AF774063

My Book

This book belongs to

Name: ______________________________

Cover Design by :
Gowri Vemuri

First Edition :
September, 2019

Author :
Gowri Vemuri

Edited by :
Ritvik Pothapragada
Siddhartha Rangavajjula
Jay Rachakonda

Questions: mathknots.help@gmail.com

NOTE : ACSL is neither affiliated nor sponsors or endorses this product.

Dedication

This book is dedicated to:

My Mom, who is my best critic, guide and supporter.

To what I am today, and what I am going to become tomorrow,

is all because of your blessings, unconditional affection and support.

This book is dedicated to the

strongest women of my life ,

my dearest mom

and

to all those moms in this universe.

G.V.

American Computer Science League (ACSL) is an international computer science competition originally founded in 1978. This organization is also an institutional member of the Computer Science Teachers Association. ACSL is on the approved activities list of the National Association of Secondary School Principals (NASSP).

ACSL consists of five divisions to appeal to the varying computing abilities and interests of students. All students at a school can take the tests but can only participate in one division. A team score is the sum of the best 3 or 5 scores in each test. Those scores can come from different students with in the team, in each contest. Prizes are awarded to top scoring students and teams based on cumulative scores after the 4th test.

The **Senior / Intermediate Division** is geared to those high school students with programming experience. Each contest consists of a 30-minute, 5- questions short answer test and a take home programming problem to be solved in 72-hours. Team scores can be based on the sum of the top 3 or top 5 scores each contest.

The **Junior Division** is geared to junior high and middle school students with no previous experience programming computers. No student beyond grade 9 may compete in the Junior Division. Each contest consists of a 30-minute 5-question short answer test and a take home program to be solved in 72-hours. Team scores are based on the sum of the best 5 scores each test.

The **Classroom Division** is open to students from all grades. It consists of a selection of the non-programming problems. As its name implies, this division is particularly well-suited for use in the classroom. Each contest consists of a 50-minute, 10-question short answer test. Team scores are based on the sum of the best 5 scores each test.
The **Elementary Division** is open to students from grades 3 - 6. It consists of non-programming problems. Four categories, one each contest, will be tested. The contest consists of a 30-minute, 5-questions test.

ELEMENTARY DIVISION	CLASS ROOM / JUNIOR DIVISION	INTERMEDIATE / SENIOR DIVISI
Elementary Computer Number Systems	Computer Number Systems	Computer Number Systems
Elementary Prefix/Infix/Postfix Notation	Recursive Functions	Recursive Functions
Elementary Boolean Algebra	What Does This Program Do? Branching	What Does This Program Do?
Elementary Graph Theory	Prefix/Infix/Postfix Notation	Prefix/Infix/Postfix Notation
	Bit-String Flicking	Bit-String Flicking
	What Does This Program Do? Loops	LISP
	Boolean Algebra	Boolean Algebra
	Data Structures	Data Structures
	What Does This Program Do? Arrays	FSA/Regular Expressions
	Graph Theory	Graph Theory
	Digital Electronics	Digital Electronics
	What Does This Program Do? Strings	Assembly Language

This book is written by computer science teachers and industry experts. Our book comprises of various practice questions in line with ACSL topics.

NOTE: ACSL is neither affiliated nor endorsed the content of this book.

This book is intended to give a hands on practice on various topics covering contest 1 and contest 2.
The questions might not replicate 100% of the actual test but are intended to reinforce the basic concepts.

NUMBER SYSTEMS

NUMBER SYSTEMS

- Decimal
- Binary
- Hexadecimal
- Octal

Decimal System:

Any number which is positionally defined in the number system is called as a decimal. All decimal numbers are with base 10.

Example: 245

$= 2 \times 10^2 + 4 \times 10^1 + 5 \times 10^0 = 200 + 40 + 5$

Binary System:

As the name implies binary numbers are only two : 1 and 0. All binary numbers are represented with base 2

Example: Convert the following to decimal system representation 0110100

64	32	16	8	4	2	1
0	1	1	0	1	0	0

$= 0X2^6 + 1X2^5 + 1X2^4 + 0X2^3 + 1X2^2 + 0X2^1 + 0X2^0$

= 0X64 + 1X32 + 1X16 + 0X8 + 1X4 + 0X2 + 0X2

= 0 + 32 + 16 + 0 + 4 + 0 + 0 = 52

Octa decimal System:

All octa decimal numbers are represented with base 8

Hexa decimal System:

All octa decimal numbers are represented with base 16

Comparison chart of various Number Systems

Decimal	Binary	Octal	Hexadecimal
0	0000	0	0
1	0001	1	1
2	0010	2	2
3	0011	3	3
4	0100	4	4
5	0101	5	5
6	0110	6	6
7	0111	7	7
8	1000	10	8
9	1001	11	9
10	1010	12	A
11	1011	13	B
12	1100	14	C
13	1101	15	D
14	1110	16	E
15	1111	17	F

1. Convert 309 into hexadecimal.

2. Convert 0x7A to binary.

3. Convert 502 into hexadecimal.

4. Convert 0x6B0 to binary.

5. Convert 532 into hexadecimal.

6. Convert the following to decimal system representation:
   110101 + 101001

7. Solve 0xAF2– 0xC2 by converting into decimal.

8. $Solve\ A_8 = 0001\ 0001\ 0001\ \ B_8 = 0100\ 0010. Find\ A_8 - B_8$

9. Solve 0xE2– 0xC4 by converting into decimal.

10. Solve A - 25 = 10 and represent the answer in decimal representation.
    A = 35

11. Solve A - 15 = 24 and represent the answer in decimal representation.
    A = 39

12. Solve $x_{16} = 3676_8$

13. $Solve\ A_8 = 0001\ 0000\ 1000\ \ B_8 = 0110\ 0000. Find\ A_8 - B_8$

14. Convert the following to decimal system representation:
010101 + 101111

15. Solve 0x1E2– 0xC4 by converting into decimal.

16. Represent E21 in binary.

17. Solve $A_{16} = 0001\ 1100\ 0001$

18. Solve $B_{16} = 1000\ 1101\ 0101$

19. Solve$B_8 = E1A$

20. Convert 0xD02 to binary.

21. Solve $B_{10} = 23D$

22. Solve $B_{16} = 0100\ 1100\ 0011$

23. Solve$B_8 = F4B$

24. Solve$B_8 = 0000\ 0001$; $A_8 = 0100\ 1000$. $Solve\ B_8 + A_8$

25. Solve$B_8 = 0001\ 0010\ 0101$; $A_8 = 0100\ 1000\ 1010$. $Solve\ B_8 + A_8$

26. Convert 0xABD to binary.

27. Solvex_8 = 1000 0101 ; y_8 = 1000 1010 . *Solve* $x_8 + y_8$

28. Solvex_8 = 0111 0101 ; y_8 = 1000 1010 . *Solve* $x_8 + y_8$

29. Solve$x_8 = 1110\ 1000$; $y_8 = 1000\ 0000\ 0000$. *Solve* $x_8 + y_8$

30. Convert 0xA47 to binary.

31. Find$B_{16} = 0111\ 1011\ 0011$

32. Find$B_{16} = 0100\ 1100\ 1100$

33. Convert 0x1E4 to binary.

34. Find$B_{16} = 0100\ 1100\ 0011$

33. Convert 0x1E4 to binary.

34. Find$B_{16} = 0100\ 1100\ 0011$

37. Convert 0x5D to binary.

38. Find$B_8 = 00111001$

39. $If\ a = 2, Find\ A_a = 1A2$

40. $If\ a = 2, Find\ A_a = 34$

41. $If\ a = 2, Find\ A_a = 68$

42. $If\ a = 2, Find\ A_a = 130$

43. $If\ a = 2, Find\ x_a = 120$

44. $If\ A_8 = 0010\ 0101\ \ B_8 = 1000\ 0000. Find\ A_8 + B_8$

45. Convert 0x49A to binary.

46. $Solve\ A_8 = 1001\ 0001\ \ B_8 = 0100\ 0001. Find\ A_8 - B_8$

47. $Solve\ A_8 = 1101\ 1001\ \ B_8 = 0100\ 0001. Find\ A_8 - B_8$

48. $Solve\ A_8 = 0001\ 0000\ 0001\ \ B_8 = 0100\ 0001. Find\ A_8 - B_8$

49. Convert 0x9C to binary.

50. $If\ a = 2, Find\ A_a = 256$

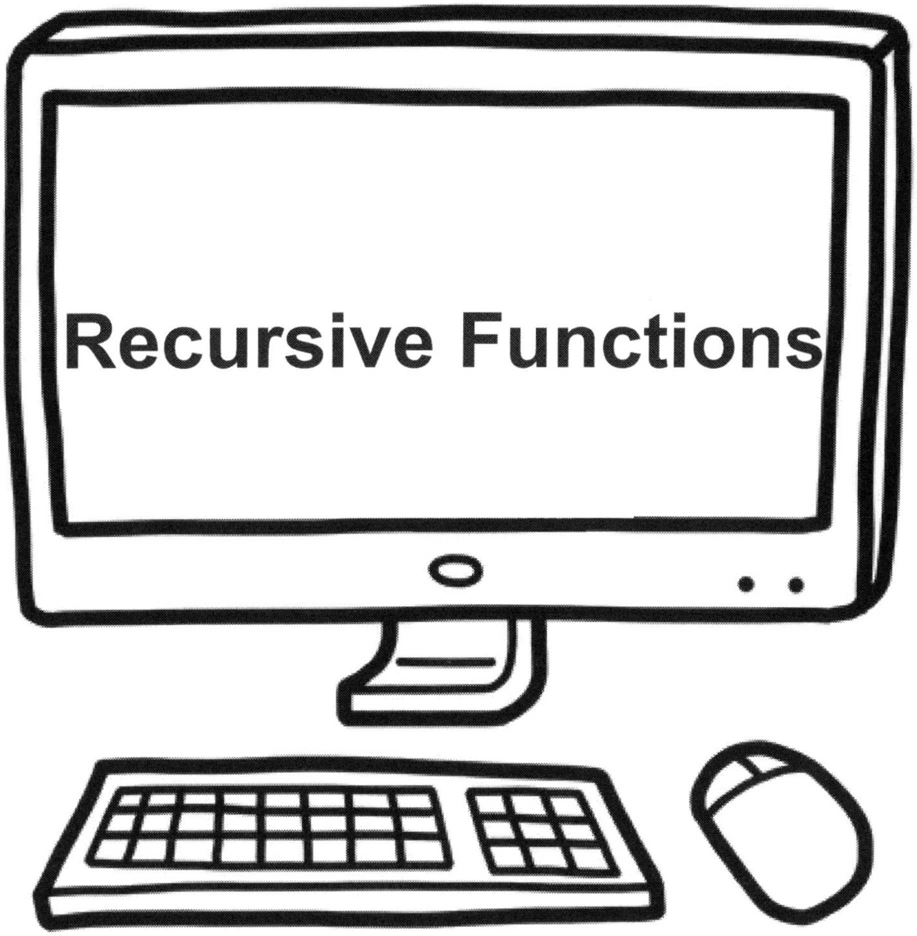
Recursive Functions

Recursive Functions

As the name says it is characterized by recurrence or repetition. This involves the repeated application of a rule, formula or a procedure based on certain given conditions to obtain successive results. These functions are widely used in programs or routines of which a part requires the application of the whole, so that its explicit call requires in general many successive executions.

Example: Any arithmetic sequence, Fibonacci numbers is a good example of recursive functions.

Sample Problem:

Find $f(5,4)$

$$f(x,y) = \begin{cases} f(x, y+1) & \text{when } x > y \\ f(x+1, y+2) & \text{when } x < y \\ -2x + \dfrac{y}{x} + xy & \text{when } x = y \end{cases}$$

Step 1:

$f(x,y) = f(5,4)$ x = 5 and y = 4 ; x > y

$f(x, y+1) = f(5, 4+1) = f(5,5)$

Step 2:

$f(x,y) = f(5,5)$ x = 5 and y = 5 ; x = y

$f(5,5) =$

$= -2(5) + \frac{5}{5} + (5)(5)$

$= -10 + 1 + 25 = 16$

1. *Find* $f(7,3)$

$$f(x,y) = \begin{cases} f(x-3, y+2) & \text{when } x > y \\ f(x+y, y+4) & x < y \\ x + \dfrac{y}{x} + xy & x = y \end{cases}$$

2. *Find* $f(8,5)$

$$f(x,y) = \begin{cases} f(x-y, y+x) & \text{when } x > y \\ x + y^2 & \text{otherwise} \end{cases}$$

3. *Find* $f(a,b)$ *where* $a > b$ *and* $a = b - a$

$$f(x,y) = \begin{cases} f(x + 3a, y + b) & when\ x < y \\ x + y^2 & otherwise \end{cases}$$

4. *Find* $f(7,4)$

$$f(x,y) = \begin{cases} f(x - y, y + 2) & when\ x > y \\ x - \dfrac{y}{x} & otherwise \end{cases}$$

5. Complete the missing values in the recursive formula of the sequence: 2,7,12, ...

$$\begin{cases} a(1) = A \\ a(n) = e(n-1) + B \\ A =?, B =? \end{cases}$$

6. *Find* $C(5)$

$$C(n) = \begin{cases} C(0) = u \\ C(1) = v \\ C(n) = C(n-2) - 2 \end{cases}$$

7. $Find\ f(7)\ if\ f(x) = 2 \times\ f(x-2) + 4\ and\ we\ know\ that\ f(3) = 10$

8. $Find\ f(8)\ if\ f(x) = 4 \times\ f(x-3) + 5\ and\ we\ know\ that\ f(2) = 10$

9. Evaluate the following recursive function:

$$Find\ f(3)\ if\ f(x) = f(x-1) + 2.x^2\ and\ we\ know\ that\ f(1) = -2$$

10. $Find\ f(4)\ if\ f(x) = f(x-2) + 3x\ and\ we\ know\ that\ f(0) = 5$

11. *Find* $C(4)$

$$C(x) = \begin{cases} C(0) = b \\ C(x) = C(x-2) + x^3 \end{cases}$$

12. *Find* $C(6)$

$$M(n) = \begin{cases} M(0) = \frac{a}{4} \\ M(n) = M(n-3) + n \end{cases}$$

13. $Find\ C(8)\ Find\ if\ f(x) = f(x-a) + a^2\ and\ we\ know\ that\ f(0) = b,$
$where\ a^2 = 2b\ and\ b = 8$

14. Complete the missing values in the recursive formula of the sequence: -4, -2, 0, . . .

$$\begin{cases} a(1) = A \\ a(n) = e(n-1) + B \\ A =?, B =? \end{cases}$$

15. *Find* $a(3)$

$$a(n) = \begin{cases} a(1) = -3 \\ a(n) = \dfrac{a(n-1)^2}{2} \end{cases}$$

16. *Find* $f(4,6)$

$$(x,y) = \begin{cases} f(x+y, y-1) & \text{when } x < y \\ f\left((x^2-75)^{\frac{1}{2}}, y\right) & \text{when } x > y \\ \left(x + \dfrac{y}{x^2}\right)^{\frac{1}{2}} & \text{otherwise} \end{cases}$$

17. *Find* $f(3,7)$

$$f(x,y) = \begin{cases} f(x+1,y) & when\ x < y \\ f(\frac{x}{4},y) & when\ x > y \\ x + \frac{xy}{4} & otherwise \end{cases}$$

18. *Find* $f(a,b)$ *where* $a^{\frac{1}{2}} = \sqrt{2}\,b$ *where* $b = 2$

$$f(x,y) = \begin{cases} f(x,y+a) & when\ x > y \\ x + \frac{y}{xy} & otherwise \end{cases}$$

19. Evaluate the following recursive functions: $Find\ f(6,4)$

$$f(x,y) = \begin{cases} f(x-3, y+2) & when\ x > y \\ f(x, y-3) & x < y \\ \dfrac{x^2}{y} & x = y \end{cases}$$

20. $Find\ f(3)\ if\ f(x) = f(x-1) + 4\ and\ we\ know\ that\ f(1) = 4$

21. *Find* $C(4)$

$$C(x) = \begin{cases} C(0) = b \\ C(x) = C(x-2) + b \end{cases}$$

22. *Find* $f(3)$

$$f(x) = \begin{cases} f(x-1)^2 & when\ x > 1 \\ 1 & x = 1 \end{cases}$$

23. Complete the missing values in the recursive formula of the sequence: 3, 8, 13, ...

$$\begin{cases} a(1) = A \\ a(n) = e(n-1) + B \\ A = ?, B = ? \end{cases}$$

24. *Find* $f(3,6)$

$$f(x,y) = \begin{cases} f(x^2 + y, y) & \text{when } x < y \\ x - y & \text{otherwise} \end{cases}$$

25. *Find* $f(1,1)$

$$f(x,y) = \begin{cases} f(x^2, y^2+1) & \text{when } x > y \\ x - \dfrac{x}{y+x} & \text{otherwise} \end{cases}$$

26. *Find* $f(5,1)$

$$f(x,y) = \begin{cases} f(x+1, 2y+1) & \text{when } x > y \\ x^2 - \dfrac{x}{y^2+x} & \text{otherwise} \end{cases}$$

27. $Find\ f(8)\ if\ f(x) = f(x-2) + \sqrt{x}\ and\ we\ know\ that\ f(2) = 3$

28. $Find\ C(5)\ if\ C(x) = C(x-2) + x\ and\ we\ know\ that\ C(1) = a$

29. Evaluate the following recursive functions:
Fibonacci numbers form a sequence defined recursively by the following equations:

$$f(x) = \begin{cases} f(0) = 0 \\ f(1) = 1 \\ f(n+2) = f(n) + f(n+1) \end{cases}$$

$$\textit{Find the fibonacci series till } f(5).$$

Write down the complete series and comment on the series.

30. Factorial of any natural number is defined by:

$$\begin{cases} f(0) = 1 \\ f(n) = n.f(n-1) \end{cases}$$

Find the Factorial of 3 using the given recursive formula.

31. *Find* $f(8)$

$$f(x) = \begin{cases} f(x-2) + \frac{a}{2} & x > 1 \\ 0 & x = 2 \end{cases}$$

32. Imagine a square park of side 100 m. Martha has to run 1.6 Km in total as per her daily workout schedule. In a recursive manner, count the total number of rounds will Martha have to run considering the running track is along the edges of the park.
Analyse the calculations and create the function from your calculations.

33. *Find* $C(6)$ *if* $C(x) = C(x-2) + \frac{x}{3}$ *and we know that* $C(2) = d$

34. *Find* $f(5)$

$$f(x) = \begin{cases} f(0) = 0 \\ f(1) = a \\ f(n+2) = f(n) + f(n+b), where\ b = 1 \end{cases}$$

35. Complete the missing values in the recursive formula of the sequence: 3, 16, 29, …….

$$\begin{cases} a(1) = A \\ a(n) = e(n-1) + B \\ A =? \;\; B =? \end{cases}$$

36. *Find* $f(3)$

$$f(x) = \begin{cases} f(x-1) - \frac{x}{2} & when\, x > 1 \\ 0 & x = 1 \\ 1 & x = 0 \end{cases}$$

37. *Find* $f(4)$

$$(x,y) = \begin{cases} f(x-2) + \frac{x^2}{2} + 1 & \text{when } x > 2 \\ 2 & x = 1,0 \\ -1 & x = 2 \end{cases}$$

38. *Find* $f(2,1)$

$$f(x,y) = \begin{cases} f(x^2, y^2 + 4) & \text{when } x > y \\ x - \frac{x}{y+x} & \text{otherwise} \end{cases}$$

39. Evaluate the following recursive functions: $Find\ f(2,3)$

$$f(x,y) = \begin{cases} f(x^2, \frac{y^2}{3}) & when\ x < y \\ x - y & otherwise \end{cases}$$

40. $Find\ f(m,n)\ where\ m > n\ such\ that\ m - n = 10$

$$f(x,y) = \begin{cases} f(\frac{x}{3}, y^3) & when\ x > y \\ x - y + 1 & otherwise \end{cases}$$

41. *Find* $f(m,n)$ *where* $m > n$ *such that* $m - n = 5$

$$f(x,y) = \begin{cases} f\left(\frac{x}{4}, 4x\right) & when\ x > y \\ \frac{x}{4} + 4x + 1 & otherwise \end{cases}$$

42. *Find* $a(4)$

$$a(x) = \begin{cases} a = -1 & at\ x = 0 \\ a(x) = a(x-2) + 2x & at\ other\ values\ of\ x \end{cases}$$

43. $Find\ f(14,3)$

$$f(x,y) = \begin{cases} f(\frac{x}{7}, y+1) & when\ x > y \\ f(x+2, y) & x < y \\ x + \frac{yx}{x^2} & x = y \end{cases}$$

44. Using the following recursive function, complete the series, complete the series considering the first element is the zeroth element,

$$\frac{1}{2}, \frac{3}{2}, \_, \_.$$

$$f(n) = 2.f(n-1) + \frac{1}{2}$$

45. Using the following recursive function, complete the series considering the first element is the zeroth element,

$$2,3,\_,\_.$$

$$f(n) = f(n-1) + n^2$$

46. *Complete the series of five elements using the data given below.*

$$f(x,y) = \begin{cases} (x + \frac{x}{a}) & \text{when } x > 3, a = 4 \\ 1 & x = 1,0 \\ 0 & x = 2 \end{cases}$$

47. Imagine a square park of side 80 m. Martha has to run 1.92 Km in total as per her daily workout schedule. In a recursive manner, count the total number of rounds will Martha have to run considering the running track is along the edges of the park.
Analyse the calculations and create the function from your calculations.

48. Using the following recursive function, complete the series,

3,_,_,_.

$$f(n) = f(n-1) + \frac{n}{2}$$

49. *Find* $f(2,6)$

$$f(x,y) = \begin{cases} f(x^2+y,y) & \text{when } x < y \\ x-y & \text{otherwise} \end{cases}$$

50. *Find* $f(5,10)$

$$f(x,y) = \begin{cases} f(2x^2,y) & \text{when } x < y \\ x-y & \text{otherwise} \end{cases}$$

What does this program do ?
Branching

What does this program do ?

Students are given small code snippets to understand and analyze the programming concepts. The goal is to get them familiarize to understand and debug the software programs written in any language.

Four concepts are emphasized as below

- Looping
- Branching
- Arrays
- Strings

1. What is the output of the program?

```
a=0
for i=1:2:20
   a=a+1
   b=a+1
   if a>4
      break
   end
   Output b
end
```

2. What is the output of the program?

```
a=1
for i=1:2:20
   a=a+i
   b=a+1
   if b>25
      break
   end
   Output b
end
```

3. What is the output of the program?

```
a=1
for i=1:2:20
   a=a+i
   b=a+1
   if b>25
      break
   end
   c=b+1
   Output c
end
```

4. What is the output of the program?

```
a=2
for i=1:2:10
   a=a+i
   b=a+1
   if b>45
      break
   end
   c=b+1
   Output c
end
```

5. What is the output of the program?

```
a=0
for i=1:2:10
   a=a+i
   b=a+1
   if b>45
      break
   end
   c=b+a
   Output c
end
```

6. What is the output of the program?

```
a=0
for i=1:2:10
   a=a+i
   b=a+1
   if b>45
      break
   end
   c=b+a
   d=c+i
   if d>25
   break
   end
   Output d
end
```

7. What is the output of the program?

```
a=1
for i=1:2:10
   a=a+i
   b=a+1
   if b>40
      break
   end
   c=b+1
   d=c+1
   if d>25
      break
   end
   Output d

end
```

8. What is the output of the program?

```
a=1
for i=1:3:15
   a=a+i
   b=a+1
   if b>40
      break
   end
   c=b+1
   d=c+1
   if d>25
      break
   end
   Output d

end
```

9. What is the output of the program if the user enters x =2329?

```
if (x > 0) || (x<0)
  Output "valid"
else if (x == 0)
  Output "invalid"
else
  Output "Enter a valid password"
```

10. What is the output of the program?

```
a=1
for i=1:3:15
    a=a+i
    b=a+1
    if b>30
        break
    end
    c=b+1
    d=c+1
    if d>30
        break
    end
    Output d

end
```

11. What is the output of the program?

```
a=1
for i=1:3:15
   a=a*i
   b=a+1
   if b>300
      break
   end
   c=b-100
   d=c+1
   Output d
end
```

12. What is the output of the program?

```
a=
for i=1:2:10
   a=a*i
   b=a+
   if b>300
      break
   end
   c=b-100
   d=c+1
   Output d

end
```

13. What is the output of the program?

```
a=1
for i=1:2:10
   a=i
   b=a+1
   if b>300
      break
   end
   c=b-100
   d=c+100
   Output d

end
```

14. What is the output of the program if s=50?

```
t = 100
s = 50
if t>s
   Output "neg"
elseif t<s
   Output "pos"
else
   Output "error"
end
```

15. What is the output of the program?

```
a=1
for i=1:5:20
   a=
   b=a+1
   if b>300
      break
   end
   c=b-100
   d=c+100
   Output d
end
```

16. What is the output of the program if age = 15?

```
age =15
if(age < 0 || age > 150)
   Output "Error."
else
   Output "Age entered is valid"
```

17. What is the value of a after the execution of the statement if a = 55 and b = 20?

```
a= (a <b) ? b : -b ;
```

18. What is the value of a after the execution of the statement if a = 4 and b = 30?

```
a= (a <b) ? b : -b ;
```

19. What is the value of a after the execution of the statement if a = 5 and b = 25?
a= (a <b) ? -b : b ;

20. What is the value of a after the execution of the statement if a = 23 and d = 32?
a= (a > d) ? d : -d ;

21. What is the output of the program?

```
for i=1:8
  a=2
  b=a*i
  c=0
  if b<11
  c=c+b
  else
     c=c-b
  end
  Output c
end
```

22. What is the output of the program?

```
for i=1:2:10
  a=2
  b=a+i
  c=0;
  if b<10
  c=c+b
  else
     c=c-b
  end
  Output c
end
```

23. What is the output of the program?

```
h=10
for i=1:2:10
  a=2
  b=a+i
  c=
  if b<10
  c=c+b
  else
     c=c-b
  end
  d=c+h
  Output d
end
```

24. What is the output of the program?

```
h=5
for i=1:2:10
  a=2
  b=a*i
  c=1
  if b<10
  c=c+10
  else
     c=c-10
  end
  d=c+h;
  Output d
end
```

25. What is the output of the program?

```
e=2
for i=1:2:10
  a=2
  b=a*i;
  c=1;
  if b<10
  c=c+10
  else
     c=a+i
  end
  d=c+e
  Output d
end
```

26. What is the output of the program?

```
e=2
for i=1:2:10
  a=2
  b=a*i
  c=1
  if b<10
  c=c+10
  else
     Output "*"
  end
  d=c+e;
  Output d
end
```

27. What is the output of the program?

```
for i=1:2:10
  a=2
  b=a*i
  %c=2
  if b<10
  c= Output "$"
  else
  c= Output "*"
  end

  Output c
end
```

28. What is the output of the program?

```
for i=1:2:10
  a=3
  b=a*i
  %c=2
  if b<15
  c=Output "*"
  else
  c=Output "$"
  end

  Output c
end
```

29. What is the output of the program?

```
for i=1:2:10
  a=3
  b=a*i
  %c=2
  if b<15
  Output "small"
  else
  Output "big"
  end
end
```

30. What is the output of the program?

```
for i=1:2:10
  a=3
  b=a+i
  c=2
  if b<10
  Output "small\n"
  else
  Output "big\n"
  end
end
```

31. What is the output of the program?

```
for i=1:2:10
   a=5
   b=a*i
   %c=2
   if b<10
   Output "*\n"
   elseif b>10
   Output "#\n"
   else
   Output "&\n"
   end
end
```

32. What is the output of the program?

```
c=zeros(5,5)
for i=1:5
    for j=1:5
        if i==j
        c(i,j)=9
        end
    end
end
Output c
```

33. What is the output of the program?

```
c=zeros(5,5)
for i=1:5
   for j=1:5
      if i==j
      c(i,j)=9
      elseif i<j
         c(i,j)=10
      else
         c(i,j)=
      end
   end
end
Output c
```

34. What is the output of the program?

```
c=zeros(5,5)
for i=1:5
   for j=1:5
      if i==j
      c(i,j)=9
      elseif i<j

         c(i,j)=i+j
      else
         c(i,j)=i-j
      end
   end
end
Output c
```

35. What is the output of the program?

```
c=zeros(5,5)
for i=1:5
   for j=1:5
      if i==j
      c(i,j)=i
      elseif i<j
         c(i,j)=i*j
      else
         c(i,j)=i-j
      end
   end
end
Output c
```

36. What is the output of the program?

```
c=zeros(4,4)
for i=1:4
   for j=1:4
      if i==j
      c(i,j)=j
      elseif i<j
         c(i,j)=i*j
      else
         c(i,j)=i/j
      end
   end
end
Output c
```

37. What is the output of the program?

```
c=zeros(4,4)
for i=1:4
   for j=1:4
      if i==j
      c(i,j)=j+i
      elseif i<j
         c(i,j)=i+j
      else
         c(i,j)=i-j
      end
   end
end
Output c
```

38. What is the output of the program?

```
c=zeros(4,4)
for i=1:4
   for j=1:4
      if i==j
      c(i,j)=j
      else
         break
      end
   end
end
Output c
```

39. What is the output of the program?

```
c=zeros(5,5)
for i=1:5
   for j=1:5
      if i>j
      c(i,j)=j+i
      else
         break
      end
   end
end
Output c
```

40. What is the output of the program?

```
c=zeros(5,5)
for i=1:5
   for j=1:5
      if i<j
      c(i,j)=j*i
      else
         break
      end
   end
end
Output c
```

41. What is the output of the program?

```
a=1
for i=1:5:20
   a=i
   b=a+1
   if b>200
      break
   end
   c=b+100
   d=c+100
   Output d

end
```

42. What is the output of the program if the age=57?

```
age= 57
if(age >= 30 && age <= 70)
   Output "You're between 30 and 70 and cannot save money on your health insurance!"
else

   Output "Contact the office, we may have deals for you"
```

43. What is the output of the program?

```
c=zeros(5,5)
for i=1:5
   for j=1:5
      if i==j
      c(i,j)=j/i
      else
         c(i,j)=i+j
      end
   end
end
Output c
```

44. What is the output of the program?

```
c=zeros(5,5)
for i=1:5
   for j=1:5
      if i<j
      c(i,j)=0
      elseif i>j
         c(i,j)=i+j
      else
         c(i,j)=1
      end
   end
end
Output c
```

45. What is the output of the program?

```
c=zeros(5,5)
for i=1:5
   for j=1:5
      if i<j
      c(i,j)=0
      elseif i>j
         c(i,j)=1
      else
         c(i,j)=i+j
      end
   end
end
Output c
```

46. What is the output of the program?

```
for i=1:2:10
  a=4
  b=a+3+i
    if b<10
  Output "*"
  elseif b>10
  Output "#"
  else
  Output "'&"
  end
end
```

47. What is the output of the program?

```
for i=1:2:10
  a=2
  b=a+i
  %c=2
  if b<10
  Output "*\n"
  elseif b>10
  Output "#\n"
  else
  Output "&\n"
  end
end
```

48. What is the output of the program?

```
c=zeros(5,5)
for i=1:5
   for j=1:5
      if i<j
      c(i,j)=10
      elseif i>j
         c(i,j)=100
      else
         c(i,j)=1000
      end
   end
end
Output c
```

49. What is the output of the program?

```
c=zeros(6,6)
for i=1:6
   for j=1:6
      if i<j
      c(i,j)=i+j
      elseif i>j
         c(i,j)=i/j
      else
         c(i,j)=i*j
      end
   end
end
Output c
```

50. What is the value of a after the execution of the statement if a = 44 and d = 30?

```
a= (a > d) ? d : -d ;
```

Pre/Post/In - fix
Notation

Prefix-Postfix Notations

Infix Notation:

P + Q

As a general way of writing expressions, Operators are written in-between their operands.

An expression such as A *(B + C) /D

is simplified as "First add B and C together, then multiply the result by A, then divide by D."

The Infix Notations use the general order of evaluation, using PEDMAS rules

Prefix Notation:

Infix, Postfix and Prefix notations are three different but equivalent ways of writing expressions. It is easiest to demonstrate the differences by looking at examples of operators that take two operands.

Infix notation: A + B. Prefix notation: + X Y

Operators are written before their operands.

An expression such as A *(B + C) /D = / (A *(B + C)) D

= / * A(B+C) D /= / * A + B C D

- The order of evaluation of operators is always left-to-right Brackets cannot be used to change this order.
- Because the "/" is to the left of the "*" in the example above, the addition must be performed before the multiplication.
- Operators act on values immediately to the left of them. For example, the "+" above uses the "B" and "C".
- Brackets can be added to make this more explicit
  (/ (* A (+ B C)) D)

SAMPLE #1:

Convert the following infix to prefix notation:

$$(A - D) + \frac{B}{C}$$

$= ((A - D)) + (\frac{B}{C})$

$= ((-AD)) + (\frac{B}{C})$

$= (-AD) + (/BC)$

$= + - AD/BC$

Postfix Notation:

Infix notation: A + B

Postfix notation (also known as "Reverse Polish notation"): A B +

Operators are written after their operands.

An expression such as A *(B + C) /D = A *(B + C) D /

= A(B+C) *D / = A B C + * D /

- The order of evaluation of operators is always left-to-right
- Brackets cannot be used to change this order.
- Because the "+" is to the left of the "*" in the example above, the addition must be performed before the multiplication.
- Operators act on values immediately to the left of them. For example, the "+" above uses the "B" and "C".
- Brackets can be added to make this more explicit

  ((A (B C +) *) D /)

Thus, the "*" uses the two values immediately preceding: "A", and the result of the addition. Similarly, the "/" uses the result of the multiplication and the "D".

In all notations, the operands occur in the same order, and just the operators have to be moved to keep the meaning correct.

SAMPLE #2:

Convert the following infix into postfix notations:

$$\frac{\boldsymbol{A+B+C\ +A^2}}{\mathbf{4}}$$

$$=\frac{((A+B+C)+A^2)}{4}$$

$$=\frac{((AB+C)+A^2)}{4}$$

$$=\frac{(ABC+A2\uparrow +)}{4}$$

$$=ABC+A2\uparrow +4/$$

1. Convert the following infix to postfix notation for an expression for "Z":

$$X^2 + \frac{Y^2}{2} - Z = 0$$

2. Convert the following infix to prefix notation:

$$A * B * \frac{(A * B)^2}{\frac{A}{B}}$$

3. Convert the following infix to prefix notation:

$$\frac{Z^{Z^{Z^{Z}}}}{Z+Z+Z+Z}$$

4. Convert the following postfix into infix:

$$AB - AC -/ B2 \uparrow C3 \uparrow /+$$

5. Convert the following postfix into infix and evaluate the answer:

$$32 + 34 - / 242 \uparrow / +$$

6. Convert the following prefix to infix and evaluate the answer:

$$+ * 42 * 5 / 42$$

7. Convert the following infix into postfix notation:

$$a = b^c + \frac{c^b}{c^{c+1}}$$

8. Convert the following infix into prefix notation:

$$\frac{A}{B} + \frac{B^B}{A - B^{B+1}}$$

9. Convert the following infix into prefix notation:

$$\left(a^{b^{c}} * \frac{c^{b^{a}}}{c+b+a}\right)$$

10. Convert the following infix into postfix notation:

$$\frac{\left(A^{B} + \frac{A-C}{B-C} + B^{A}\right)}{C}$$

11. Convert the following prefix to infix notation and evaluate the answer:

$$*\uparrow \frac{AA}{\uparrow AA} \uparrow B2, where\ A = 1\ and\ B = 2$$

12. Convert the following infix to postfix notation:

$$\frac{(\frac{1}{a}+\frac{1}{b}+\frac{1}{c})}{a*b*c}$$

13. Convert the following infix to postfix notation:

$$O = \frac{(\frac{P^2 - Q^2}{P * Q})}{(\frac{P}{Q})}$$

14. Convert the following postfix to infix notation:

$$AB + AC - /C + D -$$

15. Convert the following infix to postfix notation:

$$(\frac{A * B}{B^2}) + (\frac{C * D}{C^{3+D}})$$

16. Convert the following infix to prefix notation:

$$\frac{(A - B)^2 + (B * C)^3}{A * B * C}$$

17. Convert the following infix to postfix notation:

$$\frac{s * s + s}{t + t * t} * s * t$$

18. Convert the infix to prefix notation:

$$A^D + \frac{B}{D^B} * C^D$$

19. Convert the following postfix to infix notation:

$$A2 \uparrow B2 \uparrow + AB - 2 \uparrow +, \boldsymbol{where}\ A = 2, B = -1$$

20. Convert the following infix to postfix.

$$\frac{\left(W^2 + \frac{V * U^2}{W^{\frac{3}{2}}}\right)}{U + \frac{V}{W^{\frac{1}{2}}}}$$

21. Convert the following infix to prefix notation:

$$\left(\frac{A^2 - B^2}{A^3 + B^2} + \frac{A}{B^2}\right) * \frac{C^2}{B^3}$$

22. Convert the following postfix to infix notation:

$$\mathbf{12/32 \uparrow 3/24/* +}$$

23. Convert the following to postfix notation:

$$\frac{\alpha^2}{\beta}+\frac{\frac{\beta^\gamma}{\gamma}}{\varepsilon+\frac{2}{\alpha}}$$

24. Convert the following to prefix notation:

$$\frac{A^{\frac{B+C+D^2}{A*B*C}}}{A*B*C}$$

25. Convert the following to postfix notation for P_O :

$$\frac{P_O}{P} = \left(\frac{T_O}{T}\right)^{\frac{(\gamma-1)}{\gamma}}$$

26. Convert the three equations of motions to postfix notation. Find the postfix expression for "u" in all the equations.
HINT: The three equations of motion that defines the state of motion of any body are given by:

a) $v = u + at$

b) $s = ut + \frac{1}{2}at^2$

c) $v^2 = u^2 + 2as$

27. Convert the following postfix to infix notation. The equation obtained in the end is an equation for "Force".

$$F = GM * m * r2 \uparrow /$$

Evaluate the "Force, F" using the data given below:

$$G, Gravitational\ constant = 6.67 \times 10^{-11}$$
$$M, mass\ of\ Earth = 6 \times 10^{24}$$
$$m, mass\ of\ the\ satellite\ revolving\ around\ the\ Earth = 60\ Kg$$
$$r, distance\ between\ the\ center\ of\ Earth\ and\ the\ satellite = 7000\ Km$$

You can give an approximate answer.

28. Convert the following postfix to infix notation:

$$222 \uparrow\uparrow 23 * 2 * +22 + 222 \uparrow / + /$$

29. Convert the following to infix notation and evaluate the answer:

$$\mathbf{12/14/{+}16/{+}24 * 6 +/}$$

30. Convert the following to infix notation and evaluate the answer:

$$\boldsymbol{X2 \uparrow Y2 \uparrow X3 \uparrow /{+}YX - XY */-}$$

31. Convert the following infix to postfix notations:

$$\frac{\frac{A}{A^{A+B}}}{\frac{B}{B^{A+B}}}$$

32. Convert the following infix to prefix notation:

$$\frac{A^B + B^2 + B^A}{A * B * B}$$

33. Convert the infix to postfix notation:

$$A^2 + \frac{B^3}{A * B} * \frac{A * B}{A^3}$$

34. Convert the infix to prefix notation:

$$a * \frac{b}{b * \frac{b}{c}}$$

35. Convert the following infix to postfix notation:

$$w^2 - \frac{v^2}{w^2} + \frac{v^2}{w^2}$$

36. Convert the following infix to postfix notation:

$$\frac{A^2}{B^3} + \frac{A^B}{B^A} * (A * B)$$

37. Convert the following infix to postfix notation:

$$\frac{A^5}{(A * B)} + \frac{B^5}{\frac{A}{B} + \frac{B}{2}}$$

38. Convert the following infix to prefix notation:

$$\frac{A^2}{B^2 + C^2} * \left(\frac{A}{B}\right)^C$$

39. Convert the following infix to postfix notation:

$$\frac{x}{y}+\frac{x^2}{y^2}+\frac{x^3}{y^3}$$

40. Convert the following infix to postfix notation:

$$\left(\frac{x^3}{y}\right)^3+\left(\frac{2*x}{y}\right)^4$$

41. Convert the following infix to prefix notation:

$$\frac{w^2}{u+v}+\frac{u}{v}+\frac{u^2}{u*v}$$

42. Convert the following infix to postfix notation:

$$\frac{A^2}{2A}+\frac{B^2}{AB}+\frac{C^2}{2C}$$

43. Convert the infix to postfix notation:

$$\frac{A^{a^A}}{a * A} + \frac{A}{a^{\frac{1}{2}}} + \frac{\frac{1}{2}a}{A}$$

44. Convert the following postfix to infix and compute the final answer:

$$S = (432 \uparrow / 25 / + 912 \uparrow / 3 / -)$$

45. Convert the following postfix to infix and compute the final answer:
Note carefully the spaces between the digits as all numbers are not one-digit numbers.

$$Y = \mathbf{22 \uparrow 3 \uparrow 3/+40 \uparrow 50 \uparrow/+31\ 27/-}$$

46. Convert the following prefix to infix and compute the final answer:
Note all the numbers are one-digit numbers.

$$X = * \mathbf{3/3 + 3/33}$$

47. Convert the following prefix to infix and compute the final answer:
Note all the numbers are one-digit numbers.

$$A = * \uparrow a / 14 \uparrow / a_o + a a_o / 12$$

48. Convert the following prefix to infix and compute the final answer:

$$S = + / 3 + 4 \uparrow (-1) 7 / \uparrow 22 \uparrow 23$$

49. Convert the following postfix to infix and compute the final answer:

$$Y = x2 \uparrow y3 \uparrow /x + yxy + /*$$

50. Convert the following postfix to infix and compute the final answer:

$$Y = abba/* \, a12/\uparrow * +$$

Bit - String
Flicking

BIT STRING FLICKERING

As the name says, first let's learn about Bit and its details, then strings and Flickering

(I) Bitwise Operators

The logical operators:

(1) Binary Operators: **and** (&), **or** (|), and **xor** (⊕)
(2) Unary Operator: **not** (~ or ¬)

BINARY OPERATORS:

- **or** is a logical binary operator of each bit in each of its operands.
  The **or** of two values is 1 only if one or both values are 1 else 0

Example:

1010100 or 1100001 = 1110101

x	y	x or y
1	1	1
0	1	1
1	0	1
0	0	0
1	0	1
0	0	0
0	1	1

- **and** is a logical binary operator of each bit in each of its operands.
  The **and** of two values is 1 only if both values are 1 else 0.

Example:

1010100 and 1100001 = 1000000

x	y	x or y
1	1	1
0	1	0
1	0	0
0	0	0
1	0	0
0	0	0
0	1	0

- **xor** is a logical binary operator of each bit in each of its operands.
  The **xor** of two values is 1 if the values are different or 0
  It is equal to zero if the value is equal to 1 in both x and y

Example:

1010100 xor 1100001 = 011111

x	y	x xor y
1	1	0
0	1	1
1	0	1
0	0	1
1	0	1
0	0	1
0	1	0

UNARY OPERATORS:

- **not** is a logical unary operator that performs negation on each bit. Bit 0 becomes 1 and vice.

Example:
not 1010100 = 0101011
not 1100001 = 0101011

x	y	not x	not y
1	1	0	0
0	1	1	1
1	0	0	0
0	0	1	1
1	0	0	0
0	0	1	1
0	1	1	1

(II) Strings

Logical shifts **LSHIFT-x** meaning "ripple" the bit-string x positions in the indicated direction, to the left. Bits shifted out are lost; zeros are shifted in at the other end. The size of a bit-string does not change with shift to left. If any bit strings are initially of different lengths, all shorter ones are padded with zeros in the left bits until all strings are of the same length

x	LSHIFT- x	LSHIFT-2 x	LSHIFT-3 x
001101	011010	110100	101000
101	010	100	000
11100	11000	10000	00000
11	10	00	000

Logical shifts **RSHIFT-x** meaning "ripple" the bit-string x positions in the indicated direction, to the right. Bits shifted out are lost; zeros are shifted in at the other end. The size of a bit-string does not change with shift to right. If any bit strings are initially of different lengths, all shorter ones are padded with zeros in the left bits until all strings are of the same length.

x	RSHIFT- x	RSHIFT-2 x	RSHIFT-3 x
001101	000110	000011	000001
101	010	001	000
11100	01110	00111	00011
11	01	00	00

Circulates LCIRC-x meaning "ripple" the bit string x positions in the indicated direction. As each bit is shifted out one end, it is shifted in at the other end. The effect of this is that the bits remain in the same order on the other side of the string.

The size of a bit-string does not circulate. If any bit strings are initially of different lengths, all shorter ones are padded with zeros in the left bits until all strings are of the same length.

x	LCIRC - x	LCIRC - 2 x	LCIRC - 3 x
001101	011010	110100	101001
101	011	110	101
11100	11001	10011	00111
11	11	11	11

Circulates **RCIRC-x** meaning "ripple" the bit string x positions in the indicated direction. As each bit is shifted out one end, it is shifted in at the other end. The effect of this is that the bits remain in the same order on the other side of the string.

The size of a bit-string does not circulate. If any bit strings are initially of different lengths, all shorter ones are padded with zeros in the left bits until all strings are of the same length.

x	RCIRC - x	RCIRC - 2 x	RCIRC - 3 x
001101	100110	010011	101001
101	110	011	101
11100	01110	00111	10011
11	11	11	11

(III) Flickering

The order of precedence

High to Low: NOT; SHIFT and CIRC; AND; XOR; and finally, OR.

"Not" is a Unary operator and it is evaluated first on a single operator with a precedence of right to left

As a thumb rule Operators are evaluated from right to left

Note: when a tie situation arises with equal precedence then ONLY, they are evaluated from left to right

Example Problems:

1. (RCIRC-2 1001010) AND (RSHIFT-1 1010011)
   = (1010010) AND (0101001)
   = 0000000

2. ((RSHIFT-2 (1010010 AND 1001010)) OR (LSHIFT-1 1011001))
   = ((RSHIFT-2 (1010010 AND 1001010)) OR 0110010)
   = ((RSHIFT-2 1000010) OR 0110010)
   = (0010000) OR 0110010)
   = 0110010

1. (RCIRC-2 101011) XOR (RSHIFT-1 (RCIRC-2 111010))

2. (NOT (LSHIFT-2 (RSHIFT-3 (RCIRC-3 001111)))) OR 110000

3. (RSHIFT-2 (11110 OR 11010)) AND (RSHIFT-2 10111))

4. (RCIRC-2 (LSHIFT-2 1001010)) AND (RSHIFT-1 (LCIRC-2 1010011))

5. ((RSHIFT-2 (1010010 AND 1101011)) XOR (LSHIFT-2 1011001))

6. NOT ((0001100 XOR 1010100) AND 0011100)

7. 1111000 AND (NOT ((0000011 XOR 1010001) OR 1001100))

8. ((LCIRC-2 100000 OR LSHIFT-1 111010) AND (NOT (000001 OR 111110))

9. (NOT (RSHIFT-2 (LCIRC-3 110000))) AND (LSHIFT-3 101010)

10. (RSHIFT-3 (1010011 AND 1101000)) XOR (RSHIFT-2 1011001))

11. (LSHIFT-1 (10100 XOR (RCIRC-3 x) AND 10011)) = 01100, where x is 5-digit

12. (RSHIFT-1 (LCIRC-2 x AND 1001)) = 0110, where x is 4-digit
Give any three combinations of digits.

13. (LSHIFT-1 (LCIRC-1 x AND 1011)) = 0111, where x is 4-digit
Give any three combinations of digits.

14. (RSHIFT-1 (1011 XOR (RCIRC-2 x AND 1011)) = 0100, where x is 4-digit
Give any three combinations of digits.

15. NOT (LSHIFT-1 (RSHIFT-2 (RCIRC-3 001111 OR 100000)))

16. ((LCIRC-2 (11110 OR 10010)) XOR (LSHIFT-2 11111))

17. (RCIRC-3 1001011) AND (LSHIFT-2 1010011)

18. ((LSHIFT-2 (1110110 AND 1011011)) XOR (RSHIFT-1 1011001))

19. NOT ((00011 XOR 10101) AND 11100)

20. NOT ((x AND 1010) OR 1001) where x is a four-digit string

21. RSHIFT-2 (NOT ((01011 XOR 10101) AND 10100))

22. ((LSHIFT-3 (RSHIFT-2 (LCIRC-2 010001111))) OR (LSHIFT-2 111110010)

23. NOT ((RSHIFT-3 (LSHIFT-2 010001101)) OR (RSHIFT-2 100110010))

24. (LSHIFT-2 (10101 XOR ((LCIRC-2 x) AND 11011)))) = 11100, where x is 5 digit
Give any three combinations of digits.

25. (RSHIFT-1 (RCIRC-1 x AND 1101)) = 1110, where x is 4-digit
Give any three combinations of digits.

26. (LCIRC-1 01001001 OR LSHIFT-2 00010011)

27. ((RCIRC-3 100000 XOR RSHIFT-1 111010) OR ((000001 AND 111110))

28. (LSHIFT-1 (RSHIFT-2 (LCIRC-3 0101111))) OR (LSHIFT-2 1011110)

29. (RSHIFT-3 (LCIRC-3 011001111)) OR (LSHIFT-3 110110000)

30. (LSHIFT-3 (LSHIFT-3 (LCIRC-2 001101))) AND (RSHIFT-2 (RCIRC-1 101000))

31. If *a* equal 1 and *A* is 0. Compute:

32. (LSHIFT-2 (RCIRC-2 101111)) OR (LSHIFT-2 (RCIRC-1 101000))

33. If *a* equal 1 and *A* is 0. Compute:

34. (110011 AND (110011 OR (110011 OR (110011))))

35. (RCIRC-2 (RCIRC-1 10001)) OR (RSHIFT-2 (RCIRC-2 10101))

36. (NOT ((pqrs AND 1010) OR 1001)) OR (1000) where *pqrs* is a four-digit string

37. (RCIRC-1 110010 OR RSHIFT-2 100001) AND (NOT 001010)

38. NOT (RSHIFT-2 (LSHIFT-2 (LCIRC-2 101101 OR 101000)))

39. (RCIRC-2 (RCIRC-2 10111)) OR (LSHIFT-2 (LCIRC-3 10110))

40. If a = 101010 and b = 001100
Find RSHIFT-1 (NOT ((a XOR b) OR a))

41. (RCIRC-2 (RCIRC-1 101010)) OR (LSHIFT-3 (RSHIFT-1 100100))

42. If a = 10001 and A = 11111
Find RCIRC-1 (NOT ((a OR A) OR a))

43. (LCIRC-2 (LCIRC-2 100010)) OR (RSHIFT-2 (RCIRC-1 100010))

44. (RCIRC-2 101101)) AND (NOT (LCIRC-2 111010))

45. NOT ((00010 OR 10100) AND 00100)

46. NOT (((LSHIFT-2 (LCIRC-2 01111)) AND (LSHIFT-2 00011))

47. (LSHIFT-2 (RCIRC-2 (0010011 OR 1010010)))

48. (NOT (LCIRC-2 110001111)) OR 100000111

49. 101011 XOR (RSHIFT-4 (RCIRC-3 000011))

50. If *a* equal 1 and *A* is 0. Compute:

What does this program do ?
Loops

What does this program do ?

Students are given small code snippets to understand and analyze the programming concepts. The goal is to get them familiarize to understand and debug the software programs written in any language.

Four concepts are emphasized as below

- Looping
- Branching
- Arrays
- Strings

1. What is the output of the program if rows = 3?

```
int rows, i, j, space;
rows=3
for (i=rows; i>=rows;i- - )
for (space = i; space<rows; space++)
Output " "
for (j=1; j<=(2*i-1);j++)
Output "*"
Output "\n"
```

2. What is the output of the program?

```
int m, i, j;
m=5
for (i=1; i<=m; i++ )
for (j=1; j<=1; j++)
Output "*"
Output "\n"
```

3. What is the output if rows = 4?

```
int m, i, j
m=4
for (i=m; i>=m; i - -  )
for (j=1; j<=1; j++)
Output "*";
Output "\n"";
```

4. What is the output of the program?

```
for x = 1:4
 Output "'*\n',x"
  for y=2:4
     Output "'^',y"
  end
end
```

5. What is the output of the program?

```
for x = 1:3
 for y =1:3
    if x==y
      z(x,y) =1
    else
        z(x,y) =0
    end
  end
end
```

6. What is the output of the program?

```
for x = 1:3
 for y =1:3
    if x==y
      z(x,y) =x+y
    else
        z(x,y) =x-y
    end
  end
end
```

7. What is the output of the program?

```
for x = 1:3
 for y =1:3
    if x==y
      z(x,y) =1
    elseif x<y
       z(x,y) =0
    else
       z(x,y) = -1
    end
  end
end
```

8. What is the output of the program?

```
for x = 1:3
 for y =1:3
    if x==y
      z(x,y) =x^2
    else
      z(x,y) =y^2
    end
  end
end
```

9. What is the output of the program?

```
z = zeros(3)
for x = 1:3
 for y =1:3
    if x==y
      z(x,y) = x^2 + y^2

    else
       z(x,y) =.  x^2 + x/2
    end
  end
end
```

10. What is the output of the program?

```
A = 2
B = 1
for x = 1:3
 for y =1:3
    if x==y
      z(x,y) =A
    else
       z(x,y) =B
    end
  end
end
```

11. What is the output of the program?

```
A = 100
B = 10000
for x = 1:3
 for y =1:3
    if x<y
      z(x,y) =sqrt(A)
    elseif x>y
       z(x,y) = sqrt(B)
    else
       z(x,y) = sqrt(A+B)
    end
  end
end
```

12. What is the output of the program?

```
A = 0
B = 1
for x = 1:4
 for y =1:4
    if x<y
      z(x,y) =A
    elseif x>y
       z(x,y) = B
    else
       z(x,y) = B
    end
  end
end
```

13. What is the output of the program?

```
s = 0
for x = 1:3
 for y =1:3
    if x<y
      z(x,y) = s+x
    elseif x>y
         z(x,y) = s/y
    else
         z(x,y) = x*y
    end
  end
end
```

14. What is the output of the program?

```
s = 1
for x = 1:3
 for y =1:3
    if x>y
      z(x,y) =s+x
    elseif x<y || x==y
         z(x,y) = s/(x+y)
    end
  end
end
```

15. What is the output of the program?

```
s = 1
z = zeros(3)
for x = 1:3
 for y =1:3
    if x>y
      z(x,y) =s+x
    elseif x<y || x==y
        z(x,y) = s+x/(x+y)
    end
  end
end
```

16. What is the output of the program?

```
s = 1
for x = 1:3
 for y =1:3
    if y==1
      z(x,y) =1
    elseif y>1
        z(x,y) = s+z(x,y-1)
    end
  end
end
```

17. What is the output of the program?

```
s = 4
n = 3
for x = 1:n
 for y =1:n
    if y==1
      z(x,y) =s
    elseif y>1
        z(x,y) = s+z(x,y-1)
    end
  end
 end
```

18. What is the output of the program?

```
s = 3
n = 3
for x = 1:n
 for y =1:n
    if y==1
      z(x,y) =s
    elseif y>1 && y<n
        z(x,y) = s*z(x,y-1)
    end
  end
 end
```

19. What is the output of the program?

```
s = 3
n = 3
for x = 1:n
 for y =1:n
    if y==1
      z(x,y) =s^2
    elseif y>1 && y<n
       z(x,y) = s^2/z(x,y-1)
    end
 end
end
```

20. What is the output of the program?

```
s = 27
t = 144
n = 3
for x = 1:n
 for y =1:n
    if y==1
      z(x,y) = $\sqrt[2]{t}$
    elseif y==n
       z(x,y) = $\sqrt[n]{s}$
    elseif x==2 && y==2
       z(x,y) = 1
    else
      z(x,y) = 0
     end
 end
end
```

21. What is the output of the program?

```
s = 27
t = 144
n = 3
for x = 1:n
 for y =1:n
    if y==1
      z(x,y) = ²√t
    elseif y==n
      z(x,y) = ⁿ√s
    elseif x==2 && y==2
      z(x,y) = 1+z(x,y-1)
    else
      z(x,y) = 0
    end
  end
end
```

22. What is the output of the program?

```
s = 0
n = 4
z = zeros(4)
for x = 1:n
 for y =1:n
    if y==n or y==1
      z(x,y) = s;
    elseif y>1 && y<n
      z(x,y) = x^2 + y
    end
  end
end
```

23. What is the final value of s and x?

```
x = 1
for i = 1:3
   for j = 1:3
   s = i+j
   x =x*j
   end
end
```

24. What is the value of s?

```
y = 1
for i = 1:2:5
   for j =1:2
      x = i
      b = j
      s = x/i + y/b
   end
end
```

25. What is the final output of a?

```
x = 2
y = 3
for i = 1: x
   for j = 1:y
   t = i
   s = t
   a = s^2+t^j
   end
end
```

26. What is output of the program?

```
i=1
m=4
while i<m
  Output "*"
  i = i+1
end
```

27. What is the output of the program if rows =5?

```
int m, i, j;
m=5
for (i=m; i>=m; i - - )
for (j=1; j<=5; j++)
Output "%"
Output "\n"
```

28. What is the output of the program?

```
int m, i
int S = 0
m=3
for (i=1; i<=m; i ++ )
S = S + i;
```

29. What is the output of the program?

```
int m, i, j;
m=6
for (i=m; i>=m; i - - )
for (j=1; j<=6; j++)
Output "+"
Output "\n"
```

30. What is the output of the program?

```
int i=1;
int m=6;
while i<m
Output "$"
Output "/n"
i=i+1
```

31. What is the output of the program?

```
x = 1
for i = 1:4
   for j = 1:3
   s = i+j
   x =s*j
   end
end

Output x
```

32. What is the output of the program?

```
s = 1
for x = 1:3
 for y =1:3
   if y==1
     z(x,y) =2
   elseif y>1
      z(x,y) = s+z(x,y-1)
   end
  end
end
```

33. What is the output of the program?

```
x = 10
for i = 1:2
   for j = 1:3
   s = x+j
   x =s*j+i
   end
end
Output x
```

34. What is the output of the program?

```
for x = 1:5
 if x=1
 Output "*"
 else if x>1
    for y=2:7
     Output "@"
   end
 end
```

35. What is the output of the program?

```
for x = 1:3
   for y =1:3
    if x==y
    z(x,y)=2
   else
     z(x,y)=0
   end
  end
end
Output z(x,y)
```

36. What is the output of the program?

```
for x = 1:3
   for y =1:3
    if x==y
    z(x,y)=1
    else
    z(x,y)=x+y
   end
  end
end
Output z(x,y)
```

37. What is the output of the program?

```
for x = 1:3
   for y =1:3
    if x==y
    z(x,y)=10
    else
    z(x,y)=x-y
    end
      end
    end
    Output z(x,y)
```

38. What is the out of the program?

```
for x = 1:3
   for y =1:3
    if x==y
    z(x,y)=x^2
    else
    z(x,y)= y^2 - x
    end
   end
 end
 Output z(x,y)
```

39. What is the out of the program?

```
for x = 1:4
   for y =1:3
    if x==y   or x>y
    z(x,y)=xy+x
    else
    z(x,y)= y^2 + x^2
   end
  end
end
Output z(x,y)
```

40. What is the output of the program?

```
int x = 10
int y = 10
if ( !(x ^ y) )
   Output " x is equal to y "
else
   Output " x is not equal to y "
```

41. What is the out of the program?

```
for x = 1:4
  for y =1:4
   if x==y   or x>y
   z(x,y)=xy+y
   else
   z(x,y)= y^2 + x^2
   end
  end
 end
 Output z(x,y)
```

42. What is the out of the program?

```
int a = 12, b = 20
  Output "max = %d\n", ((a + b) + abs(a - b)) / 2)
  Output "min = %d", ((a + b) - abs(a - b)) / 2)
```

43. What is the out of the program?

```
for x = 1:4
   for y =1:4
    if x==y   or x>y
    z(x,y)=xy-y
    else
    z(x,y)= y^2 − x^2
   end
  end
end
Output z(x,y)
```

44. What is the output of the program?

```
for x = 1:4
   for y =1:4
    if x==y   or x>y
    z(x,y)=x^2+x
    else
    z(x,y)= y^2 + y
   end
  end
end
Output z(x,y)
```

45. What is the output of the program?

```
for x = 1:4
   for y =1:4
    if x==y   or x>y
    z(x,y)=x^2- y
    else
    z(x,y)= y^2 − x
    end
   end
 end
 Output z(x,y)
```

46. What is the output of the program?

```
for x = 1:4
   for y =1:4
    if x==y   or x>y
    z(x,y)=x^2
    else
    z(x,y)= y^2
    end
   end
 end
 Output z(x,y)
```

47. What is the output of the program?

```
A = 0
B = 1
for x = 1:3
 for y =1:3
    if x<y
      z(x,y) =A;
    elseif x>y
       z(x,y) = B
    else
       z(x,y) = B
    end
  end
end
Output z
```

48. What is the output of the program?

```
for x = 1:3
 for y =1:3
    if x==y
      z(x,y) = (x+y)/2
    else
       z(x,y) =x-y
    end
  end
end
```

49. What is the output of the program?

```
for x = 1:4
 for y =1:4
  if x==y   or x>y
  z(x,y)= x − y
  else
  z(x,y) y − x
  end
 end
end
Output z(x,y)
```

50. What is the output of the program?

a = 12: b = 6: c = 3: d = 1

if a > b+c then a =a*a else a = 2a

if a - b * c < a*b*c then b = b*c else b = b – c

if (a < b + c) then c = c * c else c = d + c

print $\frac{a}{b}$ + 2c + b*b+ (a+b+c+d)b -d

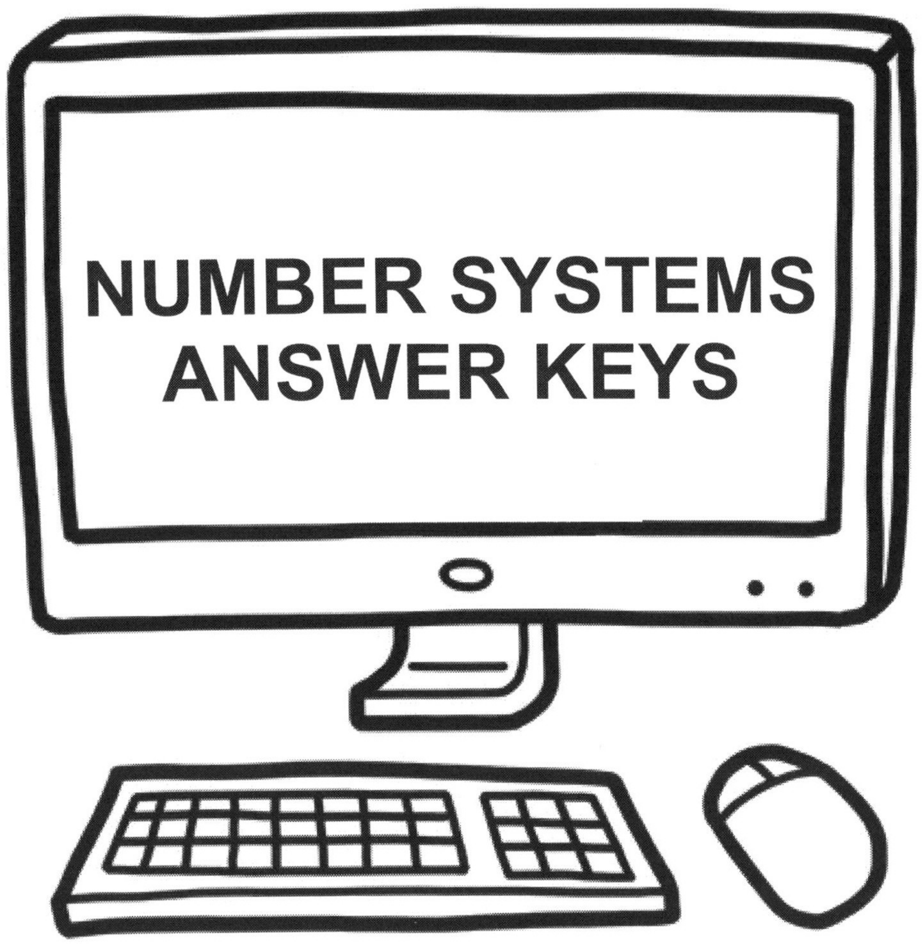
NUMBER SYSTEMS
ANSWER KEYS

1. Convert 309 into hexadecimal.

16	309	
16	19	Reminder 4
16	1	3
	0	1

=0x135

2. Convert 0x7A to binary.
   7 = 0111
   A = 1010
   The binary representation will be 0111 1010

3. Convert 502 into hexadecimal.

16	502	
16	31	Reminder 6
16	1	15
	0	1

=0x1F6

4. Convert 0x6B0 to binary.
   6 = 0110
   B = 1011
   0 = 0000
   The binary representation will be 0110 1011 0000

5. Convert 532 into hexadecimal.

16	532	
16	33	Reminder 4
16	2	1
	0	2

=0x214

6. Convert the following to decimal system representation:
110101 + 101001

32	16	8	4	2	1
1	1	0	1	0	1

=32 + 16 + 4 + 1 = 53

32	16	8	4	2	1
1	0	1	0	0	1

= 32 + 8 + 1 = 41
Therefore, 53 + 41 = 94

7. Solve 0xAF2– 0xC2 by converting into decimal.
We know A = 10, F = 15
So 0Xaf2 = (256 x 10) + (16 x 15) + (1 x 2) = 2802

And C = 12
So 0xC2 = (16 x 12) + (1 x 2) = 194
2802 – 194 = 2608

8. $Solve\ A_8 = 0001\ 0001\ 0001\ \ B_8 = 0100\ 0010. Find\ A_8 - B_8$

$$A_8 = 273\ \ B_8 = 66$$

$$A_8 - B_8 = 207$$

9. Solve 0xE2– 0xC4 by converting into decimal.
   We know E = 12
   So 0xE2 = (16 x 14) + (1 x 2) = 226

   And C = 12
   So 0xC2 = (16 x 12) + (1 x 4) = 196
   226 – 196 = 30

10. Solve A - 25 = 10 and represent the answer in decimal representation.
    A = 35

128	64	32	16	8	4	2	1
35<128 so 0	35 < 64 so 0	35> 32 35-32=3 1	3 < 16 so 0	3< 8 so 0	3< 4 so 0	3> 2 so 1 1	1= 1 1

    0010 0011

11. Solve A - 15 = 24 and represent the answer in decimal representation.
    A = 39

128	64	32	16	8	4	2	1
39<128 so 0	39 < 64 so 0	39> 32 39-32=7 1	7 < 16 so 0	7< 8 so 0	7> 4 so 3 1	3> 2 so 1 1	1= 1 1

    0010 0111

12. Solve $x_{16} = 3676_8$
    $=011\ 110\ 111\ 110_2$
    =7BE

13. $Solve\ A_8 = 0001\ 0000\ 1000\ \ B_8 = 0110\ 0000. Find\ A_8 - B_8$

$$A_8 = 264\ \ B_8 = 96$$

$$A_8 - B_8 = 168$$

14. Convert the following to decimal system representation:
010101 + 101111

32	16	8	4	2	1
0	1	0	1	0	1

=16+4+1 = 21

32	16	8	4	2	1
1	0	1	1	1	1

= 32+8+4+2+1 = 47
Therefore, 21 + 47 = 68

15. Solve 0x1E2– 0xC4 by converting into decimal.
We know E = 12
So 0xE2 = (256 x 1) + (16 x 14) + (1 x 2) = 704

And C = 12
So 0xC2 = (16 x 12) + (1 x 4) = 196
704 – 196 = 508

16. Represent E21 in binary.
=1110 0010 0001

17. Solve$A_{16} = 0001\ 1100\ 0001$

$$A_{16} = 1C1$$

18. Solve$B_{16} = 1000\ 1101\ 0101$

$$B_{16} = 8D5$$

19. Solve$B_8 = E1A$

$$14 \times 8^3 + 1 \times 8^2 + 10 \times 8^1 = 7032$$

Now converting 3610_{10} to octal

```
3610 | 8
-3610 | 451 | 8
   2   -451 | 56 | 8
          3  -56 | 7
               0
```

$= 7032$

20. Convert 0xD02 to binary.
D = 1101
0 = 0000
2 = 0010
The binary representation will be 1101 0000 0010

21. $\text{Solve}B_8 = 23D$
$$B_8 = 2 \times 8^3 + 3 \times 8^2 + 13 \times 8^1 = 1075$$

22. $\text{Solve}B_{16} = 0100\ 1100\ 0011$
$$\text{Solve}B_{16} = 4C3$$

23. $\text{Solve}B_8 = F4B$
$$\text{Solve}B_8 = F \times 8^3 + \ 4 \times 8^2 \ + B \times 8^1$$
= 7513

24. $\text{Solve}B_8 = 0000\ 0001\ ; A_8 = 0100\ 1000.\ Solve\ B_8 + A_8$
$$B_8 + A_8 = 1 + (110) = 111$$

25. $\text{Solve}B_8 = 0001\ 0010\ 0101\ ; A_8 = 0100\ 1000\ 1010\ .\ Solve\ B_8 + A_8$
$$B_8 + A_8 = (445) + (2212) = 2657$$

26. Convert 0xABD to binary.
A = 1010
B = 1011
D = 1101
The binary representation will be 1010 1011 1101

27. $\text{Solve}x_8 = 1000\ 0101\ ; y_8 = \ 1000\ 1010\ .\ Solve\ x_8 + y_8$
$$x_8 + y_8 = (135) + (138) = 273$$

28. $\text{Solve}x_8 = 0111\ 0101\ ; y_8 = 1000\ 1010\ .\ Solve\ x_8 + y_8$
$$x_8 + y_8 = (117) + (138) = 255$$

29. $\text{Solve}x_8 = 1110\ 1000\ ; y_8 = 1000\ 0000\ 0000\ .\ Solve\ x_8 + y_8$
$$x_8 + y_8 = (232) + (4096) = 4328$$

30. Convert 0xA47 to binary.
A = 1010
4 = 0100
7 = 0111
The binary representation will be 1010 0100 0111

31. $\text{Find}B_{16} = 0111\ 1011\ 0011$

$$\text{Solve}B_{16} = 7B3$$

32. $\text{Find}B_{16} = 0100\ 1100\ 1100$

$$\text{Solve}B_{16} = 4CC$$

33. Convert 0x1E4 to binary.
    1 = 0001
    E = 1110
    4 = 0100
    The binary representation will be 0001 1110 0100

34. $\text{Find}B_{16} = 0100\ 1100\ 0011$

$$\text{Solve}B_{16} = 4C3$$

35. Convert 0x00F to binary.
    0 = 0000
    F = 1111
    The binary representation will be 0000 0000 1111

36. $\text{Find}B_8 = 0001\ 0011$

$$\text{Solve}B_8 = 23$$

37. Convert 0x5D to binary.
    5 = 0101
    D = 1101
    The binary representation will be 0101 1101

38. $\text{Find}B_8 = 00111001$

$$\text{Solve}B_8 = 32 + 16 + 8 + 1 = 57$$

39. $If\ a = 2, Find\ A_a = 1A2$

$$A_2 = 0001\ 1010\ 0010$$

40. $If\ a = 2, Find\ A_a = 34$

128	64	32	16	8	4	2	1
34<128 so 0	34 < 64 so 0	34> 32 34-32=2 1	2 < 16 so 0	2 < 8 so 0	2< 4 so 0	2 = 2 so 2-2 = 0 1	0 < 1 0

0010 0010

41. $If\ a = 2, Find\ A_a = 68$

128	64	32	16	8	4	2	1
68<128 so	68 > 64 so 68-64=4	4< 32	4 < 16 so	4 < 8 so	4= 4 so	0< 2 so	0 < 1

0	1	0	0	0	1	0	0

0100 0100

42. $If\ a = 2, Find\ A_a = 130$

128	64	32	16	8	4	2	1
130>128 2	2< 64 so	2< 32	2 < 16 so	2 < 8 so	2<4 so	2= 2 so	0 < 1
1	0	0	0	0	0	1	0

1000 0010

43. $If\ a = 2, Find\ x_a = 120$

128	64	32	16	8	4	2	1
120<128 so	120 > 64 120-64= 56	56> 32 56-32=24	24 > 16 24-16=8	8= 8 so	0< 4 so	0< 2 so	0 < 1
0	1	1	1	1	0	0	0

01111000

44. $If\ A_8 = 0010\ 0101\ \ B_8 = 1000\ 0000. Find\ A_8 + B_8$

$$A_8 = 45\ \ B_8 = 200$$
$$A_8 + B_8 = 245$$

45. Convert 0x49A to binary.

4 = 0100

9 = 1001

A = 1010

The binary representation will be 0100 1001 1010

46. $Solve\ A_8 = 1001\ 0001\ \ B_8 = 0100\ 0001. Find\ A_8 - B_8$

$$A_8 = 221\ \ B_8 = 101$$
$$A_8 - B_8 = 120$$

47. $Solve\ A_8 = 1101\ 1001\ \ B_8 = 0100\ 0001. Find\ A_8 - B_8$

$$A_8 = 217\ \ B_8 = 65$$
$$A_8 - B_8 = 152$$

48. $Solve\ A_8 = 0001\ 0000\ 0001\ \ B_8 = 0100\ 0001. Find\ A_8 - B_8$

$$A_8 = 401\ \ B_8 = 101$$
$$A_8 - B_8 = 300$$

49. Convert 0x9C to binary.

9 = 1001

C = 1100

The binary representation will be 1001 1100

50. $If\ a = 2, Find\ A_a = 256$

256	128	64	32	16	8	4	2	1
256=256 1	0<128 0	0 < 64 0	0< 32 0	0 < 16 0	0 < 8 0	0< 4 0	0 < 2 0	0 < 1 0

0001 0000 0000

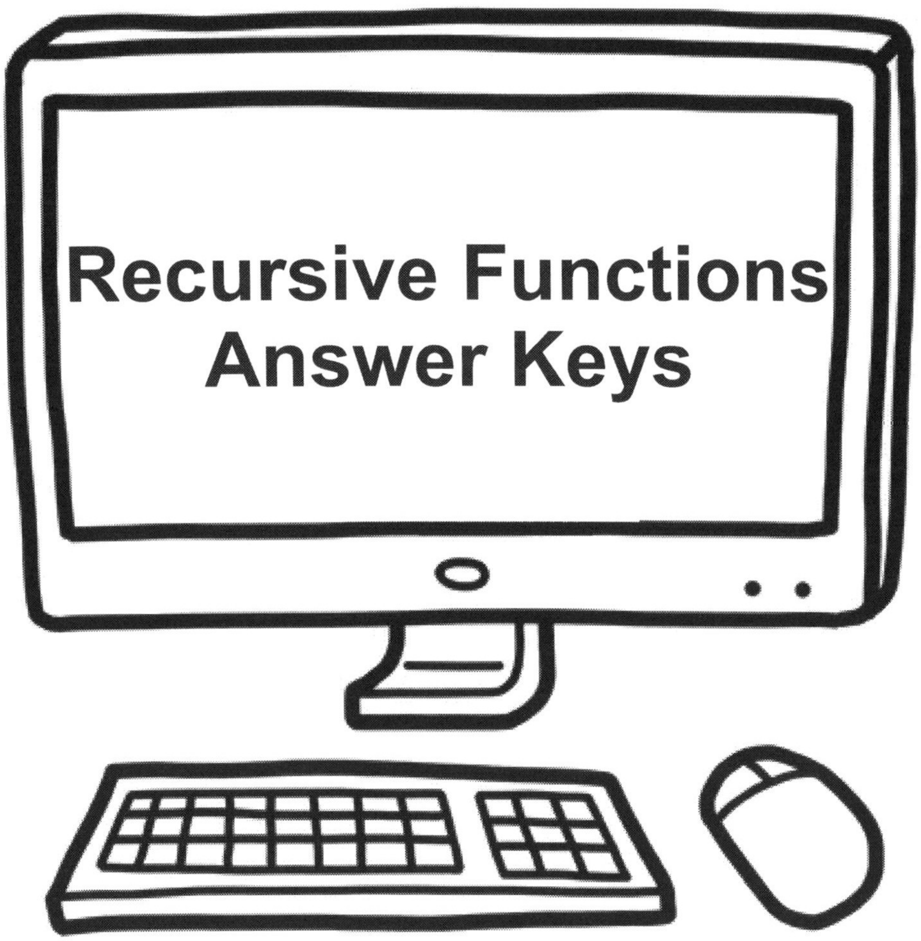
Recursive Functions
Answer Keys

1. *Find* $f(7,3)$

$$f(x,y)=\begin{cases} f(x-3,y+2) & when\; x>y \\ f(x+y,y+4) & x<y \\ x+\dfrac{y}{x}+xy & x=y \end{cases}$$

$= f(7-3,3+2) = f(4,5)$
$= f(9,9) = 9+1+81 = 91$

2. *Find* $f(8,5)$

$$f(x,y)=\begin{cases} f(x-y,y+x) & when\; x>y \\ x+y^2 & otherwise \end{cases}$$

$= f(3,13) = 3+169 = 172$

3. *Find* $f(a,b)$ *where* $a>b$ *and* $a=b-a$

$$f(x,y)=\begin{cases} f(x+3a,y+b) & when\; x<y \\ x+y^2 & otherwise \end{cases}$$

$= f(a+3a,b+b) = f(4a,2b) = 4a+4b^2 = 4a+4(2a^2) = 4a+8a^2$

4. *Find* $f(7,4)$

$$f(x,y)=\begin{cases} f(x-y,y+2) & when\; x>y \\ x-\dfrac{y}{x} & otherwise \end{cases}$$

$= f(7-4,4+2) = f(3,6)$
$= f\,(3-\frac{6}{3}) = 1$

5. Complete the missing values in the recursive formula of the sequence: 2,7,12, ...

$$\begin{cases} a(1)=A \\ a(n)=e(n-1)+B \\ A=?,B=? \end{cases}$$

The two parts of the formula should give the following information:

1. The first term (which is 2)
2. The rule to get any term from its previous term (which is 5)

A = 2, B = 5

6. $Find\ C(5)$

$$C(n) = \begin{cases} C(0) = u \\ C(1) = v \\ C(n) = C(n-2) - 2 \end{cases}$$

$C(1) = v$
$C(3) = v - 4$
$C(5) = v - 6$

7. $Find\ f(7)\ if\ f(x) = 2 \times f(x-2) + 4\ and\ we\ know\ that\ f(3) = 10$

$f(5) = 2 \times f(3) + 4$

$f(5) = 20 + 4 = 24$

$f(7) = 2.f(5) + 4$

$f(7) = 2(24) + 4$

$f(7) = 52$

8. $Find\ f(8)\ if\ f(x) = 4 \times\ f(x-3) + 5\ and\ we\ know\ that\ f(2) = 10$

$f(5) = 4.f(2) + 5 = 45$

$f(8) = 4.f(5) + 5 = 4(45) + 5 = 185$

9. Evaluate the following recursive function:

$$Find\ f(3)\ if\ f(x) =\ f(x-1) + 2.x^2\ and\ we\ know\ that\ f(1) = -2$$

$f(2) = f(1) + 2(2^2) = -2 + 8 = 6$
$f(3) = f(2) + 2(3^2) = 6 + 18 = 24$

10. $Find\ f(4)\ if\ f(x) =\ f(x-2) + 3x\ and\ we\ know\ that\ f(0) = 5$
$f(2) = f(0) + 3(2) = 5 + 6 = 11$
$f(4) = f(2) + 3(4) = 11 + 12 = 23$

11. $Find\ C(4)$

$$C(x) = \begin{cases} C(0) = b \\ C(x) = C(x-2) + x^3 \end{cases}$$

$C(2) = C(0) + 2^3 = b + 8$
$C(4) = C(2) + 4^3 = (b + 8) + 64 = b + 72$

12. *Find* $C(6)$

$$M(n) = \begin{cases} M(0) = \frac{a}{4} \\ M(n) = M(n-3) + n \end{cases}$$

$$M(0) = \frac{a}{4}$$
$$M(3) = M(0) + n = \frac{a}{4} + 3$$
$$M(6) = M(3) + 6 = \frac{a}{4} + 3 + 6 = \frac{a}{4} + 9 = \frac{a+36}{4}$$

13. *Find* $C(8)$ *Find if* $f(x) = f(x-a) + a^2$ *and we know that* $f(0) = b$, *where* $a^2 = 2b$ *and* $b = 8$

$According\ to\ the\ question, b = 8\ \ and\ a = 16^{\frac{1}{2}} = 4$
$Therefore, f(x) = f(x-4) + 16\ and\ f(0) = 8$
$f(4) = f(0) + 16 = 8 + 16 = 24$
$f(8) = f(4) + 16 = 24 + 16 = 40$

14. Complete the missing values in the recursive formula of the sequence: -4, -2, 0, . . .

$$\begin{cases} a(1) = A \\ a(n) = e(n-1) + B \\ A = ?, B = ? \end{cases}$$

The two parts of the formula should give the following information:

1. The first term (which is -4)
2. The rule to get any term from its previous term (which is +2)

A = -4, B = 2

15. *Find* $a(3)$

$$a(n) = \begin{cases} a(1) = -3 \\ a(n) = \frac{a(n-1)^2}{2} \end{cases}$$

$$a(1) = -3$$
$$a(2) = \frac{a(1)^2}{2} = \frac{9}{2}$$
$$a(3) = \frac{a(2)^2}{2} = \frac{\left(\frac{9}{2}\right)^2}{2} = \frac{81}{8}$$

16. *Find* $f(4,6)$

$$(x,y) = \begin{cases} f(x+y, y-1) & when\ x < y \\ f\left((x^2-75)^{\frac{1}{2}}, y\right) & when\ x > y \\ \left(x+\frac{y}{x^2}\right)^{\frac{1}{2}} & otherwise \end{cases}$$

$$= f(10,5) = f(5,5) = \left(\frac{26}{5}\right)^{\frac{1}{2}} = 5.2^{\frac{1}{2}} = 2.28$$

17. *Find* $f(3,7)$

$$f(x,y) = \begin{cases} f(x+1, y) & when\ x < y \\ f\left(\frac{x}{4}, y\right) & when\ x > y \\ x+\frac{xy}{4} & otherwise \end{cases}$$

$= f(4,7)$
$= f(5,7)$
$= f(6,7)$
$= f(7,7) = 7 + \frac{49}{4} = \frac{28+49}{4} = \frac{77}{4}$

18. *Find* $f(a,b)$ *where* $a^{\frac{1}{2}} = \sqrt{2}\, b$ *where* $b = 2$

$$f(x,y) = \begin{cases} f(x, y+a) & when\ x > y \\ x+\frac{y}{xy} & otherwise \end{cases}$$

According to the questions
$b = 2, a = 8, therefore$
$f(8,2) = f(8,2+8) = f(8,10)$
$= 8 + \frac{10}{80} = \frac{65}{8}$

19. Evaluate the following recursive functions: *Find* $f(6,4)$

$$f(x,y) = \begin{cases} f(x-3, y+2) & when\ x > y \\ f(x, y-3) & x < y \\ \frac{x^2}{y} & x = y \end{cases}$$

$= f(6-3, 4+2) = f(3,6)$
$= f(3,3) = \frac{9}{3} = 3$

20. $Find\ f(3)\ if\ f(x) = f(x-1) + 4\ and\ we\ know\ that\ f(1) = 4$
$f(2) = f(1) + 4 = 4 + 4 = 8$
$f(3) = f(2) + 4 = 8 + 4 = 12$

21. $Find\ C(4)$

$$C(x) = \begin{cases} C(0) = b \\ C(x) = C(x-2) + b \end{cases}$$

$C(2) = C(0) + b = 2b$
$C(4) = C(2) + b = 3b$

22. $Find\ f(3)$

$$f(x) = \begin{cases} f(x-1)^2 & when\ x > 1 \\ 1 & x = 1 \end{cases}$$

$f(1) = 1$
$f(2) = (f(2-1))^2 = (f(1))^2 = 1$
$f(3) = (f(3-1))^2 = (f(2))^2 = 1$

23. Complete the missing values in the recursive formula of the sequence: 3, 8, 13, ...

$$\begin{cases} a(1) = A \\ a(n) = e(n-1) + B \\ A =?, B =? \end{cases}$$

The two parts of the formula should give the following information:

1. The first term (which is 3)
2. The rule to get any term from its previous term (which is 5)

A = 3, B = 5

24. $Find\ f(3,6)$

$$f(x,y) = \begin{cases} f(x^2 + y, y) & when\ x < y \\ x - y & otherwise \end{cases}$$

$= f(9+6,6) = f(15,6)$
$= 15 - 6 = 9$

25. $Find\ f(1,1)$

$$f(x,y) = \begin{cases} f(x^2, y^2 + 1) & when\ x > y \\ x - \dfrac{x}{y+x} & otherwise \end{cases}$$

$= f(1^2, 1^2 + 1) = f(1,2)$
$= 1 - \dfrac{1}{1+1} = 1 - \dfrac{1}{2} = \dfrac{1}{2}$

26. *Find* $f(5,1)$

$$f(x,y) = \begin{cases} f(x+1, 2y+1) & when\ x > y \\ x^2 - \dfrac{x}{y^2 + x} & otherwise \end{cases}$$

$= f(6,3)$

$= f(7,7) = 7^2 - \dfrac{7}{7^2+7} = 49 - \dfrac{7}{56} = 48.8$

27. *Find* $f(8)$ *if* $f(x) = f(x-2) + \sqrt{x}$ *and we know that* $f(2) = 3$

$f(4) = f(2) + \sqrt{4} = 3 + 2 = 5$
$f(6) = f(4) + \sqrt{6} = 5 + 2.44 = 7.44$
$f(8) = f(6) + \sqrt{8} = 7.44 + 2.82 = 10.26$

28. *Find* $C(5)$ *if* $C(x) = C(x-2) + x$ *and we know that* $C(1) = a$

$C(3) = C(1) + 3 = a + 3$
$C(5) = C(3) + 5 = a + 3 + 5 = a + 8$

29. Evaluate the following recursive functions:
Fibonacci numbers form a sequence defined recursively by the following equations:

$$f(x) = \begin{cases} f(0) = 0 \\ f(1) = 1 \\ f(n+2) = f(n) + f(n+1) \end{cases}$$

Find the fibonacci series till $f(5)$.
Write down the complete series and comment on the series.

$f(2) = f(0) + f(1) = 1$
$f(3) = f(1) + f(2) = 2$
$f(4) = f(2) + f(3) = 3$
$f(5) = f(3) + f(4) = 5$

The complete series: 0, 1, 1, 2, 3, 5. It can be observed that the series starts from two "1"s and the numbers after that is obtained by adding the two previous numbers. This kind of series is known as Fibonacci.

30. Factorial of any natural number is defined by:

$$\begin{cases} f(0) = 1 \\ f(n) = n.f(n-1) \end{cases}$$

Find the Factorial of 3 using the given recursive formula.

$f(0) = 1$
$f(1) = 1.f(0) = 1(1) = 1$
$f(2) = 2.f(1) = 2$
$f(3) = 3.f(2) = 3(2) = 6$

31. *Find* $f(8)$

$$f(x) = \begin{cases} f(x-2) + \frac{a}{2} & x > 1 \\ 0 & x = 2 \end{cases}$$

$f(2) = 0$
$f(4) = f(2) + \frac{a}{2} = \frac{a}{2}$
$f(6) = f(4) + \frac{a}{2} = a$
$f(8) = f(6) + \frac{a}{2} = a + \frac{a}{2} = \frac{3}{2}a$

32. Imagine a square park of side 100 m. Martha has to run 1.6 Km in total as per her daily workout schedule. In a recursive manner, count the total number of rounds will Martha have to run considering the running track is along the edges of the park.
Analyse the calculations and create the function from your calculations.

$Round\ 1: 4(100m) = 400\ m$
$Round\ 2: 400 + 4(100) = 800\ m$
$Round\ 3: 800 + 4(100) = 1200\ m$
$Round\ 4: 1200 + 4(100) = 1600\ m$
The function can be written as:

$$f(1) = 4(100)$$
$$f(n) = f(n-1) + 4(100)$$

33. *Find* $C(6)$ *if* $C(x) = C(x-2) + \frac{x}{3}$ *and we know that* $C(2) = d$

$$C(4) = C(2) + \frac{4}{3} = d + 1.33$$
$$C(6) = C(4) + 2 = d + 3.33$$

34. *Find* $f(5)$

$$f(x) = \begin{cases} f(0) = 0 \\ f(1) = a \\ f(n+2) = f(n) + f(n+b), where\ b = 1 \end{cases}$$

$f(1) = a$
$f(3) = a + 1$
$f(5) = f(3) + f(4) = 2a + 3$

35. Complete the missing values in the recursive formula of the sequence: 3, 16, 29, ...

$$\begin{cases} a(1) = A \\ a(n) = e(n-1) + B \\ A =; ?\ B = ? \end{cases}$$

The two parts of the formula should give the following information:

1. The first term (which is 3)
2. The rule to get any term from its previous term (which is 13)

A = 3, B = 13

36. *Find* $f(3)$

$$f(x) = \begin{cases} f(x-1) - \frac{x}{2} & when\ x > 1 \\ 0 & x = 1 \\ 1 & x = 0 \end{cases}$$

$f(1) = 0$
$f(2) = f(1) - 1 = -1$
$f(3) = f(2) - \frac{3}{2} = -1 - \frac{3}{2} = -\frac{5}{2}$

37. *Find* $f(4)$

$$(x, y) = \begin{cases} f(x-2) + \frac{x^2}{2} + 1 & when\ x > 2 \\ 2 & x = 1{,}0 \\ -1 & x = 2 \end{cases}$$

$f(2) = -1$
$f(4) = f(2) + 8 + 1 = -1 + 8 + 1 = 8$

38. *Find* $f(2{,}1)$

$$f(x, y) = \begin{cases} f(x^2, y^2 + 4) & when\ x > y \\ x - \frac{x}{y + x} & otherwise \end{cases}$$

$$= f(2^2, 1^2 + 4) = f(4{,}5)$$
$$= 4 - \frac{4}{4+5} = 4 - \frac{4}{9} = \frac{32}{9}$$

39. Evaluate the following recursive functions: *Find* $f(2,3)$

$$f(x,y) = \begin{cases} f(x^2, \frac{y^2}{3}) & when\ x < y \\ x - y & otherwise \end{cases}$$

$= f\,(2^2, \frac{3^2}{3}) = f(4,3) = 4 - 3 = 1$

40. *Find* $f(m,n)$ *where* $m > n$ *such that* $m - n = 10$

$$f(x,y) = \begin{cases} f\,(\frac{x}{3}, y^3) & when\ x > y \\ x - y + 1 & otherwise \end{cases}$$

$= f\,(\frac{m}{3}, n^3)$

$= \frac{m}{3} - n^3 + 1$

41. *Find* $f(m,n)$ *where* $m > n$ *such that* $m - n = 5$

$$f(x,y) = \begin{cases} f\,(\frac{x}{4}, 4x) & when\ x > y \\ \frac{x}{4} + 4x + 1 & otherwise \end{cases}$$

$= f\,(\frac{m}{4}, 4n\,)$

$= \frac{m}{4} + 4n + 1$

42. *Find* $a(4)$

$$a(x) = \begin{cases} a = -1 & at\ x = 0 \\ a(x) = a(x-2) + 2x & at\ other\ values\ of\ x \end{cases}$$

$a(2) = a(0) + 2(2) = -1 + 4 = 3$

$a(4) = a(2) + 2(4) = 3 + 8 = 11$

43. *Find* $f(14,3)$

$$f(x,y) = \begin{cases} f(\frac{x}{7}, y + 1) & when\ x > y \\ f(x + 2, y) & x < y \\ x + \frac{yx}{x^2} & x = y \end{cases}$$

$= f\,(2{,}4) = f(2 + 2{,}4) = f(4{,}4)$

$= f(4{,}4) = 4 + 1 = 5$

44. Using the following recursive function, complete the series, complete the series considering the first element is the zeroth element,

$$\frac{1}{2},\frac{3}{2},\_,\_.$$

$$f(n) = 2.f(n-1) + \frac{1}{2}$$

$Given, f(0) = \frac{1}{2}; f(1) = \frac{3}{2}$

$f(2) = 2.f(1) + \frac{1}{2} = \frac{7}{2}$

$f(3) = 2.f(2) + \frac{1}{2} = \frac{15}{2}$

$Therefore, the\ series\ is\ \frac{1}{2},\frac{3}{2},\frac{7}{2},\frac{15}{2}$

45. Using the following recursive function, complete the series considering the first element is the zeroth element,

$$2,3,\_,\_.$$

$$f(n) = f(n-1) + n^2$$

$Given, f(0) = 2; f(1) = 3$

$f(2) = f(1) + 2^2 = 3 + 4 = 7$

$f(3) = f(2) + 3^2 = 7 + 9 = 16$

$Therefore, the\ series\ is\ 2,3,7,16$

46. $Complete\ the\ series\ of\ five\ elements\ using\ the\ data\ given\ below.$

$$f(x,y) = \begin{cases} (x + \frac{x}{a}) & when\ x > 3, a = 4 \\ 1 & x = 1,0 \\ 0 & x = 2 \end{cases}$$

$f(0) = 1$
$f(1) = 1$
$f(2) = \frac{5}{2}$
$f(3) = 3 + \frac{3}{4} = \frac{15}{4}$
$f(4) = 4 + 1 = 5$

47. Imagine a square park of side 80 m. Martha has to run 1.92 Km in total as per her daily workout schedule. In a recursive manner, count the total number of rounds will Martha have to run considering the running track is along the edges of the park.
Analyse the calculations and create the function from your calculations.

$Round\ 1: 4(80m) = 320\ m$
$Round\ 2: 320 + 4(80) = 640\ m$
$Round\ 3: 640 + 4(80) = 960\ m$
$Round\ 4: 960 + 4(80) = 1280\ m$
$Round\ 5: 1280 + 4(80) = 1600\ m$
$Round\ 6: 1600 + 4(80) = 1920\ m$
The function can be written as:

$$f(1) = 4(80)$$
$$f(n) = f(n-1) + 4(80)$$

48. Using the following recursive function, complete the series,

3, _, _, _.

$$f(n) = f(n-1) + \frac{n}{2}$$

$Given, f(1) = 3$

$$f(2) = f(1) + \frac{2}{2} = 3 + \frac{2}{2} = 4$$

$$f(3) = f(2) + \frac{3}{2} = 4 + \frac{3}{2} = \frac{11}{2}$$

$$f(4) = f(3) + \frac{4}{2} = \frac{11}{2} + \frac{4}{2} = \frac{15}{2}$$

49. $Find\ f(2,6)$

$$f(x,y) = \begin{cases} f(x^2 + y, y) & when\ x < y \\ x - y & otherwise \end{cases}$$

$= f(10,6)$
$= 10 - 6 = 4$

50. $Find\ f(5,10)$

$$f(x,y) = \begin{cases} f(2x^2, y) & when\ x < y \\ x - y & otherwise \end{cases}$$

$= f(50,10)$
$= 50 - 10 = 40$

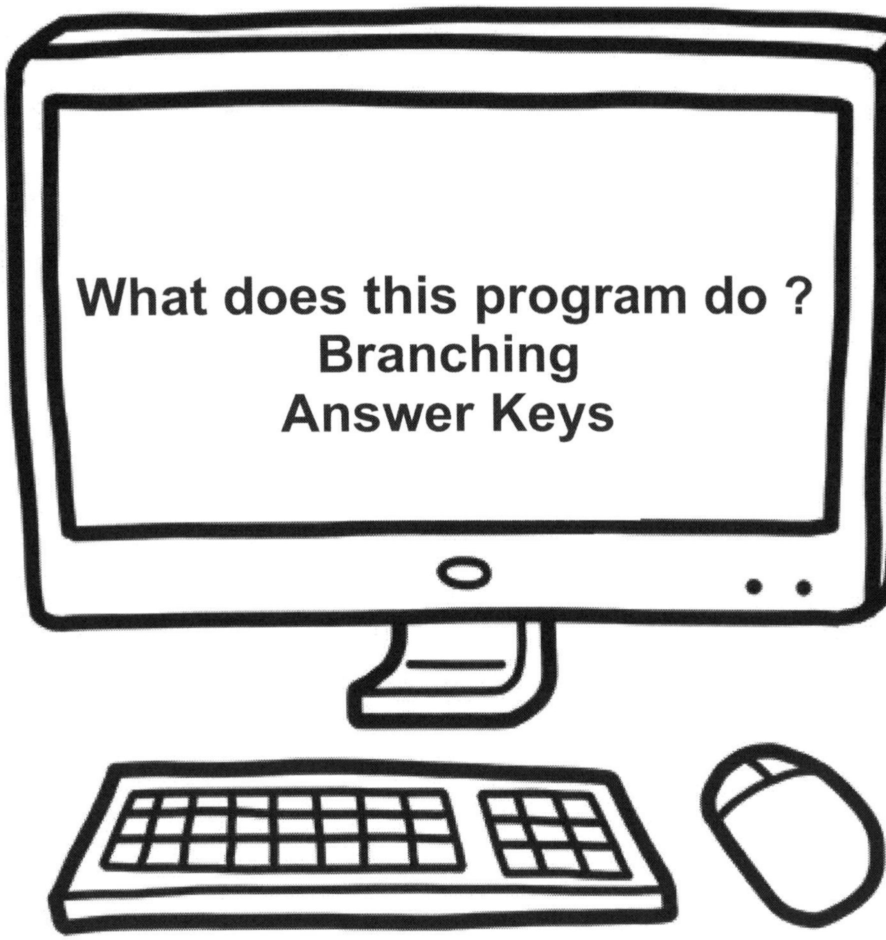
What does this program do ?
Branching
Answer Keys

1. What is the output of the program?

```
a=0
for i=1:2:20
   a=a+1
   b=a+1
   if a>4
      break
   end
   Output b
end
```

Output

2
3
4
5

2. What is the output of the program?

```
a=1
for i=1:2:20
   a=a+i
   b=a+1
   if b>25
      break
   end
   Output b
end
```

Output

3
6
11
18

3. What is the output of the program?

```
a=1
for i=1:2:20
   a=a+i
   b=a+1
   if b>25
      break
   end
   c=b+1
   Output c
end
```

Output

4
7
12
19

4. What is the output of the program?

```
a=2
for i=1:2:10
   a=a+i
   b=a+1
   if b>45
      break
   end
   c=b+1
   Output c
end
```

Output

5
8
13
20
29

5. What is the output of the program?

```
a=0
for i=1:2:10
   a=a+i
   b=a+1
   if b>45
      break
   end
   c=b+a
   Output c
end
```

Output

3
9
19
33
51

6. What is the output of the program?

```
a=0
for i=1:2:10
   a=a+i
   b=a+1
   if b>45
      break
   end
   c=b+a
   d=c+i
   if d>25
   break
   end
   Output d
end
```

Output

4
12
24

7. What is the output of the program?

```
a=1
for i=1:2:10
   a=a+i
   b=a+1
   if b>40
      break
   end
   c=b+1
   d=c+1
   if d>25
      break
   end
   Output d

end
```

Output

5
8
13
20

8. What is the output of the program?

```
a=1
for i=1:3:15
   a=a+i
   b=a+1
   if b>40
      break
   end
   c=b+1
   d=c+1
   if d>25
      break
   end
   Output d

end
```

Output

5
9
16

9. What is the output of the program if the user enters x =2329?

```
if (x > 0) || (x<0)
  Output "valid"
else if (x == 0)
  Output "invalid"
else
  Output "Enter a valid password"
```

Output

valid

10. What is the output of the program?

```
a=1
for i=1:3:15
   a=a+i
   b=a+1
   if b>30
      break
   end
   c=b+1
   d=c+1
   if d>30
      break
   end
   Output d

end
```

Output

5
9
16
26

11. What is the output of the program?

```
a=1
for i=1:3:15
   a=a*i
   b=a+1
   if b>300
      break
   end
   c=b-100
   d=c+1
   Output d
end
```

Output

-97
-94
-70
182

12. What is the output of the program?

```
a=
for i=1:2:10
   a=a*i
   b=a+
   if b>300
      break
   end
   c=b-100
   d=c+1
   Output d

end
```

Output

-97
-95
-83
7

13. What is the output of the program?

```
a=1
for i=1:2:10
   a=i
   b=a+1
   if b>300
      break
   end
   c=b-100
   d=c+100
   Output d

end
```

Output

2
4
6
8
10

14. What is the output of the program if s=50?

```
t = 100
s = 50
if t>s
   Output "neg"
elseif t<s
   Output "pos"
else
   Output "error"
end
```

Output

neg

15. What is the output of the program?

```
a=1
for i=1:5:20
   a=
   b=a+1
   if b>300
      break
   end
   c=b-100
   d=c+100
   Output d
end
```

Output

2
7
12
17

16. What is the output of the program if age = 15?

```
  age =15
  if(age < 0 || age > 150)
     Output "Error."
  else
     Output "Age entered is valid"
```

Output

Age entered is valid

17. What is the value of a after the execution of the statement if a = 55 and b = 20?

```
a= (a <b) ? b : -b ;
```

Output

-20

18. What is the value of a after the execution of the statement if a = 4 and b = 30?

```
a= (a <b) ? b : -b ;
```

Output

30

19. What is the value of a after the execution of the statement if a = 5 and b = 25?

```
a= (a <b) ? -b : b ;
```

Output

-25

20. What is the value of a after the execution of the statement if a = 23 and d = 32?

```
a= (a > d) ? d : -d ;
```

Output

-32

21. What is the output of the program?

```
for i=1:8
  a=2
  b=a*i
  c=0
  if b<11
  c=c+b
  else
     c=c-b
  end
  Output c
end
```

Output

2
4
6
8
10
-12
-14
-16

22. What is the output of the program?

```
for i=1:2:10
  a=2
  b=a+i
  c=0;
  if b<10
  c=c+b
  else
     c=c-b
  end
  Output c
end
```

Output

3
5
7
9
-11

23. What is the output of the program?

```
h=10
for i=1:2:10
  a=2
  b=a+i
  c=
  if b<10
  c=c+b
  else
     c=c-b
  end
  d=c+h
  Output d
end
```

Output

13
15
17
19
-1

24. What is the output of the program?

```
h=5
for i=1:2:10
  a=2
  b=a*i
  c=1
  if b<10
  c=c+10
  else
     c=c-10
  end
  d=c+h;
  Output d
end
```

Output

16
16
-4
-4
-4

25. What is the output of the program?

```
e=2
for i=1:2:10
  a=2
  b=a*i;
  c=1;
  if b<10
  c=c+10
  else
     c=a+i
  end
  d=c+e
  Output d
end
```

Output

13
13
9
11
13

26. What is the output of the program?

```
e=2
for i=1:2:10
  a=2
  b=a*i
  c=1
  if b<10
  c=c+10
  else
     Output "*"
  end
  d=c+e;
  Output d
end
```

Output

13
13
*3
*3
*3

27. What is the output of the program?

```
for i=1:2:10
  a=2
  b=a*i
  %c=2
  if b<10
  c= Output "$"
  else
  c= Output "*"
  end

  Output c
end
```

Output

$
$
*
*
*

28. What is the output of the program?

```
for i=1:2:10
  a=3
  b=a*i
  %c=2
  if b<15
  c=Output "*"
  else
  c=Output "$"
  end

  Output c
end
```

Output

*
*
$
$
$

29. What is the output of the program?

```
for i=1:2:10
  a=3
  b=a*i
  %c=2
  if b<15
  Output "small"
  else
  Output "big"
  end
end
```

Output

Smallsmallbigbigbig

30. What is the output of the program?

```
for i=1:2:10
  a=3
  b=a+i
  c=2
  if b<10
  Output "small\n"
  else
  Output "big\n"
  end
end
```

Output

```
small
small
small
big
big
```

31. What is the output of the program?

```
for i=1:2:10
  a=5
  b=a*i
  %c=2
  if b<10
  Output "*\n"
  elseif b>10
  Output "#\n"
  else
  Output "&\n"
  end
end
```

Output

```
*
#
#
#
#
```

32. What is the output of the program?

```
c=zeros(5,5)
for i=1:5
   for j=1:5
      if i==j
      c(i,j)=9
      end
   end
end
Output c
```

Output

9	0	0	0	0
0	9	0	0	0
0	0	9	0	0
0	0	0	9	0
0	0	0	0	9

33. What is the output of the program?

```
c=zeros(5,5)
for i=1:5
   for j=1:5
      if i==j
      c(i,j)=9
      elseif i<j
         c(i,j)=10
      else
         c(i,j)=
      end
   end
end
Output c
```

Output

9	10	10	10	10
1	9	10	10	10
1	1	9	10	10
1	1	1	9	10
1	1	1	1	9

34. What is the output of the program?

```
c=zeros(5,5)
for i=1:5
   for j=1:5
     if i==j
     c(i,j)=9
     elseif i<j
        c(i,j)=i+j
     else
        c(i,j)=i-j
     end
   end
end
Output c
```

Output

9	3	4	5	6
1	9	5	6	7
2	1	9	7	8
3	2	1	9	9
4	3	2	1	9

35. What is the output of the program?

```
c=zeros(5,5)
for i=1:5
   for j=1:5
     if i==j
     c(i,j)=i
     elseif i<j
        c(i,j)=i*j
     else
        c(i,j)=i-j
     end
   end
end
Output c
```

Output

1	2	3	4	5
1	2	6	8	10
2	1	3	12	15
3	2	1	4	20
4	3	2	1	5

36. What is the output of the program?

```
c=zeros(4,4)
for i=1:4
   for j=1:4
      if i==j
      c(i,j)=j
      elseif i<j
         c(i,j)=i*j
      else
         c(i,j)=i/j
      end
   end
end
Output c
```

Output

1.0000	2.0000	3.0000	4.0000
2.0000	2.0000	6.0000	8.0000
3.0000	1.5000	3.0000	12.0000
4.0000	2.0000	1.3333	4.0000

37. What is the output of the program?

```
c=zeros(4,4)
for i=1:4
   for j=1:4
      if i==j
      c(i,j)=j+i
      elseif i<j
         c(i,j)=i+j
      else
         c(i,j)=i-j
      end
   end
end
Output c
```

Output

2	3	4	5
1	4	5	6
2	1	6	7
3	2	1	8

38. What is the output of the program?

```
c=zeros(4,4)
for i=1:4
   for j=1:4
      if i==j
      c(i,j)=j
      else
         break
      end
   end
end
Output c
```

Output

1	0	0	0
0	0	0	0
0	0	0	0
0	0	0	0

39. What is the output of the program?

```
c=zeros(5,5)
for i=1:5
   for j=1:5
      if i>j
      c(i,j)=j+i
      else
         break
      end
   end
end
Output c
```

Output

0	0	0	0	0
3	0	0	0	0
4	5	0	0	0
5	6	7	0	0
6	7	8	9	0

40. What is the output of the program?

```
c=zeros(5,5)
for i=1:5
   for j=1:5
      if i<j
      c(i,j)=j*i
      else
         break
      end
   end
end
Output c
```

Output

0	0	0	0	0
0	0	0	0	0
0	0	0	0	0
0	0	0	0	0
0	0	0	0	0

41. What is the output of the program?

```
a=1
for i=1:5:20
   a=i
   b=a+1
   if b>200
      break
   end
   c=b+100
   d=c+100
   Output d

end
```

Output

202
207
212
217

42. What is the output of the program if the age=57?

```
   age= 57
   if(age >= 30 && age <= 70)
      Output "You're between 30 and 70 and cannot save money on your health insurance!"
   else

      Output "Contact the office, we may have deals for you"
```

Output

You're between 30 and 70 and cannot save money on your health insurance!

43. What is the output of the program?

```
c=zeros(5,5)
for i=1:5
   for j=1:5
      if i==j
      c(i,j)=j/i
      else
         c(i,j)=i+j
      end
   end
end
Output c
```

Output

1	3	4	5	6
3	1	5	6	7
4	5	1	7	8
5	6	7	1	9
6	7	8	9	1

44. What is the output of the program?

```
c=zeros(5,5)
for i=1:5
   for j=1:5
      if i<j
      c(i,j)=0
      elseif i>j
         c(i,j)=i+j
      else
         c(i,j)=1
      end
   end
end
Output c
```

Output

1	0	0	0	0
3	1	0	0	0
4	5	1	0	0
5	6	7	1	0
6	7	8	9	1

45. What is the output of the program?

```
c=zeros(5,5)
for i=1:5
   for j=1:5
      if i<j
      c(i,j)=0
      elseif i>j
         c(i,j)=1
      else
         c(i,j)=i+j
      end
   end
end
Output c
```

Output

2	0	0	0	0
1	4	0	0	0
1	1	6	0	0
1	1	1	8	0
1	1	1	1	10

46. What is the output of the program?

```
for i=1:2:10
  a=4
  b=a+3+i
    if b<10
  Output "*"
  elseif b>10
  Output "#"
  else
  Output "'&"
  end
end
```

Output

```
*
&
#
#
#
```

47. What is the output of the program?

```
for i=1:2:10
  a=2
  b=a+i
  %c=2
  if b<10
  Output "*\n"
  elseif b>10
  Output "#\n"
  else
  Output "&\n"
  end
end
```

Output

```
*
*
*
*
#
```

48. What is the output of the program?

```
c=zeros(5,5)
for i=1:5
   for j=1:5
      if i<j
      c(i,j)=10
      elseif i>j
         c(i,j)=100
      else
         c(i,j)=1000
      end
   end
end
Output c
```

Output

```
   1000      10      10      10      10
    100    1000      10      10      10
    100     100    1000      10      10
    100     100     100    1000      10
    100     100     100     100    1000
```

49. What is the output of the program?

```
c=zeros(6,6)
for i=1:6
   for j=1:6
      if i<j
      c(i,j)=i+j
      elseif i>j
         c(i,j)=i/j
      else
         c(i,j)=i*j
      end
   end
end
Output c
```

Output

1.0000	3.0000	4.0000	5.0000	6.0000	7.0000
2.0000	4.0000	5.0000	6.0000	7.0000	8.0000
3.0000	1.5000	9.0000	7.0000	8.0000	9.0000
4.0000	2.0000	1.3333	16.0000	9.0000	10.0000
5.0000	2.5000	1.6667	1.2500	25.0000	11.0000
6.0000	3.0000	2.0000	1.5000	1.2000	36.0000

50. What is the value of a after the execution of the statement if a = 44 and d = 30?

```
a= (a > d) ? d : -d ;
```

Output

30

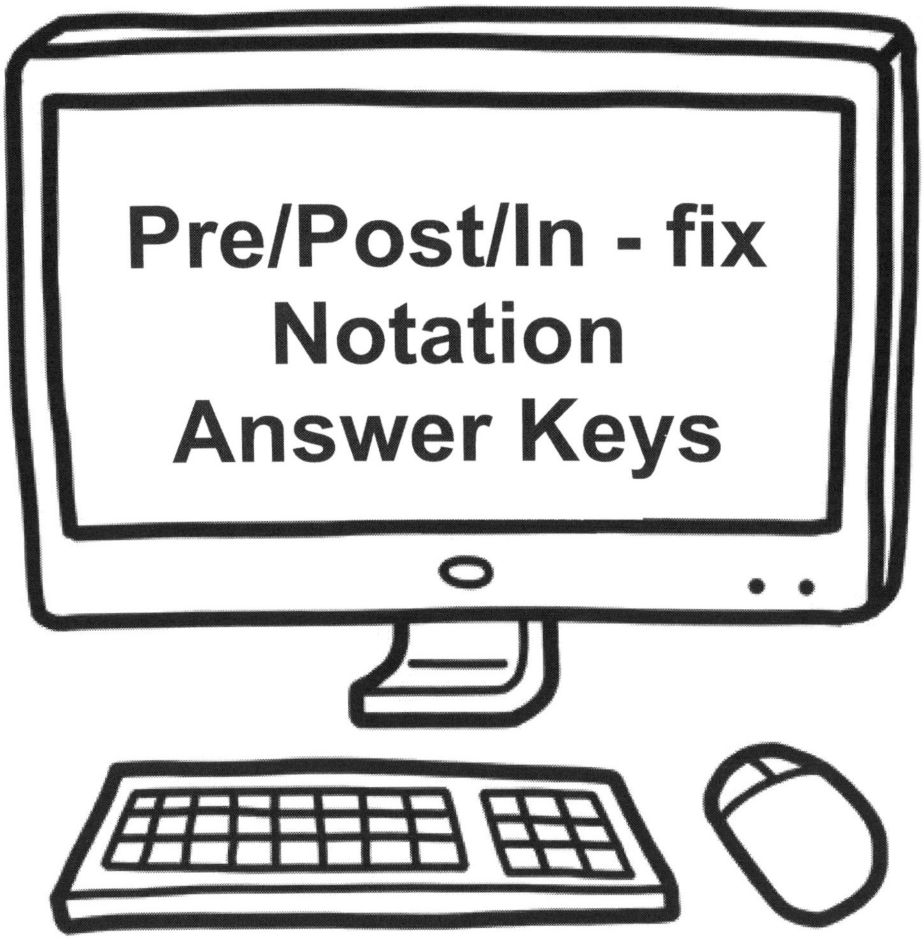
Pre/Post/In - fix
Notation
Answer Keys

1. Convert the following infix to postfix notation for an expression for "Z":

$$X^2 + \frac{Y^2}{2} - Z = 0$$

According to the question,

$$Z = X^2 + \frac{Y^2}{2}$$
$$Z = X^2 + \frac{Y2\uparrow}{2}$$
$$Z = (X2\uparrow) + (Y2\uparrow 2/)$$
$$Z = (X2\uparrow) + (Y2\uparrow 2/)$$
$$Z = X2\uparrow Y2\uparrow 2/+$$

2. Convert the following infix to prefix notation:

$$A * B * \frac{(A * B)^2}{\frac{A}{B}}$$

$$= (\,(A * B) * (\frac{(A * B)^2}{(\frac{A}{B})})\,)$$
$$= (\,(A * B) * (\frac{(* AB)^2}{(/AB)})\,)$$
$$= (\,(A * B) * (\frac{\uparrow * AB2}{(/AB)})\,)$$
$$= ((A * B) * (/\uparrow * AB2/AB))$$
$$= ** AB/\uparrow * AB2/AB$$

3. Convert the following infix to prefix notation:

$$\frac{Z^{Z^{Z^Z}}}{Z + Z + Z + Z}$$

$$= \frac{(Z^{Z^{Z^Z}})}{(((Z + Z) + Z) + Z)}$$
$$= \frac{\uparrow Z^{Z^Z} Z}{(((Z + Z) + Z) + Z)}$$
$$= \frac{\uparrow\uparrow\uparrow ZZZZ}{(((Z + Z) + Z) + Z)}$$
$$= \frac{\uparrow\uparrow\uparrow ZZZZ}{(((+ZZ) + Z) + Z)}$$
$$= \frac{\uparrow\uparrow\uparrow ZZZZ}{(+ + +ZZZZ)}$$
$$= /\uparrow\uparrow\uparrow ZZZZ + + + ZZZZ$$

4. Convert the following postfix into infix:

$$\boldsymbol{AB - AC -/ B2 \uparrow C3 \uparrow/+}$$

$= AB\text{-}AC\text{-}/B2\uparrow C3 \uparrow/+$

$=(AB\text{-}AC\text{-}/)+(CB2\uparrow C3 \uparrow)$

$$= \frac{A-B}{A-C} + \frac{B^2}{C^3}$$

5. Convert the following postfix into infix and evaluate the answer:

$$\mathbf{32 + 34 -/ 242 \uparrow/+}$$

$=(32+34\text{-}/)+(24\uparrow 2/)$

$$= \frac{3+2}{3-4} + \frac{2}{4^2} = -\frac{39}{8}$$

6. Convert the following prefix to infix and evaluate the answer:

$$\mathbf{+ * 42 * 5/42}$$

$= (*\,42) + (5 * (/42))$

$= (4 * 2) + (5 * (\frac{4}{2})) = 18$

7. Convert the following infix into postfix notation:

$$\boldsymbol{a = b^c + \frac{c^b}{c^{c+1}}}$$

$a = (b^c) + (\frac{c^b}{c^{c+1}})$

$a = (bc \uparrow) + (\frac{cb \uparrow}{cc1{+}\uparrow})$

$a = (bc \uparrow) + (cb \uparrow cc1{+}\uparrow/)$

$a = bc \uparrow\ cb \uparrow cc1{+}\uparrow/+$

8. Convert the following infix into prefix notation:

$$\boldsymbol{\frac{A}{B} + \frac{B^B}{A - B^{B+1}}}$$

$= /AB + \frac{\uparrow BB}{A - B^{B+1}}$

$= (/AB) + \frac{(\uparrow BB)}{(A - (\uparrow B + B1))}$

$= (/AB) + \frac{(\uparrow BB)}{(-A \uparrow B + B1)}$

$= (/AB) + (/\uparrow BB - A \uparrow B + B1)$

$= +/AB/\uparrow BB - A \uparrow B + B1$

9. Convert the following infix into prefix notation:

$$(a^{b^c} * \frac{c^{b^a}}{c+b+a})$$

$= (a^{b^c}) * (\frac{c^{b^a}}{c+b+a})$

$= (\uparrow a^b c) * (\frac{c^{b^a}}{c+b+a})$

$= (\uparrow\uparrow acb) * (\frac{c^{b^a}}{c+b+a})$

$= (\uparrow\uparrow acb) * (\frac{(\uparrow\uparrow cab)}{c+b+a})$

$= (\uparrow\uparrow acb) * (\frac{(\uparrow\uparrow cab)}{++cba})$

$= (\uparrow\uparrow acb) * (/\uparrow\uparrow cab ++cba)$

$=*\uparrow\uparrow acb/\uparrow\uparrow cab ++cba$

10. Convert the following infix into postfix notation:

$$\frac{(A^B + \frac{A-C}{B-C} + B^A)}{C}$$

$= \frac{((AB\uparrow) + (\frac{AC-}{BC-}) + (BA\uparrow))}{C}$

$= \frac{((AB\uparrow) + (AC-BC-/) + (BA\uparrow))}{C}$

$= \frac{((AB\uparrow AC-BC-/+) + (BA\uparrow))}{C}$

$= \frac{(AB\uparrow AC-BC-/+BA\uparrow +)}{C}$

$= AB\uparrow AC-BC-/+BA\uparrow +C/$

11. Convert the following prefix to infix notation and evaluate the answer:

$$*\uparrow \frac{AA}{\uparrow AA} \uparrow B2, where\ A = 1\ and\ B = 2$$

$= A^A * (/\uparrow AA \uparrow B2)$

$= A^A * (\frac{\uparrow AA}{\uparrow B2})$

$= A^A * (\frac{A^A}{B^2})$

$= 1^1 * (\frac{1^1}{2^2}) = \frac{1}{4}$

12. Convert the following infix to postfix notation:

$$\frac{(\frac{1}{a}+\frac{1}{b}+\frac{1}{c})}{a*b*c}$$

$$=\frac{((1a/)+(1b/)+(1c/))}{a*b*c}$$
$$=\frac{((1a/1b/+)+(1c/))}{a*b*c}$$
$$=\frac{(1a/1b/+1c/+)}{a*b*c}$$
$$=\frac{(1a/1b/+1c/+)}{(ab*)*c}$$
$$=\frac{(1a/1b/+1c/+)}{(ab*c*)}$$
$$=1a/1b/+1c/+ab*c*/$$

13. Convert the following infix to postfix notation:

$$O=\frac{(\frac{P^2-Q^2}{P*Q})}{(\frac{P}{Q})}$$

$$O=\frac{(\frac{(P2\uparrow)-(Q2\uparrow)}{P*Q})}{(\frac{P}{Q})}$$
$$O=\frac{(\frac{(P2\uparrow Q2\uparrow -)}{(PQ*)})}{(\frac{P}{Q})}$$
$$O=\frac{(P2\uparrow Q2\uparrow -PQ*/)}{(\frac{P}{Q})}$$
$$O=\frac{(P2\uparrow Q2\uparrow -PQ*/)}{(PQ/)}$$
$$O=P2\uparrow Q2\uparrow -PQ*/PQ//$$

14. Convert the following postfix to infix notation:

$$AB+AC-/C+D-$$

$$=[AB+AC-/C+]-D$$
$$=[(AB+AC-/)+C]-D$$
$$=[(\frac{(AB+)}{(AC-)})+C]-D$$
$$=[(\frac{(A+B)}{(A-C)})+C]-D$$

15. Convert the following infix to postfix notation:

$$\left(\frac{\boldsymbol{A * B}}{\boldsymbol{B^2}}\right) + \left(\frac{\boldsymbol{C * D}}{\boldsymbol{C^{3+D}}}\right)$$

$$= \left(\frac{AB *}{B2 \uparrow}\right) + \left(\frac{CD *}{B3D{+}\uparrow}\right)$$
$$= (AB * B2 \uparrow/) + (CD * B3D{+}\uparrow/)$$
$$= AB * B2 \uparrow/\ CD * B3D{+}\uparrow/+$$

16. Convert the following infix to prefix notation:

$$\frac{\boldsymbol{(A - B)^2 + (B * C)^3}}{\boldsymbol{A * B * C}}$$

$$= \frac{((A - B)^2 + (B * C)^3)}{(A * B * C)}$$
$$= \frac{((-AB)^2 + (* BC)^3)}{(A * B * C)}$$
$$= \frac{((\uparrow -AB2) + (\uparrow * BC3))}{(A * B * C)}$$
$$= \frac{(+\uparrow -AB2 \uparrow * BC3)}{(A * B * C)}$$
$$= \frac{(+\uparrow -AB2 \uparrow * BC3)}{(* AB * C)}$$
$$= \frac{(+\uparrow -AB2 \uparrow * BC3)}{(** ABC)}$$
$$= /+\uparrow -AB2 \uparrow * BC3 ** ABC$$

17. Convert the following infix to postfix notation:

$$\frac{\boldsymbol{s * s + s}}{\boldsymbol{t + t * t}} \boldsymbol{* s * t}$$

$$= \left(\left(\frac{((s * s) + s)}{(t + (t * t))}\right) * s\right) * t$$

$$= \left(\left(\frac{((ss *) + s)}{(t + (tt *))}\right) * s\right) * t$$

$$= \left(\left(\frac{(ss * s +)}{(ttt * +)}\right) * s\right) * t$$
$$= ((ss * s + ttt * +/) * s) * t$$
$$= (ss * s + ttt * +/s *) * t$$
$$= ss * s + ttt * +/s * t *$$

18. Convert the infix to prefix notation:

$$A^D + \frac{B}{D^B} * C^D$$

$= (\uparrow AD) + (\frac{B}{\uparrow DB}) * (\uparrow CD)$
$= (\uparrow AD) + (/B \uparrow DB) * (\uparrow CD)$
$= (+\uparrow AD/B \uparrow DB) * (\uparrow CD)$
$= (* +\uparrow AD/B \uparrow DB \uparrow CD)$
$= * +\uparrow AD/B \uparrow DB \uparrow CD$

19. Convert the following postfix to infix notation:

$$A2 \uparrow B2 \uparrow +AB - 2 \uparrow +, where\ A = 2, B = -1$$

$= (((A2 \uparrow)(B2 \uparrow) +) + (AB - 2 \uparrow))$
$= (((A2 \uparrow) + (B2 \uparrow)) + (AB - 2 \uparrow))$
$= ((A^2 + B^2) + (AB - 2 \uparrow))$
$= ((A^2 + B^2) + (AB -)^2)$
$= (A^2 + B^2 + (A - B)^2)$
$= (2^2 + (-1)^2 + (2 - (-1))^2) = 14$

20. Convert the following infix to postfix.

$$\frac{(W^2 + \frac{V * U^2}{W^{\frac{3}{2}}})}{U + \frac{V}{W^{\frac{1}{2}}}}$$

$$= \frac{(W^2 + (\frac{(V * U^2)}{W^{\frac{3}{2}}}))}{(U + (\frac{V}{W^{\frac{1}{2}}}))}$$

$$= \frac{(W2 \uparrow + (\frac{(VU2 \uparrow *)}{W32/\uparrow}))}{(U + (\frac{V}{W12/\uparrow}))}$$

$$= \frac{(W2 \uparrow +(VU2 \uparrow * W32/\uparrow/))}{(U + (VW12/\uparrow/))}$$

$= W2 \uparrow VU2 \uparrow * W32/\uparrow/+UVW12/\uparrow/+/$

21. Convert the following infix to prefix notation:

$$\left(\frac{A^2 - B^2}{A^3 + B^2} + \frac{A}{B^2}\right) * \frac{C^2}{B^3}$$

$$= \left(\frac{A^2 - B^2}{A^3 + B^2} + \frac{A}{B^2}\right) * \frac{C^2}{B^3}$$
$$= \left(\frac{\uparrow A2 - \uparrow B2}{\uparrow A3 + \uparrow B2} + \frac{A}{\uparrow B2}\right) * \frac{\uparrow C2}{\uparrow B3}$$
$$= \left(\left(\frac{- \uparrow A2 \uparrow B2}{+ \uparrow A3 \uparrow B2}\right) + \left(\frac{A}{\uparrow B2}\right)\right) * \left(\frac{\uparrow C2}{\uparrow B3}\right)$$
$$= ((/ - \uparrow A2 \uparrow B2 + \uparrow A3 \uparrow B2) + (/A \uparrow B2)) * (/\uparrow C2 \uparrow B3)$$
$$= (+/ - \uparrow A2 \uparrow B2 + \uparrow A3 \uparrow B2/A \uparrow B2) * (/\uparrow C2 \uparrow B3)$$
$$= * +/ - \uparrow A2 \uparrow B2 + \uparrow A3 \uparrow B2/A \uparrow B2/\uparrow C2 \uparrow B3$$

22. Convert the following postfix to infix notation:

$$\mathbf{12/32 \uparrow 3/24/* +}$$

$$= \frac{1}{2} + (32 \uparrow 3/) * (24/)$$
$$= \frac{1}{2} + \frac{3^2}{3} * \frac{2}{4} = 2$$

23. Convert the following to postfix notation:

$$\frac{\alpha^2}{\beta} + \frac{\frac{\beta^\gamma}{\gamma}}{\varepsilon + \frac{2}{\alpha}}$$

$$= \frac{\alpha 2 \uparrow}{\beta} + \frac{\frac{(\beta\gamma \uparrow)}{\gamma}}{\varepsilon + (/2\alpha)}$$
$$= \frac{\alpha 2 \uparrow}{\beta} + \frac{\frac{(\beta\gamma \uparrow)}{\gamma}}{(\varepsilon/2\alpha +)}$$
$$= \frac{\alpha 2 \uparrow}{\beta} + \frac{(\beta\gamma \uparrow)}{(\gamma\varepsilon/2\alpha +/)}$$
$$= \alpha 2 \uparrow \beta/\beta\gamma \uparrow \gamma\varepsilon/2\alpha +//+$$

24. Convert the following to prefix notation:

$$\frac{A^{\frac{B+C+D^2}{A*B*C}}}{A*B*C}$$

$$= \frac{A^{\frac{B+C+D^2}{A*B*C}}}{A*B*C}$$
$$= \frac{\uparrow A/+ + BC \uparrow D2 ** ABC}{** ABC}$$
$$= /\uparrow A/+ + BC \uparrow D2 ** ABC ** ABC$$

25. Convert the following to postfix notation for $\boldsymbol{P_O}$:

$$\frac{P_O}{P} = \left(\frac{T_O}{T}\right)^{\frac{(\gamma-1)}{\gamma}}$$

$$P_O = \left(\frac{T_O}{T}\right)^{\frac{(\gamma-1)}{\gamma}} * P$$
$$P_O = (T_O T/)^{\frac{(\gamma-1)}{\gamma}} * P$$
$$P_O = (T_O T/\gamma 1 - \gamma \uparrow) * (P)$$
$$P_O = T_O T/\gamma 1 - \gamma \uparrow P *$$

26. Convert the three equations of motions to postfix notation. Find the postfix expression for "u" in all the equations.
HINT: The three equations of motion that defines the state of motion of any body are given by:

a) $\boldsymbol{v = u + at}$
b) $\boldsymbol{s = ut + \frac{1}{2}at^2}$
c) $\boldsymbol{v^2 = u^2 + 2as}$

a) According to the question,
$u = v - at$
$= v - a * t$
$= vat * -$
$= uvat * -=$

b) According to the question,
$$s = ut + \frac{1}{2}at^2$$
$$u = \frac{(s - \frac{1}{2}at^2)}{t}$$
$$u = \frac{(s - (12/a * t2 \uparrow *))}{t}$$
$$u = \frac{(s12/a * t2 \uparrow * -)}{t}$$
$$u = s12/a * t2 \uparrow * - t/$$
$$us12/a * t2 \uparrow * - t/=$$

c) According to the question,
$v^2 = u^2 + 2as$
$u^2 = v^2 - 2as$
$u = (v^2 - 2as)^{\frac{1}{2}}$
$u = ((v2 \uparrow) - (2a * s *))^{\frac{1}{2}}$
$u = (v2 \uparrow 2a * s * -)^{\frac{1}{2}}$
$u = v2 \uparrow 2a * s * -12/\uparrow$
$uv2 \uparrow 2a * s * -12/\uparrow=$

27. Convert the following postfix to infix notation. The equation obtained in the end is an equation for "Force".

$$F = GM * m * r2 \uparrow /$$

Evaluate the "Force, F" using the data given below:

$$G, Gravitational\ constant = 6.67 \times 10^{-11}$$

$$M, mass\ of\ Earth = 6 \times 10^{24}$$

$$m, mass\ of\ the\ satellite\ revolving\ around\ the\ Earth = 60\ Kg$$

$$r, distance\ between\ the\ center\ of\ Earth\ and\ the\ satellite = 7000\ Km$$

You can give an approximate answer.

$$F = \frac{(GM * m *)}{r2 \uparrow}$$

$$F = \frac{(G * M * m)}{r^2} = \frac{6.67 \times 10^{-11} * 6 \times 10^{24} * 60}{7000^2} = 4.9 \times 10^8$$

28. Convert the following postfix to infix notation:

$$222 \uparrow\uparrow 23 * 2 * +22 + 222 \uparrow / + /$$

$$= \frac{((222 \uparrow\uparrow 23 * 2 * +))}{(22 + 222 \uparrow / +)}$$

$$= \frac{((222 \uparrow\uparrow 23 * 2 * +))}{((22+) + (222 \uparrow /))}$$

$$= \frac{((222 \uparrow\uparrow) + (23 * 2 *))}{2 + 2 + \frac{2}{2^2}}$$

$$= \frac{((222 \uparrow\uparrow) + ((23 *) * 2))}{2 + 2 + \frac{2}{2^2}}$$

$$= \frac{2^{2^2} + 2 * 3 * 2}{2 + 2 + \frac{2}{2^2}}$$

29. Convert the following to infix notation and evaluate the answer:

$$\mathbf{12/14/+16/+24*6+/}$$

$$=\frac{(12/14/+16/+)}{(24*6+)}$$

$$=\frac{(12/14/+16/+)}{(24*)+6}$$

$$=\frac{(12/14/+16/+)}{2*4+6}$$

$$=\frac{((12/14/+)+(16/))}{2*4+6}$$

$$=\frac{((12/)+(14/)+(16/))}{2*4+6}$$

$$=\frac{(\frac{1}{2}+\frac{1}{4}+\frac{1}{6})}{2*4+6}=0.06547$$

30. Convert the following to infix notation and evaluate the answer:

$$\boldsymbol{X2\uparrow Y2\uparrow X3\uparrow/+YX-XY*/-}$$

$$=X2\uparrow+(Y2\uparrow X3\uparrow/)-(YX-XY*/)$$

$$=X2\uparrow+(Y2\uparrow X3\uparrow/)-(YX-XY*/)$$

$$=X2\uparrow+(\frac{Y2\uparrow}{X3\uparrow})-(\frac{YX-}{XY*})$$

$$=X^2+(\frac{Y^2}{X^3})-(\frac{Y-X}{X*Y})$$

31. Convert the following infix to postfix notations:

$$\boldsymbol{\frac{\frac{A}{A^{A+B}}}{\frac{B}{B^{A+B}}}}$$

$$=\frac{AAAB+\uparrow/}{\frac{B}{B^{A+B}}}$$

$$=\frac{AAAB+\uparrow/}{BBAB+\uparrow/}$$

$$=AAAB+\uparrow/BBAB+\uparrow//$$

32. Convert the following infix to prefix notation:

$$\frac{A^B + B^2 + B^A}{A * B * B}$$

$$= \frac{\uparrow AB + \uparrow B2 + \uparrow BA}{(*AB) * B}$$
$$= \frac{\uparrow AB + \uparrow B2 + \uparrow BA}{(**ABB)}$$
$$= \frac{+\uparrow AB \uparrow B2 + \uparrow BA}{(**ABB)}$$
$$= \frac{++\uparrow AB \uparrow B2 \uparrow BA}{(**ABB)}$$
$$= /++\uparrow AB \uparrow B2 \uparrow BA ** ABB$$

33. Convert the infix to postfix notation:

$$A^2 + \frac{B^3}{A * B} * \frac{A * B}{A^3}$$

$$= A^2 + (\frac{B^3}{A * B} * \frac{A * B}{A^3})$$
$$= A2 \uparrow + (\frac{B3 \uparrow}{AB *} * \frac{AB *}{A3 \uparrow})$$
$$= A2 \uparrow + ((B3 \uparrow AB */) * (AB * A3 \uparrow))$$
$$= A2 \uparrow + (B3 \uparrow AB */AB * A3 \uparrow *)$$
$$= A2 \uparrow \ B3 \uparrow AB */AB * A3 \uparrow * +$$

34. Convert the infix to prefix notation:

$$a * \frac{b}{b * \frac{b}{c}}$$

$$= a * (\frac{b}{b * \frac{b}{c}})$$
$$= a * (\frac{b}{b * (/bc)})$$
$$= a * (\frac{b}{* b/bc})$$
$$= a * (/b * b/bc)$$
$$= * a/b * b/bc$$

35. Convert the following infix to postfix notation:

$$w^2 - \frac{v^2}{w^2} + \frac{v^2}{w^2}$$

$$= w2 \uparrow - \frac{v2 \uparrow}{w2 \uparrow} + \frac{v2 \uparrow}{w2 \uparrow}$$
$$= w2 \uparrow -(v2 \uparrow w2 \uparrow /) + (v2 \uparrow w2 \uparrow /)$$
$$= w2 \uparrow -(v2 \uparrow w2 \uparrow / v2 \uparrow w2 \uparrow / +)$$
$$= w2 \uparrow v2 \uparrow w2 \uparrow / v2 \uparrow w2 \uparrow / + -$$

36. Convert the following infix to postfix notation:

$$\frac{A^2}{B^3}+\frac{A^B}{B^A}*(A*B)$$

$=\frac{A2\uparrow}{B3\uparrow}+\frac{AB\uparrow}{BA\uparrow}*(AB*)$
$=(A2\uparrow B3\uparrow/)+(AB\uparrow BA\uparrow/)*(AB*)$
$=(A2\uparrow B3\uparrow/)+(AB\uparrow BA\uparrow/AB**)$
$=A2\uparrow B3\uparrow/AB\uparrow BA\uparrow/AB**+$

37. Convert the following infix to postfix notation:

$$\frac{A^5}{(A*B)}+\frac{B^5}{\frac{A}{B}+\frac{B}{2}}$$

$=\frac{A5\uparrow}{(AB*)}+\frac{B5\uparrow}{(/AB)+(/B2)}$
$=(A5\uparrow AB*/)+\frac{B5\uparrow}{(/AB/B2+)}$
$=(A5\uparrow AB*/)+(B5\uparrow/AB/B2+)$
$=A5\uparrow AB*/B5\uparrow/AB/B2++$

38. Convert the following infix to prefix notation:

$$\frac{A^2}{B^2+C^2}*(\frac{A}{B})^C$$

$=\frac{A^2}{B^2+C^2}*(\frac{A}{B})^C$

$=(/\uparrow A2+\uparrow B2\uparrow C2)*(\uparrow/ABC)$
$=*/\uparrow A2+\uparrow B2\uparrow C2\uparrow/ABC)$

39. Convert the following infix to postfix notation:

$$\frac{x}{y}+\frac{x^2}{y^2}+\frac{x^3}{y^3}$$

$=\frac{x}{y}+\frac{x^2}{y^2}+\frac{x^3}{y^3}$
$=(xy/)+\frac{x2\uparrow}{y2\uparrow}+\frac{x3\uparrow}{y3\uparrow}$
$=(xy/)+(x2\uparrow y2\uparrow/)+(x3\uparrow y3\uparrow/)$
$=xy/x2\uparrow y2\uparrow/+x3\uparrow y3\uparrow/+$

40. Convert the following infix to postfix notation:

$$\left(\frac{x^3}{y}\right)^3 + \left(\frac{2*x}{y}\right)^4$$

$= \left(\frac{x^3}{y}\right)^3 + \left(\frac{2*x}{y}\right)^4$

$= \left(\frac{x3 \uparrow}{y}\right)^3 + \left(\frac{2*x}{y}\right)^4$

$= (x3 \uparrow y/)^3 + (2x * y/)^4$

$= (x3 \uparrow y/)^3 + (2x * y/)^4$

$= (x3 \uparrow y/3 \uparrow) + (2x * y/4 \uparrow)$

$= x3 \uparrow y/3 \uparrow 2x * y/4 \uparrow +$

41. Convert the following infix to prefix notation:

$$\frac{w^2}{u+v} + \frac{u}{v} + \frac{u^2}{u*v}$$

$= \left(\frac{w^2}{u+v}\right) + \left(\frac{u}{v}\right) + \left(\frac{u^2}{u*v}\right)$

$= \left(\frac{\uparrow w2}{+uv}\right) + (/uv) + \left(\frac{\uparrow u2}{*uv}\right)$

$= (/\uparrow w2 + uv) + (/uv) + (/\uparrow u2 * uv)$

$= (+/\uparrow w2 + uv/uv) + (/\uparrow u2 * uv)$

$= + +/\uparrow w2 + uv/uv/\uparrow u2 * uv$

42. Convert the following infix to postfix notation:

$$\frac{A^2}{2A} + \frac{B^2}{AB} + \frac{C^2}{2C}$$

$= \left(\frac{A^2}{2*A}\right) + \left(\frac{B^2}{A*B}\right) + \left(\frac{C^2}{2*C}\right)$

$= (A2 \uparrow 2A */) + (B2 \uparrow AB */) + (C2 \uparrow 2C */)$

$= (A2 \uparrow 2A */B2 \uparrow AB */+) + (C2 \uparrow 2C */)$

$= A2 \uparrow 2A */B2 \uparrow AB */+ C2 \uparrow 2C */+$

43. Convert the infix to postfix notation:

$$\frac{A^{a^A}}{a*A} + \frac{A}{a^{\frac{1}{2}}} + \frac{\frac{1}{2}a}{A}$$

$= \left(\frac{A^{a^A}}{a*A}\right) + \left(\frac{A}{a^{\frac{1}{2}}}\right) + \left(\frac{\frac{1}{2}a}{A}\right)$

$= (AAa \uparrow\uparrow aA */) + (Aa12/\uparrow/) + (12/a * A/)$

$= (AAa \uparrow\uparrow aA */Aa12/\uparrow/+) + (12/a * A/)$

$= AAa \uparrow\uparrow aA */Aa12/\uparrow/+12/a * A/+$

44. Convert the following postfix to infix and compute the final answer:

$$\boldsymbol{S = (432 \uparrow /25/+912 \uparrow /3/-)}$$

$S = (432 \uparrow /25/+) - (912 \uparrow /3/)$

$S = (432 \uparrow /) + (25/) - (912 \uparrow /3/)$

$S = \frac{4}{3^2} + \frac{2}{5} - \frac{9^{\frac{1}{2}}}{3}$

45. Convert the following postfix to infix and compute the final answer:
Note carefully the spaces between the digits as all numbers are not one-digit numbers.

$$\boldsymbol{Y = 22 \uparrow 3 \uparrow 3/+40 \uparrow 50 \uparrow /+31\ 27/-}$$

$Y = (22 \uparrow 33 \uparrow /40 \uparrow 50 \uparrow /+) - (31\ 27/)$

$Y = (\frac{2^2}{3^3} + \frac{4^0}{5^0}) - (\frac{31}{27}) = 0$

46. Convert the following prefix to infix and compute the final answer:
Note all the numbers are one-digit numbers.

$$\boldsymbol{X =* 3/3 + 3/33}$$

$X = 3 * (/3 + 3/33)$

$X = 3 * (\frac{3}{+3/33})$

$X = 3 * (\frac{3}{3 + (/33)})$

$X = 3 * (\frac{3}{3+\frac{3}{3}})$

$X=\frac{9}{4}$

47. Convert the following prefix to infix and compute the final answer:
Note all the numbers are one-digit numbers.

$$\boldsymbol{A =* \uparrow a/14 \uparrow /a_o + aa_o/12}$$

$A = (\uparrow a/14) * (\uparrow /a_o + aa_o/12)$

$A = a^{\frac{1}{4}} * (\frac{a_o}{a + a_o})^{\frac{1}{2}}$

48. Convert the following prefix to infix and compute the final answer:

$$\boldsymbol{S = +/3 + 4 \uparrow (-1)7/\uparrow 22 \uparrow 23}$$

$S = +/3 + 4 \uparrow (-1)7/\uparrow 22 \uparrow 23$

$S = (/3 + 4 \uparrow (-1)7) + (/\uparrow 22 \uparrow 23)$

$S = \frac{3}{4 + (-1)^7} + (\frac{2^2}{2^3})$

$s=\frac{13}{9}$

49. Convert the following postfix to infix and compute the final answer:

$$\boldsymbol{Y = x2 \uparrow y3 \uparrow / x + yxy + / *}$$

$Y = (x2 \uparrow y3 \uparrow /) + (xyxy + *)$
$Y = (x2 \uparrow y3 \uparrow /) + (x * (yxy + /))$

$$Y = \frac{x^2}{y^3} + x * \frac{y}{x + y}$$

50. Convert the following postfix to infix and compute the final answer:

$$\boldsymbol{Y = abba / * a12 / \uparrow * +}$$

$Y = a + (bba / * a12 / \uparrow *)$
$Y = a + (bba / *) * (a12 / \uparrow)$
$Y = a + b * (ba /) * (a12 / \uparrow)$

$$Y = a + b * \frac{b}{a} * a^{\frac{1}{2}}$$

Bit - String
Flicking
Answer Keys

1. (RCIRC-2 101011) XOR (RSHIFT-1 (RCIRC-2 111010))
= (111010) XOR (RSHIFT-1 (101110))
= (111010) XOR (010111)
= 101101

2. (NOT (LSHIFT-2 (RSHIFT-3 (RCIRC-3 001111)))) OR 110000
= (NOT (LSHIFT-2 (RSHIFT-3 111001))) OR 110000
= (NOT (LSHIFT-2 000111)) OR 110000
= (NOT (011100)) OR 110000
= 100011 OR 110000
= 110011

3. (RSHIFT-2 (11110 OR 11010)) AND (RSHIFT-2 10111))
= (RSHIFT-2 (11110)) AND (00101)
= (00111) AND (00101)
= 00101

4. (RCIRC-2 (LSHIFT-2 1001010)) AND (RSHIFT-1 (LCIRC-2 1010011))
= (RCIRC-2 0101000) AND (RSHIFT-1 1001110)
= (0001010) AND (0100111)
= 0000010

5. ((RSHIFT-2 (1010010 AND 1101011)) XOR (LSHIFT-2 1011001))
= (RSHIFT-2 (1000010)) XOR (1100100)
= (0010000) XOR (1100100)
= 1110100

6. NOT ((0001100 XOR 1010100) AND 0011100)
= NOT ((1011000) AND 0011100)
= NOT 0011000
= 1100111

7. 1111000 AND (NOT ((0000011 XOR 1010001) OR 1001100))
= 1111000 AND (NOT (1010010 OR 1001100))
= 1111000 AND (NOT 1011110)
= 1111000 AND 0100001
= 0100000

8. ((LCIRC-2 100000 OR LSHIFT-1 111010) AND (NOT (000001 OR 111110))
= (000010 OR 110100) AND (NOT 111111)
= (110110) AND (000000)
= 000000

9. (NOT (RSHIFT-2 (LCIRC-3 110000))) AND (LSHIFT-3 101010)
   = (NOT (RSHIFT-2 000110)) AND (LSHIFT-3 010000)
   = (NOT 000001) AND (LSHIFT-3 010000)
   = (NOT 000001) AND (000000)
   = 000000

10. (RSHIFT-3 (1010011 AND 1101000)) XOR (RSHIFT-2 1011001))
    = (RSHIFT-3 (1000000)) XOR (0010110))
    = (0001000) XOR (0010110)
    = 0011110

11. (LSHIFT-1 (10100 XOR (RCIRC-3 x) AND 10011)) = 01100, where x is 5-digit
    Give any five combinations of digits.
    Assuming x = abcde
    RCIRC-3 abcde = cdeab
    cdeab AND 10011 = c00ab
    10100 XOR c00ab = C01ab
    LSHIFT-1 C01ab = 01100
    01ab0 = 01100
    a = 1; b = 0
    ab***, where * has to be filled with combinations of 0s and 1s.
    10000, 10111, 10100, 10011, 10001

12. (RSHIFT-1 (LCIRC-2 x AND 1001)) = 0110, where x is 4-digit
    Give any three combinations of digits.
    x = abcd
    LCIRC-2 abcd = cdab
    cdab AND 1001 = c00b
    RSHIFT-1 c00b = 0c00 = 0110
    c = 1 and therefore, *1**
    1100, 1110, 1111

13. (LSHIFT-1 (LCIRC-1 x AND 1011)) = 0111, where x is 4-digit
    Give any three combinations of digits.
    x = abcd
    LCIRC-1 abcd = bcda
    bcda AND 1011 = b0da
    RSHIFT-1 b0da = 0b0d
    0b0d = 0111
    b = 1 = d
    Therefore, *1*1 = 0101, 1101, 0111, 1111

14. (RSHIFT-1 (1011 XOR (RCIRC-2 x AND 1011)) = 0100, where x is 4-digit
Give any three combinations of digits.
x = abcd
RCIRC-2 abcd = cdab
cdab AND 1011 = c0ab
1011 XOR c0ab = C0AB
RSHIFT-1 C0AB = 0C0A
0C0A = 0100
C =1 therefore, c = 0; A = 0 therefore, a = 1
Therefore, 1*0* = 1100, 1001, 1101

15. NOT (LSHIFT-1 (RSHIFT-2 (RCIRC-3 001111 OR 100000)))
= NOT (LSHIFT-1 (RSHIFT-2 (111001 OR 100000)))
= NOT (LSHIFT-1 (RSHIFT-2 111001))
= NOT (LSHIFT-1 001110)
= NOT 011100 = 100011

16. ((LCIRC-2 (11110 OR 10010)) XOR (LSHIFT-2 11111))
= ((LCIRC-2 11110) XOR 11100)
= (11011 XOR 11100)
= 00111

17. (RCIRC-3 1001011) AND (LSHIFT-2 1010011)
= (0111001) AND (1001100)
= 0001000

18. ((LSHIFT-2 (1110110 AND 1011011)) XOR (RSHIFT-1 1011001))
= ((LSHIFT-2 1010010) XOR 0101100)
= (1001000 XOR 0101100)
= 1100100

19. NOT ((00011 XOR 10101) AND 11100)
= NOT ((10110) AND 11100)
= NOT (10100)
= 01011

20. NOT ((x AND 1010) OR 1001) where x is a four-digit string
= abcd AND 1010 = a0c0
= a0c0 OR 1001 = a0c0

21. RSHIFT-2 (NOT ((01011 XOR 10101) AND 10100))
= RSHIFT-2 (NOT (11110 AND 10100))
= RSHIFT-2 (NOT (10100))
= RSHIFT-2 01011
= 00010

22. ((LSHIFT-3 (RSHIFT-2 (LCIRC-2 010001111))) OR (LSHIFT-2 111110010)
= ((LSHIFT-3 (RSHIFT-2 000111101)) OR 111001000)
= (LSHIFT-3 000001111) OR 111001000
= 001111000 OR 111001000
= 111111000

23. NOT ((RSHIFT-3 (LSHIFT-2 010001101)) OR (RSHIFT-2 100110010))
= NOT ((RSHIFT-3 000110100) OR 001001100))
= NOT (000000110 OR 001001100)
= NOT (001001110)
= 110110001

24. (LSHIFT-2 (10101 XOR ((LCIRC-2 x) AND 11011)))) = 11100, where x is 5 digit
Give any three combinations of digits.
Assuming x = abcde
LCIRC-2 x = LCIRC-2 abcde = cdeab
= cdeab AND 11011 = cd0ab
= 10101 XOR cd0ab = Cd1aB
LSHIFT-2 Cd1aB = 11100
1aB** = 11100
a = 1; b = 0, therefore the combinations are:
11001, 11011, 11010

25. (RSHIFT-1 (RCIRC-1 x AND 1101)) = 1110, where x is 4-digit
Give any three combinations of digits.
x = abcd
RCIRC-1 abcd = dabc
dabc AND 1101 = da0c
RSHIFT-1 da0c = 0da0 = 1110
d =1, a = 1;
Therefore, 1**d can be = 1001, 1101, 1111

26. (LCIRC-1 01001001 OR LSHIFT-2 00010011)
= 10010010 OR 01001100
= 11011110

27. ((RCIRC-3 100000 XOR RSHIFT-1 111010) OR ((000001 AND 111110))
= (000100 XOR 011101) OR (000000)
= (011001) OR (000000)
= 011001

28. (LSHIFT-1 (RSHIFT-2 (LCIRC-3 0101111))) OR (LSHIFT-2 1011110)
= ((LSHIFT-1 (RSHIFT-2 1111010)) OR 1111000)
= ((LSHIFT-1 0011110) OR 1111000)
= (0111100 OR 1111000)
= 1111100

29. (RSHIFT-3 (LCIRC-3 011001111)) OR (LSHIFT-3 110110000)
= (RSHIFT-3 001111011) OR (LSHIFT-3 110110000)
= 000001111 OR 110000000
= 110001111

30. (LSHIFT-3 (LSHIFT-3 (LCIRC-2 001101))) AND (RSHIFT-2 (RCIRC-1 101000))
= (LSHIFT-3 (LSHIFT-3 110100)) AND (RSHIFT-2 (RCIRC-1 101000))
= (LSHIFT-3 100000) AND (RSHIFT-2 010100)
= (000000) AND (001101)
= 000000

31. If *a* equal 1 and *A* is 0. Compute:
= (RCIRC-3 aAaA)) XOR (RSHIFT-1 (LCIRC-2 aaAA))
= (RCIRC-3 1010)) XOR (RSHIFT-1 (LCIRC-2 1100))
= (0101) XOR (RSHIFT-1 0011)
= (0101) XOR (0001)
= 0100

32. (LSHIFT-2 (RCIRC-2 101111)) OR (LSHIFT-2 (RCIRC-1 101000))
= (LSHIFT-2 111011) OR (LSHIFT-2 010100)
= (101100) OR (010000)
= 111100

33. If *a* equal 1 and *A* is 0. Compute:
= (LCIRC-3 aAaAaaa)) XOR (LSHIFT-1 (LCIRC-1 aaAAAAA))
= (AaaaaAa) XOR (LSHIFT-1 (aAAAAAa))
= (AaaaaAa) XOR (AAAAAaA)
= 0111101 XOR 0000010
= 0111111

34. (110011 AND (110011 OR (110011 OR (110011))))
= (110011 AND (110011 OR 110011))
= (110011 AND (110011))
= 110011

35. (RCIRC-2 (RCIRC-1 10001)) OR (RSHIFT-2 (RCIRC-2 10101))
= (RCIRC-2 11000) OR (RSHIFT-2 01101)
= (00110) OR (00011)
= 00111

36. (NOT ((pqrs AND 1010) OR 1001)) OR (1000) where *pqrs* is a four-digit string
= (NOT (p0r0 OR 1001)) OR (1000)
= (NOT p0r1) OR (1000)
= (NOT p0r1) OR (1000)
= P1R0 OR 1000
= 11R0

37. (RCIRC-1 110010 OR RSHIFT-2 100001) AND (NOT 001010)
= (011001 OR 001000) AND (110101)
= (011001) AND (110101)
= 010001

38. NOT (RSHIFT-2 (LSHIFT-2 (LCIRC-2 101101 OR 101000)))
= NOT (RSHIFT-2 (LSHIFT-2 (110110 OR 101000)))
= NOT (RSHIFT-2 (LSHIFT-2 111110))
= NOT (RSHIFT-2 111000)
= NOT 001110
= 110001

39. (RCIRC-2 (RCIRC-2 10111)) OR (LSHIFT-2 (LCIRC-3 10110))
= (RCIRC-2 11101) OR (LSHIFT-2 10101)
= (01111) OR (10100)
=11111

40. If a = 101010 and b = 001100
Find RSHIFT-1 (NOT ((a XOR b) OR a))
= RSHIFT-1 (NOT ((101010 XOR 001100) OR 101010))
= RSHIFT-1 (NOT (100110 OR 101010))
= RSHIFT-1 (NOT 101110)
= RSHIFT-1 (010001)
= 001000

41. (RCIRC-2 (RCIRC-1 101010)) OR (LSHIFT-3 (RSHIFT-1 100100))
= (RCIRC-2 010101) OR (LSHIFT-3 010010)
= (010101) OR (010000)
= 010101

42. If a = 10001 and A = 11111
Find RCIRC-1 (NOT ((a OR A) OR a))
= RCIRC-1 (NOT ((a OR A) OR a))
= RCIRC-1 (NOT ((10001 OR 11111) OR 10001))
= RCIRC-1 (NOT (11111 OR 10001))
= RCIRC-1 (NOT 11111)
= RCIRC-1 00000
= 00000

43. (LCIRC-2 (LCIRC-2 100010)) OR (RSHIFT-2 (RCIRC-1 100010))
= (LCIRC-2 001010) OR (RSHIFT-2 010001)
= (101000) OR (000100)
= 101100

44. (RCIRC-2 101101)) AND (NOT (LCIRC-2 111010))
= (011011) AND (NOT (101011))
= (011011) AND (NOT 010100)
= (011011) AND (101011)
= 001011

45. NOT ((00010 OR 10100) AND 00100)
= NOT (10110 AND 00100)
= NOT (00100)
= 11011

46. NOT (((LSHIFT-2 (LCIRC-2 01111)) AND (LSHIFT-2 00011))
= NOT ((LSHIFT-2 11101) AND 01100)
= NOT (10100 AND 01100)
= NOT 00100
= 11011

47. (LSHIFT-2 (RCIRC-2 (0010011 OR 1010010)))
= (LSHIFT-2 (RCIRC-2 (1010011)))
= (LSHIFT-2 1110100)
= 1010000

48. (NOT (LCIRC-2 110001111)) OR 100000111
= (NOT 000111111) OR 100000111
= (111000000) OR 100000111
= 111000111

49. 101011 XOR (RSHIFT-4 (RCIRC-3 000011))
= 101011 XOR (RSHIFT-4 011000)
= 101011 XOR (000001)
= 101010

50. If *a* equal 1 and *A* is 0. Compute:
= (LCIRC-3 aaaAaaa) OR (LSHIFT-2 (LCIRC-2 aaAAaAA))
= (LCIRC-3 1110111) OR (LSHIFT-2 (LCIRC-2 1100100))
= (0111111) OR (LSHIFT-2 0010011)
= (0111111) OR (1001100)
= 1111111

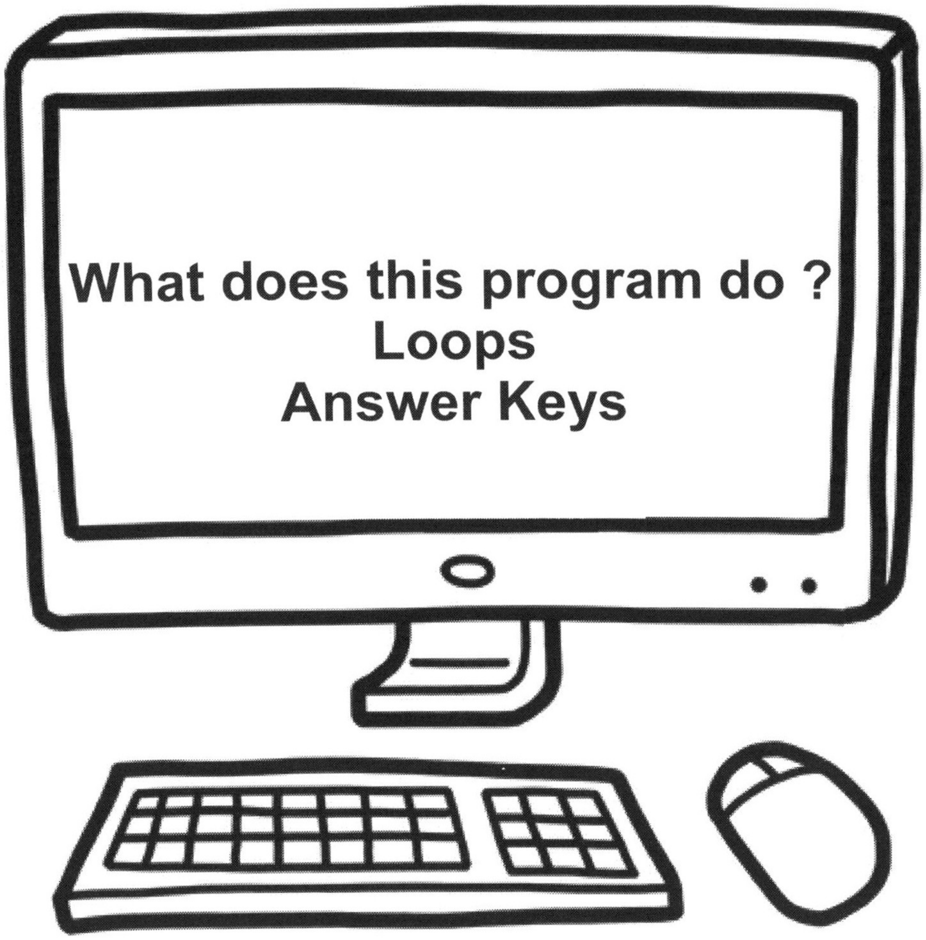
What does this program do ?
Loops
Answer Keys

1. What is the output of the program if rows = 3?

```
int rows, i, j, space;
rows=3
for (i=rows; i>=rows;i- - )
for (space = i; space<rows; space++)
Output ” “
for (j=1; j<=(2*i-1);j++)
Output ”*”
Output ”\n”
```

Output

```
***
 **
  *
```

2. What is the output of the program?

```
int m, i, j;
m=5
for (i=1; i<=m; i++ )
for (j=1; j<=1; j++)
Output ”*”
Output ”\n”
```

Output

```
*
**
***
****
*****
```

3. What is the output if rows = 4?

```
int m, i, j
m=4
for (i=m; i>=m; i - -  )
for (j=1; j<=1; j++)
Output ”*”;
Output ”\n””;
```

Output

```
****
***
**
*
```

4. What is the output of the program?

```
for x = 1:4
 Output "'*\n',x"
  for y=2:4
     Output "'^',y"
  end
end
```

Output

```
*
^^^*
^^^*
^^^*
^^^
```

5. What is the output of the program?

```
for x = 1:3
 for y =1:3
    if x==y
      z(x,y) =1
    else
        z(x,y) =0
    end
  end
end
```

Output

```
1  0  0
0  1  0
0  0  1
```

6. What is the output of the program?

```
for x = 1:3
 for y =1:3
    if x==y
      z(x,y) =x+y
    else
        z(x,y) =x-y
    end
  end
end
```

Output

```
2  -1  -2
1   4  -1
2   1   6
```

7. What is the output of the program?

```
for x = 1:3
 for y =1:3
    if x==y
      z(x,y) =1
    elseif x<y
       z(x,y) =0
    else
       z(x,y) = -1
    end
  end
end
```

Output

1	0	0
-1	1	0
-1	-1	1
0	0	0

8. What is the output of the program?

```
for x = 1:3
 for y =1:3
    if x==y
      z(x,y) =x^2
    else
      z(x,y) =y^2
    end
  end
end
```

Output

1	4	9
1	4	9
1	4	9

9. What is the output of the program?

```
z = zeros(3)
for x = 1:3
 for y =1:3
    if x==y
      z(x,y) = x^2 + y^2
    else
       z(x,y) =. x^2 + x/2
    end
  end
end
```

Output

2	4.5	9.5
2.	8	10
2.5	5.5	18

10. What is the output of the program?

```
A = 2
B = 1
for x = 1:3
 for y =1:3
    if x==y
      z(x,y) =A
    else
       z(x,y) =B
    end
  end
end
```

Output

```
    2    1    1
    1    2    1
    1    1    2
```

11. What is the output of the program?

```
A = 100
B = 10000
for x = 1:3
 for y =1:3
    if x<y
      z(x,y) =sqrt(A)
    elseif x>y
       z(x,y) = sqrt(B)
    else
       z(x,y) = sqrt(A+B)
    end
  end
end
```

Output

```
  100.4988     10.0000     10.0000
  100.0000    100.4988     10.0000
  100.0000    100.0000    100.4988
```

12. What is the output of the program?

```
A = 0
B = 1
for x = 1:4
 for y =1:4
    if x<y
      z(x,y) =A
    elseif x>y
       z(x,y) = B
    else
       z(x,y) = B
    end
  end
end
```

Output

1	0	0	0
1	1	0	0
1	1	1	0
1	1	1	1

13. What is the output of the program?

```
s = 0
for x = 1:3
 for y =1:3
    if x<y
      z(x,y) = s+x
    elseif x>y
       z(x,y) = s/y
    else
       z(x,y) = x*y
    end
  end
end
```

Output

1	1	1
0	4	2
0	0	9

14. What is the output of the program?

```
s = 1
for x = 1:3
 for y =1:3
    if x>y
      z(x,y) =s+x
    elseif x<y || x==y
       z(x,y) = s/(x+y)
    end
  end
end
```

Output

```
0.5000   0.3333   0.2500
3.0000   0.2500   0.2000
4.0000   4.0000   0.1667
```

15. What is the output of the program?

```
s = 1
z = zeros(3)
for x = 1:3
 for y =1:3
    if x>y
      z(x,y) =s+x
    elseif x<y || x==y
       z(x,y) = s+x/(x+y)
    end
  end
end
```

Output

```
1.5000    1.3333    1.2500
3.0000    1.5000    1.4000
4.0000    4.0000    1.5000
```

16. What is the output of the program?

```
s = 1
for x = 1:3
 for y =1:3
    if y==1
      z(x,y) =1
    elseif y>1
       z(x,y) = s+z(x,y-1)
    end
  end
end
```

Output

```
1   2   3
1   2   3
1   2   3
```

17. What is the output of the program?

```
s = 4
n = 3
for x = 1:n
 for y =1:n
    if y==1
      z(x,y) =s
    elseif y>1
        z(x,y) = s+z(x,y-1)
    end
  end
end
```

Output

```
4   8   12
4   8   12
4   8   12
```

18. What is the output of the program?

```
s = 3
n = 3
for x = 1:n
 for y =1:n
    if y==1
      z(x,y) =s
    elseif y>1 && y<n
        z(x,y) = s*z(x,y-1)
    end
  end
end
```

Output

```
3   9   27
3   9   27
3   9   27
```

19. What is the output of the program?

```
s = 3
n = 3
for x = 1:n
 for y =1:n
    if y==1
      z(x,y) =s^2
    elseif y>1 && y<n
        z(x,y) = s^2/z(x,y-1)
    end
  end
end
```

Output

```
9   1   9
9   1   9
9   1   9
```

20. What is the output of the program?

```
s = 27
t = 144
n = 3
for x = 1:n
 for y =1:n
    if y==1
      z(x,y) = ²√t
    elseif y==n
       z(x,y) = ⁿ√s
    elseif x==2 && y==2
       z(x,y) = 1
    else
      z(x,y) = 0
    end
  end
end
```

Output

```
12  0  3
12  1  3
12  0  3
```

21. What is the output of the program?

```
s = 27
t = 144
n = 3
for x = 1:n
 for y =1:n
    if y==1
      z(x,y) = ²√t
    elseif y==n
       z(x,y) = ⁿ√s
    elseif x==2 && y==2
       z(x,y) = 1+z(x,y-1)
    else
      z(x,y) = 0
    end
  end
end
```

Output

```
12   0  3
12  13  3
12   0  3
```

22. What is the output of the program?

```
s = 0
n = 4
z = zeros(4)
for x = 1:n
 for y =1:n
    if y==n or y==1
      z(x,y) = s;
    elseif y>1 && y<n
       z(x,y) = x^2 + y
     end
  end
end
```

Output

```
   0    3    4    0
   0    6    7    0
   0   11   12    0
   0   18   19    0
```

23. What is the final value of s and x?

```
x = 1
for i = 1:3
    for j = 1:3
    s = i+j
    x =x*j
    end
end
```

Output

s = 6, x = 216

24. What is the value of s?

```
y = 1
for i = 1:2:5
    for j =1:2
       x = i
       b = j
       s = x/i + y/b
    end
end
```

Output

s = 1.5

25. What is the final output of a?

```
x = 2
y = 3
for i = 1: x
   for j = 1:y
   t = i
   s = t
   a = s^2+t^j
   end
end
```

Output

a = 12

26. What is output of the program?

```
i=1
m=4
while i<m
  Output "*"
  i = i+1
end
```

Output

```
*
*
*
```

27. What is the output of the program if rows =5?

```
int m, i, j;
m=5
for (i=m; i>=m; i - - )
for (j=1; j<=5; j++)
Output "%"
Output "\n"
```

Output

```
%%%%%
%%%%
%%%
%%
%
```

28. What is the output of the program?

```
int m, i
int S = 0
m=3
for (i=1; i<=m; i ++  )
S = S + i;
```

Output

S = 6

29. What is the output of the program?

```
int m, i, j;
m=6
for (i=m; i>=m; i - - )
for (j=1; j<=6; j++)
Output "+"
Output "\n"
```

Output

```
++++++
+++++
++++
+++
++
+
```

30. What is the output of the program?

```
int i=1;
int m=6;
while i<m
Output "$"
Output "/n"
i=i+1
```

Output

```
$
$
$
$
$
```

31. What is the output of the program?

```
x = 1
for i = 1:4
    for j = 1:3
    s = i+j
    x =s*j
    end
end

Output x
```

Output

21

32. What is the output of the program?

```
s = 1
for x = 1:3
 for y =1:3
    if y==1
      z(x,y) =2
    elseif y>1
        z(x,y) = s+z(x,y-1)
    end
  end
end
```

Output

```
   2   3   4
   2   3   4
   2   3   4
```

33. What is the output of the program?

```
x = 10
for i = 1:2
   for j = 1:3
   s = x+j
   x =s*j+i
   end
end
Output x
```

Output

593

34. What is the output of the program?

```
for x = 1:5
 if x=1
 Output "*"
 else if x>1
    for y=2:7
     Output "@"
  end
end
```

Output

```
*
@@@@@@
@@@@@@
@@@@@@
@@@@@@
```

35. What is the output of the program?

```
for x = 1:3
   for y =1:3
    if x==y
    z(x,y)=2
    else
     z(x,y)=0
    end
   end
 end
 Output z(x,y)
```

Output

```
  2  0  0
  0  2  0
  0  0  2
```

36. What is the output of the program?

```
for x = 1:3
   for y =1:3
    if x==y
    z(x,y)=1
    else
    z(x,y)=x+y
    end
   end
 end
 Output z(x,y)
```

Output

```
  1  3  4
  3  1  5
  4  5  1
```

37. What is the output of the program?

```
for x = 1:3
   for y =1:3
    if x==y
    z(x,y)=10
    else
    z(x,y)=x-y
    end
      end
    end
    Output z(x,y)
```

Output

```
  10  -1  -2
   1  10  -1
   2   1  10
```

38. What is the out of the program?

```
for x = 1:3
   for y =1:3
    if x==y
    z(x,y)=x^2
    else
    z(x,y)= y^2 - x
   end
  end
end
Output z(x,y)
```

Output

1	3	8
-1	4	7
8	1	9

39. What is the out of the program?

```
for x = 1:4
   for y =1:3
    if x==y   or x>y
    z(x,y)=xy+x
    else
    z(x,y)= y^2 + x^2
   end
  end
end
Output z(x,y)
```

Output

2	5	10
4	6	13
6	9	12
8	12	16

40. What is the output of the program?

```
int x = 10
int y = 10
if ( !(x ^ y) )
   Output " x is equal to y "
else
   Output " x is not equal to y "
```

Output

x is equal to y

41. What is the out of the program?

```
for x = 1:4
   for y =1:4
    if x==y   or x>y
    z(x,y)=xy+y
    else
    z(x,y)= y^2 + x^2
    end
   end
 end
 Output z(x,y)
```

Output

2	5	10	17
3	6	13	20
4	8	12	25
5	10	16	20

42. What is the out of the program?

```
int a = 12, b = 20
  Output "max = %d\n", ((a + b) + abs(a - b)) / 2)
  Output "min = %d", ((a + b) - abs(a - b)) / 2)
```

Output

max = 20
min = 12

43. What is the out of the program?

for x = 1:4
  for y =1:4
    if x==y or x>y
    z(x,y)=xy-y
    else
    z(x,y)= $y^2 - x^2$
    end
  end
end
Output z(x,y)

<u>Output</u>

0	3	8	15
1	2	5	12
2	4	6	7
3	6	9	12

44. What is the output of the program?

for x = 1:4
  for y =1:4
    if x==y or x>y
    z(x,y)=x^2+x
    else
    z(x,y)= $y^2 + y$
    end
  end
end
Output z(x,y)

<u>Output</u>

2	6	12	20
6	6	12	20
12	12	12	20
20	20	20	20

45. What is the output of the program?

```
for x = 1:4
  for y =1:4
   if x==y  or x>y
   z(x,y)=x^2 - y
   else
   z(x,y)= y^2 − x
   end
  end
end
Output z(x,y)
```

Output

0	3	9	15
3	2	7	14
8	7	6	13
15	14	13	12

46. What is the output of the program?

```
for x = 1:4
  for y =1:4
   if x==y  or x>y
   z(x,y)=x^2
   else
   z(x,y)= y^2
   end
  end
end
Output z(x,y)
```

Output

1	4	9	16
4	4	9	16
9	9	9	16
16	16	16	16

47. What is the output of the program?

```
A = 0
B = 1
for x = 1:3
 for y =1:3
    if x<y
      z(x,y) =A;
    elseif x>y
       z(x,y) = B
    else
       z(x,y) = B
    end
  end
end
Output z
```

Output

1	0	0
1	1	0
1	1	1

48. What is the output of the program?

```
for x = 1:3
 for y =1:3
    if x==y
      z(x,y) = (x+y)/2
    else
       z(x,y) =x-y
    end
  end
end
```

Output

1	-1	-2
1	2	-1
2	1	3

49. What is the output of the program?

```
for x = 1:4
 for y =1:4
  if x==y   or x>y
  z(x,y)= x − y
  else
  z(x,y) y − x
  end
 end
end
Output z(x,y)
```

Output

0	1	2	3
1	0	1	2
2	1	0	1
3	2	1	0

50. What is the output of the program?

```
a = 12: b = 6: c = 3: d = 1
if a > b+c  then a =a*a else a = 2a
if a - b * c < a*b*c  then b = b*c else b = b – c
if (a < b + c)  then c = c * c else c = d + c
print a/b + 2c + b*b+ (a+b+c+d)b -d
```

Output

if a > b+c then a =a*a else a = 2a

12>6+3 So, a= 12*12 =144

Now, a=144, b = 6: c = 3: d = 1

if a - b * c < a*b*c then b = b*c else b = b – c

144-18<144*6*3 so, b=6-3=3

Now, a=144, b = 3: c = 3: d = 1

if (a < b + c) then c = c * c else c = d + c

144<6 so,c=1+3

Now, a=144, b = 3: c = 4: d = 1

print $\frac{a}{b}$ + 2c + b*b+ (a+b+c+d)b-d

$\frac{144}{3}$ + 2(4) + 3*3+ (144+3+4+1)3 -1 = 48+8+9+152*3-1

=65+456-1

= **520**

Made in the USA
Middletown, DE
05 December 2020

26314809R00163

Proceedings of WORDS'96

The Second Workshop on Object-Oriented Real-Time Dependable Systems

February 1–2, 1996

Laguna Beach, California, USA

Sponsored by

IEEE Computer Society Technical Committee on Distributed Processing

IEEE Computer Society Press

Los Alamitos, California

Washington • Brussels • Tokyo

IEEE Computer Society Press
10662 Los Vaqueros Circle
P.O. Box 3014
Los Alamitos, CA 90720-1264

IEEE Computer Society Press Order Number PR07570
IEEE Order Plan Catalog Number 96TB100036
Library of Congress Number 96-83326
ISBN 0-8186-7570-5
Microfich ISBN Number 0-8186-7572-1

Additional copies may be ordered from:

IEEE Computer Society Press
Customer Service Center
10662 Los Vaqueros Circle
P.O. Box 3014
Los Alamitos, CA 90720-1264
Tel: +1-714-821-8380
Fax: +1-714-821-4641
Email: cs.books@computer.org

IEEE Service Center
445 Hoes Lane
P.O. Box 1331
Piscataway, NJ 08855-1331
Tel: +1-908-981-1393
Fax: +1-908-981-9667
misc.custserv@computer.org

IEEE Computer Society
13, Avenue de l'Aquilon
B-1200 Brussels
BELGIUM
Tel: +32-2-770-2198
Fax: +32-2-770-8505
euro.ofc@computr.org

IEEE Computer Society
Ooshima Building
2-19-1 Minami-Aoyama
Minato-ku, Tokyo 107
JAPAN
Tel: +81-3-3408-3118
Fax: +81-3-3408-3553
tokyo.ofc@computer.org

Editorial production by Penny Storms
Cover by Alex Torres
Printed in the United States of America by KNI, Inc.

The Institute of Electrical and Electronics Engineers, Inc.

Table of Contents

Message from the Workshop General Co-Chairs

Kane Kim

Raymond Paul

Building on the success of the first WORDS held in October 1994, WORDS '96 was a step up in invigorating a technology drive to integrate three major computer system engineering technologies (CSETs): Object-Oriented CSET, Real- Time CSET, and Dependable CSET. This drive toward establishing an integrated technology, which may be called the Object-Oriented Real-Time Dependable System (ORDS) technology, cannot be exempt from the following law: success in any important technology drive depends upon leading efforts of innovation forces in industry. In this respect, WORDS '96 was an important event. A majority of participants who filled the limited 50 seats were invited presenters and a substantial segment of those invited presenters came from industry.

WORDS '96 served well as a serious forum for exchanging newly recognized research issues, advanced formulations that show promise, and research progress reports. The exchanged materials were of conceptual, theoretical, innovative design, or experimental nature, and represented technological or scientific advances. Most of the presented papers were polished further after the workshop and assembled into this proceedings. WORDS '96 continued the policy adopted in WORDS '94 of welcoming contributions that represent significant advances in any two of the three component technology fields rather than focusing on those contributions that cut across all three component technologies.

Many participants marveled at the splendid nature of the location of WORDS '96. Lively technical exchanges were nicely balanced out by the relaxing atmosphere provided by the beautiful natural surroundings. The success of this workshop was largely due to the efforts of all participants. However, we would like to recognize the more than usual efforts of the Program Committee co-chairs, Phillip Sheu and Edgar Nett; the Program Committee members; the local arrangements co-chairs, Eltefaat Shokri and Luiz Bacellar; other committee chairs, Larry Peterson, Charlie Jung, Chittur Subbaraman, and Taehyung Wang; and the registration officers, Astrid Boetel and Yuseok Kim. It was a great pleasure to be a part of the team who organized the second workshop in the WORDS series. We feel it is a safe bet to predict that the next WORDS (to be held in 1997) will be an even more exciting event!

Workshop Chairs and Program Committee

Workshop General Co-Chairs

Kane Kim
University of California, Irvine (UCI), USA
kane@ece.uci.edu

Ray Paul
Office of the Under Secretary of Defense, USA
paulra@acq.osd.mil

Program Co-Chairs

Phillip C-Y. Sheu
UCI, USA
sheu@ece.uci.edu

Edgar Nett
GMD, Germany
nett@gmd.de

Local Arrangements Co-Chairs

Eltefaat Shokri, *SoHaR, USA*
Luiz Bacellar, *UCI, USA*

Registration Co-Chairs

Chittur Subbaraman, *UCI, USA*
Taehyung Wang, *UCI, USA*

Publicity Committee Co-Chairs

Larry Peterson, *USN NRaD, USA*
Charles Jung, *IBM, Boca Raton, USA*
Lonnie Welch, *NJIT, USA*

Program Committee

Alan Burns, *York University, U.K.*
Jen-Yao Chung, *IBM, T.J. Watson, USA*
Michel Gien, *Chorus Systems, France*
Katsumi Kawano, *Hitachi, Japan*
Moon H. Kim, *Konkuk University, Korea*
Hermann Kopetz, *Technische Universitat Wien, Austria*
Tom Lawrence, *USAF Rome Laboratory, USA*
Insup Lee, *University of Pennsylvania, USA*
James Lee, *ITRI, Taiwan*
K.J. Lin, *UCI, USA*
W.L. McCoy, *USN NSWC, USA*
C.J. Paul, *IBM, Austin, USA*
Santosh Shrivastava, *University of Newcastle upon Tyne, UK*
Richard Soley, *OMT, USA*
Hide Tokuda, *Keio University, Japan*
Anand Tripathi, *NSF, USA*
Susan Urban, *Arizona State University, USA*
Doug Wells, *OSF, USA*

Session 1

Real-Time Object Models and Languages

Session Chair:
Radu Popeseu-Zeletin
Technical University of Berlin and GMD-FOKUS, Germany

A Node as a Real-Time Object

H. Kopetz
S. Poledna
Technical University of Vienna, Austria
hk@vmars.tuwien.ac.at

Abstract: *In real-time systems the temporal performance of the implementation is a key issue. Since this temporal performance depends not only on the functions and structure of the application software, but also on the services of the operating system, and the architecture and performance of the hardware, we argue that a node, i.e., a complete hardware software unit, is a proper unit of abstraction when building large distributed fault-tolerant real time systems.*

Key words: distributed systems, real-time systems, object orientation, fault tolerance, replica determinism

1 Introduction

In a hard real-time environment the properties in the temporal domain are as important as the properties in the value domain. The object oriented design techniques will only be successful in the area of real-time systems if the design and analysis of the temporal properties of an object oriented system are supported by the object technology. The issues in the design of performance sensitive objects are yet to be studied [Pereira, 1994].

In this workshop contribution we deal with the design of distributed fault-tolerant real-time systems in embedded applications. These application are characterized by high production numbers (e.g., the computer control of a television set) and by a foremost concern for cost-effective and correct solutions. Some of these applications are safety critical and must not have a single point of failure, e.g., the control of the braking system in an automobile. In these applications fault-tolerant computer systems must be deployed. We focus on the temporal properties of the computer system, assuming that the correct operation in the value domain has been established.

The rest of this paper is organized as follows. In section two we discuss the temporal aspect of a client-server interaction in the object model and emphasize the temporal obligations of the server and the client. In section three we discuss the issues that have to be dealt with when establishing an upper bound for the execution time of the server object. The subseqeunt sections deals with the obligations of the client and discusses the desired properties of real-time communication protocols. Finally, we investigate the issues of replica determinism of the server.

2 Composability

The most widely cited advantage of Object Technology is the fact that objects can be used as components [Pancake, 1995] when designing a larger system. The driving force behind the object-oriented technology is thus the reuse of subsystems (objects) in widely differing contexts. To achieve this goal a complex subsystem is replaced by an understandable abstraction [Corbett & Avrunin, 1994] that captures the essential properties of this subsystem and hides the irrelevant details of the subsystem behind a simple and stable external interface. The essential properties of each subsystem can then be verified in isolation. The correctness of the whole system is established by combining the verification arguments for each subsystem. It is important to ascertain that the subsystems do not interact via *hidden* interfaces. Otherwise uncontrolled interactions between the subsystems via these hidden interfaces can invalidate the arguments that are the basis for reasoning about the correctness of a composition.

In a hard real-time system it is not possible to abstract form the temporal performance of a subsystem as an irrelevant detail. Therefore the temporal properties of the objects must be part of the interface specification. These temporal properties depend not only on the software module that implements the object, but also on the characteristics of the operating system and the performance of the target hardware. Any resource sharing of different objects on the same target hardware must be analyzed with great scrutiny in order to detect hidden interactions between the objects that effect the temporal performance.

We assume a hardware structure as depicted in Fig. 1. A node of the distributed systems is considered a smallest replaceable unit (SRU). A set of SRUs forms a fault-tolerant unit (FTU). The individual SRUs are interconnected by means of a replicated broadcast bus. If a node fails, the failure is masked by the other nodes of the FTU and the service is provided without any interruption. Since the communication system is replicated as well, the failure of a single bus is also tolerated. This guarantees that there is no single point of failure in the system.

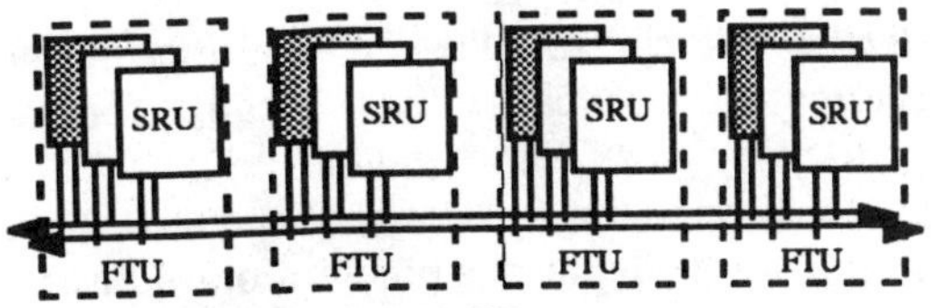

Fig. 1: System Structure

The fault-tolerant system as depicted in Fig. 1 applies active replication as its fault-tolerance strategy. This presupposes that the operation of the SRUs that form an FTU are replica determinate, i.e., they visit the same states at about the same time (c.f. section "Replica Determinism").

Let us now analyze the temporal performance of a real-time transaction between a stimulus from the environment and a response to the environment by making use of the client-server model. In the client-server model a request (a message) from a client object to a server object causes after some time a response from the server. This response can be a state change of the server object and/or the sending of a message. There are three temporal parameters characterizing such a client-server interaction: The maximum response time *resp* that is expected by the client and given in the specification, the maximum execution time of the server *maxt* that is determined by the implementation, and the minimum time between two successive requests by the client *mint*. It is important to note that only one of these parameters, namely *maxt*, is completely in the sphere of control of the server and that the minimum time between two successive request *mint* is completely in the sphere of the client. In a hard real-time environment the implementation must guarantee, that

$$maxt < resp$$

holds as long as the server respects the minimum time between two requests, *mint*.

If in the given execution environment $maxt \ll resp$, i.e. the performance of the object is orders of magnitude faster than required by the particular application in the worst case condition, i.e. the hardware is "over dimensioned", then performance concerns are not an issue. However this situation is exceptional in embedded systems, because of the cost pressure in a market economy. If *maxt* is in the same order of magnitude as *resp*, then a careful analysis of the temporal properties of an object, i.e. *maxt*, is required to assure that the system will always meet the temporal requirements, provided the client object observes its obligation to issue request only at a rate less then $1/mint$.

3 Temporal Properties of Objects

In a hard real-time system the most important temporal property of the server is the maximum execution time between the stimulus and the response. We first discuss the maximum execution time analysis of simple sequential objects and subsequently of more complex concurrent objects.

Simple Objects

The simplest object we can envision consists of a single nonpreemptive sequential task that runs on dedicated hardware without requiring any operating system services. The maximum execution time of such a task depends on the source code of the task, the properties of the object code generated by the compiler, and the characteristics of the target hardware. In this section we investigate the required effort for the analytical establishment of a tight maximum execution time bound of such a simple task.

Source Code Analysis: The first problem concerns the calculation of the maximum execution time of a program written in a higher level language under the assumption that the maximum execution times of the basic language constructs are known. Most source programs do not contain enough information for the determination of the maximum execution time. Consider a simple WHILE statement that controls the iterations of a loop. Without knowing the exact data, it is not possible to establish the maximum number of iterations in such a construct. It is even known that this problem, in general, is undecidable. The problem can be solved by restricting the language and by instrumenting the program with preprocessor statements that provide this required additional information, e.g., the maximum number of iterations of every loop etc.[Puschner & Koza, 1989]. If tight bounds are to be found then the available semantic information about the program must be expressed in some form and entered into the maximum execution time analysis tool. An example of such an elaborate tool is given by [Puschner, 1993].

Compiler Analysis: The next problem concerns the determination of the maximum execution time of the basic language constructs of the source language under the assumption that the maximum execution time of the machine language commands are known. For this purpose the code generation strategy of the compiler has to be analyzed. Most compilers are designed to optimize the *average* performance, but not the maximum execution time. Therefore the determination of the maximum execution time of basic source language constructs can be a tedious task, if not impossible. In our institute a new compiler for the language MODULA/R [Vrochticky, 1992] (a real-time derivative of MODULA) with predictable maximum execution time constructs was developed [Vrchoticky, 1994]. This compiler is being used for developing the application software for the MARS [Kopetz et al., 1989] architecture.

Hardware Analysis: The final problem concerns the determination of the maximum execution time of the commands of the target hardware. If the target hardware is a processor without cache and pipelining, then the execution time of the hardware instructions are contained in the hardware documentation. If a modern processor with cache and pipelining is used, then the execution time of a hardware instruction depends on a number of additional parameters, such as the interruption of the pipeline by an interrupt or the need to reload the cache. The analytical determination of the maximum execution time of modern processors is the subject of intensive research see, e.g. [Lim & al., 1994].

This discussion shows that the analytical calculation of the maximum execution time for a simple sequential task which does not make use of operating system service is only possible under very restricting assumptions: It requires an instrumented source program with a restricted language that guarantees termination. Furthermore, the code generation strategy of the compiler needs to be predictable and the architecture of the processor must be accessible to analysis. We feel that, at least at present, all these assumptions can only be satisfied in a research environment but not for industrial software construction.

Concurrent Objects

We now turn to maximum execution time analysis of concurrent objects with shared resources and other task dependencies. In this scenario two types of interactions have to be analyzed, the direct interactions caused by the intended task dependencies (mutual exclusion, precedence) and the indirect interactions caused by the competition for shared resources (e.g., the cache). In addition, the time required by the operating system services has to be scrutinized.

Direct Interactions: In the restricted context of the rate monotonic assumptions [Liu & Layland, 1973] the task dependencies introduced by critical regions can be analyzed according to the generalized rate monotonic theory developed by [Sha, Rajkumar, & Sathaye, 1994]. This examination requires the analysis of the complete system under the assumption that the client observes its obligation, i.e., that the tasks are activated with a minimum period between two consecutive activations that is larger than *mint*.

Indirect Interactions: The indirect interactions of concurrent tasks are caused by contention for resources that are used by more than one task (e.g., cache) and by the request of operating system services to resolve these conflicts. If for example, the preemption of tasks can be eliminated by an a priori analysis of the application semantics, then considerable coordination services from the operating systems, such as the queue management around semaphores protecting critical regions, can be eliminated.

The analytic determination of the maximum execution time between a stimulus to and a response from a server that executes a number of dependent concurrent tasks is beyond the present state of practice. A disciplined approach tries to combine a number of different methods to solve this important real-time system problem:

(i) The provision of a restricted architecture that reduces the interactions between the tasks by providing a static structure that can be analyzed a priori and eliminates wherever possible the need for explicit synchronization actions that require context switches and operating system services.

(ii) The analysis of subproblems (e.g., maximum execution time analysis of the source program) such that some of the questions can be answered analytically.

(iii) The realization of experiments with a concrete implementation to gather the missing data by measurement.

(iv) The informed testing of the implementation to validate the assumptions and to measure the safety margin between the specified response time and the actual execution time.

The maximum execution time of a server object is thus determined after the server software is bound to the server hardware. In our view the concept of a 'real-time software object' has only marginal utility since it is not possible to give sound analytical results for the temporal properties of such an object. To guarantee fault-tolerance it is therefore necessary to enforce this obligation explicitly by proper methodologies [Poledna, 1995b].

4 A Node as a RT Object

According to [Wegner, 1987] p.168 'Objects are autonomous entities that respond to message or operations and share a state'. This definition of an object supports the view that a node of a distributed real-time system can be considered as an object. A node as a RT object is an encapsulated software/hardware unit with physical interfaces to the outside world.

Simple Nodes

The simplest case is a node with only a single interface to the communication system as depicted in Fig. 1. The communication protocol that controls the interface to the communication system transports the requests from the client object to the server object and returns the responses from the server object to the client. The obligation of the client to maintain a minimum interval between requests can be delegated to the communication protocol. If a time-triggered protocol such as TTP [Kopetz & Grünsteidl, 1994] is used in the communication system then the protocol enforces the client obligation. The regularity and clairvoyance of the protocol makes it possible to specify precisely the temporal properties of the external object interface such that the temporal properties can be validated in isolation. The autonomy of the protocol assures the composability of the architecture.

Interface Nodes

A real-time control system has to collect data from sensors and has to deliver setpoints to the actuators in the environment. For this purpose some of the nodes have to contain an Input/Output interface to the environment. In many situations the designer has the freedom to plan an I/O interface based on sampling or to implement an interrupt-driven I/O interface.

In an interface based on sampling the values of the state variables in the process are interrogated periodically under the control of the interface node. The temporal control of the interface node remains within the interface node and the I/O subsystem has no further obligation to fulfill.

If the I/O interface is interrupt-driven, then some aspects of the temporal control of the interface nodes are delegated to the I/O subsystem. The interface node can only meet its temporal specification if the I/O subsystems observes its obligation, i.e., it does not generate interrupts with a frequency higher than *1/mint*. The critical role of this interface obligation is sometimes not considered appropriately.

5 Replica Determinism

Hard real-time systems are often subject to dependability requirements since the cost of a missed deadline is very high. Fault-tolerance of these systems has to be guaranteed. In real-time systems fault-tolerance is best supported by active replication. With this technique a client broadcasts its request to a set of replicated server objects. The server group executes all requests in parallel. Active replication, however, is based on a very restricting consistency criterion for the replicated server objects: Fault-free replicated servers are required to deliver identical responses to service requests within a time interval of ε time units. This criterion is called *replica determinism* [Poledna, 1995a].

It might seem trivial at the first moment that, given an identical sequence of service requests, replicated servers will produce identical responses. Unfortunately, this is not the case. Mechanisms such as on-line scheduling, preemptive scheduling, uncoordinated access to clocks, reading of replicated sensors, timeouts and others are causing non-deterministic behavior of replicated servers. It is therefore necessary to enforce replica determinism by appropriate methodologies.

It is important to note that replica determinism is not a local property of a server object but a system global property. Replica determinism is closely related to areas such as communication, synchronization, failure handling, redundancy preservation and scheduling. Achievement of replica determinism depends on the system architecture and the implementation of server objects. The requirements for deterministic behavior of servers are most commonly decomposed into the following basic abstractions [Schneider, 1986, Cristian, 1991]

- **Membership:** Every non-faulty client has timely and consistent information on the set of functioning servers objects which constitutes a group. (This information can be hidden behind the abstraction of the communication service.)
- **Agreement:** Every non-faulty server in a group receives the same service requests within a given time interval.
- **Order:** Explicit service request as well as implicit service request, which are introduced by the passage of time, are processed by non-faulty servers of a group in the same order.

It is well known that all these three requirements can only be fulfilled in a timely manner in a synchronous system [Fischer, Lynch, & Paterson, 1985]. A system is said to be synchronous if servers are synchronous and the communication is bounded. That is, there exists an a priori known constant Δ such that every server takes at least one processing step within Δ real-time steps. Furthermore, service requests sent between clients and servers arrive at their destination within Φ real-time steps, for some fixed Φ.

Correct function of a synchronous system depends therefore on the a priori knowledge of the parameters Δ and Φ. The reason for this is that the algorithms which are implementing membership, agreement, and order are based on the assumption that the parameters Δ and Φ hold. Therefore, time-outs can be used to discriminate between correct and faulty behavior. This in turn has the consequence that the correct functionality of the fault-tolerance mechanism depends on the timeliness of servers and communication.

Again, this discussion of the fault-tolerance mechanism leads us back to the problem of temporal performance of real-time objects and their composability. To establish the bounds for Δ and Φ the maximum execution time between a stimulus and its response of an object needs to be determined. This determination, however, can only be done after the server software has been bound to the hardware. It is therefore impossible to guarantee fault-tolerance for 'real-time software objects' independently of the hardware, the operating system, and communication services.

6 Conclusion

Objects are providing a powerful abstraction which cleanly separates the interface aspects from implementation issues. One main advantage of this separation is seen in the ability to build complex systems by a composition of simpler subsystems or objects. Composability, however, can only be achieved if it is possible to establish correctness criteria for the whole system by combining the correctness arguments of the individual subsystems. For hard real-time systems it is necessary that an real-time object obliges to its specified temporal properties, since the correct function of real-time systems depends on the delivery of timely responses. Object-oriented design techniques can only be applied successfully to real-time systems if the temporal properties of objects can be guaranteed.

Fault-tolerance in real-time systems is implemented by active replication. This requires that replicated server objects behave deterministically, i.e., they deliver identical responses to service requests within a fixed time interval. Achievement of replica determinism under real-time constraints, however, is only possible in a synchronous system. By definition, a synchronous system is able to guarantee bounded maximum execution times and communication delays. It follows therefore that fault-tolerance can only be guaranteed if real-time objects oblige to their temporal specification.

The most important temporal property of a server object is the maximum execution time between a service request sent to the object and its response. According to the current state of practice this temporal parameter is determined by careful analysis combined with measurements on a concrete implementation. Separating the software from the hardware will lead to a loss of the measured temporal parameters and thus to a loss of the basis for reasoning about the temporal properties of real-time objects. Therefore, according to our view, the proper abstraction of a real-time object is a node where software and hardware are bound together and a proper communication interface that guarantees the client's temporal obligation is provided.

This discussion should not be understood to imply that object-oriented system construction [Meyer, 1988] is not considered to be of value for real-time system design. It just presents the view that the concept of a 'real-time software object' is of marginal utility.

Acknowledgments

This work has, in part, been supported by the ESPRIT LTR Project DEVA.

References

Corbett, J. C., & Avrunin, G. S. (1994). *Towards Scalable Compositional Analysis* ACM Software Engineering Notes, *19* (5), 53-61.

Cristian, F. (1991). *Understanding Fault-Tolerant Distributed Systems* Comm. ACM, *34* (2), 57-78.

Fischer, M., Lynch, N., & Paterson, M. (1985). *Impossibility of Distributed Consensus with one Faulty Processor* Journal of the ACM, *32* (2), 374-382.

Kopetz, H., Damm, A., Koza, C., Mulazzani, M., Schwabl, W., Senft, C., & Zainlinger, R. (1989). *Distributed Fault-Tolerant Real-Time Systems: The MARS approach* IEEE Micro, *9* (1), 25-40.

Kopetz, H., & Grünsteidl, G. (1994). *TTP-A Protocol for Fault-Tolerant Real-Time Systems* IEEE Computer, *24* (1), 22-66.

Lim, S. S., & al., e. (1994) An Accurate Worst Case Timing Analysis for RISC Processors. *Real-Time Systems Symposium RTSS 94*, (pp. 97-108). San Juan, Puerto Rico: IEEE Computer Society.

Liu, C. L., & Layland, J. W. (1973). *Scheduling Algorithms for Multiprogramming in a Hard-Real-Time Environment* J. of the ACM, *20* (1), 46-61.

Meyer, B. (1988). *Object-Oriented Software Construction* .Prentice Hall,

Pancake, C. M. (1995). *The Promise and the Cost of Object Technology: A five-year Forecast* Comm. ACM, *38* (10), 33-49.

Pereira, C. E. (1994) Enhancing Real-Time Object-Oriented Analysis Methods. *Sixt Euromicro Workshop on Real-Time Systems*, (pp. 104-115). Vaesteraas, Sweden: IEEE Press.

Poledna, S. (1995a). *Fault-Tolerant Real-Time Systems, The Problem of Replica Determinism* .Kluwer Academic Publishers, Hingham, Mass, USA.

Poledna, S. (1995b). *Tolerating Sensor Timing Faults in Highly Responsive Hard Real-Time Systems* IEEE Transactions on Computers, *44* (2), 181-191.

Puschner, P. (1993). *Zeitanalyse von Echtzeitprogrammen*. PhD, Technical University of Vienna.

Puschner, P., & Koza, C. (1989). *Calculating the Maximum Execution Time of Real-Time Programs* Real-Time Systems, *1* (2), 159-176.

Schneider, F. B. (1986) A Paradigm for Reliable Clock Synchronization. *Proc. Advanced Seminar Real-Time Local Area Networks*, (pp. 85-104).

Sha, L., Rajkumar, R., & Sathaye, S. S. (1994). *Generalized Rate-Monotonic Scheduling Theory: A Framework for Developing Real-Time Systems* Proceedings of the IEEE, *82* (1), 68-82.

Vrchoticky, A. (1994). *The Basis for Static Execution Time Prediction*. PhD, Technical University of Vienna.

Vrochticky, A. (1992). *MODULA R Language Definition* No. 2/92). Technical University of Vienna.

Wegner, P. (1987) Dimensions of Object-Based Language Design. *Proc. of OOPSLA 87*, (pp. 168-179). ACM Press.

A Logic-Based Requirements Language for the Specification and Analysis of Real-Time Systems

Jeffrey J. P. Tsai†
Dept. of EECS
University of Illinois at Chicago
Chicago, IL 60607
tsai@eecs.uic.edu

Thomas Weigert
Softw. Tech. Ctr.
Motorola, Inc.,
Schaumburg, IL 60196
weigert@comm.mot.com

ABSTRACT

The applications of real-time systems have dramatically increased over recent years. Requirements for real-time systems need to specify timing constraints. In this paper, we augment the FRORL *requirements specification language with constructs specifically geared towards capturing requirements of real-time systems. We give meaning to these constructs through temporal logic. The non-functional aspects of a specification such as temporal properties asserted of a real-time system and the consistency of timing constraints can be verified through model checking and graph-theoretical algorithms.*

1 Introduction

The Software development life cycle begins with the development of requirements specifications. Requirements are acquired in a continuous dialog with the user. During the requirements acquisition process, software engineers are forced to make assumptions. Some of these assumptions may later be withdrawn and new assumptions may be added due to a better understanding of the user needs. An added assumption may invalidate consequences of earlier assumptions. We refer to the fact that consequences do not grow monotonically with the assumptions in a specification as the non-monotonicity property. Requirements specifications will be more straightforward if the language used to capture requirements possesses the non-monotonicity property.

In [1], we presented a formal requirements specification language, FRORL (Frame-and-Rule Oriented Requirements specification Language). The applications of real-time systems, (such as avionics, patients monitoring, and military command and control systems), have dramatically increased recently. Beyond purely functional requirements, specifications for real-time systems must specify timing constraints, i.e., bounds on the execution time of the specified system. Further, specifications for real-time systems often include safety and lifeness properties.

In this paper, we augment FRORL with constructs specifically geared towards the specification of real-time systems. FRORL continues to rely on the concepts of frames and production rules for its surface syntax. We distinguish between two aspects of a FRORL specification: Parts of a specification are concerned with producing the functionality of the specified system. Other parts of a specification are concerned with stating constraints imposed on the system, due to its being embedded into a real-time environment and its need to exchange information with other processes. We refer to the former as the functional aspects of a specification, and to the latter as the time-related aspects of the specification.

The functional aspects of a FRORL specification are given their meaning through a non-monotonic extension of Horn-clause logic. The time-dependent aspects of a FRORL specification are given a semantics through temporal logic.

Various properties of a specification, like reachability, reversibility, liveness, consistency, etc., are of interest to the software engineer. Specifications constructed in FRORL may be subject to analysis tools which attempt to determine whether such properties hold for a given specification. Through this checking process, the user, with the aid of the system, goes through a sequence of iterations of specification modification and verification until the specification is assumed to be correct. FRORL employs (i) resolution refutation to determine reachability, reversibility, liveness, consistency, synchronic distance, and bounded fairness; (ii) model checking to verify temporal properties; and (iii) graph-theoretical algorithms to establish the consistency of timing constraints [2].

†. This work is supported in part by Fujitsu Network Switching of America, Inc.

0-8186-7570-5/96 $5.00 © 1996 IEEE

2 Overview of FRORL

The FRORL requirements specification language builds on Horn-clause logic for its clear semantics and the ability of extensive verification of the specification it provides. FRORL was designed to satisfy desiderata on specification languages common to the software engineering literature, such as

- Freedom from implementation concerns
- Naturalness of representation
- Verification
- Validity checking
- Ease of modification and construction
- Soundness and completeness

The FRORL modeling primitives are *objects* and *activities*. Each object in the model stands for some entity in the domain being modeled. Changes taking place in the domain and relations between entities of the domain are represented as activities in the requirements model. Each object and activity has certain properties, assumptions, or constraints associated with it. The information related to objects and activities is expressed using a frame representation.

The modeling constructs of FRORL are given their meanings in terms of the non-monotonic logic presented in [1]. The FRORL syntax is only surface syntax; reasoning about specifications is performed at the level of non-monotonic Horn-clause logic. In [3] we showed how the constructs of the FRORL surface syntax can be translated into predicates of the underlying logic.

Objects represent entities in the domain being modeled. An object can also be viewed as a data structure or a data type in a software system. Each object has attributes or properties. These details of an object are stated in slots of the object frame.[1]

There are two kinds of slots in each object frame, abstract relations and attributes. The abstract-relation slot indicates the hierarchical relationship a frame bears to its parent frame. Generalization (a_kind_of) allows a class to capture common characteristics shared by a group of individuals. Instantiation (an_instance_of) allows to obtain an individual from a prototype (class). Aggregation (a_part_of) allows an entity to be constructed from its constituent entities. An object may bear an abstract relation to more than one parent object (multiple inheritance). The attribute slots represent particular properties of an object. Each property may have a value defined in the frame (which may either be a single object, or a list of objects which is represented in Prolog list notation). The value of an object may also be empty, in which case it will either be defined by an object which descends from the given frame in the inheritance hierarchy, or it will be defined through later computation.

object: subscriber_line_controller
    a_part_of: exchange
    id:
    line:
    alloc_out:
    alloc_in:
    call_line:

This example shows the subscriber_line_controller object. It bears an aggregation relationship to the exchange object. It has various attributes that are expressed as slots. None of the slots have initial values.

FRORL refers to changes taking place in the domain as *activities*. Each activity is again represented as a frame. The activity frame has four attribute slots: part, precondition, action, and alt_action (alternative action).

The part slot states the objects participating in an activity. It contains the class to which the objects participating in the activity belong. The precondition slot describes constraints or facts associated with the activity. (A fact is some assertion about the domain; as such it either postulates the existence of objects or activities that do not have preconditions associated with them.) The action slot contains sequences of references to activities. If the precondition is satisfied, then the action stated in the action slot will be executed. If the precondition is not satisfied the sequence of actions that may be stated in the alternative action slot will be executed. Actions and preconditions are stated as a comma-separated list of (Prolog-style) predicates. Any of the slots may be empty.

Activities are passed values as arguments, or they may be passed uninstantiated variables which will receive values as a result of performing the activity. Variables may be instantiated to lists and refer to instances of objects. It is also possible to refer to the object that is the value of an attribute slot (through dot-notation).

3 Modeling real-time systems

A real-time system is typically comprised of a collection of processes which communicate with one another through shared variables and/or by message passing over a communication channel. A set of state variables and a set of events are associated with each process.

[1] In FRORL expressions begin with lower-case letters. Expressions with an initial upper-case letter are implicitly universally quantified variables.

From a functional point of view, we can treat each process as an independent activity which occasionally outputs events or waits to input events.

However, the correctness of a real-time system depends not solely on the functional result of the computation, but also on the time at which the results are produced. Often, the processes composing a real-time system are coupled to processes or events of the environment. The system's correctness hinges also on meeting the timing constraints imposed by the environment.

A periodic process requires actions to be taken at fixed intervals. It usually performs a cyclic function, such as collecting data values at predetermined times or performing routine maintenance chores. Sporadic processes typically respond to external stimuli or events. Their behavior is not necessarily periodic or predictable. They occur because the real-time computer system must interact with its environment, such as commands from human operators or changes in a monitored sensor's value.

Timing constraints may specify when and for how long a given process has to execute, how quickly it needs to respond to the triggering event, and for how long a process is willing to wait for a message.

Within FRORL, processes are modeled as activities. The frame representation is extended to capture the constraints important for specifying real-time systems. Timing constraints are classified into maximum (no more than a certain length of time may elapse between two events), minimum (no less than a certain length of time may elapse), and durational timing constraints. Timing constraints are asserted in slots added to standard activity frames.

To ensure the repetitive nature of a *periodic activity*, we associate a cyclic value with an activity to represent its period. Such an activity frame expresses periodic processes in an external manner (rather than expressing the period as internal to the activity) to ensure that the timing constraints of periodic activities are enforced by the system and do not depend on the execution of the processes themselves.

activity: poll_link_layer(*Llc*, *Mupex*)
  parts: *Llc*: link_layer, *Mupex*: app_layer
  period: 15ms
  precond: receive(*Llc*.in, *Val*)
  actions: send(*Mupex*.out, *Val*)
  alt_actions: send(*Mupex*.out, dummy_val())

This specification describes a periodic process poll_link_layer which should be executed every 15 milliseconds—we use units such as seconds (s), milliseconds (ms), microseconds (us), etc., as multipliers in the definition of time constraints. Every 15 milliseconds this activity will attempt to receive an input val from the input channel in of llc. If it can obtain an input, it will execute the specified actions and pass the input value on to mupex.out. If no input is available, a default value dummy_val() is constructed and passed on to the application layer.

Note that actions or alternative actions might represent a single activity (process) or a set of activities (processes). In the latter case, the periodic timing constraint will constitute a single deadline for the specified composite task.

Sporadic activities are initiated like non-timed activities and do not contain a period attribute. Timing constraints are classified into maximum and minimum timing constraints, and durational constraints which are combinations of both. Two attribute slots, max_constraint and min_constraint, capture constraints on the time duration between the occurrence of two events. Activities are constrained regarding the time lag allowed between the occurrence of the event designated by the from slot and the event designated by the to slot. For a maximum timing constraint, no more than the given length of time may elapse between the occurrences of two events. No less than the given length of time may elapse between the occurrence of two events in a minimum timing constraint. Note that the starting event or the stopping event (or both) may be left unspecified. If the starting event is missing, then the timing constraint holds from the moment of the initial invocation of the activity. If the stopping event is missing, the activity functions as a pure timeout operation.

activity: first_digit(*Contr*)
  parts: *Contr*: subscriber_line_controller
  max_constraint: 40s
  to: receive(*Contr*.line, *Digit*)
  actions: send(*Contr*.line, stop_tone),
    send(*Contr*.alloc_out, *Digit*),
    collect_digits(*Digit*)
  alt_actions: send(*Contr*.line, timeout_tone),
    send(*Contr*.alloc_out, release)

The first_digit activity has an associated maximum timing constraint of 40 seconds from the start of this activity to the receipt of a digit on the terminal line of the subscriber line controller. If no digit is received within the specified time, processing of the actions is abandoned and the alternative actions are executed. The alternative actions release the resources that had been acquired by the subscriber line controller.

Durational timing constraints [4, 5] are modeled by both maximum and minimum timing constraints. For durational timing constraints, the time between the starting event of an activity and the stopping event of an activity is precisely constrained.

4 Temporal logic

We had earlier indicated that a FRORL specification is given its semantics by a non-monotonic extension to Horn-clause logic. Strictly, this is true only for the functional aspects of a FRORL specification. Above we have discussed extensions to FRORL that allow to represent real-time systems. Some of these extensions concern the functional aspects of a specification: Activities can be treated as processes which communicate via message passing or shared variables. For example, an activity with an associated maximum timing constraint is interpreted as interacting with a timer process which signals violation of the timing constraint and causes alternative behavior in response.

However, the timing constraints as well as the concurrent execution of activities also induce non-functional aspects that essentially rely on the concept of time. We define the semantics of the non-functional, time-dependent aspects of a FRORL specification through temporal logic.

Temporal logic is a language for both describing sequences of situations over time and for reasoning about them. In this context, we consider sequences of situations as representing the successive states a program runs through during its execution. This interpretation allows us to use temporal logic to specify and verify time-dependent programs. (This was first pointed out by Pnueli [6], where he argued that the then *en vogue* notions of predicate transformers could not handle real-time systems which do not have a final state.)

Due to these non-functional, time-dependent aspects, it is difficult to implement real-time systems correctly. However, an incorrectly executing real-time program might have grave consequences. Verification of real-time programs has received much attention and consequently, many different approaches to the verification of such systems have been presented in the literature. For an overview of the use of temporal logic in the specification and verification of real-time systems, see [7, 8], but in particular the excellent survey by Emerson in [9].

Temporal logic allows, along with the standard vocabulary of first-order logic, the use of temporal modalities to form sentences, such as (provided that P and Q are sentences) $\bigcirc P$, $\Diamond P$, $\Box P$, and $P \mathcal{U} Q$. Intuitively, "$\bigcirc P$" means that "P holds in the next state," "$\Diamond P$" means that "P will eventually hold," "$\Box P$" is read as "P will hold from now on" (or "P always holds," but note that this operator looks only into the future and does not extend into the past). Lastly, "$P \mathcal{U} Q$" means that "P will hold until Q holds." Depending on the types of states over which we interpret the sentences of temporal logic, we get different types of temporal logics.

The logic presented in this paper is an extension of the modal μ-calculus first presented by Kozen [10], who took a modal logic with labels and added a least fix-point operator. $\mu z \cdot P[z]$ is intended to be the least fix-point of P, considered as a functional on its free propositional variable z. The introduction of a fix-point operator enriches the language allowed by the above modal systems in important ways: Standard branching-time temporal logic cannot express certain fairness assumptions crucial for real-time systems [11]. Standard temporal logics cannot express certain modular properties of concurrent programs (e.g., "P holds at all even moments" [12]). Furthermore, most modal approaches deal with temporal aspects of a system purely from a qualitative viewpoint. Quantitative aspects, such as the demand that a system respond to an event within a certain fixed time, have only been dealt with in an *ad hoc* manner. The extension to the modal μ-calculus presented in this paper allows us to reason about quantitative aspects which are crucial for real-time systems.

We refer to our extension of the modal μ-calculus as RTμ. Sentences in RTμ are formed in the standard way: We assume an infinite supply of propositional constants P, Q, etc., and propositional variables x, y, and so on. Provided that P and Q are sentences, $\neg P$, $\bigcirc P$, and $P \wedge Q$ are sentences. As explained earlier, the intuitive reading of "$\bigcirc P$" is that "P holds at the next time." Furthermore, if P is a sentence and z a propositional variable, then $\mu z \cdot P[z]$ is a sentence, where by "$P[z]$" we mean that z occurs free in P. We pose the additional restriction that $P[z]$ must be syntactically monotonic in the propositional variable z, i.e., all free occurrences of z in $P[z]$ must fall under an even number of negations. Intuitively, "$\mu z \cdot P[z]$" represents the least fix-point of $P[z]$.

Other operators are introduced by definition: $P \vee Q = \neg(\neg P \wedge \neg Q)$, $P \Rightarrow Q = \neg P \vee Q$, and $\nu z \cdot P[z] = \neg \mu z \cdot \neg P[^z/_{\neg z}]$; the latter denotes the greatest fix-point (as usual $P[^z/_{z'}]$ means that all free occurrences of z in $P[z]$ have been replaced by z').

Consider the following sentences containing fix-points.

$$\mu z \cdot P \vee \bigcirc z$$
$$\nu z \cdot P \wedge \bigcirc z$$

Intuitively, the first sentence means that either P holds now, or P holds in the next state, or $\ldots$, up to the fix-point. This sentence says that P will eventually hold. The second sentence expresses the meaning of "always": P holds now, and P holds in the next state, and so on. The above sentences express the meaning of $\Diamond P$ and $\Box P$ of standard temporal logic. One may wonder why one formulation uses the least fix-point whereas the other formulation uses the greatest fix-point. The former requires P to be true in some finite time whereas the latter is concerned with infinite behavior (P is required to hold along infinite computations). $P \,\mathcal{U}\, Q$ is just as easily expressed:

$$\mu z \cdot Q \vee (P \wedge \bigcirc z).$$

We consider the three sentences presented above as definitions of the corresponding sentences of standard temporal logic. A sentence that cannot be expressed in standard temporal logic is

$$\nu z \cdot P \wedge \bigcirc\bigcirc z,$$

which is the formulation of "P holds every other time" of [12]. For a final, more complicated example, consider

$$\nu z \cdot \mu x \cdot (P \vee \bigcirc x) \wedge \bigcirc z$$

Let the above sentence be abbreviated by Φ. If we unfold the fix-point, we obtain

$$(P \vee \bigcirc \mu z \cdot (P \vee \bigcirc z) \wedge \bigcirc \Phi) \wedge \bigcirc \Phi$$

The second conjunct tells us that the next state must again satisfy Φ. Concerning the first conjunct, either P holds, or $\mu z \cdot (P \vee \bigcirc z) \wedge \bigcirc \Phi$ holds in the next state. In the latter case, we can unfold the fix-point. Since we are dealing with a least fix-point, this sentence must be true in a finite length of time. Thus, P must eventually hold. Taken together, Φ expresses that P eventually holds, and that it will do so infinitely often, i.e., that P holds infinitely often.

The temporal operators discussed so far allow qualitative claims regarding the temporal aspects of a theory. For example, $\Diamond P$ says that *eventually* P will hold. We add the $\mu^{\leq k} z \cdot P[z]$ operator that allows us to express that $P[z]$ will hold after a certain length of time k (where k is a natural number). Intuitively, k corresponds to the maximum number of applications of the functional corresponding to $P[z]$. Again, we can define the dual $\nu^{\leq k} z \cdot P[z]$. We relativize the abbreviations $\Diamond$, $\Box$, and $\mathcal{U}$ to k analogously. Further, we introduce by definition the complementary notions $\mu^{=k} z \cdot P[z]$ and $\mu^{\geq k} z \cdot P[z]$ to the bounded fix-point operator introduced above. Intuitively, the former states that P will be true after precisely k time steps, whereas the latter states that P will be true no earlier than k time steps. For example,

$$\Box(\neg P \vee \mu^{\leq k} z \cdot Q \vee \bigcirc z)$$

states that P always results in Q within a bounded length of time k. This sentence could be used to assert that a system must respond to an event P within a certain length of time k with the action Q.

To define the semantics of RTμ we assume a set of propositional constants *Prop*, and a set of propositional variables *Var*. Further, we are given a finite, nonempty set $\mathcal{S}$ of states. Let $\mathcal{L}$ be a mapping from $\mathcal{S}$ into 2^{Prop}. We can view $\mathcal{L}$ as going from $\mathcal{S}$ into *Prop* into the truth values, i.e., as a sequence of choices of truth values for the propositional constants in the theory. Furthermore, let a valuation be a mapping from *Var* into $2^{\mathcal{S}}$. A valuation assigns a set of states to each propositional variable. Lastly, we have a transition function $\mathcal{R}$ from $\mathcal{S}$ into $2^{\mathcal{S}}$. $\mathcal{R}$ tells us which states can be reached in a step of the computation from each given state. We call the triple $\langle \mathcal{S}, \mathcal{R}, \mathcal{L} \rangle$ a structure.

We define a mapping from $2^{\mathcal{S}}$ to $2^{\mathcal{S}}$:

Definition 1 *Let $\mathcal{M}$ be a structure, $\mathcal{V}$ a valuation, and Φ a sentence of* RTμ*, possibly containing the free propositional variable z. The* temporal theory operator *$\mathcal{T}_{\mathcal{M},\mathcal{V},\Phi}$ is a mapping from $2^{\mathcal{S}}$ to $2^{\mathcal{S}}$ defined by*

$$\mathcal{T}_{\mathcal{M},\mathcal{V},\Phi}(s) = \Phi_{\mathcal{M}}^{\mathcal{V}[z:=s]}$$

for $s \in 2^{\mathcal{S}}$.

(By $\mathcal{V}[z := s]$ we mean the valuation that agrees with $\mathcal{V}$, except that it assigns s to z.)

The power set of states, $2^{\mathcal{S}}$, forms a complete lattice under the partial order of set inclusion.[2] The temporal theory operator is monotonic, due to the restriction on the fix-point operator that it must be syntactically monotonic in its free variable. Therefore, by the Knaster-Tarski fix-point theorem we know that a greatest lower bound and a least upper bound exist for this functional.

Next, we define the *approximation from below* to the least fix-point of the temporal theory operator in terms of its ordinal powers.

Definition 2 *Let $\mathcal{M}$ be a structure, $\mathcal{V}$ a valuation, and Φ a sentence of* RTμ*, possibly containing the free propositional variable z. Let $\mathcal{T}_{\mathcal{M},\mathcal{V},\Phi}$ be the temporal theory operator. The approximation from below to the least fix-point of the temporal theory operator is defined as*

2. The top element is the set of states, the bottom element is the empty set; union and intersection form the upper and lower bounds, respectively.

(i) $\mathcal{T}_{\mathcal{M},\mathcal{V},\Phi}\uparrow_0 = \emptyset$

(ii) $\mathcal{T}_{\mathcal{M},\mathcal{V},\Phi}\uparrow_k = \mathcal{T}_{\mathcal{M},\mathcal{V},\Phi}(\mathcal{T}_{\mathcal{M},\mathcal{V},\Phi}\uparrow_{k-1}) = \Phi_{\mathcal{M}}^{\mathcal{V}[z:=\mathcal{T}_{\mathcal{M},\mathcal{V},\Phi}\uparrow_{k-1}]}$

(iii) $\mathcal{T}_{\mathcal{M},\mathcal{V},\Phi}\uparrow_\omega = \bigcup_{k\geq 0} \mathcal{T}_{\mathcal{M},\mathcal{V},\Phi}\uparrow_k$, if ω is the limit ordinal.

The approximation from below reaches the least fix-point at the limit. We interpret sentences of RTμ relative to structure $\mathcal{M} = \langle \mathcal{S}, \mathcal{R}, \mathcal{L}\rangle$ and valuation $\mathcal{V}$.

Definition 3 *Let Φ be a sentence of* RTμ *possibly containing free propositional variables. Given a structure $\mathcal{M} = \langle \mathcal{S}, \mathcal{R}, \mathcal{L}\rangle$ and a valuation $\mathcal{V}$, the semantic value of Φ, $\Phi_{\mathcal{M}}^{\mathcal{V}}$ is defined as*

$$\begin{aligned}
\Phi_{\mathcal{M}}^{\mathcal{V}} &= \{s \in \mathcal{S} \mid \Phi \in \mathcal{L}(s)\}, \text{ if } \Phi \in Prop \\
\Phi_{\mathcal{M}}^{\mathcal{V}} &= \mathcal{V}(\Phi), \text{ if } \Phi \in Var \\
\Phi_{\mathcal{M}}^{\mathcal{V}} &= \phi_{\mathcal{M}}^{\mathcal{V}} \cap \psi_{\mathcal{M}}^{\mathcal{V}}, \text{ if } \Phi \text{ is } \phi \wedge \psi \\
\Phi_{\mathcal{M}}^{\mathcal{V}} &= 2^{Prop} - \phi_{\mathcal{M}}^{\mathcal{V}}, \text{ if } \Phi \text{ is } \neg\phi \\
\Phi_{\mathcal{M}}^{\mathcal{V}} &= \{s \in \mathcal{S} \mid \exists t \in \phi_{\mathcal{M}}^{\mathcal{V}} \cdot t \in \mathcal{R}(s)\},
\end{aligned}$$

if Φ is $\bigcirc\phi$. The semantic value for the μ operator is given in terms of the fix-point of $\Phi_{\mathcal{M}}^{\mathcal{V}[z:=s]}$, where Φ is $\mu z \cdot \phi$:

$$\Phi_{\mathcal{M}}^{\mathcal{V}} = \text{glb}\{s \in 2^{\mathcal{S}} \mid \mathcal{T}_{\mathcal{M},\mathcal{V},\phi} s \subseteq s\}.$$

We define the semantic value for $\mu^{\leq k} z \cdot P[z]$ in terms of the ordinal powers of the temporal theory operator.

$$\Phi_{\mathcal{M}}^{\mathcal{V}} = \mathcal{T}_{\mathcal{M},\mathcal{V},\phi}\uparrow_k, \text{ if } \Phi \text{ is } \mu^{\leq k} z \cdot \phi, 1 \leq k < \omega$$

Definition 4 *A (closed) sentence Φ of* RTμ *is true in structure $\mathcal{M} = \langle \mathcal{S}, \mathcal{R}, \mathcal{L}\rangle$ in state s*

$$\mathcal{M}, s \models \Phi \textit{ if and only if } s \in \Phi_{\mathcal{M}}^{\mathcal{V}},$$

for some valuation $\mathcal{V}$. We call a structure $\mathcal{M}$ that meets this condition a model of Φ.

5 Temporal aspects of a FRORL specification

In [1], we presented object and activity frames, and their translations into the underlying non-monotonic logic. In this section, we show how the temporal constructs introduced into FRORL in this paper are translated into constructs of the underlying logics.

Translation of the temporal constructs yields, on one hand, sentences of the non-monotonic base logic of FRORL representing the functional behavior of the time-dependent constructs. These are used along with the rest of the specification both for execution as a prototype and for validation. Note that the concurrency of processes is irrelevant when considering the functional aspects of the specification. For example, the spawning of a concurrently executing activity (such as that of the timer that might signal violation of a timing constraint) is merely another conjunctive goal among the actions performed by an activity. Input via message passing is merely the use of a variable that had been bound earlier. Naturally, for observation of the functional behavior of a specification through operational interpretation the fact that several activities do interact does make a difference. For purposes of operational interpretation the surface syntax of FRORL provides special connectors indicating parallel execution of activities, mutual exclusion between events, communication between activities, and so on. Again, none of these are given any logical meaning.

On the other hand, we obtain assertions in temporal logic used in verification of the time-dependent behavior of the specification. The time-related aspects of the specification receive their meaning through the temporal logic presented earlier. Time-dependent information is extracted from the specification and expressed in a form amenable to formal verification techniques, such as the approaches described later.

As far as the functional aspect is concerned, a periodic activity represents a periodically scheduled process. Preconditions, actions, and alternative actions are translated as usual (see [1]). A recursive call to the activity is added as additional conjunct to both actions and alternative actions, preceded by an invocation of a timer activity. Again, from the point of view of the non-monotonic logic giving meaning to the functional aspects of the specification, the specification states that after evaluation of either the actions or the alternative actions the activity is recursively evaluated (the call to the timer activity always evaluates to true). From the point of view of operational interpretation a copy of the activity is spawned and delayed until after expiration of the respective timer process.

The activity also expresses a timing constraint for periodic processes. Let p be the time period associated with the activity K, and $t_1 \ldots t_n$ be arguments to K. The activity is converted into the following sentence of the RTμ calculus.

$$\begin{aligned}
&(\mu^{=p} z \cdot \mathsf{K}(t_1 \ldots t_n) \vee \bigcirc z) \wedge \\
&\quad \Box(\mathsf{K}(t_1 \ldots t_n) \Rightarrow \\
&\qquad \bigcirc\mu^{=p-1} z \cdot \mathsf{K}(t_1 \ldots t_n) \vee \bigcirc z)
\end{aligned}$$

The first conjunct states that activity K is executed in exactly p time-steps. The second conjunct states that each time K has been executed, it will again be scheduled to execute in exactly p time-steps.

When translating the functional aspects of an activity with an associated maximum timing constraint p, we add the spawning of a timer process, corresponding to the stated timing constraint, to the preconditions of the frame. An additional process is created which watches for a violation of the timing constraint. A third clause is created should the evaluation of the precondition fail, as described in [1]. Successful evaluation of the preconditions causes execution of the starting event E_1 and simultaneous spawning of a timer with duration p. An additional clause detects the timeout generated by the timer and leads to execution of the alternative actions (as it would if the preconditions of the activity had failed). If the timing constraint is not violated, the actions are executed as usual.

A maximum timing constraint also generates a timing assertion which is expressed in RTμ as

$$\mathsf{E}_1 \, \mathcal{U}^{\leq p} \, \mathsf{E}_2$$

(where E_1 and E_2 are the events between which the maximum timing constraint is asserted). This sentence of RTμ states that E_1 holds now but E_2 will become true within p length of time.

The situation is analogous for minimum timing constraints. Instead of creating a timer that raises an alarm upon timeout, we create a timer which raises an alarm if accessed before the timeout period.

The accompanying timing assertion induced by a minimum timing constraint is

$$\mathsf{E}_1 \, \mathcal{U}^{\geq p} \, \mathsf{E}_2.$$

6 Verification of FRORL specifications

FRORL employs various strategies when analyzing requirements specifications. The analysis is performed at the level of the underlying logic, rather than at the level of the FRORL surface syntax of frames and rules.

In [1] we have presented analyses possible by exploiting the operational interpretation of FRORL specifications. Dynamic properties of a specification can be verified through the construction of resolution refutations from the (clauses of a) theory derived from the specification and a chosen goal clause. *Reversibility* determines whether a clause of a specification can always be resumed. *Liveness* ensures that every goal in a specification is resumable from any other reachable goal. *Consistency* analysis determines whether contradictions occur between the clauses of a FRORL specification. *Synchronic Distance* is a measure of the mutual dependence among rules (that is, their relevance to one another). Two clauses of a FRORL specification are in a *bounded-fair* relation, if there is a bound on the number of times one is invoked while the other is not.

6.1 Analysis through model checking

Sentences of the temporal calculus RTμ express properties of specifications. If the structure corresponding to a given specification is a model for a sentence Φ of RTμ, we say that the specification has the property Φ. For a detailed discussion of specification properties of interest see [13] or [14].

Given a sentence Φ of RTμ, we want to be able to determine whether this sentence holds: Given a particular structure $\mathcal{M}$, is $\mathcal{M}$ a model of Φ, i.e., is there a state $s \in \mathcal{S}$ such that $\mathcal{M}, s \models \Phi$? To answer this question is referred to as *model checking*.

Alg.1 determines whether a given structure $\mathcal{M}$ is a model for a sentence Φ.

Our algorithm proceeds in stages. For each subsentence of Φ, we determine whether it holds in a state of the structure. We begin by examining the propositional constants in Φ. These are true at all states in the structure in which they are contained. We then straightforwardly apply the rules given in Def.3. After considering the finitely many subsentences of Φ, we have obtained the set of states at which Φ holds in the structure. If this set is empty, the structure is not a model for Φ.

Algorithm 1 (Model checking)

```
input structure M = ⟨S, R, L⟩, sentence Φ
output set of states in which Φ holds
types S', S'', S_z: set of states
case syntactic form of Φ in
    P (constant): S' ← {s ∈ S | P ∈ L(s)}
    z (variable): S' ← S_z
             ¬P: S' ← S − Model checking(M, P)
          P ∧ Q: S' ← Model checking(M, P) ∩
                      Model checking(M, Q)
             ○P: S'' ← Model checking(M, P)
                 S' ← {s ∈ S | ∃t ∈ S'' · t = R(s)}
        μz · P[z]: S_z ← { }
                 repeat
                     S' ← S_z
                     S_z ← Model checking(M, P[z])
                 until S' = S_z
     μ^{≤k} z · P[z]: count ← k; S_z ← { }
                 repeat
                     S' ← S_z
                     S_z ← Model checking(M, P[z])
                     count ← count − 1
                 until S' = S_z ∨ count = 0
```

The treatment of sentences of the form $\mu z \cdot P[z]$ and $\mu^{\leq k} z \cdot P[z]$ relies again on the Knaster-Tarski fix-point theorem and Kleene's proof that ordinal approximations to the greatest lower bound and least upper bound, respectively, do in fact, reach the fix-points:

$$\mu z \cdot P[z]^{\mathcal{V}}_{\mathcal{M}} = \text{glb}\{\mu^{\leq n} z \cdot P^{\mathcal{V}}_{\mathcal{M}} \mid n < \omega\}.$$

Alg.1 is similar to one presented by Emerson and Lei [15]. They have shown, for a slightly different version of the μ-calculus, that the complexity of this algorithm is a polynomial of a degree related to the alternated nesting of μ and ν operators. It is important to note that most (or all) modalities that have proven useful for discussing real-time programs can be expressed by nesting depths of degree 1 or 2. Therefore, the above model checking algorithm has a small polynomial time complexity.[3]

Alg.2 determines whether a given property Φ holds for a specification P. Since the specification itself does not contain temporal modalities, the polynomial depends solely on the structure of the sentence to be checked (i.e., on the property to be verified).

Pixley [16] mentions that structures with up to 10^{200} states can currently be handled by sophisticated model checking techniques that increase the number of storage elements through binary decision diagrams. Using scaling he has been able to sequentially compare machines with about 700 latches.

Algorithm 2 (Verify temporal property)

input property Φ, specification P
Convert P into the corresponding structure M
if Model checking$(M, \Phi) \neq \{\}$
then report "P satisfies Φ"
else report "P does not satisfy Φ"

6.2 Timing constraints consistency analysis

The correctness of a real-time system depends not only on the logical result of the computation, but also on the time at which the results are produced. An important aspect of the correctness of a real-time system is the consistency of the various timing constraints imposed on the individual components of a system. The system will not function correctly if even two of the timing constraints cannot jointly be met.

Timing constraints consistency analysis is concerned with real-time related aspects of the specification only. We may, therefore, safely omit all those clauses of a specification not related to stating timing-specific requirements.

Step 1: Generate directed timing constraints graph (DTCG) and the corresponding directed timing constraints matrix (DTCM). Based on the timing related parts of a FRORL specification, we create a DTCG by letting the vertices be the events mentioned in the specification, and the arrows (with numeric values of constraints associated as weights) be the timing constraints that hold between two events. The arrows point in the direction of the timing constraint (e.g., for a maximum timing constraint between event e_1 and event e_2 the arrow goes from e_1 to e_2, and vice versa for a minimum timing constraint). (We assume there is no timing constraint on events, so the DTCG is a directed graph with n vertices and no looping edges.)

The directed timing constraints graph is then converted into the directed timing constraints matrix DTCM by letting the cells in the matrix hold either the weight of the arrow between two events, or 0, if there is no arrow between two events. Thus, the size of the graph is reduced by eliminating redundant arrows and vertices.

Detection of inconsistencies between timing constraints relies on the following theorem (which can be shown by basic inequality reasoning).

Theorem 1 *For a cycle in a directed timing constraints graph (DTCG), if the sum of the weights of the arrows in the cycle is positive, then the timing constraints in the cycle are inconsistent.*

Step 2: Transform DTCM to canonical form. Based on Thm.1, we shall locate all the cycles in a DTCM and compute the timing constraints on the cycles. First, we eliminate all events and arrows from the DTCM which cannot contribute to a cycle, i.e., all events (and their connected arrows) from which either no arrow originates or at which no arrow ends.

Step 3: Locate timing constraints contradictions. Now find cycles with the sum of the weight of the contained arrows being positive, and compute the overall timing constraints values.

Alg.3 outputs all directed cycles in the timing constraint detection matrix and their corresponding sums of the weights associated with each cycle. According to Thm.1, we locate the cycles with positive overall weights to detect inconsistent timing constraints in a requirements specification.

3. Emerson [14, 15] has pointed out that the performance of model checking can be improved further by relying on memoization when constructing the set of states in which a property holds (obviously, identical subsentences need to be checked only once). Furthermore, the propositional variables bound by fix-point operators induce simultaneous monotonicity. He also shows that the number of iterations necessary to evaluate the fix-point operator is bounded by the number of states in the structure.

Algorithm 3 (Timing Constraints Detection)

/*Let *sumtc* be an accumulator for the overall weight of the cycle through the canonicalized DTCM.*/
input *path* is a list of events $e_1, \ldots, e_n$ in a specification
$start \leftarrow 1$
(i) $F \leftarrow 0$
(ii) $i \leftarrow start + 1$
(iii) **if** $i > n$
  then goto (v)
  else if $\exists i \leq n \cdot tc_{\text{last}(path),e_i} = 0$
    then goto (iv)
    else if $e_i \in path$
      then goto (iv)
      else if e_i is forbidden from last(*path*)
        then goto (iv)
        else $k \leftarrow k + 1$
          $path \leftarrow path \cup e_i$
          $sumtc \leftarrow sumtc + tc_{\text{last}(path),e_i}$
          goto (ii)
(iv) $i \leftarrow i + 1$
  goto (iii)
(v) **if** $tc_{\text{last}(path),start} \neq 0$
  then $sumtc \leftarrow sumtc + tc_{\text{last}(path),e_i}$
    report "weight of *path* is *sumtc*"
  else if $k = 1$
    then goto (vi)
    else clear the list of vertices forbidden from
      last(*path*) and make last(*path*) forbidden
      from one_but_last(*path*)
      remove last(*path*) from *path*
      $k \leftarrow k - 1$
      goto (ii)
(vi) $start \leftarrow start + 1$
  $path \leftarrow path \cup e_{start}$
  if $start = n$
  then goto (v)
  else if $start < n$
    then goto (i)

7 References

[1] J. Tsai, T. Weigert, and H. Jang, "A hybrid knowledge representation as a basis of requirement specification and specification analysis," *IEEE Trans. Software Engineering*, vol. 18, no. 12, pp. 1076–1100, 1992.

[2] F. Jahanian and A. Mok, "Safety analysis of timing properties in real-time systems," *IEEE Trans. Software Engineering*, vol. 12, pp. 890–904, Sept. 1986.

[3] J. Tsai and T. Weigert, *Knowledge-Based Software Development for Real-Time Distributed Systems.* Singapore: World Scientific Publ., 1993.

[4] B. Dasarathy, "Timing constraints of real-time systems: Constructs for expressing them, methods of validating them," *IEEE Trans. Software Engineering*, vol. 11, pp. 80–86, Jan. 1985.

[5] J. Tsai, K. Fang, H. Chen, and Y. Bi, "A noninterference monitoring and replay mechanism for real-time software testing and debugging," *IEEE Trans. Software Engineering*, vol. 16, pp. 897–916, Aug. 1990.

[6] A. Pnueli, "The temporal logic of programs," in *Proc. 18th Ann. IEEE Symp. Foundations of Comp. Sci.*, pp. 46–57, 1977.

[7] P. Wolper, "On the relation of programs and computation to models of temporal logic," in *Temporal Logic in Specification* (B. Banieqbal, H. Barringer, and A. Pnueli, eds.), no. 398 in Lect. Notes in Comp. Sci., pp. 75–123, Berlin: Springer Verlag, 1987.

[8] A. Pnueli, "Applications of temporal logic to the specification and verification of reactive systems: A survey of current trends," in *Current Trends in Concurrency, Overviews and Tutorials* (J. de Bakker, W. de Roever, and G. Rozenberg, eds.), no. 224 in Lect. Notes in Comp. Sci., pp. 510–584, New York: Springer Verlag, 1986.

[9] E. Emerson, "Temporal and modal logic," in *Handbook of Theoretical Computer Science* (J. van Leeuwen, ed.), vol. B, pp. 995–1072, Amsterdam: Elsevier Science Publ., 1990.

[10] D. Kozen, "Results on the propositional μ-calculus," *Theoretical Computer Science*, no. 27, pp. 333–354, 1983.

[11] E. Emerson and J. Halpern, "'Sometimes' and 'not never' revisited, on branching versus linear time temporal logics," *J. ACM*, vol. 33, no. 1, pp. 151–178, 1986.

[12] P. Wolper, "Temporal logic can be more expressive," in *Proc. 22nd Ann. IEEE Symp. Foundations of Comp. Sci.*, pp. 340–348, 1981.

[13] F. Kröger, *Temporal Logic of Programs.* No. 8 in EATCS Monographs on Theoret. Comp. Sci., Berlin: Springer Verlag, 1987.

[14] E. Emerson, A. Mok, A. Sistla, and J. Srinivasan, "Quantitative temporal reasoning," in *Computer Aided Verification* (E. Clarke and R. Kurshan, eds.), no. 531 in Lect. Notes in Comp. Sci., pp. 136–242, Berlin: Springer Verlag, 1990.

[15] E. Emerson and C.-L. Lei, "Efficient model checking in fragments of the propositional mu-calculus," in *Proc. Symp. Logic in Comp. Sci.*, (Cambridge), pp. 267–278, 1986.

[16] B. Plessier and C. Pixley, "Formal verification of a commercial serial bus interface," in *Proc. of 14th Annual IEEE International Conf. on Computers and Communications*, (Phoenix), March 1995.

Real Time Formal Specification using VDM^{++}

Jan van Katwijk
Delft University of Technology
Netherlands
Email: J.vanKatwijk@twi.tudelft.nl

Eugène Dürr
CAP Gemini Sogeti
Netherlands

Stephen Goldsack
Imperial College London,
United Kingdom

Abstract

VDM^{++} [1] *is a formal Object Oriented Specification language, derived from VDM. It extends VDM by providing object-orientation and concurre.cy features. The use of the language is supported by design guidelines and a toolset. The latter offers graphical representations, syntactic and semantic checking, pretty printing and code generation.*

In this paper we address real-time extensions as being developed for the language.
Keywords: *Software Engineering, Real-time Object-Oriented processing,*

1. Introduction

In this paper we report on our work on extending VDM^{++} in the area of real-time specifications. VDM^{++} is meant to be a notational vehicle for modeling and development of systems. Being born from VDM, the language provides a rich set of data structures for building abstract models of the system to be built. To support the development process, data and operation reification techniques, as known in plain VDM, are extended with specific *Object-Oriented* refinement steps (termed "annealing"). The development of a system from a high-level model consists of repeatedly applying these annealing steps. A large collection of more or less standardized reification steps is being developed.

The language has shown to be successful as specification language in a number of commercial project. One of the major successes of the language was its use in the Combicom project [10], [11].

In this paper we discuss extensions to the VDM^{++} language that we are currently working on to support real-time issues in modeling and development.

[1] The VDM++ work was partially funded by the European Information Technology Research Programme, Esprit III under number 6500.

The structure of this paper is as follows. In section 2 we briefly discuss the main characteristics of VDM^{++} , in section 3 we discuss the way action/event modeling is included in the language, and in section 4 we illustrate the typical real-time features of the language by an example. Finally, in section 5 we discuss some conclusions and future work.

2. VDM++, a brief overview

VDM^{++} [3, 1] is a formal specification language based on VDM-SL [5, VDM Standard] and extended in an object-oriented fashion with elements from Smalltalk [4]. VDM^{++} provides a wide range of constructs such that a user can formally specify (concurrent) systems in an object-oriented fashion.

The language was originally created by Eugène Dürr. It has been developed into a full, mature language as as part of the ESPRIT-III project Afrodite (project number 6500). During the Afrodite project, the language was established by developing a complete definition. Furthermore, a toolset was designed and implemented to provide commonly used object-oriented graphical representations (OMT) as representation of VDM^{++} specifications. Apart from a toolset aiming at support in building a specification, a language interpreter was developed (obviously only interpreting the execuatble elements in the language), while furthermore a C++ prototype code generator was developed with which (parts of) VDM^{++} specifications can be transformed into C++ code. Recent developments of the language are related to extensions for real-time object-oriented specification.

In VDM^{++} a complete formal specification, also called a *Model*, comprises a set of *class* specifications and an optional specification of a *workspace*.

A class specification in VDM^{++} has the following components :

Class header In the class header one specifies the name of the class, and its possible relationship with parent classes. *is subclass of* clause achieves (multiple) inheritance from other classes.

Instance Variables The state of an object is made up from the values of the instance variables of the object. Instance variables are variables of simple types, VDM-SL typeconstructors as sets, sequences and maps and object references (*clientship relation* denoted by @Classname) can be declared here. The state space of objects of a given class can be constrained by the specification of an *invariant* clause, a relation over the instance variables of the objects of the class. Similarly, the set of allowable initial values of the instance variables of the objects of a given class can be specified using an *initial* specification.

Methods The methods belonging to the objects of a class can be defined several ways:

- *implicitly*, by specifying the pre- and postconditions of the method. As in VDM-SL , the precondition refers to values of state elements before invocation of the method, the post condition may refer to values of state elements as they were before invocation as well as as they are after invocation of the method.

```
class example is
    s: sequence of T;
    method my_try ()
        pre not is_empty (s)
        post is_inverse (~s, s)
...
```

  As in VDM-SL , a reference to a name preceded by a tilde in the post condition indicates a reference to the value prior to invocation of the method.

  Roughly speaking, the VDM-SLsemantics apply, i.e. the operation is only defined when the precondition is satisfied, while loosenessin the post condition is allowed.

- *explicitly*, using functional and imperative constructs, although this style is not encouraged in the specification of high-level models.
- as *not yet defined*, A specification as *not yet defined* is useful when the environment of an object demands the existence of a method, but at the same time – at that stage of development – it is not considered necessary to present its exact definition.

Invocation of a method of a client object is considered to be a *statement*. The invocation syntax is

```
expression ! methodname (parameters)
```

For a well-formed specification, the *expression* yields an object of a class that contains the method *methodname*.

Controlled Inheritance An inheritance reduction clause can restrict visibility of methods obtained from the superclass to the external user of an object.

Auxilary Reasoning A class might have an *Auxilary Reasoning* part. In such a part, axioms, properties, and invariants required to perform correctness proofs and constraints for other depending classes can be specified here. References to internal states of client objects are limited to read-only accesses. The format is not constrained, the contents of the auxiliary reasoning part are not processed by the available tools.

Outline of an example

```
class Train
   instance variables
       speed : N;
       power : N;
       direction: (Forward, Backward);
       inv speed = F (power)
   methods
       set_power () == not yet defined
       get_speed () == not yet defined
end Train
```

The example class defines objects of class *Train*, used in a later section of this paper. The state of the train consists of the power supplied to drive the train, and the direction of the transmitted speed. The instance variable speed is related to the power by the invariant P.

A system specification is completed by the description of a *workspace*. The workspace mechanism has the role of the 'main procedure' in other languages. Usually, a workspace has a special method called the 'initial-method'. Its role is to create the objects of which the system is initially composed, and to establish their topological relations. A workspace object is implicitly created at the start of the pseudo execution of the specified system and its initial method is invoked by *Deus ex Machina*. The (pseudo) execution of parallel executing objects is started from here.

3. Real-time specification and VDM^{++}

3.1. Introduction

The VDM^{++} paradigm states that building a high-level model of the system should result in a model of the software system, together with its environment. The model is closed, i.e. it contains all interactions between the submodel of the software system and the submodel of the environment.

Elements in the environment are usually represented by objects, as is the model of the software system itself. Objects therefore are considered to model/simulate active world entities, as a result, objects need some form of behaviour.

In VDM^{++} we have chosen to make an object active by allowing a class to contain a specification of a *thread*. The thread definition has the form

```
thread
   thread specification
```

A thread specification is either a procedural thread, or a declarative one. The declarative form specifies a periodic action, the execution of a method, to be executed with a given period.

$periodic(\Delta T)(methodname)$

It is assumed that for each thread a processor is available for the execution of methods.

The procedural form a the thread is not used throughout this paper and will not be discussed here.

Apart from objects being active by letting them have threads, we needed the notion of time. In VDM^{++} , a global clock is assumed, that can be read by anyone. It's value can be read by refering to the implied clock variable now.

Since, time is considered to be a continuous variable with infinite accuracy, expressions using references to time in a system cannot not use strict equality.

Introducing time this way, allows the functionality of methods to depend on time (at least to a certain extent).

Consider as an example

```
class example is
  .......
  method exec_time () returns d: real;
  post (d = time - ~time)
```

Since $time$ is considered to be a state element, its value before and after invocation of a method can be queried.

In the example an implicit method is specified. The post condition of the method states that the returned value (d), will be equal to the difference between the value of $time$ as it holds after the invocation of the method is finished and the value of $time$ as it holds before invocation. I.e., the result of invocation of the method will be the duration of its execution.

3.2. VDM^{++} and events

The paradigm in modeling real-time systems states that the state of the system acts as the interface between the controlling and the controlled system. Each object in the system inspects the state and reacts upon changes in this state. Each change in the state can be considered to be an event. More formally speaking, an event is the marker in time at which a predicate yields **true**.

Being in an object-oriented world however, where hiding is one of the fundamental issues, only a part of the whole system state is directly visible. This is fundamentelly different than e.g. in VDM-SL where the whole state is visible from each operation.

Some attention to what can be part of the predicate defining an event is therefore inevitable. Elements to be taken into account when defining events are:

1. an *observable* transition of the state of system. Within an OO specification visibility of the state is typically limited to the values of the instance variables of the current object, the values of instance variables of parent objects and the values of instance variables in the worksapce. Visibility of change is therefore typically limited to a change in the value of these variables.

2. the transition of the systems' state resulting from handing a method request or method invocations. Here the changes are made observable through the introduction of *synchronisation variables*. VDM^{++} has the following built-in events:

   - the request to execute a method. $req(methodname)$ will become **true**, raising an event, as soon as a request to execute the method $methodname$ is issued;
   - the actual activation (start) of a method, $act(methodname)$ will be **true**, raising an event, as soon as the invocation of the method is a fact;
   - the finalisation of a method, $fin(methodname)$ will be **true**, raising an event, as soon as the invication of the method $methodname$ terminates.

   The default available variables #req, #act and #fin record these transitions as counters. Notice that the events will be raised independent of the method of invocation of a method. I.e., although a method invocation in a periodic thread is not visibly by an explicit call, the same events apply.

Requirements for event handling in a specification notation are (i) the ability to specify events, (ii) the ability to specify the system's reaction upon an event, and (iii) the ability to specify temporal constraints on the occurrences of events and reactions. In our VDM^{++} extension, we offer the capability specifying within the thread part of a class specification a connection between an action to the occurrence of an event, using a statement of the form:

```
whenever condition also
      [from delay ] ==>  predicate
```

Such an action specification tells that whenever *condition* becomes **true**, the system should realize *predicate* within *delay* units of time.

Obviously, having the notion of time available, one might raise an alarm at a specified time

```
whenever  now > ThisAfternoon
also from delta ==> fin (methodname)
```

This specification states that an event is raised as soon as a certain calendar time is reached. On raising this event, the execution of the method *methodname* should be finished within delta units of time.

Generalizations are straightforward:

```
whenever P (now, period) also
from delta ==> fin (methodname)
```

Suppose that P is a predicate that yields **true** when

$$now\ MOD\ period < Small\ Value$$

I.e., in this case, we emulate a periodic obligation.

An obvious well formedness condition is that

$$Duration(methodname) < period$$

The ability to refer to 'state' of method invocations allows us to specify e.g. the upperbound of the execution of a particular invocation. Consider as an example

In this specification, an upperbound for the execution of *methodname* is given. As soon as *methodname* is invoked, the specification requires the implementation to ensure that the finalization of the invocation of *methodname* is not later than delta1 units of time.

```
whenever  req (methodname)
also from delta1 ==> fin (methodname)
```

In this specification, it is indicated that the execution of a method *methodname* should be terminated within *delta*1 units of time after a request to its invocation is made.

```
whenever #act (methodname) >
         #fin (methodname)
also from delta ==> .....
```

This example shows a statement in which an unspecified predicate (indicated by '...', is to become **true** within delta units of time from #act > #fin becoming **true**.

It is assumed that the performance "consumption" for the evaluation process of the various conditions is taken into account when writing the specification. The existence of the "from delta" construct acknowledges that in the real world computations and reactions takes time and cannot be executed in zero time. Omitting such a restriction specifies a requirement that is hard to fullfil: the predicate acting as consequent in the implication is to be made **true** in zero units of time.

If sensor based systems are involved, the interface between the controlled system and the controlling system may contain variables that can be assumed to be changed by the environment.

E.g.

$$whenever\ \ temp > t$$

can be used. In a later refined model, a sensor object will probably be introduced and the temperature observation mechanism will be specified in more detail.

3.3. Discussion

The semantics of having more than a single *whenever* clause within a single thread deserves some attention. Current semantics assume that multiple *whenever* clauses within such a thread exclude each other. In the event that two or more events become **true** at the same moment, only one of the clauses will be executed further. It is the responsibility of the specifier to ensure that all specified *whenever* clauses are indeed schedulable within their deadlines. Consider as an example

```
class example
   .....
   methods
   ....
thread
   whenever Pa also within delta1 ==> Qa
   whenever Pb also within delta2 ==> Gb
   ...
end example
```

Assume furthermore that, in order to satisfy Qa a method need to be invoked that takes Ex_1 units of time. Assume furthermore that, in order to satisfy Qb a method need to be invoked that takes Ex_2 units of time.

If, at any moment Pa and Pb become **true** at the same time, a necessary condition is that

$$Ex_1 + Ex_2 < min(delta1, delta2)$$

In general, however, two or more conditions in different objects may become **true** at the same time, raising different exceptions within different threads. If the actions, associated with the handling of these events, refer to the same state variables, the result is, formally speaking, undefined. It will depend upon the speed and the (arbitrary) order in the execution of the components, race conditions might occur.

Safe specifications can therefore not use instance variables in an updating mode, unless extensive proof is delivered about non interference to these variables at all times. Read only access is safe. We are considering a language extension which allows users to add their own sync variables

(variables which can be updated atomically) similar to the currently built-in #req, #act and #fin.

It is interesting to compare the event/action approach, as followed here, to the enabling approach. In the latter, at any state *an* operation will be selected that has a fullfilled precondition. No queuing of 'enabled' operations will take place. In our approach, queuing will occurr indeed. Whever two or more events are raised the associated operations will be queued for execution.

An interesting question was raised in [13]. In this paper, the notion of real-time object (RTO.k) was introduced. Such real-time objects would form the basic building blocks of real-time systems. The structure of an RTO, obviously, resembles the structure of the VDM^{++} class. However, in RTO.k objects, the methods are partitioned into two classes. Elements of one class are exported and made visible to cleints of the objects, elements of the other class are kept invisible to the external world. It is with elements of this latter class that structures, comparable to our *whenever* constructs are built.

Whether or not additional constraints should be put on the methods that can be used within thread specification is subject to further research.

Notice that an action/event model has some intrinsic problems, since the conditions leading to raising an event may have been changed whenever the action, scheduled by the event, is executed. Actions that are invoked as a response upon the raising of an event should establish whether or not the conditions, causing the event to be raised, still exist.

4. A case study: the railroad controller

Effectively evaluating formal specification notations is not always easy. In our research, we use a single example for the comparison of different specification notations. In [12], a brief comparison is given of the use of formal specification notations for a simple train controller. In this section, we briefly discuss the use of VDM^{++} as vehicle for the specification of the same controller.

The specification is sufficiently realistic to get a good insight in the possibilities and limitations of specification notations. It involves both data and control issues. The railroad itself is to be represented as some (rather complex) data structure. Physical devices (i.e. switches and trains) have to be manipulated as functions of the time and system state as well. Furthermore, the controller is subject to temporal constraints. Each train has deadlines w.r.t. its arrival time on specified positions, while furthermore, reactions of the system on occurrences of events should be within specified times.

Figure 1. Fragment of railroad

4.1. The problem

A (toy) railroad system is built up from a variety of (connected) rail elements (straight elements, bowed elements, crossings and switches. Rail elements have a fixed length of app 23 cm, they are grouped into physical blocks, *a block is connected to hardwired devices. Through a computer (serial) interface, commands can be given:*

- *status enquiry commands (i.e. blocks being occuppied or not).*
- *switch setting/resetting commands.*
- *train commands, setting speed (between 0 and 30 cm/sec), setting lights, a horn and a direction (in our model, we ignore the horn and the lights).*

The controlled behaviour of the train should conform to requirements *set for the train movements by a specified mission, one for each train. Missions, and therefore train routes, are expressed in terms of* tracks, *where a track is a path over a block.*

The problem is to specify (and subsequently design and implement) a software controller with which the behaviour of the system, i.e. train movements and switch status, can be controlled. Following obervations made by many researchers in the field, we require the model to be closed, *i.e. encompassing both a model for the controller and for the controlled entities.*

A train drives according to a sequence of *commands*. A command indicates what a train is required to do on a given track, i.e. a *PASS* command indicates that the train should drive over the specified track and should have reached the end of it no later than the specified time. (A track represents a block, i.e. a group of connected rail elements, together with an indication of an entry and exit. Blocks, entries and exits are encoded as naturals. A track is encoded as a triple (*block, entry, exit*).)

Consider for example the following specification of a small mission

```
INIT   (1, 1, 2)
START  (1, 2, 1), 1000, 1010
PASS   (2, 1, 2), 1020
PASS   (3, 1, 2), 1030
STOP   (4, 1, 2), 1035
```

```
START  (4, 2, 1), 1035, 1040
PASS   (3, 2, 1), 1050
PASS   (2, 2, 1), 1060
STOP   (1, 2, 1), 1070
```

After being initialized (*INIT*) on track (1, 1, 2), the train starts at time 1000 on track (1, 2, 1) (in our encoding, track (1, 2, 1) is the reverse from (1, 1, 2)). The train is forced to move such that it leaves that track no later than 1010. The subsequent track on the route, (2, 1, 2), should be passed such that the train reaches the end of the track no later than 1020, etc. The track (2, 1, 2) being the logical successor of track (1, 2, 1) is obviously a well-formedness constraint on the railroad.

4.2. A VDM^{++} solution

The structure of the specification In [9] the architecture of the solution is given in great detail. Due to space constraints, we limit the presentation of the solution here to a discussion of some details.

The specification contains the following classes:

- The class *Train*. Objects of this class model the physical trains, as driving on the physical railroad;
- The class *RailRoad*. A single object of this class models the railroad on which the trains are driving;
- The class *TrainPosition*. Objects of this class model the relation between the trains and the railroad, i.e. they model the positions of the trains.
- The class *TrainController*. Objects of this class (one for each train) model the controller function to be executed in order to have the associated train show the required behaviour. The objects of this class contain the sequences of commands, to be executed for the associated trains.
- The class *AllocationTable*. In order to enforce that no two trains ever appear on the same track, a track is assigned to eat most one train. A train only drives on allocated tracks. Management of the tracks and their association to trains is maintained in a single object of class *AllocationTable*.

Specification of train positions Train positions relate trains to the railroad. A position can be derived, given a previous position, the structure of the railroad and the driving characteristics of the train.

```
class TrainPosition
   instance variables
      railroad: @RailRoad
      train   : @Train
      position: Position
   methods
      set_position (p: Position) == not yet defined
      get_position (p: Position) == not yet defined
      new_position () == not yet defined
   thread
      period (100 ms) (new_position)
end TrainPosition
```

The behaviour of each actual train in the system is modeled by an instance of the class *TrainPosition*. The class relates the trains (models of which are maintained in objects of class *Train*) to the actual railroad (modeled in an object of class *RailRoad*).

The thread ensures that the position information will be updated once every 100 milli seconds. Taken the speed of the train into account (max 30 cm/sec) the maximum difference between two position updates is less than or equal to 3 cm.

The railroad object, referred to from the class *TrainPosition* provides information on the structure of the railroad. Given a position on the railroad, and given a distance, a new position (taking into account the state of the switches) is computed.

Train objects contain state information on the trains they represent.

Outline of the controller To each individual train a controller object is associated that manages that train. The controller object contains an encoding of the mission of the train and essentially executes a finite state machine for that train. The FSM is given in figure 2.

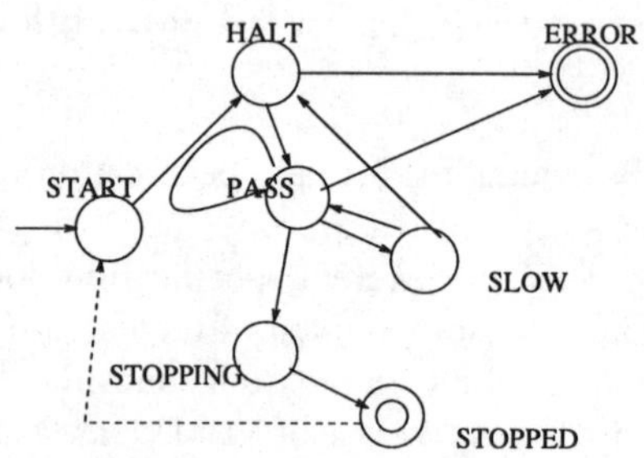

Figure 2. The state diagram of the controller

The states and the transitions in the FSM are determined by (i) the command for the train to be executed, (ii) the position of the train and (iii) the current time.

1. A train starts in state START as soon as the starting time is reached.
2. In this state, the next track on the route of this train is computed, and the train is set in state HALT.

3. A train in state PASS is a running train. Such a train remains running at least until it enters the next track on its route. If the command to be performed on this next track is a *PASS* command, an attempt will be made to allocate the next track on the route. If this attempt fails, the train will slow down (and enter state SLOW), if the attempt succeeds the state PASS will be reentered. If the next command is *STOP* command, state STOPPING will be entered.

4. In state STOPPING, the train will move forward until it reaches the middle of the track and take state STOPPED and halt. In this state, the only valid next command is a *START* command.

5. In state SLOW, the train will continue to attempt to allocate the required resource. If the train reaches the middle of the track *before* being able to allocate this resource it enters state HALT and the train halts. If the train does obtain the required resource, state PASS will be (re)entered, and the train will move with the speed required to meet its temporal constraint.

6. In state HALT, the train will continue its attempts to allocate the resource during a period of time. If however, this period is finished, the train will enter state ERROR and stops functioning. Otherwise, it will enter state PASS and it continues.

Modeling the control for the trains is then modeling the automaton in a single class.

Objects of the class contain a thread that will continuously react upon events and issue commands when reacting upon them.

```
Class TrainController is
  instance variables
    mission: Instructions*
    train: @TrainPosition
    TrackTable   : @AllocTable
    CurrentMode  : (HALT, STOPPING,
                    STOPPED, PASS,
                    SLOW, START)
    CurrentTarget : Position
       -- The target of
       -- the current command
    SlowTime     : Time
    inv .....
    init ......
  methods
    ....
  Thread
    ....
    whenever     -- Transition Pass to Pass
       CurrentMode = PASS and
       TrackOf (CurrentTarget) #
         TrackOf (PositionOf (train)) and
       WantsToContinue (hd Instructions) and
       CanAllocateNextTrack (hd Instructions,
                                     TrackTable)
       also from 100 ==>
         CurrentMode = PASS and
         CurrentTarget =
           TargetFrom (hd Instructions) and
           act (train.
                set_power (ComputePower (
                TargetFrom (hd Instructions))))
       and mission = tl ~mission

    whenever     -- Transition Slow to Pass
       CurrentMode = SLOW and
       now < SlowTime + deadline and
       CanAllocateNextTrack (hd Instructions,
                                       Tracktable)
       also from 500 ==>
              CurrentMode = PASS and
              CurrentTarget =
                  TargetFrom (hd Instructions) and
              act (train. set_power (
                      ComputePower (
               TargetFrom (hd Instructions)))) and
              mission = tl ~mission
    ....
  end TrainController
```

For each transition in the FSM the pre- and postconditions can be written down easily in a functional style. Notice however that the postcondition contains the *act* predicate. The conforming implementation is to ensure that the method set_power is activated with the correct parameter, within 100 resp. 500 milli seconds after the predicate became **true**, raising the event.

Specification and implementation The complete specification of the railroad controller in VDM^{++} took less than 400 lines. This is roughly comparable to the sizes of specifications of the same controller in other notations. As in other notations, it excludes the specification of the railroad data structures, which were 'borrowed' from the plain VDM specification of the same railroad (see [9]).

The specification is, due to its structuredness, somewhat larger than a similar specification in plain VDM-SL .

The next step in the development has been the translation of this highly structured specification into Ada-83 code. The translation was performed by simply transforming each (active) class into a task type, with an instantiation for each object. Although the translation is in general a fairly simple one, two problems occur:

- the 'model' used in the VDM^{++} specification is an action/event model. I.e. 'some' event occurs, as a result an action is scheduled. The Ada model is, however, a synchronous one. The translation is therefore somewhat obscured since essentially an action/event model is 'implemented' in Ada.

  It is interesting that a translation into Ada-95 where asynchronicity can be obtained by using protected objects, which is underway, turns out to be straightforward.

- systematically dealing with temporal constraints in the implementation is not always trivial. The semantic gap between the *also form* part of the VDM^{++} whenever clause and the delay alternative in the Ada selective wait construct is fairly large. Ada (and Ada-95 as well in that respect) do not provide much support for a direct translation. Neither Ada nor Ada-95 provide support for mapping temporal constraints on scheduling directives.

5. Results and Conclusions

VDM^{++} has shown to be a valuable modeling tool in the area of industrial systems. The first phase of the project, establishing the language and a supporting toolset has been completed. Current research aims at

- establishing the real-time elements of the language,
- investigating and formalizing the annealing steps in the systems development.

Our experiences in the design and the useage of VDM^{++} have shown that there is a significant amount of tension between *useability* of the notation and the degree of *formalism* in the notation. Our choices have been infuenced by the desire to develop a notation that can be used as an engineering's tool. I.e. we have been aiming at a suitable notation, sufficiently defined, that allows engineers from a variety of disciplines to specify high level models of the system to be build. Less emphasis has been given to formal verifyability of properties of particular models.

Although we still have a long way to go w.r.t. the formal verifyability of properties of given models, we have paid attention to a rigorous approach to refinement (annealing).

The real-time extensions we are working with, are presented in this paper. The structure is similar to extensions earlier proposed for VDM-SL , however, the object-orientedness of the notation has influenced the design.

Current work involves handling the delay between the event being raised and the associated action to be invoked and the problem of verification.

6. References

References

[1] Dürr, E (1994), The use of Object-Oriented Specifications in Physics. PhD Thesis, Utrecht University, Utrecht, The Netherlands, 1994, ISBN 90-393-0684-2.

[2] Dürr, R and N. Plat Editors (1995). VDM^{++} Language Reference Manual, Afrodite (ESPRIT-III project number 6500) document AFRO/CG/ED.LRM/V10, Cap Volmac.

[3] Dürr, E and J. van Katwijk, (1992), VDM^{++} - A formal Specification Language for Object-Oriented Designs. In: Georg Heeg, Boris Magnusson, Editors *technology of Object-Oriented Languages and Systems*, pp 63 - 78. Prentice hall International, Proceedings of Tools Europe '92.

[4] Goldberg, A., Editor, (1984) Smalltalk-80, The Interactive Programming Environment. Addison Wesley Publishing Company.

[5] Jones, C. (1990). Systematic Software Development using VDM (2nd edition). Prentice Hall International.

[6] Mahoney, B. and Hayes, I. Using Continuous Real Functions to Model Time Histories. In Bailes, P. Editor, *Proceedings of the 6th Australian Software Engineering Conference (ASWEC91)*, pp 257-270, Australian Computer Society.

[7] Eugène Dürr, Stephen Goldsack and Nico Plat. *Rigorous Development of Concurrent Object Oriented Systems*. Tutorial(MM5) at the Tools Europe '94 Conference, March 7-10,Versailles France. In : Technology of Object Oriented Languages and Systems, Editors. B Magnusson,B.Meyer, J.Nerson, J.F. Perrot TOOLS 13, ISBN 0-13-350539-1, Prentice Hall, UK, (page 515)

[8] W. Cellary, E. Gelenbe, T. Morzy, *Concurrency Control in Distributed Database Systems* , Studies in Computer Science and AI nr. 3, North-Holland, Elseviers Science Publishers, Amsterdam, 1988.

[9] Ton Biegstraaten, Klaas Brink, Jan van katwijk, Hans Toetenel. A simple railroad controller: A case study in real-time specification. Technical Report 94-86. Reports of the Faculty of Technical Mathematics and Informatics. Delft University of Technology, Delft 1994.

[VDM Standard] ——. *VDM Specification Language: Proto-Standard(Draft)*. Document N-246(I-9), BSI IST/5/-/19 and ISO/IEC JTC1/SC22/WG19, December 1992.

[10] *Cap Gemini CombiCom Team*, CombiCom Internal Deliverable S.1.X,
High Level Formal specification of the CombiCom architecture, CombiCom Consortium.

[11] *Cap Gemini CombiCom Team*, CombiCom Internal Deliverable S.3.X,
Final Formal specification of the CombiCom architecture, CombiCom Consortium.

[12] J. van Katwijk, W.J. Toetenel. Comparing formal specifications by measuring. Proceedings of the second International Workshop on Real-Time Computing Systems and Applications. IEEE 1995, ISBN 0-8186-7106-8, pp 184 - 190.

[13] K.H.Kim, C. Subbaraman, Y. Kim. Imitation of RTO.k Objects using PCD Components in C++. In *these proceedings*.

7. About the authors

- Dr Ir E.H.Dürr is Technical Manager at Cap Gemini Sogeti, and Assistant Professor at Utrecht University, both in the Netherlands.
- Prof. J.van Katwijk, is professor at the Faculty of Mathematics and Informatics of the Technical University of Delft, The Netherlands
- Prof S. Goldsack is emeritus Professor at the Department of Computing, at Imperial College in London U.K.

Schedulability and Safety Analysis in the Graphical Communicating Shared Resources*

Hanêne Ben-Abdallah
Dept. of Computer and Information Science
University of Pennsylvania
Philadelphia, PA 19104

Young Si Kim
Electronics and Telecommunications
Research Institute
Taejon, Korea

Insup Lee
Dept. of Computer and Information Science
University of Pennsylvania
Philadelphia, PA 19104

Abstract

Graphical Communicating Shared Resources is a formal language for the specification and analysis of real-time systems, including their functional, temporal and resource requirements. GCSR supports the explicit representation of system resources and priorities to arbitrate resource contentions. These features allow a designer to examine resource inherent constraints and to experiment with various resource allocations and scheduling disciplines in order to produce a more dependable specification. In addition, GCSR has a precise operational semantics and notions of equivalence that allow the execution and formal analysis of a specification. In this paper, we show how to model a scheduling discipline and verify schedulability and safety properties within GCSR. We illustrate our method through a mobile robotic application.

1 Introduction

The Communicating Shared Resources (CSR) paradigm is an ongoing project at the University of Pennsylvania to build a formal framework for the development of real-time systems. It is based on the premise that the timed behavior of a real-time system is affected not only by the time its components take to execute and synchronize, but also by delays introduced due to the scheduling of actions that compete for shared resources. One of the objectives of the CSR paradigm is therefore to provide a rigorous framework in which the run-time resource requirements of a real-time system can be specified together with its functional requirements. The integration of the two types of system requirements has several advantages. One advantage is that designers can formally specify and rigorously analyze and test various design alternatives before implementation. Another is that designers can consider resource-induced constraints during the design stage of the development cycle; this in turn allows them to develop more realistic design solutions.

Within the CSR paradigm, we developed the Graphical Communicating Shared Resources (GCSR) [1] for the following reasons. One reason is that, similar to most other formal methods, the CSR formalisms developed so far (CCSR [6] and ACSR [9, 2]) tend to be not easily accessible to practitioners, in part, because of their mathematical, textual notations that often produce obtuse specifications. Another reason is that popular graphical languages for real-time systems (Statecharts [7], Modechart [8] and Communicating Real-time State Machines [13]) do not support explicit notions of resources. While it might be possible to encode the run-time resource requirements as boolean conditions using formalisms without explicit notions of resources, the resulting specification is often complex and hard to modify. (For an example, the reader is referred to the description of the Modechart toolset MT [5] where several tasks of a missile system are modeled together with a cpu allocation.)

In addition, the development of the GCSR language is motivated by two other concerns: one is to support the modular, hierarchical, and thus, scalable specification of real-time systems; the other is to benefit from the analysis techniques [3, 10] and toolset [4] developed for the Algebra of Communicating Shared Resources (ACSR), timed process algebra that is developed within the CSR paradigm.

Statecharts and Modechart support modularity through their notions of state and mode, respectively; they, however, lack a notion of scopes for variables since they allow a variable to be used in any state and mode. Furthermore, Statecharts and Modechart support hierarchy through state and mode nesting, but allow edges to cross state and mode boundaries. Such a feature can be compared to *goto* statements in high-level programming languages as it may create unstructured specifications. Also, there are graphical ambiguities in describing when control moves out of a state and mode: does control abort the execution of

*This research was supported in part by NSF CCR-9311622, NSF CCR-9415346, AFOSR F49620-95-1-0508, ARO DAAH04-95-1-0092, and NSF SBR 8920230.

0-8186-7570-5/96 $5.00 © 1996 IEEE

the source mode, or does it wait for its completion? GCSR remedies these weaknesses by offering several types of nodes and edges. In GCSR, the visibility scope of communication events, which reflect potential dependencies between nodes, can be limited. Furthermore, GCSR disallows edges from crossing node boundaries and offers instead two types of edges that graphically distinguish between control transfer due to an interrupt versus an exception, i.e., involuntary versus voluntary release of control.

The semantics of GCSR can be defined either directly as a labeled transition system or through a translation to the ACSR process algebra. As shown in [1], the second way of defining the semantics of GCSR allows a designer to interchangeably use the graphical and textual notations, for example, to specify the high-level view of a system graphically and fill the details of components textually. In addition, since we already have developed VERSA [4], a toolset for the specification, syntactic manipulation, execution, and equivalence checking of ACSR specifications, we can use VERSA to analyze GCSR specifications by an automated translation from GCSR to ACSR. We have extended VERSA with a graphical user interface that allows the drawing, syntactic checking, and automated translation of GCSR specifications.

In this paper, we describe the GCSR language and illustrate how the explicit representation of resources and priorities allows a designer to model static priority based scheduling disciplines and to experiment with different resource allocations without major changes to a given design solution. We use a robotic application to illustrate schedulability and safety analysis in GCSR. The analysis is based on verifying deadlock freeness.

2 Mobile Robot

The system consists of a robot that autonomously navigate in an environment while avoiding obstacles. The robot is equipped with sensors, e.g., cameras, to monitor the environment and actuators to react upon detecting an obstacle by rotating a set of wheels to move the robot. The robot also has two dedicated processors, *cpu*1 and *cpu*2, to process collected sensory data and to determine and execute navigation strategies.

The software architecture, which is inherited from the system functionality, consists of three components: an obstacle detection and avoidance planning strategy, a normal motion planning strategy, and a motion execution strategy. The obstacle detection and avoidance planning strategy is a periodic task with period $d1$ and computation time $c1$. The normal motion planner is also a periodic task with period $d2 > d1$ and computation time $c2$. The motion execution strategy synchronizes its computation with the other two components through two communication events: *obs* when an obstacle is detected, and *loc* when a safe, target location to which the robot must move is set.

Since there are three tasks and only two processors, it is obvious that the behavior of the system will depend on the resource allocation scheme and how resource contention is resolved. A design solution that explicitly incorporates these decisions gives the designer more confidence in the solution before the implementation stage. We will use the above mobile robotic example to illustrate how we can model it in GCSR to describe the behavior of the system, the resource allocation, as well as resolution of resource contention through priorities. In addition, we will show how to verify the safety of a design and the schedulability of tasks that share a processor under a specific scheduling discipline.

3 The GCSR Language

The GCSR paradigm is based on the view that a real-time system consists of a set of communicating components, called *processes*, that execute on a finite set of serially reusable resources and synchronize with one another through communication channels. The use of shared resources is represented by timed *actions*, and synchronization is supported by instantaneous *events*. The execution of an action is assumed to take nonzero time units with respect to a global clock, and to consume a set of resources during that time. The execution of an action is subject to the availability of the resources it uses. Contention for resources is arbitrated according to the priorities of competing actions. To ensure uniform progress of time, processes execute actions synchronously. Time can be either dense or discrete. However, for implementation reasons, we consider discrete time only.

Unlike an action, the execution of an event is instantaneous and never consumes any resource. Processes execute events asynchronously except when two processes synchronize through matching event names, i.e., channels.

Graphically, a GCSR process is represented by a finite set of *nodes* that are connected with directed *edges*. Figures 2 and 3 show the graphical GCSR symbols. Before describing the details of these symbols, we first use the motion executor component of the mobile robot to illustrate the GCSR symbols.

3.1 Example

Figure 1 shows one specification of the motion executor component for the robot example. The behav-

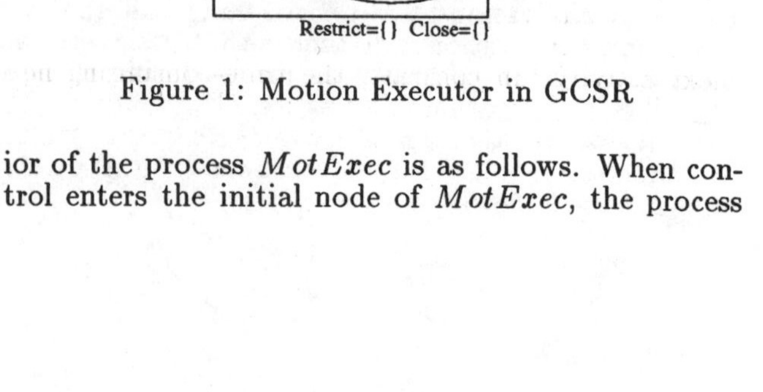

Figure 1: Motion Executor in GCSR

ior of the process *MotExec* is as follows. When control enters the initial node of *MotExec*, the process starts behaving like the process *Wait*. The *Wait* recursively consumes time and no resources. Within the process *MotExec*, the execution of the *Wait* process can be interrupted in two ways which correspond to the event-labeled edges out of the node *Wait*: one is by receiving the event $(obs?, 2)$ and another by receiving the event $(loc?, 1)$. Receiving the event *obs* has a higher priority than receiving the event *loc*. This reflects the fact that avoiding an obstacle is more critical than *normal* motion.

Once the event $(obs?, 2)$ is received, control instantaneously moves to the target node of the edge. In this node, control remains for *to* time units while using the *cpu*1 resource at priority level *po*. This node represents the robot avoiding an obstacle detected in its environment. After *to* time units, control moves back to the initial node where the robot is ready to execute a navigation strategy.

If event $(loc?, 1)$ is received, while the event $(obs?, 2)$ was not present, control instantaneously moves to the target node of the corresponding event-labeled edge. This node is a *compound* node where the nested GCSR process represents the navigation strategy in the absence of obstacles. Control starts at the initial node of the nested process which is an *instantaneous* node. At this node, an instantaneous decision must be made whether to idle or use the *cpu*1 resource at priority level *pw*. As we will see in Section 4, the choice between the two actions is determined by the availability of the *cpu*1 resource. The execution of the nested process can be stopped in three ways: the first way is by the reception of the event $(obs?, 2)$ at which time control moves to the node where the avoidance strategy is executed; the second way is by a *timeout* after control has remained *tw* time units inside the compound node; and the third way is when the nested process *relinquishes* control by sending the event $(done!, 1)$ at which time control moves through the edge labeled with $(done?, 1)$ back to the initial node of the *MotExec* process. The first two types of control transfer do not depend on the nested process; the third on the other hand is initiated by the nested process itself.

3.2 GCSR Nodes and Edges

Figure 2 shows the graphical symbols for the five types of GCSR nodes.

The *Resource* attribute of a *time-consuming* node is a set of (resource, priority) pairs, with the restriction that each resource is listed once. The *Name* attribute of a *reference* node refers to the name of a GCSR process. The *Restrict* and *Close* attributes of a *compound* node are sets of event names and resource names, respectively.

The motivation for various node symbols in GCSR is a succinct and scalable representation of the different system activities and components. The instantaneous node requires that no delay is allowed before the next activity. In contrast, the time-consuming node describes a time consuming action. Furthermore, the Resource attribute of a time-consuming node explicitly describes the required resources for the system action and makes it easy to modify any resource requirement to reflect different resource allocation and scheduling disciplines.

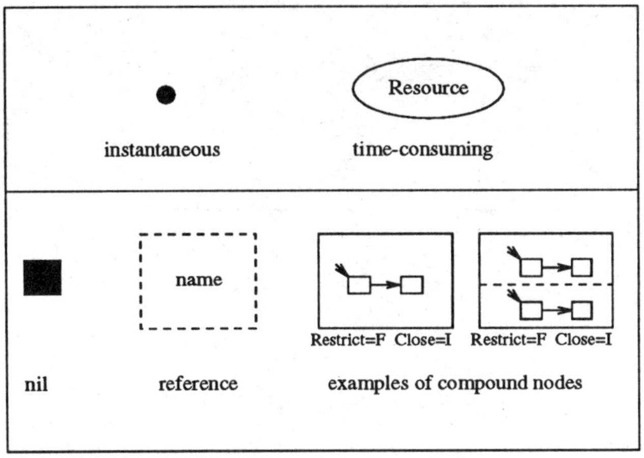

Figure 2: GCSR nodes

The nil node describes a halting process, i.e., end of system execution. The reference node allows the decomposition of a large specification into subspecifications which eases the visual structuring of such a specification. On the other hand, the compound node visually distinguishes a system action from a system component. It is essential in supporting scalable and modular specifications since it allows a designer to 1) group GCSR processes into a higher level entity, 2) connect several GCSR processes that are executed sequentially, and 3) reflect the fact that system components execute in parallel. In addition to the structural modularity, compound nodes also provide for semantic modularity by encapsulating dependencies through their Restrict and Close attributes. The Restrict attribute identifies a set of events that are local to the GCSR nested processes inside the node; they are not visible outside the node boundaries. The Close attribute identifies a set of resources that are reserved for the nested GCSR processes, even if their actions do not explicitly request them.

GCSR nodes can be connected with edges to describe sequential execution. GCSR offers four types of edges shown in Figure 3. The distinct symbols for exception edges and the other edges are motivated by the desire to support a structured, hierarchical specification in which edges do not cross node boundaries and to graphically distinguish two types of control flow: one that is externally controlled by an interacting process and one that is triggered internally through voluntary release of control by raising an *exception*. The second type of control flow is specified using an exception edge. Control moves to the destination node of an exception edge when the process at the source node executes an exception event that labels the exception edge. The transfer of control through an exception edge allows synchronization between a process inside a compound node with an outside node and thus emulates a transition between nodes at different levels of nesting.

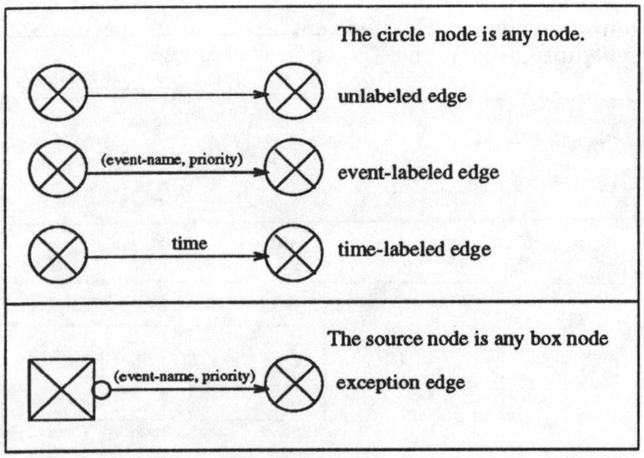

Figure 3: GCSR edges

3.3 Informal Semantics

Intuitively, the behavior of a GCSR process consists of a sequence of execution steps each of which represents either a communication event or a time and resource consuming action. For example, in Figure 4 (a), once execution reaches the instantaneous node, the system *sends* the event named e at priority 1 and then execution moves instantaneously to the target node of the event-labeled edge from where execution continues. On the other hand, in Figure 4 (b) execution remains in the time-consuming node for 3 time units while using simultaneously the *cpu*1 resource at priority level 1 and the *cpu*2 resource at priority level 2; after three time units in the source node, execution moves to the target node of the time-labeled edge.

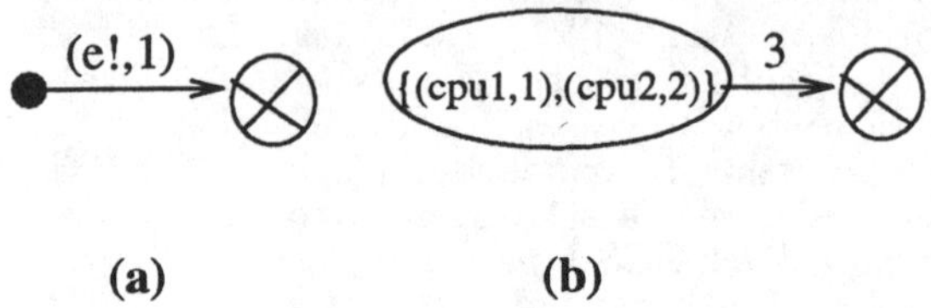

Figure 4: GCSR partial specifications (part 1)

In addition to this basic notion of sequential execution in GCSR, several processes can be combined through compound nodes to describe a large system where processes execute in sequence or in parallel. For example, in Figure 5 (a), once execution reaches the compound node, the GCSR process P is executed for at most t time units, after which execution moves to the target node of the time-labeled edge. The execution of P can also be terminated in two other ways. One way is when the target node of the unlabeled edge is enabled, the execution of P is aborted and execution moves to the target node. Another way of terminating the execution of P is through an exception raised by P which sends the event a. At this time, the execution moves to the target node of the exception edge, labeled with the *receive* event a at priority p. The difference is that, in the former the enabling of the edge is caused by a process other than P, whereas in the latter the enabling of the edge is caused by the process P itself.

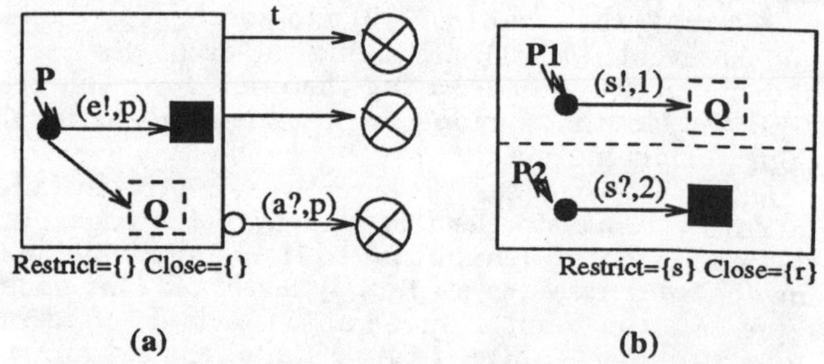

Figure 5: GCSR partial specifications (part 2)

GCSR processes can also be combined in parallel by nesting them inside a compound node. When control reaches a compound node, it simultaneously enters all the initial nodes of its nested GCSR processes. Control can move through event-labeled edges in different nested processes in an interleaved fashion for unrestricted events, but synchronously in any two nested processes for restricted events. In addition, if control spends time in one nested process, then control in all the remaining nested processes must be in time-consuming nodes where it can spend time. This forces synchronous time passage between parallel processes. Furthermore, since resources are assumed to be serial, the set of resources used in all the time-consuming nodes that simultaneously have control must be disjoint. For example, in Figure 5 (b) the processes $P1$ and $P2$ execute in parallel. Furthermore, $P1$ and $P2$ communicate privately through the event named s and use the resource r exclusively; that is, if the compound node is combined in parallel with another GCSR process, this latter can not communicate with $P1$ and $P2$ through the event s and will not have access to the resource r unless it requires it at a higher priority.

The overall behavior of a GCSR process can be formally described as a labeled transition system where each transition represents an execution step with the label being a communication event or a time and resource consuming action. In the translation of a GCSR process to a labeled transition system, there are two main difficulties that stem from handling the time-labeled edges and the unlabeled edges. The time-labeled edges require a notion of local *clock* associated with each time-labeled edge to keep track of how long control remained inside the source node. The local clocks of edges in parallel components are decremented each time the global clock "ticks". When the local clock of an edge reaches zero, control moves to the source node of the edge. The unlabeled edges do not participate in a *visible* activity, i.e., communication event of time and resource consumption. We therefore had to define the semantics at two levels: the first level basically prunes out the unlabeled edges and produces the sequential execution steps in a GCSR process; the second level uses the first level transitions to define the parallel execution and synchronization within a GCSR process.

As seen in the previous examples, there might be several execution steps that are simultaneously possible. The selection among such execution steps is done via a notion of priority. Informally, the selection among two execution steps that are simultaneously possible is either nondeterministic, or is arbitrated according to the following three rules for selecting an execution step α over an execution step β: 1) α and β are events with the same label and α has a higher priority; 2) α and β are actions and α uses a subset of the resources used in β at priorities no lower than in β and with at least one resource in α at a higher priority; or 3) α is a τ event (i.e., synchronization of two events) with a non-zero priority and β is a time-consuming action. (The technical reasons behind these rules are rather complicated and outside the scope of this paper [9].)

The second way of defining the semantics of GCSR is through a translation to ACSR: each GCSR process is translated to an ACSR process. The translation is consistent with the direct semantics, in the sense that the corresponding ACSR process has an equivalent labeled transition to the one generated directly from the GCSR process. This direct translation which we have implemented in the GCSR toolset allowed us to use the ACSR toolset VERSA to execute and analyze GCSR specifications.

To define precisely the semantics of GCSR, either directly or through ACSR, we imposed syntactic restrictions on how edges can connect nodes. Detailed description of the syntax of *well-defined* GCSR processes and the GCSR-ACSR correspondence can be found in [1].

4 The Mobile Robot in GCSR

Figure 6 shows the high level structure of the mobile robot example in GCSR. In this first solution, the components in charge of planning motion strategies, *ObstDetect* and *MotPlan*, have been allocated on *cpu*2. The motion execution component, *MotExec*, is allocated on *cpu*1.

Figure 7 shows the detailed specification of the *MotExec* process; this is a simplified version of the one in Figure 1. Figure 8 shows the detailed specifications of the *ObstDetect* and *MotPlan* processes. Each of these latter processes consists of two processes that execute in parallel and coordinate their activities with the synchronization events *start* and *end*. The process *ObsDetTask* (*MotPlanTask*) represents the actual task that computes the obstacle detection and avoidance (normal motion planing) strategy. The second process *ObsDestSched* (*MotPlanSched*) serves as a dispatcher for the task: it instantiates the task by sending the event *start* and waits for its period to finish at which time it informs it to stop by sending the event *end* and instantiates it again for the following period. Note that at the end of the task period, the scheduler moves to an instantaneous node from which it can only send the event *end*; in particular it can not consume time. At this time, however, the task might not be able to receive the event *end* if it

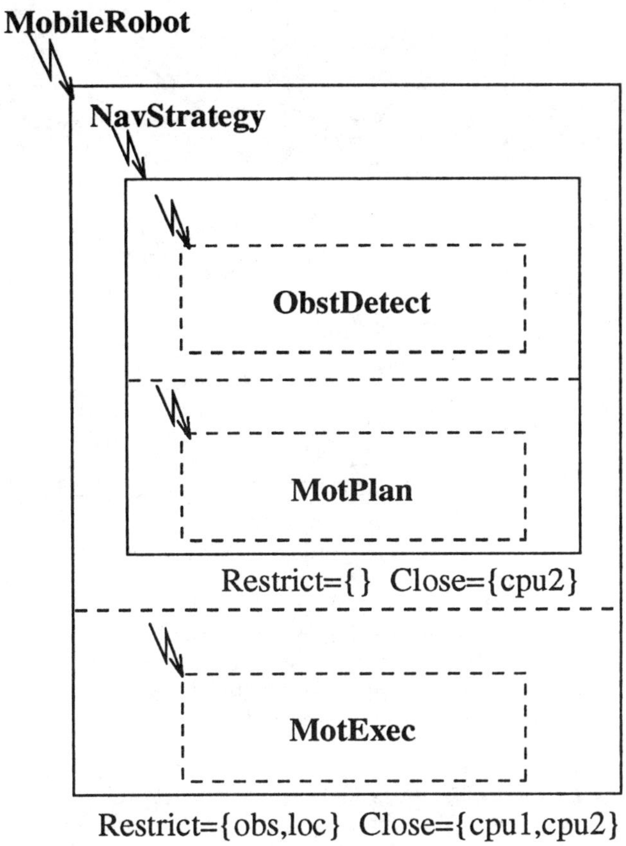

Figure 6: Mobile Robot Architecture in GCSR

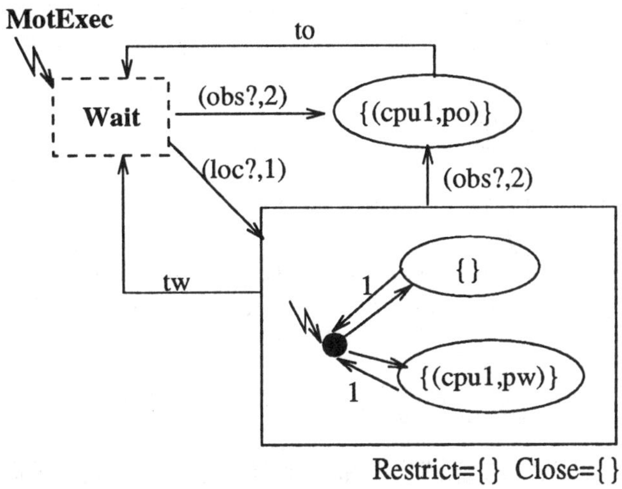

Figure 7: Motion Executor in GCSR

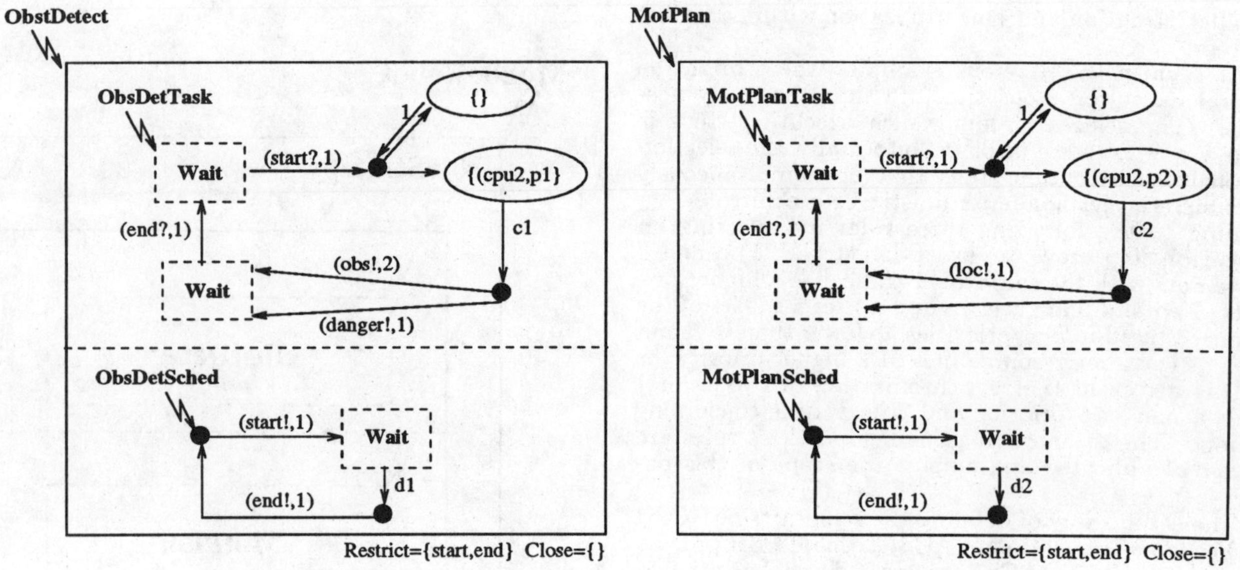

Figure 8: Obstacle Detection (left) and Motion Planer (right) in GCSR

is in any of its time-consuming nodes. By operational semantics of the parallel execution and event restriction, the system deadlocks at this instant, i.e., does not have any more execution steps.

Each of the two tasks, *ObsDetTask* and *MotPlanTask*, is modeled such that it waits for the event *start*, after which it can either acquire the *cpu*2 resource or idle. According to the preemption relation, acquiring the resource has a higher priority than idling. In addition, the task with the higher priority will acquire the *cpu*2 resource first. Once a task acquires the *cpu*2 resource, it executes without preemption until the end of its computation time. It then tries to signal the motion executor process the result of its computation (i.e., *obs* and *loc*, respectively) and waits for the event *end* to go back to the initial state of waiting.

Note that in the *ObsDetTask* process, after using the *cpu*2 resource for *c*1 time units, the process can either send $(obs!, 2)$ or $(danger!, 1)$. The first event when synchronized will have a higher priority. However, if the motion executor process is not ready to receive the *obs* event, the process will send the *danger* event, to indicate that an obstacle was detected but the motion planer is not aware of it and thus there is a potential safety violation: the robot may collide with the obstacle. We will use this event during the analysis stage in the next section to verify the safety of the system.

In the *MotPlanTask* process, the event *loc* is not as critical to send. This is modeled by the unlabeled edge which represents the possibility of skipping the synchronization with the motion executor on the event *loc*.

5 Analysis in GCSR

Formal analysis and rigorous testing of a real-time specification before implementation is essential, especially for real-time systems that are embedded in safety critical applications. Analysis includes checking that a design is valid with respect to a (requirements) specification, and that a specification satisfies certain properties such as safety and schedulability of system processes, in the presence of resource sharing.

A real-time system property is usually expressed in terms of the (timed) occurrences of certain events. For example, a safety property can be formulated as "bad events never occur" or as "a bad event never occurs within t time units after some event ". In the presence of communication and resource sharing, deadlock freeness can be considered as a safety property. Furthermore, schedulability analysis in GCSR can be turned into verifying deadlock freeness; that is, a real-time system can be modeled such that if one process misses its deadline, the system deadlocks [4].

The operational semantics of GCSR allows two types of analysis: execution to test a specification for "unintended" behaviors, and verification of properties through either state space exploration or behavioral equivalence checking. Within the CSR paradigm, there are two notions of behavioral equivalence that are defined over the prioritized labeled transition system. The first equivalence relation is based on strong bisimulation [12], $\sim_\pi$, which ensures that equivalent processes match each other's execution steps, i.e., events and time-consuming actions. The second is based on weak bisimulation, $\approx_\pi$, which ensures that equivalent processes match each other's execution steps for non-τ events but allows one process to make τ execution steps that an equivalent process does not match. Based on the prioritized strong and weak bisimulations, we have defined two corresponding equivalence relations that abstract out certain resource names [10]. These relations relate processes based on their behaviors while ignoring the usage of certain resources. It turns out that these equivalence relations are extremely useful since during the specification phase, resource identity may not be relevant, while it is important during the design phase.

In this paper, we use state space exploration to verify the safety and schedulability of a design and search for deadlocked states in the prioritized labeled transition system of the design. We note that an analysis based on state space exploration (as well as equivalence checking) is feasible only when the processes involved have transition systems with a *finite* number of states. All of the processes we used in the mobile robot example are finite.

To verify the safety of our design, we proceed in two steps: we first build a simple, abstract *tester* process that basically either idles or synchronizes on the "bad" event and terminates, i.e., becomes the Nil process. We then run the tester process in parallel with the given design and force synchronization on all events. Figure 9 (a) shows the verification setting, where the * in the Restrict attribute indicates all events; Figure 9 (b) shows a tester process for the safety of the mobile robot example where the event *danger* is the "bad" event. The semantics of the Parallel operator guarantees that the verification process *SystemVerifier* will reach a deadlocked state *if and only if* the *Tester* process enters its Nil node. In this case, we can conclude that the "bad" event in deed can happen and the design violates the safety property.

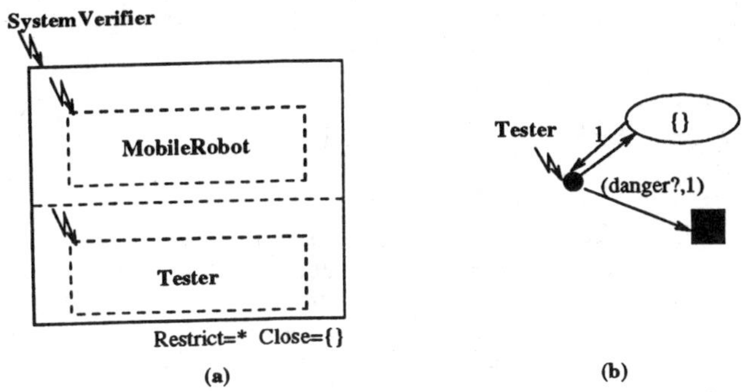

Figure 9: (a) System Verification Process; (b)Tester Process For Safety of The Mobile Robot

We next verify the schedulability and safety of two design solutions for the mobile robot example. The first design uses the resource allocation of the previous section and the rate monotonic scheduling discipline [11] to arbitrate the usage of the *cpu*2 resource between the *ObstDetect* and *MotPlan* processes. The second design shortens the period of the *ObstDetect* process and requires a different resource allocation as we find out that the two processes are no longer schedulable on *cpu*2 under the rate monotonic scheduler. The second design illustrates the ease of changing resource requirements and priorities in GCSR.

5.1 Analysis of the Mobile Robot: the First Design

In the first design of the mobile robot example, the timing constraints are as specified in Table 1. In addition, to arbitrate the usage of the *cpu*2 resource between the *ObstDetect* and *MotPlan* processes, we use rate monotic and assign priority 2 ($p1 = 2$) for the *ObstDetect* and priority 1 ($p2 = 1$) for *MotPlan*, since the period of the first is larger than that of the second. Since the *MotExec* process runs alone on *cpu*1, any non-zero priority for its resource consuming actions will be acceptable; we set $po = 1$ and $pw = 1$.

Table 1: Timing Assumptions for The Mobile Robot

MotExec	*ObstDetect*	*MotPlan*
to : 3	$c1$: 3	$c2$: 2
tw : 6	$d1$: 5	$d2$: 6

Schedulability Analysis. The two processes which share the *cpu*2 resource are schedulable if the process *NavStrategy* never deadlocks. We used the automated GCSR to ACSR translation and VERSA to verify that this process has no deadlocked state. VERSA supports a command that shows statistics about the transition system of a process, e.g., number of states, deadlocked states, and edges.

Safety Analysis. The safety requirement for the mobile robot is that it does not collide into a detected obstacle. There are two potential ways to violate this safety requirement: one is when the motion planer does not execute the corresponding motion avoidance strategy and instead keeps on navigating; another is when the motion planer misses to receive the event *obs* from the obstacle detection process, *ObstDetect*.

In our design, the first scenario can be detected by the flow of control within the *MotPlan* process right after receiving the event (*obs*?, 2). It is straightforward to check that the first scenario does not happen. The second scenario can be detected by the occurrence of the event (*danger*!, 1) which is sent by the *ObstDetect* when it can not communicate with the *MotPlan* process right after finishing its computation.

To verify that the event *danger* never occurs in our design, we restricted the event *danger* in the *SystemVerifier* process of Figure 9; all other events in *MobileRobot* are already restricted. We then used the automated translation within the GCSR toolset to translate all the processes and used VERSA to verify that *SystemVerifier* has no deadlocked state, and thus, our *MobileRobot* design is safe.

5.2 Analysis of the Mobile Robot: the Second Design

In this design, the obstacle detection strategy must be invoked more frequently to detect the position of mobile targets in the environment. The new period of the *ObstDetect* process is $d1 = 4$. With the

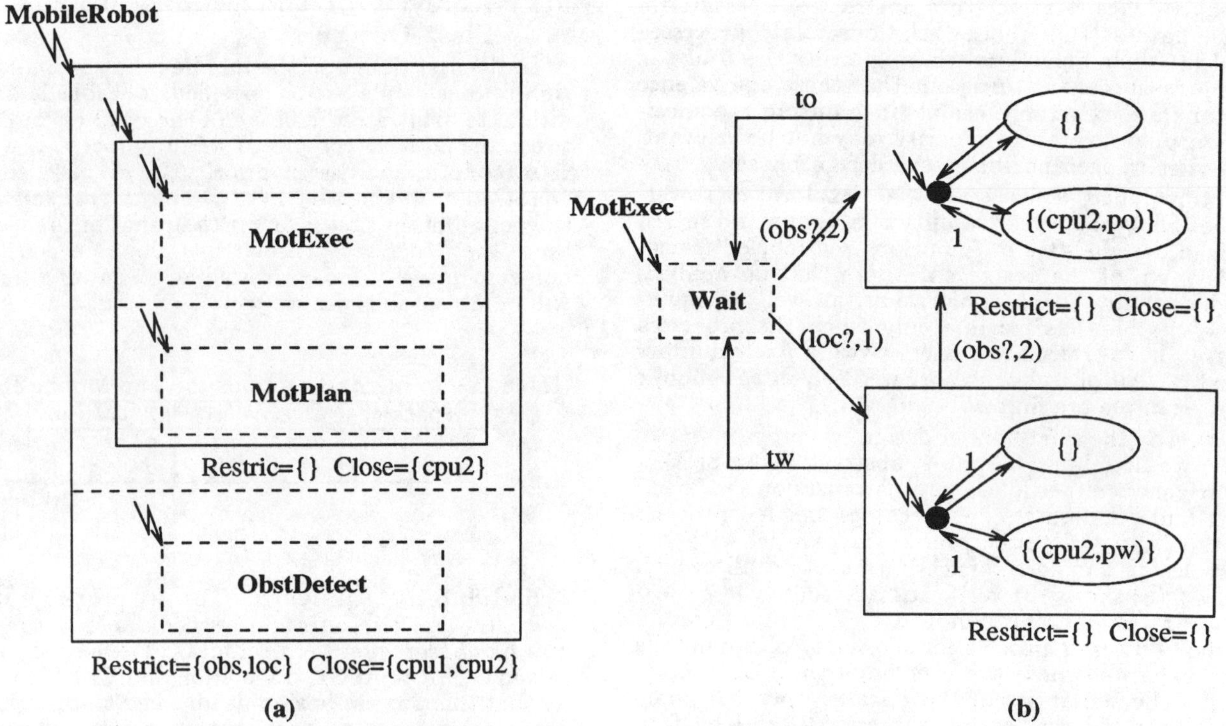

Figure 10: (a) Mobile Robot with New Resource Allocation; (b) Preemptable Motion Executor

shorter period, the *ObstDetect* and *MotPlan* processes are no longer schedulable on the *cpu*2 resource using rate monotonic scheduling. VERSA produced a sample of execution steps that lead to the deadlocked state. In these execution steps, the process *NavStrategy* reaches a deadlocked state at time 12 when the second period of the *MotPlanTask* process expires without finishing its execution. This latter was delayed for 5 time units during its second period since the *ObstDetTask* was using the *cpu*2 resource during the end and beginning of its third and fourth periods. In other words, at time 12, control is inside the time-consuming node with the Resource attribute $(cpu2, p2)$ (see Figure 8) and the *MotPlanTask* therefore can not receive the $(end?, 1)$ event from its scheduler.

To get a schedulable design, we relocate the *ObstDetect* process to *cpu*1and move the *MotExec* to the *cpu*2. The new architecture is show in Figure 10 (a). To reflect the new resource allocation in the detailed specifications of the components involved, we only need to change the Resource attributes of their time-consuming nodes. More specifically, the time consuming node of the *ObsDetTask* process now has a Resource attribute of $\{(cpu1, p1)\}$ and both time-consuming nodes in the *MotExec* process must have their *cpu*1 resource changed to *cpu*2.

With the new resource allocation, the system is however not safe any more: this is due to the fact that the *MotPlan* process may be running on the *cpu*2 resource when the *ObstDetect* process sends the event $(obs!, 2)$ to the *MotExec* process. This later can not preempt the *MotPlan* process which will have control in the time-consuming node with Resource attribute $\{(cpu2, p2)\}$. To fix the safety violation, there are two necessary changes: One is to make one of the processes *preemptable*; and the second is to adjust their priorities so that the most critical action is executed first.

Since the *MotExec* process is an aperiodic process, we elected to make it preemptable. This requires to change the time-consuming action with the Resource attribute $(cpu2, po)$ with a compound node where the nested process describe the behavior of either using the *cpu*2 resource or idling. Figure 10 (b) shows the modified process. Furthermore, to ensure preemption according to the importance of the actions, we assign priorities as follows: When the *MotExec* process executes an obstacle avoidance strategy, it must run at the highest priority $po = 3$; the *MotPlan* process executes at a lower priority $p2 = 2$; and finally, when the *MotExec* process executes normal navigation, it uses the processor at priority $pw = 1$. With this arbitration, the *MotPlan* process acquires the *cpu*2 resource if the *MotExec* process does not need it to avoid an obstacle. It however can preempt the *MotExec* process when executing a normal motion strategy.

We used the automated GCSR to ACSR translation to verify that the new resource allocation and arbitration produces a safe design.

6 Conclusion

We have presented the Graphical Communicating Shared Resources (GCSR) languages. The GCSR language supports the specification of resources and priorities in an intuitive way that produces easy to understand and modify specifications. These features allow a designer to incorporate resource requirements early in a development phase and experiment with different resource allocations to eliminate unimplementable or undependable solutions. In addition, the GCSR operational semantics makes it possible to execute a GCSR specification and verify its correctness through state exploration and various notions of behavioral equivalence relations. We have illustrated how to conduct schedulability analysis and verify safety properties with GCSR through a mobile robot example. The analysis technique we used is based on detecting deadlocks.

One might argue that integrating too fine details about the resource requirements with the functional requirements of a real-time system might render the design process unwieldy. It would be therefore fruitful to develop a software engineering methodology where the resource requirements are added incrementally. The incremental integration of the resource requirements should preserve certain properties between different levels of design abstractions, e.g., deadlock freeness. We are currently working on a notion of refinement that will support such a notion of incremental specification of resource requirements. This notion of refinement also provides for a top-down and modular development of a GCSR specification. Here, a designer starts with an abstract description that contains minimal details about the resource requirements, communications, and system components, which are represented as GCSR actions. The designer then gradually adds the details of the resource requirements, implementation related communications, and internal structures of system actions.

References

[1] Hanêne Ben-Abdallah, Insup Lee, and Jin-Youg Choi. A graphical language with formal semantics for the specification and analysis of real-time systems. In *IEEE Proceedings of Real-Time Systems Symposium (RTSS' 95)*, Pisa, Italy, December 1995.

[2] P. Brémond-Grégoire, I. Lee, and R. Gerber. ACSR: An Algebra of Communicating Shared Resources with Dense Time and Priorities. In *Proc. of CONCUR '93*, 1993.

[3] Jin-Youg Choi, Insup Lee, and Hong-Liang Xie. The specification and schedulability analysis real-time systems using acsr. In *IEEE Proceedings of Real-Time Systems Symposium (RTSS' 95)*, Pisa, Italy, December 1995.

[4] Duncan Clarke, Insup Lee, and Hong-Liang Xie. VERSA: A Tool for the Specification and Analysis of Resource-Bound Real-Time Systems. *Journal of Computer and Software Engineering*, 3(2), April 1995.

[5] P.C. Clements, C.L. Heitmeyer, B.G. Labaw, and A. T. Rose. MT: A toolset for specifying and analyzing real-time systems. In *Proc. of the Real-Time Systems Symposium*, pages 12–22, Raleigh-Durham, North Carolina, December 1-3 1993.

[6] R. Gerber and I. Lee. A Layered Approach to Automating the Verification of Real-Time Systems. *IEEE Trans. on Software Eng.*, 18(9):768–784, 1992.

[7] D. Harel. Statecharts: A visual formalism for complex systems. *Science of Computer Programming*, 8:231–274, 1987.

[8] F. Jahanian and A. K. Mok. Modechart: A specification language for real-time systems. *IEEE Transactions on Software Engineering (to appear)*, November 1989. IBM Tech Report RC 15140.

[9] I. Lee, P. Brémond-Grégoire, and R. Gerber. A Process Algebraic Approach to the Specification and Analysis of Resource-Bound Real-Time Systems. *Proceedings of the IEEE*, pages 158–171, Jan 1994.

[10] Insup Lee, Hanêne Ben-Abdallah, and Jin-Young Choi. A process algebraic method for the specification and analysis of real-time systems. In Constance Heitmeyer and Dino Mandrioli, editors, *Formal Methods for Real-Time Computing*, chapter 7. John Wiley & Sons, Chichester, January 1996. (To appear).

[11] C.L. Liu and J.W. Layland. Scheduling algorithms for multi-programming in a hard-real-time environment. *Journal of the ACM*, pages 46 – 61, January 1973.

[12] R. Milner. *A Calculus for Communicating Systems*. LNCS 92, Springer-Verlag, 1980.

[13] A. C. Shaw. Communicating Real-Time State Machines. *IEEE Transactions on Software Engineering*, 18(9):805–816, September 1992.

Session 2

Real-Time O.S. Architectures and Mechanisms

Session Chair:
Michel Gien
Chorus Systems, France

Paths: Programming with System Resources in Support of Real-Time Distributed Applications

Franco Travostino, Ed Menze, Franklin Reynolds

Open Software Foundation Research Institute
11 Cambridge Center, Cambridge, MA 02142, USA
{travos, fdr}@osf.org, menze@paws.com[1]

Abstract[2]

Real-time distributed applications often rely upon underlying communication services to provide different types of guarantees, such as forward progress and bounded delivery time. Providing such guarantees requires the careful management of several resources, including memory, CPU and network bandwidth. This paper describes how our Object-Oriented Communication Subsystem, CORDS, was enhanced to manage these resources. A new abstraction, the path, provides an end-to-end context for ***programmable*** *control over system resources. Paths can be thought of as naming distinct virtual channels through a CORDS protocol graph.*

Although our work is motivated by hard real-time distributed applications (requiring deterministic guarantees), the new notion of "programming with system resources" also applies to scenarios outside the hard real-time realm. Preliminary empirical results demonstrate the effectiveness of CORDS' control over system resources.

1. Introduction

The growing family of real-time distributed applications covers a vast and diverse problem domain. A large number of real-time distributed applications, like multimedia applications (i.e., cu_c_me, MPEG players, etc.), rely upon underlying communication services for very limited end-to-end guarantees, such as statistical guarantees on the delivery time; the applications themselves can, in fact, compensate for omission failures, timing failures and delivery time jitter in the communication layers with adaptive algorithms, buffering techniques and controlled degradation of quality to the end user. A common communication paradigm for these applications is a continuous data stream (with episodes of bursty traffic).

There are real-time distributed applications, however, that need deterministic end-to-end guarantees (i.e., bounded delivery time, bounded delivery time jitter, etc.). These applications cannot "share the pain" with other "best-effort" applications (that is, applications that do not need any guarantee beyond the system's notion of fairness). Often, these are mission critical control applications: typical scenarios are factory floors, airplanes and combat systems. The applications are carefully designed with an emphasis on synchronization operations among parties, rather than stream or bulk data transfers; end-to-end guarantees must apply to these synchronization handshakes, which we model as remote procedure calls (RPCs). In most cases, end-to-end guarantees must hold within a fairly limited geographic perimeter.

Our work is motivated by the second type of distributed application, known as "hard real-time" distributed applications. These applications depend on system resources (memory, CPU cycles and network bandwidth) in several ways: resource shortages can induce omission failures, performance failures and episodes of overload that jeopardize the distributed application and/or the real-time host system as a whole.

We have created an Object-Oriented Communication Subsystem which provides a fine level of programmable control over system resources and supplies hard real-time applications with a broad range of deterministic guarantees, from a simple forward-progress guarantee to a bound on delivery time jitter. Our design has been centered around medium-sized networks (e.g., 100 nodes) with a very simple topology, either a bus or a ring. Although multi-hop networks introduce new problems, we expect the fundamental results of this work to extend to complex topologies as well. We have performed most of our development on ethernets and are considering FDDI and Myrinet [3] for deployment.

1. Ed Menze can currently be reached at EMASS, Inc., 10949 E. Peakview Ave., Englewood, CO 80111, USA.

2. This research was supported in part by the Advanced Research Projects Agency (ARPA) and the Rome Laboratory of the Air Force Materiel Command (AFMC).

0-8186-7570-5/96 $5.00 © 1996 IEEE

Although our immediate focus is limited in terms of target applications and guarantees, our approach is fairly general and can serve well in several communication scenarios. Outside of the hard real-time realm, we have exploited or we plan to exploit programmable system resource control in different ways:

- "Soft" real-time distributed applications: Even though these applications can "share the pain" during episodes of performance degradation, they can still benefit from a combination of statistical and deterministic guarantees, rather than statistical guarantees only. For example, every channel can be guaranteed a minimal amount of memory, while the remaining memory is overbooked. Thus, applications can operate under less conservative working assumptions without compromising real-time behaviors.

- Real-time node-liveness and membership: A protocol that relies on timeouts to determine liveness runs the risk of prematurely declaring "dead" a node which is simply overloaded and slow in producing a liveness signal (or replying to queries). If we grant privileges to the "liveness channels", in the form of preallocated memory and favorable scheduling attributes, we will reduce the number of "false failures" being called, with less noise associated with recoveries from "false failures".

- Receiver-based congestion control over reliable interconnects: Even in environments where the hardware is reliable, some of the communication layers beneath the application can still drop incoming packets during episodes of memory shortage. The advantages of having the combination of hardware and communication layers immune from omission failures are significant: clients avoid arming retransmission timeouts and can efficiently recycle packets and packet-related data structures immediately after the packet emission (when bits are still "hot" in cache). Both of these techniques make a difference at gigabit speeds.

- Protection from denial of service security attacks: One or more network connections may maliciously consume most or all of the network resources in a server, such as communication buffers and I/O bandwidth: these resources are, in fact, typically overbooked and assigned on a FCFS based policy. With the proposed control over system resources, we can certainly complement the traditional overbooking model with advisory quotas that may trigger the enforcement of security policies towards fairness and pseudo-random behaviors.

The next section describes our O-O Communication Subsystem. Section 3 introduces a new abstraction, the path, that plays a pivotal role in our work with system resources. Section 4 and 5 describe how memory, CPU cycles and network bandwidth are reserved or allocated, and how paths are established and used. Experimental results (Section 6) provide a quantitative analysis of the benefits of using paths in an environment close to our design target.

2. The O-O Communication Subsystem

CORDS (Communication Objects for Real-time Dependable Systems) is a modular and highly configurable O-O Communication Subsystem. CORDS grew out of our work [15] with the x-kernel [5][8], a protocol framework and toolkit created at University of Arizona. X-kernel was developed as a vehicle for research into problems posed by the development of complex network communication protocols. One of the key ideas was the application of software engineering practices from other fields, such as object orientation, to the design and implementation of protocols. In the x-kernel, protocols and communication subsystems are implemented as graphs of protocol objects (Figure 1). Protocol objects are passive; they are either static ("protocols") or dynamic ("sessions"). The latter are meant to capture the run-time view of the protocol graph; the resulting web of sessions unambiguously define access to logical resources, i.e., the next object upstream/downstream, the channel's Maximum Transfer Unit (MTU), etc. Data move from object to object in the form of messages; messages tend to be associated with threads which "shepherd" the message through the protocol graph via object invocation.

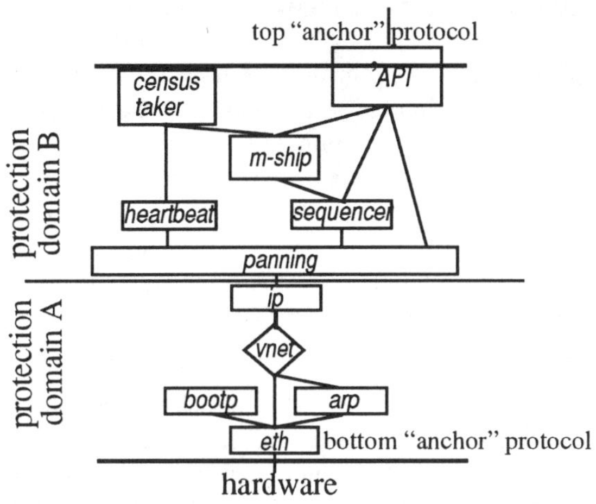

Figure 1: A x-kernel/CORDS protocol graph: *protocol objects (11) interact in two different protection domains (e.g., user and kernel) to form a group communication server.*

CORDS builds upon x-kernel's concept, architecture and code. The first task in the CORDS project was to adapt the original x-kernel (version 3.2) to take advantage of the

Mach-derived [1] OSF MK real-time micro-kernel [12], both as a new kernel subsystem and as (a collection of) user space tasks. In this phase, the x-kernel was modified to provide real-time programming rules (e.g., eager allocation), failure semantics (e.g., memory allocation failures), improved thread management and simpler interactions with the OSF MK primitives. In spite of these incompatible changes, modifying existing "off-the-shelf" x-kernel protocols to run under CORDS is a fairly straightforward proposition [9]. On a similar note, in spite of the OSF MK adaptations it is conceivable to run CORDS on platforms other than OSF MK.

From the foundation represented by the new framework, CORDS has evolved in two orthogonal directions. On the protocol side, we have implemented several new protocols which complement the original set of protocols from the x-kernel. These include protocols which form a communication fabric of multicomputers and clusters, a distributed clock synchronization service with an upper bound on client/master deviation, and various group communication facilities (from unreliable multicast to atomic broadcast). Our protocol graphs are designed to operate over unreliable, asynchronous media (e.g., ethernet).

At the same time, we have refined the framework to support a strict control over system resources through the new concept of programming with system resources, as presented in the rest of the paper. We validate our accomplishments in this direction by exploring real-time behaviors in distributed, dependable systems using the large set of CORDS protocols available.

3. Moving along three-dimensions

In our effort to control system resources, we resisted making ad-hoc changes to protocols and drivers in a given protocol stack. This solution, in fact, does not lend itself to code reuse (new protocols need similar modifications), it pollutes protocol code with various dependencies (from machine dependencies to OS dependencies) and produces protocols whose fate is to grow by accretion (as a result of new system features such as a different interface to memory allocators).

In CORDS, we wanted to establish an interface to various system resources and to define semantics for failures introduced by lack of resources. With these rules, we extended the new framework, the interface to protocols and the protocols themselves for programming with system resources.

Communication subsystems based on frameworks, such as UNIX STREAMS [13] and the x-kernel, have introduced the notion of "programming with protocols". This concept is best viewed as a bi-dimensional space where the developer is free to move: in one dimension, the developer can code new protocols, well supported by the framework; on an orthogonal dimension, the developer can import "off-the-shelf" protocols into the protocol stack, to assume roles needed in a specific configuration. With CORDS, we wanted to move one step beyond the benefits of "programming with protocols" and provide the developer with a third programming dimension. With this new dimension, the developer can select and commit system resources for individual end-to-end channels, or virtual circuits, as appropriate, on a per-channel basis.

Our "programming with system resources" revolves around a new first class abstraction, the path. With a path, we model an end-to-end channel and we capture all the system resources for which the end-to-end channel has exclusive use (Figure 2).

Paths transcend the traditional dichotomy between connection-oriented and connection-less traffic: the mechanisms for path set-up and path identification do not depend, in fact, on a particular protocol or delivery semantic. Thus, paths enable control over system resources even when the protocol stack is as simple as a UDP/IP stack.

The actual process of creating paths and committing resources to paths typically takes place at the uppermost protocol (called "top anchor protocol") in the CORDS protocol stack, in accordance with the end-to-end argument; this is the place with the highest exposure to the application-specific needs, and this is where the Quality of Service (QoS) negotiation takes place.

With paths, we eliminate an undesirable side effect of object decomposition typical of object-oriented architectures such as CORDS', i.e., the higher the level of decomposition, the higher the risk of losing the context that objects (protocols and sessions) need to operate effectively. From the uppermost protocol to the data sink, all the objects that compose a CORDS protocol graph can now refer to paths in order to manipulate system resources with the proper context.

Paths can be manipulated both at compile-time and at run-time. We support a broad spectrum of scenarios, from the simplest "embedded" uses of paths (compile-time only) to sophisticated applications that operate on paths throughout their instantiations to obtain different guarantees at different times (e.g., a control system for a plane during takeoff, cruise and landing).

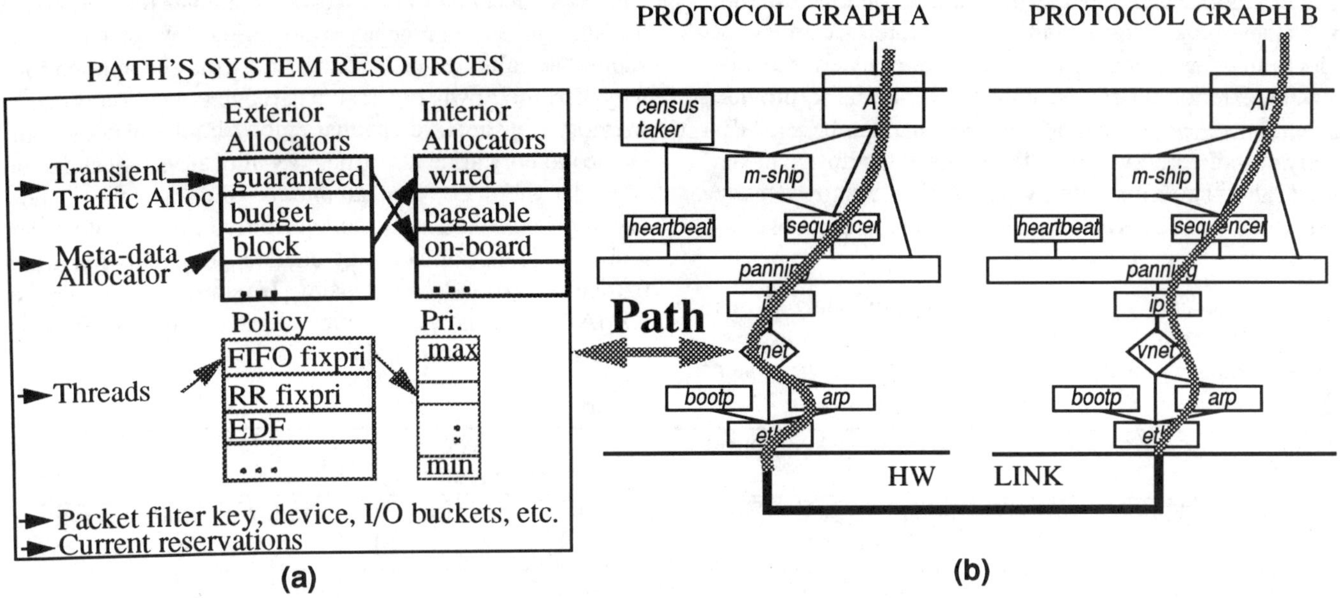

Figure 2: A path: *with a path, we capture all the system resources (a) for which the end-to-end channel (b) has exclusive use.*

4. Programming with system resources

4.1. From paths to memory

Traditional memory allocators do not distinguish between policy and mechanism, with the undesirable result that developers are presented with a few fixed policy-mechanism combinations. We recognize the need for multiple independent policies (i.e., from overbooking to deterministic memory allocations), mechanisms (i.e., pageable, wired and on-board memory allocation) and failure semantics (i.e., block, wait and/or retry upon allocation failure).

We also observe that memory allocated within a communication protocol stack is used either for transient data (i.e., MTUs to and from the network) or for meta-data (i.e., structures for protocol control blocks and connection state). In CORDS, we want to be able to allocate memory for transient traffic and meta-data with different mechanisms and policies.

In some scenarios, for instance, we can imagine carrying some transient traffic with memory that resides on the network controller; in this case, we would want to use the host virtual memory for any allocation of meta-data. Protecting from allocation failure of meta-data usually turns out to be a very simple task: it is fairly easy, for example, to use eager allocation and re-use of meta-data to make the code robust to allocation failures (i.e., a timeout started at connection set-up re-arms itself without allocating any data). Conversely, manipulating transient data is typically more complex and prompts for sophisticated allocation semantics such as reservations: the allocation cannot fail if it takes place above the reliability boundary of the protocol stack, or if it takes place beneath the reliability boundary and we are keen about real-time behaviors (no omission failures introduced by memory shortages). For these reasons, we wanted our channel abstraction, the path, to describe distinct methods for accessing memory for transient traffic and meta-data.

We consider memory allocators as a stack of allocation modules; each module contributes to memory allocation with a policy or a mechanism. Several discussions within the x-kernel community, a design straw man [11] and a prototype implementation at University of Arizona were synthesized into the current CORDS implementation of general purpose composite memory allocators. A memory allocator stack includes one or more "interior allocators" with an "exterior allocator" on top that provides the interface for memory operations (Figure 3). Interior allocators interface with

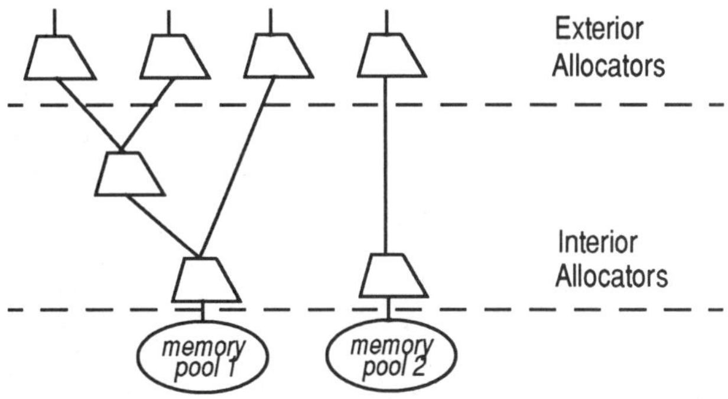

Figure 3: CORDS' general purpose composite memory allocators.

memory pools and provide memory accounting for the pools they serve. They are designated to interface only with pools and other allocators (either interior or exterior). Conversely, exterior allocators, at the top of the stack, provide allocation semantics and an interface that can be used directly or indirectly (via the path interface) to allocate memory (Table 1). A running system typically shows many instantiations of external allocators and only one or two instantiations of interior allocators.

Protocols can take advantage of two additional features of CORDS memory allocators: call-backs and transfers. Protocols and applications that use CORDS may register call-back functions to be invoked whenever the composite memory allocator reaches a specified low-water mark: should the call-back be invoked, protocols and applications may start their own aggressive garbage collection process. Memory transfers are enabled among homogeneous composite memory allocators (memory allocators that share the same derivation tree from an interior allocator); thus, memory may be debited and credited from/to the allocators that maintain quotas, without any data copying. This feature is particularly useful when using Direct Memory Access (DMA) capable network devices that have little or no packet filter facilities on-board.

Name	Class	Description	Type of memory	Failure semantics
Guaranteed	Exterior	It pre-allocates memory in response to reservations.	Interior's	No failures within reservations. Interior's for excess
Budget	Exterior	"Soft" reservations (no pre-allocation)	Interior's	Interior's
Block	Exterior	It rounds up allocation requests to the next largest block size configured.	Interior's	Interior's
Kernel's Interior	Interior	Kernel-only optimized version of interior allocator.	Wired	It returns an error without blocking
User Interior.	Interior	An interior allocator built upon libc.	Pageable	Interior's, if any. Otherwise, malloc() semantics.

Table 1: Exterior and interior allocators used in CORDS

A path is typically configured to use a composite allocator that guarantees a certain amount of memory for its incoming data from the network: aggressive demultiplexing among paths in the device driver makes sure that we use reserved memory without any dependency on other memory. Such deterministic behavior is certainly appealing in a real-time environment; however, we must worry about memory (especially kernel wired memory) being wasted because of aggressive allocation to the exclusive benefit of a single path. Using existing technology, we have envisioned a few ways to limit the waste of memory inherent to deterministic memory allocation:

- Buffer coloring: We leverage the notion of path in the case where all incoming traffic to a specific path has a size smaller than the network MTU (i.e., a path expects RPC replies whose size is only 16 bytes). Thus, the eager memory allocation takes place in multiples of the packet data unit expected by the path rather than the generic network MTU[3].
- Priority encoding: We can imagine that, as part of the end-to-end protocol, user data is organized in a set of data packets with various degrees of redundancy and priority. Deterministic allocation applies to the "mandatory" group of data packets, whereas for the remaining data we resort to traditional overbooking techniques.
- Limited overbooking: We are exploring relaxed forms of deterministic memory allocation. For instance, well-behaved controlled paths which aren't likely to require large amounts of memory simultaneously can share one branch of a composite allocator hierarchy, and therefore overbook a small set of individual reservations; they would still be protected from unknown, uncontrolled "best-effort" traffic.
- Quick reservations: Given a path and its needs, it may happen that we only need the memory allocation guarantees to hold for a limited amount of time. As the mechanisms for manipulating memory and reservations are not expensive, we can imagine changing the allocation semantics and guarantees often during the lifetime of a path.

3. The developer is responsible for following the new advisory allocation limit, unless a new exterior allocator is provided to limit path's memory allocations to a single allocation size.

4.2. From paths to CPU cycles

The scheduler of a real-time kernel may offer various scheduling policies, such as FIFO and round robin fixed priority, rate monotonic, earliest deadline first, Locke's best-effort[4] [7], capacity reservation, and calendar scheduling. While fixed priority is by far the most common, the other policies are useful in a variety of real-time scenarios. In addition, real-time kernels are usually preemptable with some degree of granularity, depending on the techniques used (either locks or explicit preemption points).

Unfortunately, popular network implementations (e.g., NetBSD) ignore much of this real-time scheduling support. For instance, their use of socket queues for outbound traffic defeats any notion of priority. Similarly, applications are accustomed to an interface oblivious to priorities, such as "select", for inbound traffic.

A primary objective for CORDS is to leverage the existing real-time scheduling support in the host OS for path scheduling and Application Programming Interfaces (CORDS does not mandate a specific API, like sockets). CORDS is multithreaded and employs a "one thread per message" model; in some cases (i.e., when using kernel threads), such a model is relaxed in order to minimize the number of context switches, and the same thread serves a number of packets within a unique path. While output packets are usually processed using the context of a thread "stolen" from the CORDS client application, input packets rely upon CORDS threads for upstream processing all the way to the application.

CORDS allows applications and protocols to select the number of threads and their scheduling attributes (policy and priority) for an individual path. For these selections to be effective, they need to apply as soon as packet processing begins. In the in-kernel portion of CORDS, for instance, we decide within the host device interrupt handler which thread to wake up. This allows us to avoid unnecessary context switches (from a generic thread to the right thread for this path) and we avoid processing any privileged packets with improper generic or unidentified scheduling attributes. For this reason, our model calls for a lightweight packet filter that is integrated within the device driver and provides path demultiplexing. Once the proper thread for a path begins executing, it pulls the incoming packet from a queue where it was left by the interrupt handler. Every path has its own input queue, which is operated with a FIFO policy. If, in spite of managing system resources, episodes of overload occur, we have two options to deal with the overrun of the input queue: as part of the interface to a path, it is possible to specify whether to dispose of the most-recently-arrived or the longest-lived member of the queue. The latter choice is well suited to applications that exchange data packets whose value decays over time.

4. This notion of "best-effort" is different from the "best-effort" used in communication literature to indicate unprivileged traffic.

It is worth noting that the framework does not impose any form of queueing on outbound or inbound traffic: such queues could generate episodes of priority inversion. The sole exception to this model is the input queue mentioned above: this queue, however, is path specific and collects traffic with homogeneous attributes (same path, hence same priority and privileges). As for protocols, it is up to the protocol implementor to pursue priority-inversion-free algorithms.

Given two or more paths whose threads have different scheduling attributes, the problem of preemption arises. CORDS is SMP- and preemption-safe; locking, however, is still coarse, and we have several potential points of priority inversion. We can deal with these using traditional priority boosting mechanisms.

With paths and their interface to threads and scheduling, as described above, we have the opportunity to implement path-centric optimizations for multiprocessing on UMA/NUMA machines, such as affinity path scheduling and path-wise locking. The former is intended to optimize cacheing and cache affinity; the latter aims to minimize lock contention by locking whole paths, rather than locking the individual protocol objects. In another x-kernel-based environment, both these approaches have been thoroughly analyzed [14][18].

We are also interested in exploring the notion of the utility function of a path. This function quantifies the relative marginal benefit of receipt of the next packet on a particular path. As a consequence of an end-to-end protocol based on priority encoding (see Section 4.1), we may encounter paths that achieve most of utility with, say, only 60% of fragments received. It would be nice to factor the path's utility function into the thread scheduling function that some scheduling policies (such as [7]) define: should overload occur, the paths that have accomplished a satisfactory portion of their mission would have their priority dropped, with consequent withdrawal from the competition for further CPU cycles.

4.3. From paths to network bandwidth

As we have seen in the previous two sections, programming with memory and CPU resources builds a strong case for a demultiplexing layer, or packet filter, in input: this facility must operate on all the input traffic, with the goal of isolating paths and their system requirements. The packet filter ought to be located as low as possible in the protocol stack. Ideally, we would like the packet filter to be integrated in the link protocol and the Media Access Control (MAC): this way, the network device itself is proactive in distinguishing between privileged and best-effort traffic. Unfortunately, programmable network devices are not common; in most cases, we must resort to layering the packet filter at the host driver level, with the critical portion of the filter executing within the driver interrupt handler.

Packet filters are wellknown in the literature [10][2]. In the real-time realm, a key requirement for packet filters is the ability to introduce the same latency regardless of the size of the filter function's range (in our case, the number of established paths). For pure development purposes, we have implemented the simplest version of a packet filter: we overload the protocol type field of a ISO 8802-2 header with non-standard values that mark privileged traffic and, within privileged traffic, which path the incoming packet belongs to. Our primitive packet filter is intrusive and, among other things, prevents us from interoperating with protocol stacks that are not based on CORDS (i.e., within routers); nevertheless, such a packet filter allows us to evaluate the utility of paths and let us measure the performance of the resulting system.

So far, we have addressed problems concerning incoming traffic and input queues. For output traffic, we need the capability for a path to advertise the shape of its traffic (i.e., MTUs per second). With the aggregation of shapes of all paths, the system can perform admission control tests at path setup time (e.g., is the system able to satisfy all these guarantees?), and it can schedule I/O according to paths' priorities. To this extent, each path can be terminated in output with a leaky bucket [16] that is regulated to inject data at a rate monitored by the system.

Aperiodic traffic patterns, such as most of the RPCs used by the distributed applications in our design target, are hostile to the notion of traffic shape. We give a path with aperiodic traffic the opportunity to send immediately, regardless of the backlog of network packets queued at the host. Due to the aperiodic nature of the traffic, the limited payload bay (i.e., hundreds of bytes) of our RPCs and the fairly small number of paths configured in our systems, we have not yet observed conflicts among aperiodic paths even with the slowest network controllers (i.e., 10 Mb/s ethernet cards).

Similar to what we have done for input traffic, we would like to influence the handling of outgoing data by network controllers through powerful interfaces to controllers and/or micro-programming of the controllers. There can be, in fact, several queues interposed between the transmission from the host and the emission from a network controller: Network Interface Controller queues, ATM framer queues, etc. Without control, any of these queues may be responsible for priority inversions. As we have observed in the case of input traffic, it is rather uncommon to gain this type of access with modern network controllers.

Ultimately, the link protocol itself must be friendly to the notion of privileged traffic. Ethernet's CSMA/CD is unable to handle privileged traffic: upon collisions, in fact, the emission is re-scheduled with an exponential back-off. Researchers have complemented CSMA/CD with variations of the link protocol that take privileged traffic into account (e.g., [17]).

Conversely, FDDI gives developers an easier handle on privileged traffic: through changes to station management variables, such as token rotation, it is possible to control the behavior of the link protocol. For ATM, researchers are still investigating the use of the CLP bit for a bi-modal priority scheme.

5. Path set-up

In general, paths are set up by the CORDS top anchor protocol upon request from the CORDS client applications: the top anchor protocol is, in fact, responsible for connecting to the CORDS abstractions and, among other things, it maps an application's QoS classes into paths and requests for resources. The protocols underneath the top anchor protocol are free to make additional resource requests on the paths passed to them. Examples of protocols making such requests include TCP (congestion control facilities trigger resource reservations), IPv6 and RSVP (flow set-up across routers and switches), and RTP (standard encoding of real-time flow information in data packets).

The operations of defining a path and committing system resources to a path are performed in several steps. An arbitrary fraction of these steps (optionally, all steps) can be specified at compile time through a description language: as part of the CORDS build process, utilities interpret the description language and generate code to be executed at CORDS boot time. The remaining steps are performed at arbitrary times, after the protocol graph is set up and operational.

The first step is to allocate a path structure and a unique path ID. The path ID space can be managed in different ways. For example, it can be statically partitioned among applications, and individual IDs are assigned through application-specific end-to-end protocols. Conversely, the path ID space may an exclusive resource of a node; facilities are provided to map incoming path IDs into local path IDs, and vice versa (much like VCI pairs in an ATM hop). The same mapping considerations for trans-node paths apply to those paths that cross protection domain boundaries local to a node (from user to user or from kernel to user): paths' continuity across boundaries are guaranteed via the path ID.

The second step is to populate the newly created path with capabilities to system resources. In case of memory, for instance, we have seen how we must specify exterior and interior allocators that produce the desired set of policies, mechanisms and failure semantics.

The third step is to instantiate system resources for exclusive use of the path. Given the system resources selected in the previous step, we proceed to reserve memory (or to pre-allocate memory, depending on the type of memory that the path manipulates), to spawn threads and to set-up the packet filter and the output I/O slots (if applicable).

At this point, the path is operational. As a new output message is generated, the message is bound to a path (a "default" path addresses the needs of unprivileged, uncharacterized traffic): such a binding will accompany the message in its journey through the protocol graph, from top to bottom. Should the message require additional system resources (i.e., expand the message structure with more data or meta-data), the binding to the path leads to the correct set of resources. At the receiving end, the binding between message and path is restored by the packet filter in input.

A CORDS object (session) may need to allocate system resources (i.e., meta-data) when there are no current messages available to extract the path from. In order to sustain the presence of paths along the whole protocol graph, even in absence of messages, we bind every object to the path used for its creation. Due to the multiplexing of several upper objects into a single lower object, it may happen that an object processes messages whose path differs from its own path. It is up to the protocol developers to handle this potential discrepancy in a protocol-sensitive way.

6. Experimental results

We have observed UDP/IP round-trip times between two nodes[5] directly connected by a private ethernet link. From the density function of UDP/IP round-trip times, we can analyze both maximum delivery time and delivery time jitter. The UDP/IP round-trips play the role of the synchronization handshakes (RPCs) which nodes will perform within a mission critical control system.

UDP/IP round-trips take place in the presence of background traffic, such as heavy host workload and sustained network traffic (memory-to-memory bulk data transfers). We are particularly interested in variations of the density function of UDP/IP round-trips, depending on whether or not a path with privileged system resources is used to protect UDP/IP round-trips from the interference of background traffic. For convenience of testing, the UDP/IP round-trips as well as the background traffic are generated by a CORDS protocol graph that runs in the OSF MK kernel; data samples are retrieved from the kernel in a non-intrusive way.

The interference between UDP/IP round-trips and background traffic has two components: a host component and a network link component. Our path implementation must address and minimize the host component, with the use of thread scheduling attributes, memory and I/O privileges. As for the network link component, our action is severely limited by ethernet's CSMA/CD: the presence of collisions prevents us from finalizing our analysis of density functions into a reasonable upper bound guarantee for an UDP/IP round-trip time.

In both Figure 4a and 4b, the solid line defines the density function of UDP/IP round-trip samples when a path with privileged system resources is used (pre-allocated memory buffers, highest priority for FIFO fixed priority scheduling, and highest I/O privilege). Conversely, the dotted line represents the density function of UDP/IP round-trip samples when the UDP/IP round-trips do not take advantage of paths. The sole difference between the tests captured in Figures 4a and 4b is the amount of background traffic (respectively two and four concurrent 16K memory to memory transfers initiated and sustained from both nodes). The experimental results show that the use of paths produces a much more favorable density function, i.e., the delivery time jitter is significantly reduced. In addition, the results show that, in fact, paths shield the UDP/IP round-trips from background traffic (minimal variations to the solid line between the two tests).

The overhead imposed by using paths is limited to the latency of the packet filter which eagerly demultiplexes incoming traffic. Due to the minimal packet filter presented in Section 4.3, the path overhead that we measured is negligible.

7. Related work

Our work with system resources is largely orthogonal to research programs that focus on real-time communication channels (e.g., [4][6]). With CORDS, its notion of paths and an underlying real-time OS (such as OSF MK), we are well positioned to import protocols and algorithms from those research programs in order to exploit real-time communication over multi-hop networks.

5. 486s 50 Mhz; clock resolution: 1 usec.

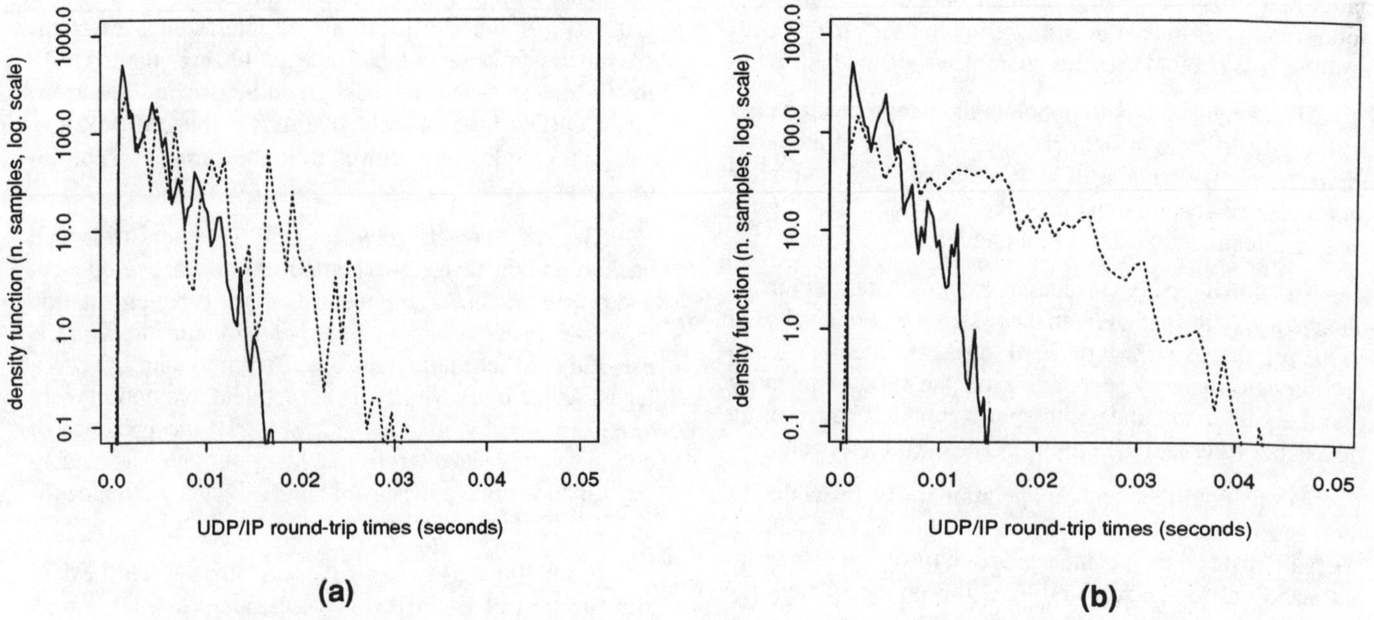

Figure 4: Density functions of UDP/IP round-trip times: *the use of paths with privileged resources (solid lines) limits interference with simulated background traffic (two (a) and four (b) concurrent 16K memory to memory transfers initiated and sustained from both nodes).*

8. Summary and conclusions

Our O-O Communication Subsystem, CORDS, establishes a new programming dimension: programming with system resources, such as memory, CPU and network bandwidth. Through a first class abstraction, the path, it is possible to select and commit system resources for the exclusive use of individual communication channels. In addition, CORDS' framework conforms to real-time design principles: meta-data can be eagerly allocated and re-used; queues responsible for priority inversions have been eliminated and framework invocations have bounded execution time. Preliminary experimental results show CORDS' high potential for addressing the needs of hard real-time distributed applications. The programmable control over system resources offers several opportunities outside of the hard real-time realm: from soft real-time to receiver-based congestion control, live group membership, protection from denial of service attacks, parallel execution on UMA/NUMA architectures, and multi-hop networks.

Acknowledgments

The authors would like to thank Prof. Larry L. Peterson and his colleagues at the University of Arizona who originally developed the x-kernel and who worked with us to enhance its support for real-time systems. The authors would also like to thank our colleagues at the OSF Research Institute whose contributions were many and invaluable.

References

[1] M. Accetta, R. Baron, W. Bolosky, D. Golub, R. Rashid, A. Tevanian, and M. Young, "Mach: A New Foundation for UNIX Development," *Proceedings of the 1986 Summer USENIX Conference*, Atlanta, GA, 1986.

[2] M.L.Bailey, B.Gopal, P.Sarkar, M.A.Pagels and L.L.Peterson, "PathFinder: A Pattern-Based Packet Classifier," *Proceedings of the 1st Symposium on Operating System Design and Implementation*, November 1994.

[3] N. Boden et al., "Myrinet: a Gigabit-per-Second Local Area Network," *IEEE Micro*, pp. 29-36, February 1995.

[4] D.Ferrari, "Real-Time Communication in an Internetwork," *Journal of High Speed Networks,* Vol. 1, No. 1, pp. 78-103.

[5] N.C.Hutchinson and L.L.Peterson, "The x-kernel: An architecture for implementing network protocols," *IEEE Trans. on Software Eng.*, 17(1):64-76, Jan. 1991.

[6] D.D. Kandlur, K.G.Shin and D.Ferrari, "Real-time Communication in Multi-hop Networks," *IEEE ICDCS-11*, May 1991.

[7] C.D.Locke, "Best-Effort Decision Making for Real-Time Scheduling," *Doctoral thesis, CMU-CS-86-134*, Carnegie Mellon University, 1986.

[8] S.W.O'Malley and L.L.Peterson, "A dynamic network architecture," ACM Trans. Comput. Syst., 10(2):110-143, May 1992.

[9] E. Menze, F. Travostino, "The x-kernel-based communication subsystem for OSF Mach: changes to the x-kernel 3.2 API," *OSF Research Institute TR*, May 1995.

[10] J.C.Mogul, R.F.Rashid, M.J. Accetta, "The Packet Filter: An Efficient Mechanism for User-Level Network Code," *Proceedings of the 11th Symposium on Operating System Principles*, November 1987.

[11] L.L.Peterson, "Resource Allocation in the x-kernel (Working Draft)," *Unpublished Draft*, November 1993.

[12] F.Reynolds, "The OSF Real-Time Micro-kernel," *OSF Research Institute TR*, July 1995.

[13] D.M. Ritchie, "A stream input-output system," *AT&T Bell Laboratories Technical Journal*, 63(8):311-324, Oct. 1984.

[14] J.D.Salehi,J.F.Kurose and D.Towsley, "The Performance Impact of Scheduling for Cache Affinity in Parallel Network Processing," IEEE 4th International Symposium on High-Performance Distributed Computing, Pentagon City, VA, Aug. 1995.

[15] F.Travostino, F.Reynolds, "An O-O Communication Subsystem for Real-Time Distributed Mach," *IEEE Proceedings of the Workshop on Object-Oriented Real-Time Dependable Systems*, Dana Point, CA, 1994.

[16] J.S.Turner, "New directions in Communications (or Which Way to the Information Age)," *IEEE Communications*, Vol. 24, No. 10, October 1986, pp. 8-15.

[17] C.Venkatramani, T.Chiueh, "Design, Implementation, and Evaluation of a Software-based Real-Time Ethernet Protocol," *Proceedings of ACM SIGCOMM '95*, Cambridge, MA, 1995.

[18] D.J.Yates, E.M.Nahum, J.F.Kurose and D.Towsley, "Networking Support for Large Scale Multiprocessor Servers," *Proceedings of the ACM Sigmetrics Conference on Measurement and Modeling of Computer Systems,* Philadelphia, PA, May 1996

Trademarks

A Client-Server Approach to DCE Inter-operability

Patrick M. Hurley
Rome Laboratory
RL/C3AB 525 Brooks Rd.
Rome, New York 13441
hurleyp@rl.af.mil

Terrance A. Stedman
Rome Laboratory
RL/C3AB 525 Brooks Rd.
Rome, New York 13441
stedmant@rl.af.mil

Abstract

Advances in distributed computing technology have lead to the development of numerous Distributed Computing Environments (DCEs). Most support a common set of services necessary for implementing distributed applications, while also featuring some unique properties or additional services which distinguish them from the rest. Each product is well suited to serve the needs of application developers within the constraints of a specific and limited problem domain; however, none can claim to be the most effective and efficient choice for all applications. The goal of this project was to investigate the feasibility of combining the attributes of multiple DCEs. This affords the application developer the choice of using the best features and services from any of the constituent DCEs. This paper describes the client-server approach that was used to accomplish this goal.

1. Introduction

Distributed computing environments (DCEs) support several key attributes that are essential for the development and execution of Command and Control (C2) applications. These attributes include heterogeneity, resource availability, concurrent programming, fault tolerance, and resource management. Many DCEs claim to support all or some of these key attributes. In reality, however, no one DCE can easily and efficiently address all the needs of every application. The needs of large applications that encompass a diverse set of requirements, such as C2 applications, are especially difficult to satisfy with any one DCE. The reason for this is that many DCEs are designed using vastly different design methodologies that make them better suited for different types of problems.

The Cronus/Isis (CRISIS) project is an attempt to combine attributes from multiple DCEs to allow the application designer more freedom when designing complex applications. Specifically, the objective of the CRISIS project is to design and demonstrate the inter-operability of two structurally different DCEs. This was accomplished by building an application that spans two mature yet different DCEs (Cronus and Isis). The rationale for doing this was to utilize the strengths of each distributed computing environment, where a certain DCE's design methodology is better suited for that part of the application.

This paper is organized as follows: In section 2 a Cronus overview is presented. An overview of the Isis distributed computing environment is presented in section 3. Section 4 compares several mechanisms provided by Cronus and Isis. The CRISIS architecture and two examples are discussed in section 5. Section 6 summarizes the CRISIS work as well as how it benefits current and future technologies.

2. Cronus overview

Cronus is a distributed computing environment that supports heterogeneous computer systems interconnected on a high-speed local area network (LAN) or wide area network (WAN) [1]. Cronus was developed by BBN Systems and Technologies Incorporated to support distributed command and control applications, under the sponsorship of the U.S. Air Force (Rome Laboratory). The present version of Cronus provides support for such diverse systems as Sun workstations running Sun UNIX, DEC machines running VMS or ULTRIX, HP machines running HP-UX, and several parallel machine architectures. Cronus currently supports applications written in the following languages: C, C++, FORTRAN, and Common Lisp.

The Cronus distributed computing environment is based on the *object model*. An object consists of state information maintained in an object database and a collection of rules that govern how this state information may be examined or changed. Each rule represents an operation on the object. Objects and their associated operations are managed by object managers. Operations on objects are invoked by client programs or by other object managers.

Cronus object types define how the objects are to be used and implemented. Types are made up of operation code, operation interfaces, and data structures that specify the representation of the various objects. Types are also placed in a hierarchy structure that allows new types to be created as subtypes of existing ones.

Cronus consists of services, clients, and the Cronus kernel. Services (a service consists of one or more object managers) implement both system and application functions. Current system services provided by Cronus include an authentication service, a symbolic naming service (global), a network configuration service, a directory service, and an object type definition service. Clients within Cronus are processes that use services. The Cronus kernel itself resides above the native operating system and is primarily responsible for transmitting synchronous or asynchronous operation invocations from clients to services. The Cronus kernel makes the network appear transparent. For example, a client on one host in a given Cronus configuration can invoke an operation on an object type whose manager resides on another host in the network (Figure 1).

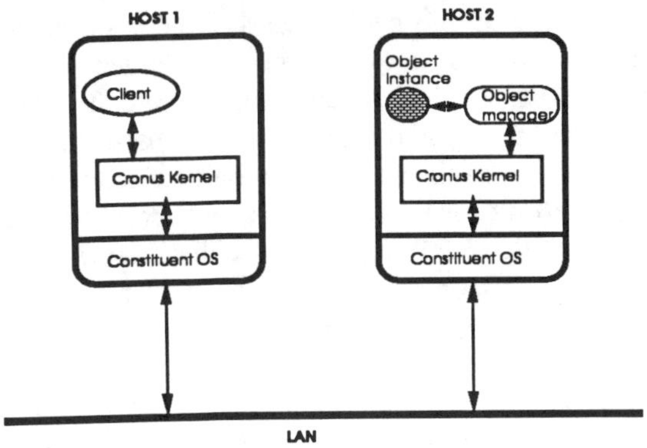

Figure 1. Cronus Communication.

By using an object-oriented approach, Cronus is able to provide a variety of high level abstractions such as: (1) Management of replicated data (on disk); (2) Parallelism, by splitting a computation among several machines; (3) Monitoring and reporting the status of a computation; (4) Dynamic reconfiguration from failures; (5) Support for multi-domained applications; and (6) Support for multi-clustered networks.

3. Isis overview

Isis is a distributed computing environment that provides high level tools which support the development of fast reliable distributed applications. Isis provides mechanisms that support heterogeneous computer systems interconnected on a high-speed local area network (LAN) or wide area network (WAN). Isis was sponsored in part by the Defense Advanced Research Projects Agency (DARPA) and developed at Cornell University, Ithaca, New York. It is currently available commercially through Isis Distributed Systems, Inc.

The Isis distributed computing environment is based on the concept of *process groups* [4][5]. Process groups are a lightweight programming construct. A single process can belong to any number of groups and there is minimal overhead associated with joining and leaving groups. Groups have a hierarchical namespace, much like a file system namespace, and permit flexible, location transparent addressing. Process groups are also capable of spanning across multiple machines. Isis provides mechanisms for communicating atomically with a group, as one might do to inform its members of some event, or to issue them a request of some sort. Such a communication takes the form of a *multicast*, in which one or all the members of the group receive the message, and zero or more respond (depending on the needs of the particular application) [3].

An important strength of the Isis environment lies in its support for implementing *virtual synchrony* [2]. Virtual synchrony permits the programmer to design a distributed program for execution in a simplified environment, wherein all processes observe events simultaneously and therefore in the same order. Events such as communication with a group or detection of failures are atomic in a virtual synchronous setting: all group members receive a message (or observe failures) if any does [3].

Using process groups and virtual synchrony, Isis is able to provide a variety of high level abstractions such as: (1) Management of replicated data (in memory or on disk); (2) Parallelism by splitting a computation among several machines; (3) Coordination of an external action; (4) Synchronization of concurrent actions (such as when several processes share a resource that only one can use at a time); (5) Monitoring the status of a computation, process or computer, and triggering user-programmed defined actions should the status change; and (6) Dynamic reconfiguration from failures to include the integration of a recovered machine into the operational system, restarting of the services that should run at that location and the ability to bring the services up-to-date concerning the active state of the system [3].

The Isis distributed computing environment currently supports applications written in C, C++, FORTRAN, ADA and Common Lisp. Isis runs on (and between) SUN, DEC, HP, GOULD, NEXT, and APOLLO equipment, and currently requires the UNIX or MACH operating system. However, ports to other operating systems such as AIX and VMS are being considered.

4. Cronus and Isis mechanisms

Cronus and Isis provide support for several key distributed computing attributes. These attributes include distributed processing, fault tolerance, real time, limited adaptive resource management, heterogeneity (to include languages, operating systems and machine architectures) and support for multi-domained applications. In the paragraphs that follow each of these characteristics will be described further, and the extent to which Cronus or Isis provide inherent support for these capabilities will be compared.

4.1. Distributed processing

Distributed processing may be defined as concurrent processes cooperating towards common goals where each process is likely to have incomplete and/or inconsistent global state information. Distributed processing provides many desirable attributes to include: *Redundant Processing* - several processes performing the same task perhaps on different data; *Concurrent Processing* - work is divided among several processes, each of which will contribute to the final result; *Resource Sharing* - sharing data, memory, CPUs, disks, and other physical devices for a common goal, or because a host is deficient in the resources it requires; *Reliable Communication* - reliable communication between processes whether they are on the same host or on different hosts connected by LANs and WANs; and *Heterogeneity* - the inter-operability of languages, operating systems and machine architectures.

Cronus and Isis both support the attributes of distributed processing to some degree. However, different characteristics are evident due to the vastly different design methodologies of Cronus and Isis. For example, both systems provide reliable communications between processes but a characteristic of Isis is to also preserve message order. Therefore, if your application needs reliable ordered communications, Isis should be used because Cronus provides no inherent mechanisms to preserve order.

4.2. Fault tolerance

Fault tolerance may be defined as the ability of a system or component to perform its function, despite the presence of hardware or software faults. Fault tolerant mechanisms must also be able to detect and/or recover from hardware and software faults. Faults fall into three classes hardware, software, and communication. Two examples of hardware faults are: (1) the complete loss of a host/service; and (2) degradation in the performance of a host/service due to overload conditions. Software faults include the following types of faults: (1) algorithmic faults; (2) software component faults (service loss); and (3) timing faults. Finally, two examples of communication faults are: (1) loss of connectivity (local area network or wide area network); and (2) overloaded or congested communications.

Cronus and Isis both provide and/or support a limited number of fault tolerant mechanisms. *Replication* - Replication is defined as multiple copies of a resource maintained on different hosts to improve availability or survivability; *Triple Modular Redundancy (TMR)* - TMR is defined as three processors running the same code on the same data and voting on the result to mask out an error by any one processor; and *N-Version Programming* - N-Version Programming is defined as N versions of a program (using different algorithms) operating on the same data and voting on the results. The only one of these fault tolerant mechanisms that is directly supported by Cronus and Isis is replication. The other mechanisms, however, can be supported but are more application specific.

Both Cronus and Isis directly support replication. The replication mechanisms within Isis are more flexible and dynamic than those found within Cronus. For example, Isis requires no pre-defined and pre-compiled read or write quorum as does Cronus. However, Cronus replication mechanisms are better able to maintain consistency during and after a network partition. This is because Cronus does have pre-defined read and write quorums. If a distributed application becomes partitioned due to a network failure, Cronus will allow the partition containing a majority of the application copies (i.e., managers) to continue execution pending a recovery from the network failure. Upon recovery of the network, Cronus provides guarantees to insure all copies of the application are brought up to the current (consistent) state. The choice to use Cronus or Isis once again depends on the needs of the particular application.

4.3. Real time

"Real time may be defined as a system or mode of operation in which computation is performed during the actual time that an external process occurs, so that the computation results can be used to control, monitor or respond in a timely manner to the external process" [8]. Using this definition, a system with real time response capability would be able to incorporate scheduling and resource decisions based on the current conditions of both system and environment. At present, neither Cronus nor Isis has the ability to comprehensively support such real time computations. However, if only part of an application has a real time requirement then an integration similar to the CRISIS integration, but with real time system support, would address this issue. This shows the strength of the CRISIS environment, i.e. an application programmer is not stuck with any one technology but can exploit different technologies to solve his problems.

4.4. Resource management

Adaptive resource management may be defined as the ability of a system to change its internal state to be consistent with its external environment. Adaptive resource management should allow for both the static mapping of application components within the system as well as the ability to dynamically react to changes in the runtime requirements that could not possibly be anticipated. Therefore, a base for adaptive resource management should be supported to improve throughput and provide some level of fault avoidance. This base should also support both static and dynamic migration of processes and data.

Adaptive resource management mechanisms can be used to automatically adapt to evolving resource availability. Therefore, in the event of failures, the system may automatically instantiate or migrate application functionality within and among the surviving system resources. Isis provides some support for adaptive resource management that includes “recycling” idle workstations for remote use and managing a pool of "compute servers". It can also manage a reliable replicated service by ensuring that some desired number of copies of the service are always running, despite machine failures. Cronus, on the other hand, provides no such mechanisms [9].

4.5. Multi-domained support

Support for multi-domained applications is also desirable. This would permit controlled access to resources (e.g., computers and data) in a flexible and efficient manner. Cronus supports multi-domained applications by a mechanism called *clusters*. A cluster is a set of hosts grouped together into a single administrative unit. Each cluster is autonomous, and therefore responsible for its own administration and control. No host is permitted to be a member of more than one cluster. Clusters allow boundaries to be erected between organizations, but can selectively allow or deny foreign clusters access to services that they support. This feature allows an administrator to selectively choose the services that will be exported from one Cronus cluster to another, thereby limiting the remote computing/communications burden. Isis process groups, by default, provide a much less sophisticated mechanism (compared to Cronus) to support multi-domained applications. Process groups differ from clusters in that a member of one group can join any number of other groups. Process groups also have very little access control for joining a group, and once they become a member they can gain access to all the group services.

5. The CRISIS architecture

The design goals of the CRISIS architecture are very straightforward. CRISIS must demonstrate the inter-operability of two structurally different DCEs (Cronus and Isis) in a form that is easy to use. In doing so, it must not inhibit the functionality of either DCE.

In order to accomplish this goal, CRISIS capitalized on the fact that both DCEs selected support the *client-server* model. In this model, clients make invocations to a population of servers which can be resident on several nodes in the system. The servers are passive entities waiting to service an invocation from a client. Therefore, by using this client-server model, an Isis client can be embedded inside a Cronus application to allow an Isis server to be called from Cronus. The reverse is also possible. A Cronus client can be embedded in an Isis application to allow a Cronus server to be called from Isis.

This is a fairly simple approach to the integration of two complex systems. However, this form of integration preserves the integrity of both DCEs and is fairly simple to use, thus meeting the design goals. A description of the testbed and two examples follow to illustrate the use of such a system.

5.1. The Rome Laboratory testbed

Rome Laboratory currently has an in-house distributed computer systems research, development and evaluation laboratory called the *Distributed Systems Environment* (DISE). Among the capabilities available within the DISE is the ability to demonstrate key distributed systems attributes utilizing various DCEs e.g. Cronus [1], Isis [4], and OSF/DCE [11].

The hardware configuration used consisted of three Sun workstations (SPARCstation 20’s) connected on an ethernet LAN. The operating system used was Solaris 1 (BSD v4.1.3). The DCEs used were Cronus v3.0 and Isis v3.0.8.

5.2. The inventory example

To test out the feasibility of the CRISIS architecture, a simple Cronus manager called the inventory manager was used. The inventory manager is a sample manager distributed with the Cronus software. It simulates an inventory control system which provides operations to create items, add and subtract from the quantity of an item, and delete items.

5.2.1. An Isis client within a Cronus manager. The first case is to embed Isis client code inside of a Cronus manager. Below is the canonical representation of the Cronus inventory item type.

cantype ITEM
representation is Item: record
Description: ASC
annote "A string that describes the item.";
Count: U16I
annote "The number of items in the inventory.";
end ITEM
annote "This datatype stores all the information about an item in the inventory.";

So, an inventory item is a record containing a description string (e.g. hammer) and an unsigned 16-bit integer representing the current quantity of the item in the inventory.

A typical operation on an inventory item is to add to the quantity of that item. This operation is defined in the inventory manager as a generic operation called "AddtoItem". This operation is defined in the inventory manager's typedef file in the manner shown below:

```
generic operation AddToItem(
   Description:  ASC;
   Add:      U16I;
)
returns(
   Count:  U16I;
   Description:  ASC;
)
   annote "This operation adds new stock to an inventory
item.";
```

So, operation AddToItem takes a description of an item and the number of these items to add to the database as input and outputs the new number of these items that are in stock and the description of the item as well. For example, if there are currently 5 hammers in the inventory and AddToItem("hammer", 5) is invoked, the new count of hammers in stock would be 10.

The AddToItem operation was originally written in Cronus. It was implemented in Isis to demonstrate the capability of Cronus to communicate with an Isis server. To do this, the portion of the Cronus AddtoItem operation that actually performed the add was stripped out and replaced with Isis client code. This client code makes a call to an Isis server to perform the add. Isis then returns the description and the new count to Cronus.

In order to make these Isis calls from Cronus, several include files are required in the Cronus operations code, as below:

```
#include "isis.h"
#include "isis_errno.h"
```

These are the standard include files in Isis server code as well.

Below is a pseudocode version of the Cronus AddtoItem operation with the embedded Isis client code.

```
AddtoItem(...)
begin
   retrieve item from Cronus object database;
   join Isis process group;
   call Isis task invAddToStock(item_name, count,
      addamt);
   leave Isis process group;
   update item count with result returned from Isis;
   store the updated object in the Cronus object database
end
```

Lastly, here is pseudocode for the Isis AddToStock task as defined in the Isis server code.

```
AddToStock(message)
begin
   get item_name, count, and addamt from message
      received from Cronus;
   add addamt to count;
   return a message to Cronus consisting of the
      item_name and the new count;
end
```

It should be readily apparent that the Isis AddToStock task simply adds an amount to a count and returns the new count and description. While this is a very trivial operation, it nevertheless demonstrates that a Cronus manager can issue a call to an Isis server and receive results back.

This example is executed by first starting the Cronus inventory manager and the Isis server. Using either a Cronus client or TROPIC (TRansportable Operation Interface for Cronus), the AddToItem operation can be invoked. For example, entering "on {inv} AddToItem hammer 10" through tropic will invoke the Cronus AddToItem operation which in turn will call the Isis AddToStock task.

5.2.2. A Cronus client within an Isis server. Next comes the second case: embedding Cronus client code inside of an Isis server. Demonstrating this case was slightly more difficult because the Cronus inventory manager had to first be converted into Isis server code. After that was accomplished, the AddToStock task as defined in the Isis server code was modified to call a Cronus operation to perform the add. This is quite similar to what was done in the first case.

Below is pseudocode for the Isis AddToStock routine with the embedded Cronus client code.

```
AddToStock(message)
begin
    get item from Isis store;
    invoke the Cronus operation AddToStock(item_name,
        count, addamt);
    update the inventory item in the Isis store with the
        new count;
end
```

In order for the Isis server to make Cronus calls, like the AddToStock call above, the following Cronus include files are required in the Isis server code file.

```
/* Cronus includes */
#include <stdio.h>
#include <cronus.h>
#include <getqual.h>
#include <futures.h>
#include "invatmgr.h"
```

Pseudocode for the Cronus AddToStock operation which is called by the Isis task AddToStock follows below.

```
AddToStock(...)
begin
    add addamt to count;
    return new count and item_name to Isis;
end
```

Again, this Cronus operation merely adds a numeric quantity to an inventory item and returns the new quantity and description to Isis. Regardless of its simplicity, it demonstrates that not only can Cronus managers communicate with Isis servers, but Isis servers can talk to Cronus managers as well.

This example is executed by starting the Isis server and the Cronus manager. By issuing Isis calls of the form

```
cbcast(group_addr, ADD, "%s%d", "hammer", 25, 1,
    "%d",answer);
```

the Isis AddToStock task is called which in turn calls the Cronus operation AddToStock to perform the add. The Cronus operation then returns the new quantity and description back to Isis.

Since communication is possible in both directions between Isis and Cronus, it seems likely that these two distinctly different DCEs can be used together to create a single distributed application. This distributed application could take advantage of the features unique to both Cronus and Isis, and thus enhance its capability.

5.3. The JDL example

The inventory example above proved the inter-operability of Cronus and Isis for a simple example. To test this inter-operability on a more complicated example, the Target Filter (targfil) Manager component of the Joint Directors of Laboratories (JDL) experiment was selected.

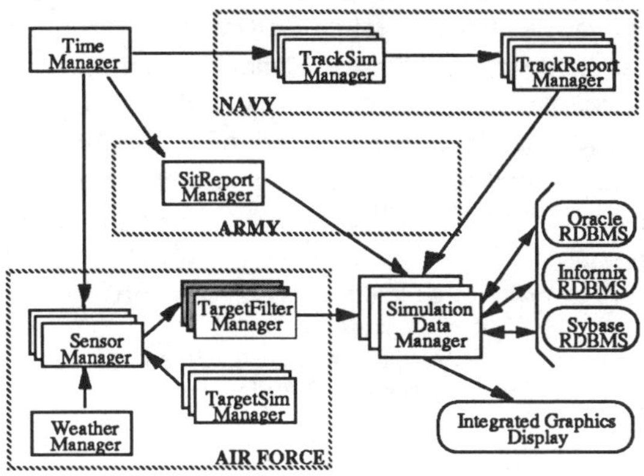

Figure 2. JDL Diagram.

The JDL experiment simulates a joint tracking and targeting system. Each service (Army, Navy, and Air Force) provides an application, which is integrated into the overall, distributed environment. Common software elements, user interfaces, data storage devices, and synchronization mechanisms are shared between the three services. The application is constructed of 10 components: synchronization (timing), target simulation, track simulation, track reporting, sensor simulations, target filtering, weather information, situation reporting, data management, and graphical interfaces (see Figure 2). The Target Filter manager is a Cronus manager that receives target detections from sensor managers, filters out false detections and sends real detections to a Simulation Data Manager.

Originally, all of the JDL managers were written in Cronus. The goal was to implement the Filter Detections operation as an Isis server and have the JDL simulation run as before. That is, the modification to the simulation would be transparent to the user and both Cronus managers and Isis servers would be used in the simulation.

It is important to note that the Cronus Filter Detections routine is not entirely replaced by the Isis server code. The header of the operation is left intact, along with the three standard Cronus operation parameters: a pointer to an operation independent header, a pointer to the operation's input arguments, and an output parameter for returning values. This was intentionally done to facilitate communication with the other Cronus managers in the simulation, most notably the Sensor Managers and SimData manager. If the entire Cronus Filter Detections routine was removed, the Sensor Manager code would have to be modified to be capable of calling the Isis Filter Detections task instead of the Cronus operation. The goal was to keep the modifications isolated to the Target Filter manager to minimize the impact on other JDL managers. Keeping a "stub" of the Cronus Filter Detections operation allows this to be done. The Sensor Managers communicate with the Cronus Filter Detections operation as before. The difference is that the "guts" of the Cronus Filter Detections operation have been removed and replaced with a call to the Isis Filter Detections task (see Figure 3).

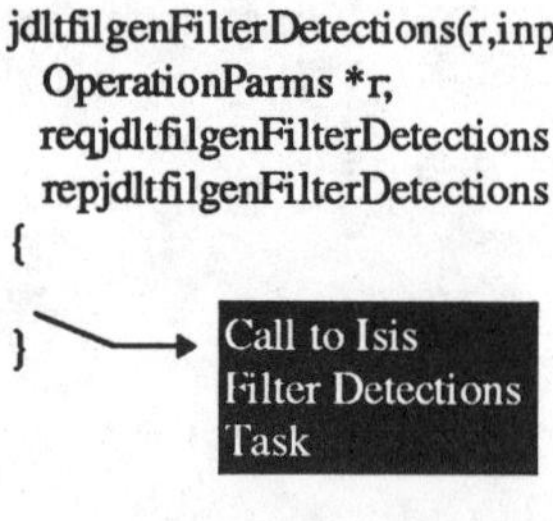

Figure 3. Cronus Stub.

The first part of this Cronus operation converts the array of DETECTIONDATA into an array that can be handled by Isis. The reason for this is the nested structures within the DETECTIONDATA structure. There is no easy way in Isis to pass an array of structures with nested structures. The workaround was to "flatten" the DETECTIONDATA structure. That is, all of the nested structures are removed and each field is inserted in a new IDD (Isis Detection Data) structure. So, the first part of the code copies the appropriate DETECTIONDATA information into the "flattened" IDD structure. An array of IDD structures can be passed to Isis with no problem at all. Once the array of IDD structures is passed to Isis, the Isis Filter Detections task converts it back into an array of DETECTIONDATA, filters out false target detections, and stores the results with the Cronus Simulation Data Manager.

The next section of the routine joins the Isis process group and issues the call to the Isis Filter Detections task shown below:

nresp=cbcast(group_addr,FILTER,"%X",Idd,
input->dimensions.Detections, ALL,"%d",&test);

Idd is the flattened array of DETECTIONDATA, and input->dimensions.Detections is the number of detections in the current sensor sweep. After this call, the rest of the routine simply leaves the Isis process group and frees memory allocations.

In summary, the Cronus Filter Detections routine converts the data into a form that can be used by Isis, joins the Isis process group, makes the call to Isis to perform the target filtering, and then leaves the Isis process group and frees memory.

The first part of the Isis FilterDetections routine converts the flattened array of IDD type back into an array of DETECTIONDATA, array "list". After this is done, the array of targets are filtered. New and/or modified targets are updated with the Cronus Simulation Data Manager. The actual update is performed by a call to the Cronus operation UpdateSimObject.

One important point needs to be made about the following line in the Isis server code:

i = jdlobj_declare();

Without this declaration, the CanTypeToLen call fails, and thus the Isis Filter Detections task fails as well. The reason this call is needed is that a DETECTIONDATA object is a subtype of type JDLOBJ which in turn is a subtype of type OBJECT. In the inventory example, an INVENTORY object was a direct subtype of OBJECT. To quote from the Cronus 3.0 Programmer's reference manual: "All cantypes defined by type object (obj), are already declared so there is no need to call obj_declare(). Whenever using routines that access any standard Cronus cantypes, the user must declare the cantypes first." This is accomplished above by calling the jdlobj_declare routine.

So, in summary, Isis received the array of detections from Cronus, filtered out the false detections, and made the appropriate Cronus calls to store the real target information in a Cronus object database. The entire JDL demonstration runs as if never modified. So the

Cronus/Isis integration is a success in this more complicated example as well.

The JDL example above shows the inter-operability of Cronus and Isis. However, it does not show how the unique features of Cronus and Isis were combined to produce a greater capability than either DCE alone. To do this, the Isis Resource Manager is configured to keep a minimum number of copies of the Isis FilterDetections service running at all times. This is superior to the Cronus replication mechanism because Isis will automatically restart the FilterDetections service even after failures occur to maintain the desired number of copies. With Cronus, manual intervention is required to restart a failed service.

6. Summary

The CRISIS project focuses on one serious limitation of today's distributed computing technologies. This limitation is that no one DCE can easily and efficiently address all the needs of every application. The reason for this is that many DCEs are based on vastly different design methodologies that make them better suited for different types of problems. With the technology available today, why should application designers settle for one particular DCE that does not totally fulfill their needs? The CRISIS project addresses this problem by using the client-server model to integrate two different DCEs. This method was fairly simplistic and preserved the integrity of both DCEs.

What is the future of CRISIS? It is useful for legacy software. For example, a Cronus developer can use existing Isis services in the development of applications, thus saving time and money. Similarly, existing Cronus services can be used in the development of Isis applications.

An environment like CRISIS would also be useful as the micro-kernel architecture matures. As DCEs are layered on top of the micro-kernel, CRISIS-like solutions can act as the glue to integrate them and allow them to inter-operate. An application designer could then use several different DCEs to solve a problem and not make the compromises that are associated with choosing just one DCE.

The Object Management Group (OMG) is currently developing a Common Object Request Broker Architecture (CORBA) specification that is trying to address this inter-operability concept. Both Cronus and Isis are undergoing development to become CORBA compliant. Hopefully CORBA compliant DCEs will eventually support heterogeneous distributed computing environments. Until then, efforts like CRISIS demonstrate that this inter-operability is attainable today without undue complexity.

7. References

[1] Berets, J.C., N. Cherniack, and R. Sands, "Introduction to Cronus", BBN Technical Report, Jan. 1993.

[2] Birman, Kenneth P. and Thomas A. Joseph, "Exploiting Virtual Synchrony in Distributed Systems", Proceedings of the 11th ACM Symposium on Operating Systems Principles, pp. 123-138, Nov. 1987.

[3] Birman, Kenneth P. and Keith Marzullo. "A Brief Overview of the ISIS Distributed Programming Toolkit and the Meta Distributed Operating System", March 1989.

[4] Birman, Kenneth P., "The Process Group Approach to Reliable Distributed Computing", Department of Computer Science, Cornell University, July 1991.

[5] Cooper, Robert C. B., Bradford B. Glade, Kenneth P. Birman, and Robert van Renesse, "Light-Weight Process Groups", Department of Computer Science, Cornell University, 1992.

[6] "Cronus Programmer's Reference Manual: Release 3.0", BBN Systems and Technologies, Dec. 1992.

[7] Hurley, Patrick M. and Scott M. Huse, "The Survivable Distributed Computing Environment", Rome Lab Technical Report RL-TR-94-29, June 1994.

[8] IEEE STD 610.12-1990, "IEEE Standard Glossary of Software Engineering Terminology", 1990.

[9] "Isis Distributed Toolkit User Guide and Reference Manual", Isis Distributed Systems Inc., 1992.

[10] "Annual Technical Report", Joint Directors of Laboratories (JDL) Tri-Service Distributed Technology Experiment Working Group, 1992.

[11] Johnson, Brad C. "A Distributed Computing Environment Framework: An OSF Perspective", OSF Technical Paper, June 1991.

Implementation of a Transparent RPC for Distributed Object-Oriented Systems

O.S. van Roosmalen and W.J.M. Lemmens

Department of Computing Science
Eindhoven University of Technology
e-mail:wsinonno@win.tue.nl

Abstract

In this paper we describe a remote procedure call (RPC) mechanism that is used to implement the object oriented programming language Deal. Deal (Dedos Application Language) is a C++-based language that supports real-time and concurrency and is developed in the context of the Dependable Distributed Operating System (Dedos) project at Eindhoven University of Technology. An essential aspect of the presented RPC mechanism is that it helps to hide distribution from the programmer during program construction and provides the Deal programmer with the abstraction of a single object space. The design of the (C++) classes that implement the RPC mechanism is described in detail.

1. Introduction

The work presented in this paper is performed in the context of the Dependable Distributed Operating System (Dedos) project that is being carried out at Eindhoven University of Technology [1]. It concerns a library of C++ classes that is employed to implement an RPC mechanism between software objects in a distributed system. This RPC mechanism was developed as part of the implementation of the C++-based concurrent real-time language Deal (Dedos Application Language) and the class library is to be used in conjunction with the Deal preprocessor. Although design of the mechanism is the topic of this paper, it is necessary that we describe Dedos and the development steps for its applications in more detail to give some understanding of the motivation for the development of the present RPC mechanism and the requirements on it.

The Dedos project is aimed at supporting the construction of real-time systems for process control and covers the implementation of an execution platform, the implementation of a programming language and the formulation of a design and implementation method for such systems. Dedos is intended as a test bed for methods and techniques in the field of dependable real-time computing. An important aspect of Dedos is the dual dependability paradigm: an application is divided in a hard real-time (HRT) part with critical real-time requirements and a soft real-time (SRT) part with more relaxed real-time constraints. Deal is intended to become a language that can be applied in both regimes. Essential in the HRT domain is that timeliness of that part of an application must be guaranteed. To accomplish this, system construction techniques are used that employ the results of static (off-line) analysis of on-line resource utilization. Often worst case resource usage of HRT tasks must be anticipated and a pessimistic resource reservation is necessary. In case HRT tasks do not require all reserved resources, such resources may be allocated to SRT tasks dynamically for as far as no unwanted interference with the HRT part results.

The desire to construct a concurrent (hard) real-time language that (1) has as high an abstraction level as non-real-time languages, (2) promotes reuse and extendibility of programs and (3) enables the off line analysis and scheduling of applications to guarantee timeliness, has resulted in the definition of the Deal language as described in [2]. The first and second desire are reflected in the choice to make Deal object-oriented and to construct programs without a priori knowledge of the hardware configuration. That is, object distribution over the hardware platform is not an issue dealt with at the programming stage. The third desire is taken into account by dividing application construction into two phases. (1) The programming phase: reusable program components (classes) are constructed that have a well defined (real-time) semantics. (2) The system generation phase. In this last phase all the program components are analyzed and a feasibility test is performed to inspect whether the semantics of the program can be implemented on the hardware configuration at hand. An object distribution and a schedule per processor are generated and the program components are linked into a set of loadable modules, one or more per processor. If the application is executed using the off-line generated schedules, the timeliness is guaranteed (under a particular fault hypothesis).

The following requirements were formulated for distribution and remote object communication in Deal.

0-8186-7570-5/96 $5.00 © 1996 IEEE

- Distribution transparency: programs with the contained class definitions should contain nor require any reference to object locations. The programmer must be able to view the system as a single space of communicating objects.
- Uniform object model: no distinction should exist between classes of objects that can be distributed and classes of objects that cannot. In principle an allocation decision must be possible for all objects.
- Suitability for hard real-time: implementation should be efficient, applications are closed and must be analyzable and schedulable off line. Some restrictions on C++ language features may be introduced if they do not affect the object-oriented nature of the language.

These requirements are different from those satisfied by other object communication paradigms. E.g. many approaches make a distinction between ***heavy-weight*** and ***light-weight*** objects, where the former type supports remote communication and can be executed concurrently [3]. In Deal no such distinction is made. A second difference with other approaches [4] is that Deal systems are closed. This is related to the third, most important, difference with other approaches: Deal is a real-time language and should enable off-line analysis to guarantee timeliness of applications.

2. Concurrency in Deal

Because distribution and concurrency are very dependent issues, concurrency in Deal is briefly described. Concurrency is introduced in Deal through the possibility to define special member functions, called activities. An activity implies the creation of a separate thread of control during the construction of a class instance. This thread starts its execution at the designated member function. More than one activity can be defined in a class. In addition every activity can (synchronously) invoke operations of objects that happen to be in scope. Two categories of objects can therefore be distinguished: local objects, which are only accessible by the one activity that created it, and global objects, which can possibly be accessed by more than one activity (intra-object concurrency). This concurrency model, which ties activities to objects, requires that a distinction is made between constructors for global objects and constructors for local ones: One cannot allow the start of an activity in a local object because local objects are removed from the system when a program block is exited, leaving the activity without a parent object.

Consistency of global objects is an issue that is left to the programmer. Deal offers some language concepts for concurrency control: objects can be declared atomic. This implies that invocations to the object are (virtually) serialized. To enable synchronization of timing of activities, special system classes, events (semaphore-like constructs) and guards (which are like condition variables) are introduced in Deal .

3. Language restrictions related to HRT

The requirement that HRT programs must be analyzable and schedulable off-line has some consequences for the language definition. Behavior of the activities must be extracted from a program and this behavior must be sufficiently predictable. Active objects, atomic objects and events are of crucial importance to activity behavior and information on these objects must be extracted for the scheduler. Also, resource usage must be efficient and predictable. Therefore, global objects (active objects, atomic objects, events are always global) are only created at system initialization. For HRT programs we have introduced the following restrictions on the Deal syntax:

- no loops with unbounded number of iterations,
- recursion is prohibited,
- objects in global scope are created with global-object constructors,
- no dynamic instantiation of objects except at program initialization: the C++ keyword `new` may not be employed, except in the global-object constructors. (As a consequence, normal constructors must receive all initialization values for pointer attributes via their parameter list.)

The last two restrictions are relevant for the distribution of objects.

4. Realization of distribution

In an HRT application it is neither efficient nor useful to allow for allocation of a local object on different processing nodes than the one where the creating object resides. Therefore, allocation decisions are made by the off-line scheduler only on global objects. Information on these global objects is generated by the Deal preprocessor and passed to the scheduler. The latter determines a global schedule and one local schedule per node. The allocation implied by the global schedule is used to generate one main file per node in which the objects allocated to the node are created, and stub classes for the remote objects are inserted. For every class a stub class is generated by the preprocessor that has an interface identical to the class of the remote object but implements the behavior by passing the request to the remote object. In each main file the remote objects are transparently replaced by stubs via the dynamic binding mechanism. Class definitions linked to a main file need not be altered and do not depend on the object distribution. Each main file is compiled and linked with the classes and stubclasses it requires to form a loadable module. Such a module creates the object space residing on a node, with light-weight activities (processes) as concurrent entities. All intra-node communication consists of efficient regular procedure calls.

The amount of remote (i.e. inter-node) communication within the system could possibly be reduced through caching of object information in stub-objects. Although efficiency is a requirement, we have decided not to employ caching for the sake of simplicity. All operations on an object are passed, via the stubs, to the object's location (i.e. the node where the object resides) and are performed on the actual object. The result of an operation is subsequently returned.

5. Distribution and language restrictions

The mentioned approach for realizing distribution is particularly simple to implement if some additional restrictions on the Deal language are introduced:

1. Simple types (e.g. char, int, float) and structs may not be global variables.
2. The use of pointers to simple types is prohibited.
3. Classes may not offer public attributes.
4. Public member functions are always virtual.

The first three restrictions are a consequence of the desire to make remote communication necessary only for function calls, not direct access of data. These restrictions coincide with the tendency in object-oriented programming to hide information in objects, keeping attributes private and only accessible via getter and setter functions. This proper object-oriented programming practice is enforced in Deal.

6. Properties of the underlying platform

Dedos and Deal are implemented on the Eindhoven Multi Processor System (EMPS) hardware and kernel [5]. The EMPS real-time kernel offers, among other things, a process model and primitives to implement a remote procedure call mechanism. E.g. it offers mailboxes for the handling of messages and system functions to send and receive both requests and replies via such mailboxes. EMPS processes are essentially light weight and may share address space. To obtain the Deal activity model each activity is mapped onto an EMPS process and all processes on each node share address space. Therefore, they act within the same object space. At each node a mailbox is kept that can store the requests intended for the node. This mailbox keeps track of the process pair involved in a particular request/reply instance. Replies can thus be returned to the remote process that sent the request. On each node there is a pool of EMPS processes that handle the requests deposited in the mailbox. Which one of the processes out of this pool handles the request is irrelevant. An activity that enters the node from a remote location is carried by an arbitrary process from the pool. When a process is finished with the remote request it returns to the pool of processes and awaits the arrival of a new message in the mailbox. The process pool is of fixed size to avoid the overhead of dynamic process creation. The off-line program analysis yields the required size of the pool for each node.

Deal is presently implemented through a preprocessor that delivers C++ programs. Typical Deal aspects are implemented through a class library and EMPS system calls. The present setup is meant to gain experience with the concepts proposed for Dedos and Deal in the realm of distributed real-time computing. It is our intention to construct a proper language and language compiler when the concepts and their implementation have been tested and are sufficiently stable.

7. Implementation of the RPC mechanism

In this section we describe the classes that implement the remote procedure calls. The Deal programmer will define classes that will have instances (objects) at run time. An object that invokes an operation of another object is referred to as the *client* object. The invoked one is called the *server object.* If the client and server reside on the same node, in the same address space, the invocation can be a simple procedure call. However, when this is not the case, the remote procedure call mechanism has to be applied. Before we go into a more detailed description of the design of this mechanism, we describe the approach in brief.

To enable communication between a client object and a remote server *x*, a so-called client-stub object (also called proxy) *xclient* is created at the client's node and a server-stub object *xserver* is created at the server's node. The *xserver* object is created as soon as the existence of the server object is announced to other nodes. This can either be at system initialization or when a reference to the object is part of the parameter list of a remote call. The client stub *xclient* is created when a node learns about the existence of the server x. Also here, this may happen at initialization or through the passing of references with remote calls. The objects *x*, *xclient*, and *xserver* all have the same interface signature so that the *xclient* can transparently replace *x*. Instead of implementing the functionality of *x* directly, *xclient* constructs an internode message that contains all information related to the call. This translation of the call-data into a message is called marshalling. The message is sent to the mailbox at the node where *x* resides. For a mailbox there is a handler object that allocates a process from the process pool to execute the request. Upon execution the message content is analyzed, and a normal procedure call to the server-stub object *xserver* is performed. This server stub invokes the appropriate member function of the actual server *x*. The invocation sequence is reversed on return from the call and a reply message is returned to the client-stub *xclient* which returns control to the client object from which the call originated. The client-stub object *xclient* keeps an identifier of the destination mailbox and the server object *x*. On an invocation it performs the following tasks:

- It packs the procedure call information (destination object, member function identification, parameters) into a message.
- If a reference to a near object *y* is to be passed in the message it finds or creates a server stub *yserver*.
- It sends the message to the remote mailbox and awaits the reply.
- When the reply message arrives it unpacks the message data.
- If a reference to a remote object *z* is returned with the reply, it finds or creates the client stub *zclient* and uses its pointer in the reply.
- It returns to the caller.

As is apparent from this description, object references may be passed between nodes. A method for identifying remote objects is required. We introduced the *universal reference* that can be used to refer to objects anywhere in the application. Essentially it consists of a mailbox identifier and an identification of the object on the remote node. Each node keeps track of the client stubs created on behalf of global remote objects and the corresponding universal references. Because all global objects and their classes are known at compile time, efficient methods can be devised for locating client stubs given a universal reference.

Typical actions taken by mailbox handler and the server stub *xserver* on arrival of a message are:

- One of the processes executing the handler is awoken to carry out the handler's procedure for the incoming message.
- The handler extracts the identification of the server-stub object *xserver* from the message. A standard procedure of the server stub is called. The remainder of the message is passed with the call.
- The standard procedure extracts the member function identification from the message and calls this member-function of *xserver.*
- The member function extracts the parameters from the message, finds or creates client stubs for pointer parameters to remote objects and makes the call to the actual destination function of the server object *x.*
- If a pointer to an object is the result of the call, its universal reference is inserted in the reply by *xserver* on return of the call to *x.*
- The reply message is returned via the mailbox by the mailbox handler.

The server-stub object fulfills a crucial role in unmarshalling the request data and marshalling the reply. Because the unmarshalling can only be carried out if there is sufficient information on the types of the parameters that are passed, it has to be carried out by an object that has information on the signature of the server object, i.e. the server-stub object. The server stub has the same members as the server object but does not use the parameters in the member's implementation.

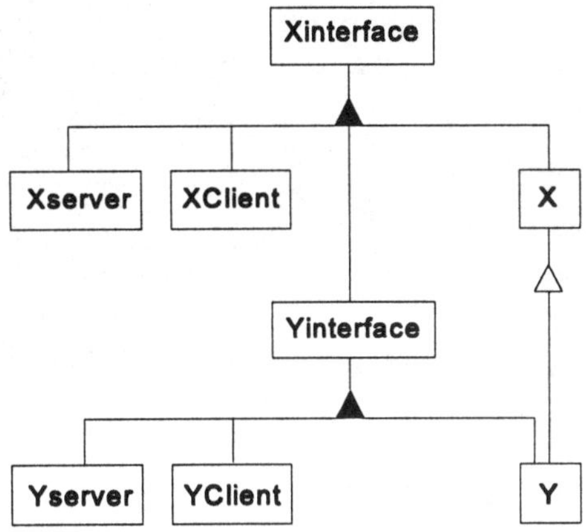

Figure 1

Some of the classes introduced to implement this RPC protocol are described in figures 1 and 2 using the OMT notation [6]. For every application class *X*, the preprocessor generates a number of classes as in Fig. 1. The class *Xinterface* is constructed which essentially contains the same public member functions as *X*, except that they are not implemented. The static type (class) of all pointers in the application is changed by the preprocessor into the corresponding interface class. The classes *Xserver* and *Xclient*, also generated by the preprocessor, are derived from *Xinterface.* The reason behind this is that now a pointer referring to a global object can be transparently changed to refer to a client stub. Thus, stubs can be inserted without altering the code of client object's class: an instance of *Xclient* replaces and instance of *X*, the server object, at the client node and an instance of *Xserver* handles the communication at the node of the server object. The inheritance graph is constructed as depicted in figure 1 to provide the client and server stubs with the same virtual-member tables for their interface sub-objects, so that they can exchange member-function identifiers using the pointer-to-member construct in C++ [7].

If an application class has descendants, like *X* has the derived class *Y* in Fig. 1, the inheritance chain is copied in the interface classes. This must be done to obtain an assignment compatibility between pointers of the interface classes that is identical to that for pointers of the original application-classes. Although many instances of repeated inheritance result because of this scheme, the dominance rules for function binding [7] are such that always the function implementations defined by the application classes are selected in calls to the application objects.

A more detailed diagram of the aforementioned classes is given in Fig. 2. Assuming there is an application class *X* with *dothis()* and *dothat()* public members, other classes will have

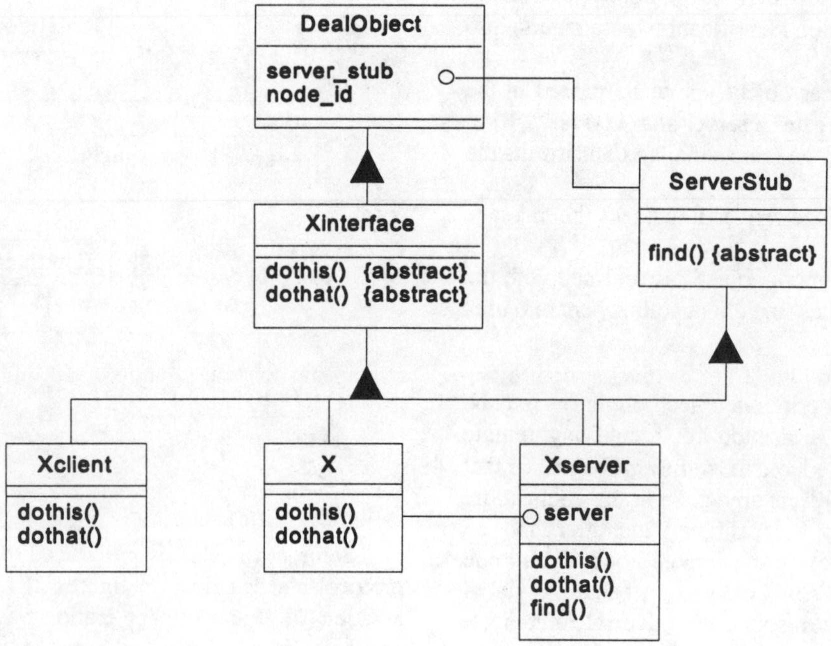

Figure 2

a structure as sketched in this figure. The classes *X*, *Xclient* and *Xserver* will inherit a universal-reference attribute from the class *DealObject*. This attribute consists of a node-id and a reference to a server stub. Instances of the class *X* and instances of the client-stub class, *Xclient*, will hold a reference to their corresponding server stub. The client stub will use the node id to obtain a mailbox port enabling communication with the server object. On every call the server-stub reference is passed with the call and the mailbox handler uses this reference to locate the server stub and to call the *find()* function that extracts the pointer-to-member and the parameter from the call message. The server stub holds a pointer to the server to which it delegates each call.

8. Conclusions

We have presented the design of the RPC implementation for the Deal language. This design is particularly tuned to the requirements on Deal programs. Important requirements are: efficiency, transparency and predictability of RPC implementation. Every object to which a pointer is obtained can, in principle, be at a remote location when called. We have used the efficiently implemented dynamic binding mechanism in C++ to resolve the desitination of each call. As a consequence only the dynamic binding overhead is incurred on every call.

The presented mechanism is currently being implemented on the Dedos platform, and has been tested to a large extend. Its performance depends very much on the facilities offered by this platform. We will publish some figures in a longer publication where we will also discuss other aspects of the implementation in more detail.

References

1. D.K Hammer *et al.*, "DEDOS: A Distributed Real-Time Environment", *IEEE Journal of Parallel and Distributed Technology*, Vol. 2, Nr.4, Winter 1994.
2. O.S. van Roosmalen, "Deal: an Object Oriented Language for Distributed Real-Time Systems", *Int. Workshop on Object-Oriented Real-Time Dependable Systems (WORDS)*, Dana Point, CA, October 1994.
3. Takashio et al. , "An Object-Oriented Programming Language for Distributed Real-Time Systems", *proceedings of OOPSLA'92, ACM Press* October 1992, p276.
4. A.B. Schill and M.U. Mock, "DC++: distributed object-oriented system support on top of OSF DCE", *Distr. Syst. Engin. 1(1993)112.*
5. G.J.W. van Dijk, "The Design of the EMPS Multiprocessor Executive for Distributed Computing", Ph.D. Thesis, *Eindhoven University of Technology,* Eindhoven, The Netherlands, 1993.
6. J. Rumbaug et al., "Object Oriented Modelling and Design", *Prentice-Hall*, Englewood Cliff, NY, 1991.
7. B. Stroustrup, "The C++ programming Language", second edition, *Addison-Wesley Publishing Company*, Reading Massachusetts, 1993.

The DREAM Library Support for PCD and RTO.k programming in C++

K. H. (Kane) Kim, Chittur Subbaraman, and Yuseok Kim
Dept. of Electrical and Computer Engineering
University of California, Irvine

Abstract: In order to realize real-time computing in the form of a generalization of non-real-time computing and yet allowing system engineers to confidently produce certifiable RTCS's for safety-critical applications, we have established a real-time object-oriented (OO) structuring approach called the RTO.k object structuring scheme in recent years. Also, a model of an operating system kernel which can support both conventional real-time processes and RTO.k objects with guaranteed timely services, has been formulated. The model has been named the DREAM kernel. The DREAM kernel supports three powerful building blocks of process-structured real-time distributed programs: real-time process, concurrent-read-&-exclusive-write (CREW) shared data structure monitor for intra-node inter-process communication, and data field channels (DFC's) which are real-time logical multicast channels for inter-node and intra-node inter-process-group communication. Programs composed of these components are called PCD programs. A DREAM library that consists of a collection of specific C++ classes provides a user-friendly interface to DREAM kernel services for real-time application programs. An implemented prototype of the DREAM kernel encapsulated by the DREAM library supports not only well-structured PCD programming but also efficient programming of RTO.k objects in C++. The essence of PCD and RTO.k object programming in C++ with the aid of the DREAM library is discussed.

Index terms: GT design, DREAM library, C++ PCD programming, C++ RTO.k programming

1. Introduction

A few years ago we became convinced that it was time to start vigorously pursuing the following idealistic real-time computing paradigms [Kim95b]:

(1) *General-form design style*: Future real-time computing must be realized in the form of a generalization of the non-real-time computing, rather than in a form looking like an esoteric specialization.

(2) *Design-time guarantee of timely service capabilities of subsystems*: To meet the demands of the general public on the assured reliability of future real-time computing systems (RTCS's) in safety-critical applications, it is inevitable to require the system engineer to produce design-time guarantees for timely service capabilities of various subsystems (which will take the form of objects in object-oriented system designs).

The motivating factors behind these paradigms which may be called the GT (General-form Timeliness-guaranteed) design are the newly improved hardware economy and component reliability which provide impetus in expanding the real-time computing application field.

To realize the idealistic GT design, a powerful structuring scheme capable of dealing with all practically useful real-time and non-real-time computing requirements must be established. In the last several years, there has been a growing trend of research activities aimed for extending the conventional OO-structuring approaches to support RTCS design [Att91, Ish92, Kim94b, Kop90, Tak92]. However, most of those works have not been aimed for supporting the design-time guarantee of timely service capabilities of objects, which is one of the fundamental parts of GT design. In recent years, the first co-author and his collaborating researchers together have established a real-time OO-structuring approach with the purpose of facilitating GT design. The new object structuring approach is called the RTO.k object structuring scheme [Kim94a, Kim94b]. The RTO.k object structuring scheme, though an extension of the conventional OO-structuring approaches, has several unique features of fundamental nature which will be reviewed later in Section 2.

Until a highly abstract programming language for RTO.k object structuring and efficient execution facilities for such objects become available, the most cost-effective approach to practicing the RTO.k object structuring appears to be to program RTO.k objects by use of conventional concurrent and distributed process structuring languages supported by real-time operating systems. The first co-author recently formulated a model of an operating system kernel which can support real-time processes with guaranteed timely services. The model is named the DREAM (Distributed Real-time Ever Available Micro-computing) kernel. The DREAM kernel supports three powerful building blocks of process-structured real-time distributed programs:
(1) real-time processes subject to various activation and synchronization requirements,
(2) shared data structure monitors called concurrent-read-&-exclusive-write (CREW) monitors &,
(3) real-time multicast logical (RML-) channels called data field channels (DFC's).
These three basic components fully represent the general facilities for concurrent and distributed program structuring. Programs composed of these three components are called PCD (process-CREW-DFC) programs. In fact, the DREAM kernel can support both process-structured real-time application software (i.e., PCD programs) and RTO.k object structured application

software. Components of an RTO.k object are treated as special types of PCD components. The key emphasis in formulating the DREAM kernel was in realization of guaranteed timely service capabilities with minimal loss of hardware utilization. The DREAM kernel can thus be viewed as a model of a general-purpose timeliness-guaranteed OS kernel.

In addition, the authors have designed a library of functions that eases not only PCD programming but also RTO.k object programming in C++. This library called the DREAM Library is a collection of several C++ classes, each class functioning as an interface between a PCD or RTO.k object program and a specific DREAM kernel service call (KSC) group. The DREAM library hides various details of parameter passing between application and the DREAM kernel and thus offers a user-friendly interface to the application programmer. An implemented prototype of the DREAM kernel encapsulated by the DREAM library supports not only well-structured PCD programming but also efficient programming of RTO.k objects in C++.

Several prototype implementations of the DREAM kernel, have been produced by the authors to run on networks of PC's connected by Ethernet. An earlier prototype kernel has been used to run an RTO.k structured non-trivial defense C^3 application, together with an RTO.k structured real-time simulator of the application environment.

The paper starts in Section 2 with a discussion on the essence of the RTO.k object structuring scheme. The major features of the DREAM kernel offered to application programmers are briefly presented in Section 3. The essence of PCD program structuring in C++ with the support of the DREAM library is discussed in Section 4. Section 5 then discusses the programming of RTO.k objects in C++ with the support of the DREAM library. The next section, Section 6, looks at the various advantages and limitations of the PCD based structuring scheme for RTO.k objects. The paper concludes in Section 7 with discussions on the issues to be resolved via future research.

2. Overview of the RTO.k Object Structuring Scheme

An initial abstract framework of the *RTO.k object model*, also called the time-triggered real-time object (TT-RTO) model, came out of the attempt by the first co-author and Hermann Kopetz at the Technical University of Vienna to find a proper extension of the basic object model which is highly cost-effective in development of *hard-real-time* application systems. Based on the initial abstract framework formulated in late 1980's [Kop90], a concrete syntactic structure and execution semantics was developed in recent years [Kim94a, Kim94b].

The basic structure of an RTO.k object is depicted in Figure 1. It is an extension of the conventional object

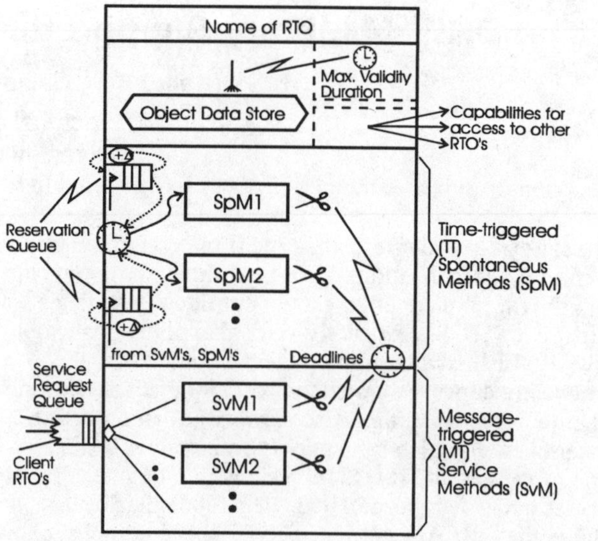

Figure 1. Structure of the RTO.k Object Model (Adapted from [Kim94b])

model(s) and four most important extensions are the following:

(a) *Two clearly separated groups of methods*:

For some methods of an RTO.k object, a real-time clock serves as the mechanism for triggering the method executions as the clock reaches some values specified at design time and such methods are called time-triggered (TT-) methods, also called the spontaneous methods (SpM's), and *clearly* separated from the conventional service methods (SvM's) triggered by messages from clients. The two types of methods in an RTO.k object are different not only in the way their executions are triggered but also in that "actions to be taken at real times *which can be determined at the design time* can appear only in SpM's".

Therefore, actions of the type "at constant-clock-value do S" or the type "sleep-until constant-clock-value" can appear only in SpM's. Incorporation of SpM's means introducing the potential for the following two new types of concurrent executions of object methods in addition to the potential for concurrent executions of SvM's that exist in conventional objects:

(Type I) Concurrency among SpM executions: This concurrency is specified in an implicit but natural manner, e.g., two SpM's designed to be triggered at 10 am.

(Type II) Concurrency between SpM executions and SvM executions.

(b) *Basic concurrency constraint (BCC)*:

In order to dramatically reduce the designer's efforts in guaranteeing timely service capabilities of RTO.k objects, the execution rule which prevents conflicts between SpM's and SvM's is incorporated. Basically, *activation of an SvM triggered by a message from an external client is allowed only when potentially conflicting SpM executions are not in place.* To be exact, when a message-triggered SvM is not free of conflict with an SpM in accessing the same portion of the object data space (ODS), execution of the former method (SvM) must not be allowed in a time zone earmarked for a TT-execution of the latter method (SpM). This restriction is called the basic concurrency constraint (BCC). Therefore, SpM's are given higher priorities for execution over the SvM's. Note that this BCC does not impose any restriction on concurrent execution of SpM's or concurrent execution of SvM's. Therefore, executions of SpM's are not disturbed by SvM executions and triggering times of SpM's are fixed at the design time. At least this makes it very easy to analyze the execution time behavior of SpM's. For example, if a statement of the type "at 10am do S" appears in an SpM, its reliable execution can be easily assured.

(c) For each execution of a method of an RTO.k object, a *deadline* is imposed.

(d) Real-time data contained in an RTO.k object become invalid after the interval called the maximum validity duration passes.

The first two features (a) and (b) mentioned above make the RTO.k object model clearly distinguished from other proposed real-time object models [Att91,Ish92, Tak92].

Triggering times for SpM's must be fully specified as constants during the design time. Those real-time constants appear in the first clause of an SpM specification called the autonomous activation condition (AAC) section. An example of an AAC is

"for t = from 10am to 10:50am every 30min
start-during (t, t+5min) finish-by t+10min"

which has the same effect as

{"start-during (10am, 10:05am)
finish-by 10:10am",
"start-during (10:30am, 10:35am)
finish-by 10:40am"}.

A provision is also made for making the AAC section of an SpM contain only candidate triggering times, not actual triggering times, so that a subset of the candidate triggering times indicated in the AAC section may be dynamically chosen for actual triggering. Such a dynamic selection occurs when an SvM within the same RTO.k object requests future executions of a specific SpM. The AAC specifying candidate triggering times rather than actual triggering times starts with a declaration "if-demanded".

An underlying design philosophy of the RTO.k object model is that an RTCS will always take the form of a network of RTO.k objects. RTO.k objects interact via calls by client objects for service methods in server objects. The caller may be an SpM or an SvM in the client object. In order to facilitate highly concurrent operations of client and server objects, non-blocking (sometimes called asynchronous) types of calls (i.e., service requests) can be made to SvM's.

The designer of each RTO.k object provides a guarantee of timely service capabilities of the object by indicating the *deadline for every output* produced by each SvM (and each SpM which may be executed on requests from SvM's) in the specification of the SvM (and some relevant SpM's) advertised to the designers of potential client objects. Before determining the deadline specification, the server object designer must convince himself/herself that with the object execution engine (hardware plus operating system) available, the server object can be implemented to always execute the SvM such that the output action is performed within the deadline. Again, the BCC contributes to major reduction of these burdens imposed on the designer.

The RTO.k object model is effective not only in the multiple-level abstraction of real-time (computer) control systems under design but also in the accurate representation and simulation of the application environments. In fact, it enables uniform structuring of control computer systems and application environment simulators [Kim94a] and this presents considerable potential benefits to the system engineers.

3. Major Features of the DREAM Kernel

An essential building-block for construction of timeliness-guaranteed real-time application systems is a *timeliness-guaranteed operating system* which together with the hardware platform forms an execution engine that provides guaranteed timely services to concurrent and distributed real-time application software. If the operating system lacks such guaranteed timely service capabilities, then timely service capabilities of real-time applications cannot be ensured. Existing commercial operating systems lack the capability for providing guaranteed timely services required by a great deal of real-time application software. The DREAM kernel has been formulated as a model of an operating system kernel that can support both PCD-based real-time processes and RTO.k objects with guaranteed timely services. An experimental prototype of the DREAM kernel was implemented in the DREAM laboratory of UC Irvine. The experimental version runs on a network of PC's connected via Ethernet and equipped with i486 or Pentium processors, DOS-BIOS device drivers, and the Packet Ethernet driver. This section briefly discusses some major features of the DREAM kernel facility offered to real-time C++ PCD and C++ RTO.k program designers.

3.1 Guaranteed worst-case response

One of the most distinguishing features of the DREAM kernel is that it provides guaranteed timely services to real-time applications with minimal loss of hardware utilization.

The guaranteed timely service capabilities of the DREAM kernel is realized by adopting a unique approach for the layering of its components [Kim95b]. The kernel consists of five layers L0 through L4 in total.

This approach is based on the organizational principle called the time-leasing machine layering, which is of fundamental nature. Under the time-leasing machine layering principle, the bottom layer (L0 in the DREAM kernel) owns the full power of the hardware machine. So, the bottom layer uses the hardware machine at its own will. The remainder of the hardware machine time after the bottom layer use of the hardware machine, is "leased" to the next upper layer (L1 in the DREAM kernel). The time leasing relationship is recursive, i.e., it is maintained through all the layers that compose the kernel (in the DREAM kernel layers L0-L4). Moreover, there is a tight bound on the amount of machine time that each layer uses. This sometimes implies a bound on the frequency of certain types of interrupts that are accepted.

3.2 Exploitation of parallelism within the kernel by use of kernel-threads

In order to maintain a high degree of hardware utilization, parallelism is exploited extensively within the DREAM kernel through the introduction of kernel-threads. The kernel-thread, or thread for short, is an active concurrency unit operating inside the DREAM kernel. The set of threads is fixed at the operating system loading time. All the threads share the same address space and they are strictly *periodic threads* which make the analysis of their worst-case execution behavior a trivial task. Currently, four basic kernel-threads are used in the DREAM kernel. Additional application-specific custom kernel-threads can be introduced if fast response activities specific to the application must be supported.

4. DREAM Library and DREAM Kernel Support for PCD Programming in C++

As mentioned in Section 1, the DREAM kernel supports the execution of both process-structured and RTO.k object structured real-time application software. This section focuses on the construction of real-time PCD programs in C++ and the support offered by the DREAM library and DREAM kernel.

4.1 Basic components of process-structured real-time concurrent and distributed PCD programs

In Section 1, we briefly mentioned the PCD components. Let us now consider each of these components in some detail.

(1) Processes: These are real-time processes with various synchronization and activation requirements. In this paper, these are also called *application processes* (AP's) although in reality, some of these processes may play the roles of system management processes, e.g., processes managing I/O. The processes may frequently impose deadlines on the kernel for execution of their next computation-segments.

(2) CREW (concurrent-read-&-exclusive-write) monitors: The CREW monitor is a shared data structure monitor and an extension of the *monitor* defined in [Bri77] in that the former possesses the *readers-writers* semantics (i.e., concurrent-read-&-exclusive-write semantics) instead of the exclusive-read-&-exclusive-write semantics associated with the latter.

(3) Data field channels (DFC's) in the HU-DF scheme [Kim95a]: The *HU-DF (HU data field) scheme* for inter-process-group communication is an extension of the original *data field* scheme developed by Mori and other researchers in Hitachi, Ltd. [Mor93]. The essence of the data field scheme is to facilitate dynamic creation of *logical multicast channels* called data field channels (DFC's) and dynamic connection of processes to the DFC's in such a way that the idiosyncrasies of the physical communication networks are transparent to the process designer. If the physical communication facility has the broadcast capability, then a logical multicast channel is facilitated by making all processing nodes using the channel to broadcast (through the physical communication facility) messages with the headers containing the ID of the channel called the content code. The processing nodes connected to the logical channel can see all the messages coming through the physical broadcast facility but will pay attention only to those messages containing relevant content codes. The HU-DF scheme differs from the original data field scheme in that the former allows dynamic flexible connection of processes to the logical channels and supports not only conventional *event messages* but also *state messages* which are based on the distributed replicated memory semantics (a variation of the state message semantics in [Kop89]).

The three basic components (process, CREW and DFC's) represent fully general facilities for concurrent and distributed program structuring. Therefore, any operating system kernel that supports these three components along with various I/O operations can be viewed as a general purpose kernel.

In addition to these three program components, a PCD program may also use semaphores for inter-process communication and synchronization. Also, a PCD program may use ports for inter-process communication apart from the DFC's [Ras89]. A port is a message communication channel with only a single AP as receiver. An AP may notify its intention to the underlying kernel to function as a receiver to a specific port. In such a case, the kernel makes sure that no other AP's are granted the receive rights to that port. The DFC based communication facility is more powerful than the port based communication facility since the former supports group communication whereas the latter supports only point-to-point communication.

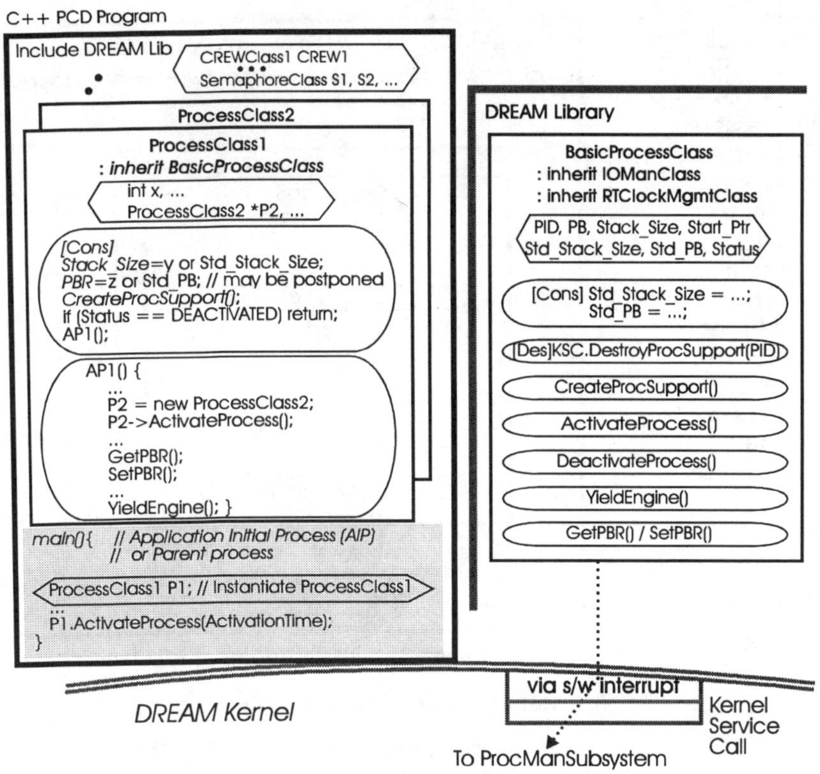

Figure 2. An application process constructed with DREAM library support

4.2 Construction of C++ PCD programs with the support of the DREAM Library

Application programmers may construct each PCD program as a collection of C++ objects and the main function which is treated as the body of the special process called the Application Initial Process (AIP). The PCD program may request the services of the DREAM kernel during its execution by making kernel service calls (KSC's) to the DREAM kernel. An interface to the DREAM kernel which is friendly to the PCD programmer has been constructed. This interface is called the DREAM library. The DREAM library is a collection of several C++ classes, each class functioning as an interface between a PCD program and a specific KSC group. For instance, the *BasicProcessClass* of the DREAM library provides a PCD program with the interface to KSC's related to AP management. A DREAM library class houses a set of pre-defined data structures and member functions all of which are well-known to the PCD programmer. In order to pass parameters to the DREAM kernel, in most cases the PCD programmer just needs to initialize the appropriate data structure(s) of the pertinent DREAM library class and then call the appropriate member function of this class. This member function is responsible for packaging the input parameters supplied by the PCD program, making the correct KSC, and possibly receiving output parameters from the DREAM kernel and saving it in the appropriate data structure defined in the class. The PCD program can then retrieve the appropriate output parameters.

Figure 2 shows the construction of an AP in a PCD program. The user first defines a *ProcessClass* which *inherits* the *BasicProcessClass* of the DREAM library. The BasicProcessClass provides a user-friendly interface to the programmer wishing to make KSC's related to AP management. The AP management KSC's include creating and destroying the kernel support for an AP, activating an AP, deactivating an AP, obtaining and setting priority values for the AP, etc. The BasicProcessClass defines default values for the process stack size and priority. The user may use the default value for the stack size and process priority or may redefine these values. Once the application-specific ProcessClass is defined, the user creates an instance of a ProcessClass inside the AIP. The AIP is the starting point of the entire application and is responsible for initializing the various child processes located in the node. When an instance of a ProcessClass is created, kernel support for this AP is automatically created since the constructor of the ProcessClass contains the appropriate DREAM library call (*CreateProcSupport*). Since the ProcessClass inherits the BasicProcessClass, it can access the capabilities provided by the BasicProcessClass. Figure 2 also shows how parameters are passed from the PCD program to the DREAM library.

In addition to the AP class, the user may also define an AP group class that encapsulates a group of communicating AP's within a node. The AP's in the AP group class would use semaphores and/or CREW monitors to communicate with each other.

In order to construct a CREW monitor, the PCD programmer typically defines a *CREWClass* which inherits the *CREWMgtClass* provided by the DREAM Library. The CREWMgtClass provides the PCD programmer with a user-friendly interface to KSC's for creating and destroying CREW support and obtaining and releasing READ-WRITE or READ-ONLY access rights to the CREW monitor. When an instance of a CREWClass is created, kernel support for the CREW monitor is automatically created since the constructor of the CREWMgtClass (inherited by the CREWClass) makes the appropriate KSC.

The DFC's provide an alternate means of communication among AP's within a local node and the main means of communication among a group of AP's located in different nodes distributed in a network such as a Local Area Network (LAN). For highly efficient communication, AP's within a local node are usually designed to use CREW monitors. However, this would prevent efficient migration of these AP's to other nodes. In order to design an AP in a PCD program to use the DFC facilities for sending and receiving event messages the designer may create an instance of the *DFCClass* provided by the DREAM library. The DFCClass provides an interface to KSC's such as creating and destroying DFC support, connecting an AP to and disconnecting an AP from a DFC, sending, forwarding, and receiving event messages via DFC, etc. When an AP connects itself to a DFC, it basically joins the multicast group of AP's which use that DFC for multicasting. An AP may also send and read state messages [Kim95a] which are shared among the AP's connected to the same DFC. State messages contain up-to-date information about the state of some entity in the system, for instance, the temperature of a room. The DREAM library offers support for the creation, sending and receiving of state messages through DFC's, etc.

As we mentioned in the previous subsection, a PCD program may also use semaphores and ports for interprocess communication. Similar to the DFCClass, the DREAM library provides a *PortClass* and a *SemaphoreClass* with the appropriate data structures and member functions.

4.3 Real-time processes and scheduling

The DREAM kernel supports the execution of real-time application processes (AP's) with various synchronization and activation requirements. The AP's may frequently impose deadlines on the kernel for execution of their next computation-segments. The AP management subsystem of the DREAM kernel facilitates the static and dynamic creation and destruction of AP's, static and dynamic setting of process priority values, voluntary yield of the execution engine by an AP, etc.

The DREAM kernel provides a flexible platform to easily test out a variety of real-time AP scheduling policies. In the current prototype version of the DREAM kernel, v.D2, the real-time AP scheduling policy incorporated is the *priority bracket (PB) scheduling* policy. The PB scheduling policy is quite powerful in that it can simulate a variety of other scheduling policies such as round robin, rate monotonic and other fixed priority scheduling policies, deadline-driven scheduling policies, etc. Nevertheless, the priority bracket scheduling policy is just one of many AP scheduling policies that can be incorporated into the DREAM kernel.

5. DREAM Library and DREAM Kernel Support for RTO.k Programming in C++

In the previous section, we have seen the construction of C++ PCD programs with the DREAM library and DREAM kernel support. We now discuss the construction of C++ RTO.k programs with the DREAM library and DREAM kernel support.

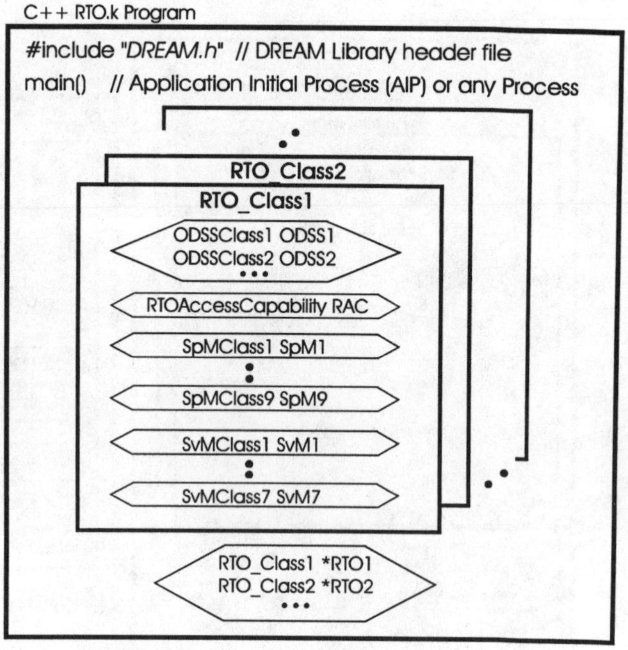

Figure 3. The definition of RTO class

5.1 Definition of an RTO.k object

As discussed in section 4.2, the application program begins execution from the AIP. The C++ RTO.k programmer typically defines an RTO.k class as a C++ class and creates an instance of the RTO.k class inside the AIP. Figure 3 shows such RTO.k classes defined as C++ classes. In this figure, the *RTO_Class1* represents one RTO.k class. It consists of

(1) instances of user-defined *ODSSClass*es which represent the various ODSS's of the RTO.k object,

(2) an instance of a user-defined *RTOAccessCapabilityClass*; Client RTO's need to obtain capabilities for accessing server RTO's before requesting for service. The access capability will be provided to a client RTO in the form of the service request queue (SRQ) ID that should be used to request for service. The RTOAccessCapabilityClass facilitates a client RTO to obtain capabilities for accessing server RTO's, and

(3) instances of user-defined *SpMClass*es and *SvMClass*es. These classes represent the SpM's and SvM's of this RTO.k object.

In subsequent subsections, we will describe the details on the user-defined classes mentioned in this subsection.

5.2 Overview of the approach for composing an RTO.k object by using PCD program components

The DREAM kernel treats components of RTO.k objects as special types of PCD components, i.e., special types of processes, CREW monitors, and DFC's. Figure 4 depicts the mapping relationship between components of an RTO.k object and the corresponding components of an equivalent PCD-program.

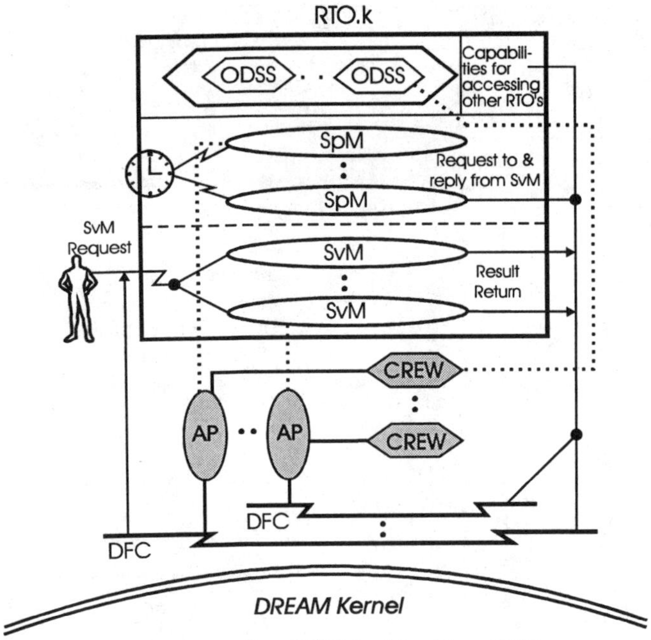

Figure 4. Mapping of an RTO.k object to a PCD program

As shown in Figure 4,
(1) both SpM's and SvM's, are treated as application-specific program bodies of processes (called method execution processes or MEP's),
(2) access paths (i.e., SRQ's) to SvM's as DFC's created by the SvM's, and
(3) result-return paths to clients as DFC's created by the clients,
(4) ODS segments (ODSS's) as special CREW monitors,
(5) capabilities for accessing other RTO's as DFC ID's.
Note that this way of utilizing DFC's facilitates the transparency of object locations.

An RTO.k object method should be assigned to an MEP before the method can be executed. This assignment may be specified by the application programmer or may be left unspecified in which case the DREAM kernel will perform this function. We distinguish the following two types of RTO.k object method executions:
(1) An RTO.k object method executes on an MEP that is bound (by design) to this method
(2) An RTO.k object method executes on an MEP that is not bound to any method (i.e. the MEP is unbound).

When an MEP is bound to an object method, a permanent connection is established between the object method and the MEP. In this case, the corresponding object method will always execute on the same MEP to which it is bound. Also, in this case, we distinguish whether a single method is mapped to an MEP or a set of methods is mapped to the MEP. If only a single object method is bound to an MEP, then the MEP always executes just this method. In other words, the MEP is *permanently dedicated* to the object method. In contrast, a set of methods may share the execution facility offered by the same MEP. However, at any time the MEP is only *temporarily dedicated* to any one of the methods executing on it. One important requirement for sharing an MEP among several object methods is that these object methods cannot be active concurrently. This requirement has been imposed in order to reduce the complexity of implementation. The RTO.k designer who wishes to use the shared mapping facility should ensure that the methods that share the same MEP do not have timing requirements that necessitate their concurrent execution. The shared mapping of object methods onto the same MEP has the advantage of increased resource utilization (e.g., memory space) and would facilitate a large number of object methods to be executed on the same local node.

When a certain execution run of an RTO.k object method is assigned by the DREAM kernel to an unbound MEP, the MEP is *temporarily dedicated* to this method just for this run. The next execution run of this object method may be assigned to a different MEP. Thus in the case of an unbound MEP, no permanent connection exists between a method and the unbound MEP. SpM's are typically executed on bound MEP's. SvM's may be executed on bound or unbound MEP's, but should be executed on unbound MEP's when the pipelined execution of the SvM is enabled.

5.3 Construction of ODSS

In this section, we consider the construction of the Object Data Space Segment (ODSS) of an RTO.k object as a special CREW monitor. The ODSS of an RTO.k object may be concurrently accessed by multiple methods of the same RTO.k object. Hence a mechanism for concurrency control is necessary to maintain consistency within the ODSS and among the ODSS and its users. The use of a CREW monitor to construct an ODSS seems to be a natural choice.

The C++ RTO.k programmer can start designing the ODSSClass by inheriting the DREAM library class, *BasicODSSClass*. This BasicODSSClass in turn inherits the *CREWMgtClass* of the DREAM library. Due to this inheritance, the BasicODSSClass can provide the interface to kernel services such as: creating and destroying ODSS (CREW) support, obtaining and releasing ODSS access rights, etc. The following sample program shows the use of the ODSS class:

```
class ODSSClass:public BasicODSSClass {
    int x,...; // private data area
  public:
    void Method1(); // a member function
};
void ODSSClass::Method1() {
    // Obtaining RW access rights to the CREW
    // monitor by calling the member function of the
    // CREWMgtClass of the DREAM library
    EnterCREW_RW();

    ...; // access the data area
```

```
        // Releasing RW
        // access rights to
        // the CREW
        // monitor
        ExitCREW_RW();
    }
    // Instantiation of the
    // ODSS class
    ODSSClass ODSS1;
```

5.4 Construction of RTO access capabilities

The RTO access capability section specifies the capabilities of the methods in this RTO.k object to call the services of other RTO objects. The access capability will be provided to a client RTO.k in the form of the DFC ID that should be used to request for execution of the corresponding SvM. In the approach adopted by the DREAM kernel, the creator of a server RTO.k should inform all potential clients about the creation of certain services by broadcasting the service-request DFC ID's of the SvM's of the server RTO.k via a special global broadcast DFC. At a client site, the client RTO.k, during initialization, makes a KSC to obtain the DFC ID of the external server method.

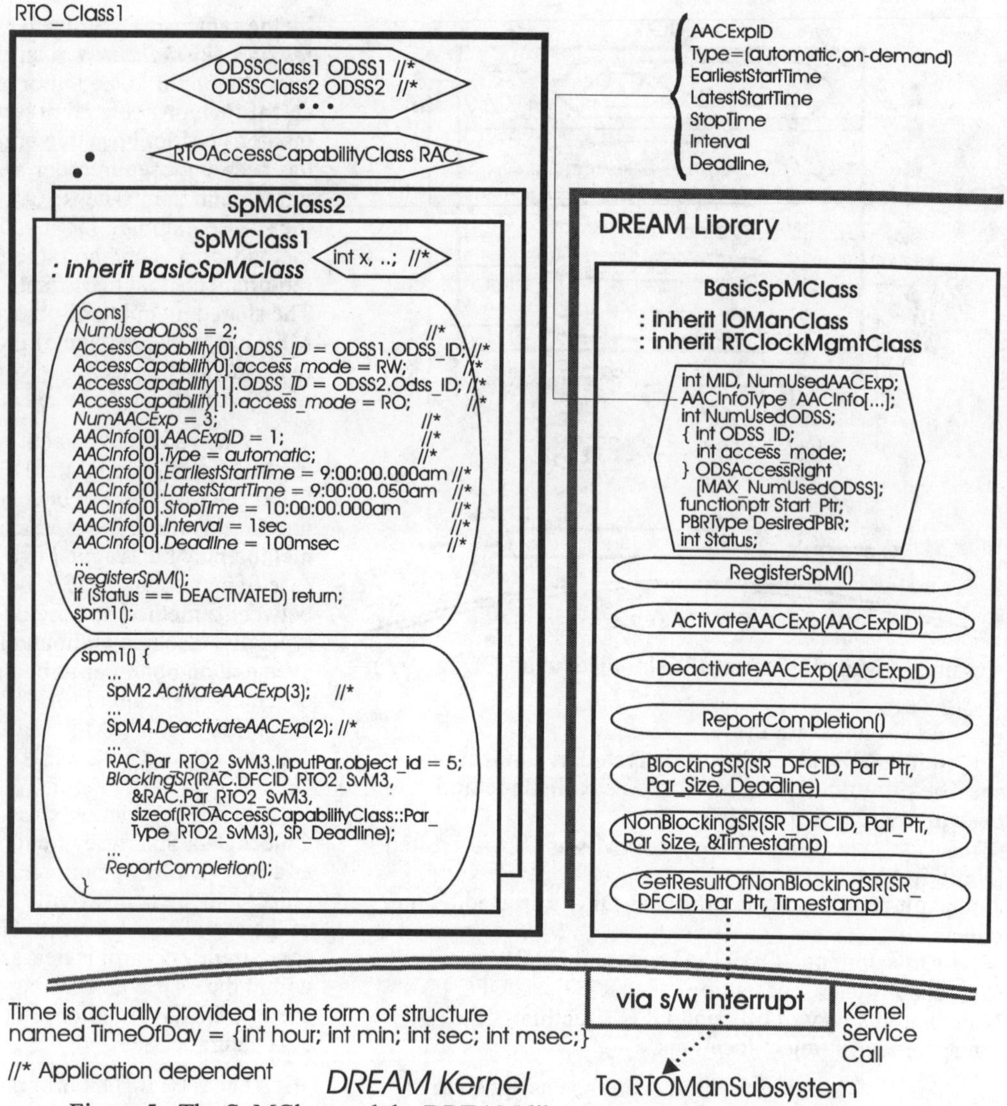

Figure 5. The SpMClass and the DREAM library support for SpM management

The RTO access capabilities are managed by a DREAM library class called the *BasicRTOAccessCapabilityClass*. The C++ RTO.k programmer can start designing an RTOAccess-CapabilityClass by inheriting the BasicRTOAccess-CapabilityClass. The BasicRTOAccessCapabilityClass provides the interface to the KSC for obtaining the capability for accessing an SvM in another RTO.k object. This KSC RegisterRTOAccessCapability() obtains the DFC ID of the SvM whose symbolic name is specified as input parameter. It is a good practice to define inside the RTOAccessCapabilityClass the parameters for each service request to be sent from this RTO. The following illustrates construction of RTO access capabilities:

```
class RTOAccessCapabilityClass: public
BasicRTOAccessCapabilityClass
{
  public:
    // parameter for access to SvM3 of RTO2
        struct ParType_RTO2_SvM3 {
            ...;
        } Par_RTO2_SvM3;
        // Service-request DFC ID of SvM3 of RTO2
        int DFCID_RTO2_SvM3;
        RTOAccessCapabilityClass(); // constructor
};
RTOAccessCapabilityClass::
        RTOAccessCapabilityClass()
{
        RegisterRTOAccessCapability
        ("RTO2", "SvM3", &DFCID_RTO2_SvM3);
}
// Instantiation of the RTOAccessCapability class
RTOAccessCapabilityClass RAC;
```

5.5 Construction of SpM's

As we mentioned in section 5.2, SpM's may be bound to MEP's or may be left unbound. In case the

SpM's are bound to MEP's, these MEP's may be generic or custom-tailored for the SpM. In the case of *generic MEP's*, the process priority, process stack size, etc. are fixed to some pre-defined standard values. On the other hand, these values may be custom-tailored for *custom MEP's*. Each SpM may request the services of one or SvM's through the appropriate service-request DFC. The result-returns from the SvM's are obtained through a unique reply DFC associated with the SpM. This unique reply DFC is allocated to an SpM by the DREAM kernel during the SpM initialization (registration) time. An SpM client may make two types of service requests [Kim94a]:
(1) *blocking* service requests in which the client is blocked after making a service request until the result is obtained from the server or until a pre-specified timeout expires.
(2) *non-blocking* service requests in which the client proceeds after making a service request and retrieves the result of the request later during its execution.

The DREAM library supports the registering of the SpM with the DREAM kernel, mapping of the SpM to an MEP, blocking and non-blocking service request calls, etc. Figure 5 shows the DREAM library support offered for SpM construction and management. The C++ RTO.k programmer defines an *SpMClass* inside an *RTOClass*. The SpMClass inherits the *BasicSpMClass* of the DREAM library. As shown in figure 5, the constructor of the SpMClass initializes several data items of the BasicSpMClass and registers the SpM with the DREAM kernel. The data items initialized include: number and ID's of the ODSS's accessed by this SpM along with the respective access modes, autonomous activation condition (AAC) information, start pointer to the function body of this SpM, etc.

After registering the SpM with the kernel, the user may create an MEP and bind the SpM to the MEP. This binding is typically done inside the AIP using DREAM library calls. The SpM can be permanently bound to one MEP or can be one among several SpM's that are mapped to the same MEP (shared binding), or left unbound at the end of the initialization. All unbound methods are mapped to (unbound-free) MEP's at the activation time of the method by the DREAM kernel.

5.6 Construction of SvM's

SvM's are *usually not bound* to any MEP unless the pipelined execution is explicitly disabled. When the SvM is activated, the kernel maps the method to an unbound-free MEP and executes it. If the service request frequency for an SvM is sufficiently high, the designer may decide to execute the SvM in a pipelined fashion. In other words, when multiple service requests arrive, each service request is handled by a separate invocation of the same SvM. In such cases, the kernel maps each SvM invocation to a different MEP. Each SvM is associated with a unique service-request DFC through which it may receive service requests. This service-request DFC is assigned to the SvM by the DREAM kernel during its registration time. Each SvM invocation is associated with a unique reply DFC through which it may receive replies from server methods to which it sent requests. The ID's of these reply DFC's are determined by the DREAM kernel at the time of the activation of the SvM invocation and are hidden from the RTO.k application programmer. The service requests from a client method to an SvM carry the ID of the reply DFC to which the SvM should send the reply back.

SvM itself may be a client of some other SvM(s), the SvM client may make both blocking and non-blocking service requests. In addition, an SvM may also pass a client request to another SvM by using a *client-transfer* call [Kim94b]. The latter SvM may again pass the client request to another SvM. This chaining sequence may repeat until the last SvM in the chain returns the results to the client. The main motivation behind such a client transfer call stems from the basic concurrency constraint which requires an execution of an SvM to be made only if a sufficiently large time window between potentially conflicting SpM's opens up. Hence, in certain situations a highly complicated SvM may never be executed due to lack of a wide enough time window. One way to get around this problem is to divide the SvM into multiple smaller SvM's, SvM1, .. SvMx. A client can then call each smaller SvM each time. Calling each smaller SvM incurs communication overhead of transmitting a request to the smaller SvM and obtaining the results. Eliminating most of such communication overhead is the motivation behind an arrangement in which the client calls the first SvM and the latter passes on the client to another SvM and so on until the last SvM of the chain returns the results to the client.

The DREAM library supports the registering of the SvM with the DREAM kernel, receiving service requests from clients, returning results to client methods, blocking and non-blocking service requests to other SvM's, and client transfer calls. The RTO.k programmer can start defining an SvMClass by inheriting the *BasicSvMClass* of the DREAM library. The constructor of the SvMClass should initialize several data items defined in the BasicSvMClass and then register the SvM with the DREAM kernel. The initialized data items include: the name of this method and the owner RTO.k, number and ID's of the ODSS's accessed by this SvM along with the respective access modes, method deadline, whether or not pipelining is allowed, start pointer to the function body of this SvM, etc.

Upon registering an SvM, the name of the SvM and the owner RTO.k along with the service-request DFC ID (assigned by the DREAM kernel) are broadcast to all nodes which may contain potential clients. The potential client RTO.k needs to obtain the capability for accessing the server method before they can request the service. Upon arrival of a service request, the DREAM kernel recognizes the request, binds and dedicates one SvM invocation to an unbound-free MEP, and activates the MEP for execution. It also connects this MEP to the appropriate reply DFC through which the SvM invocation may receive replies from SvM's to which it sent requests.

6. Advantages and Limitations of RTO.k Object Execution using PCD Components

The execution of RTO.k objects using PCD components offers several advantages to the system designer. These include:

(1) The use of DFC's for inter-RTO communication makes the RTO.k objects to be location transparent and allows easy and efficient migration of RTO.k objects. There exists no difference if communicating RTO.k objects are located in the same node or in different nodes.

(2) General purpose kernels (such as DREAM kernel) supporting various types of concurrent and distributed application software can be utilized for the execution of RTO.k objects.

(3) The shared binding of several object methods to an MEP provides for very efficient resource utilization and facilitates the execution of a large number of object methods on the same local node.

Among the limitations of this type of structuring include:

(1) Efficient implementation of DFC is possible only if the underlying communication network supports efficient broadcast capability.

(2) The use of DFC's for communication among RTO's within a local node may not be as efficient as a shared-memory form of communication. However, in the latter case, it is difficult to realize the location transparency of RTO's.

Even though the DREAM library eases the efforts for structuring C++ RTO.k objects, it still imposes burdens on the application designer of dealing with low-level details in redundant forms which could be automatically deduced from a high-level abstract and yet accurate representation of RTO.k objects. An extension of the C++ language to support abstract RTO.k programming is the most natural path to follow in this direction. The authors are currently developing one such extension named the C++T language.

7. Conclusion

The imitation of RTO.k objects using PCD program components facilitates the execution of RTO.k objects on a general-purpose kernel such as the DREAM kernel that views RTO.k objects as PCD program components. The DREAM library considerably reduces the programming efforts through use of the advanced inheritance features of C++. The DREAM kernel offers guaranteed timely services to real-time applications. These suite of tools would go a long way in our efforts to realize the GT design paradigms. One meaningful direction to pursue is to develop the C++T language, the extension of C++ language, that would provide a natural means of accurately representing RTO.k objects.

Acknowledgment: The research work reported here was supported in part by US Navy, NSWC Dahlgren Division under Contract No. N60921-92-C-0204, in part by the California Transportation Department via the UCI Institute for Transportation Studies, in part by Hitachi, Ltd, in part by IBM, in part by Postech, and in part by ETRI.

References

[Att91] Attoui, A. "An Object Oriented Model for Parallel and Reactive Systems", *Proc. IEEE CS 12th Real-Time Systems Symp.*, 1991, pp. 84-93.

[Bri77] Brinch Hansen, P., 'The Architecture of Concurrent Programs', Prentice-Hall, 1977.

[Ish92] Ishikawa, Y., Tokuda, H., and Mercer, C. W., "An Object-Oriented Real-Time Programming Language", *IEEE Computer*, October 1992, pp. 66-73.

[Kim94a] Kim, K.H. and Kopetz, H., "A Real-Time Object Model RTO.k and an Experimental Investigation of Its Potentials", *Proc. 1994 IEEE CS Computer Software and Applications Conf. (COMPSAC)*, Nov. 1994, Taipei, pp.392-402.

[Kim94b] Kim, K.H. et al., "Distinguishing Features and Potential Roles of the RTO.k Object Model", *Proc. 1994 IEEE CS Workshop on Object-oriented Real-time Dependable Systems (WORDS)*, Oct. 94, Dana Point, pp.36-45.

[Kim95a] Kim, K.H., Mori, K., and Nakanishi, H., "Realization of Autonomous Decentralized Computing with the RTO.k Object Structuring Scheme and the HU-DF Inter-Process-Group Communication Scheme", *Proc. 1995 IEEE CS Int'l Symp. on Autonomous Decentralized Systems (ISADS)*, April 1995, Phoenix, pp.305-312.

[Kim95b] Kim, K.H. et al., "A Timeliness-Guaranteed Kernel Model - DREAM Kernel and Implementation Techniques", *Proc. 1995 Int'l Workshop on Real-Time Computing Systems and Applications (RTCSA 95)*, Tokyo, Japan, Oct. 1995, pp.80-87.

[Kop89] Kopetz, H. et al., "Distributed Fault-Tolerant Real-Time Systems: The Mars Approach", IEEE Micro, Feb. 1989, pp. 25-39.

[Kop90] Kopetz, H. and Kim, K.H., "Temporal Uncertainties in Interactions among Real-Time Objects", *Proc. IEEE CS 9th Symp. on Reliable Distributed Systems*, Huntsville, AL, Oct. 1990, pp.165-174.

[Mor93] Mori, K., "Autonomous Decentralized Systems: Concept, Data Field Architecture, and Future Trends", *Proc. IEEE CS Int'l Symp. on Autonomous Decentralized Systems (ISADS 93)*, Mar.93, Kawasaki, Japan, pp. 28-34.

[Ras89] Richard Rashid et al., "A Foundation for Open Systems", *Proc. the IEEE Second Workshop on Workstation Operating Systems (WWOS2)*, September 1989, pp.109-113.

[Tak92] Takashio, K., and Tokoro, M., "DROL: An Object-Oriented Programming Language for Distributed Real-Time Systems", *Proc. OOPSLA*, 1992, pp. 276-294.

TaskPair-Scheduling with Optimistic Case Execution Times - An Example for an Adaptive Real-Time System

Martin Gergeleit, Hermann Streich
GMD – German National Research Center for Information Technology (GMD)
Research Division SET.RS (Responsive Systems)
Schloß Birlinghoven, D-53754 St. Augustin, Germany

Abstract

This paper first describes a model for an adaptive real-time system, that consists of four components: the real-time application, a monitor that observes the behaviour of the application, a dynamic resource manager that controls the execution of the application based on the data collected by the monitor, and finally an adaptation manager that uses the input from the monitor to implement long-term adaptation strategies. In the second part of the paper, a concrete adaptation mechanism is introduced. This mechanism extends TaskPair scheduling [14], an on-line, fault-tolerant scheduling method developed at GMD within the context of the DIRECT project [3]. Task Pair scheduling is extended with the notion of Optimistic Case Execution Time (OCET), that describes the time that a task needs in most of the cases for successful completion. As this time may depend on the system state and the environment, the system observes the actual resource consumption during execution and adapts future resource requests accordingly.

1. Introduction

Real-world systems operate under real-time conditions, they are inherently distributed and dynamic. In order to reflect these properties the controlling hard- and software for computer-based systems which monitor, control, or simulate real world processes must provide adequate means to cope with dynamics and with the effects of decentralisation. A typical application, e.g. a control system, has to provide high availability in the presence of failures or during maintenance activities or has to provide fail-safe behaviour. Traditional static system structures based on a total a-priori knowledge about the complete system do not suffice any longer, but a new generation of computer-based control systems is needed instead, based on a more dynamic view of a real-time system.

The project DIRECT at GMD works on solutions for this problem by developing an integrated approach for self-adapting real-time systems. This paper describes a generic model for an adaptive real-time system and as a realisation of this model a concrete adaptation mechanism which adapts the Optimistic Case Execution Time (OCET). The OCET is the time that a task needs in most of the cases for successful completion. The OCET expresses a duration in which a specified amount (e.g. 90%) of such executions will succeed. As this time may depend on the system state and the environment, the system observes the actual resource consumption during execution and adapts future resource requests accordingly. Our adaptation mechanism is based on the TaskPair scheduling [14], a method that performs on-line guarantees for real-time-applications. Different variants of a basic scheduling algorithm can be combined with this method. One variant is applicable for real-time applications with unknown worst-case execution time (WCET) or large WCET compared to the average execution time.

2. A Generic Model for an Adaptive Real-time System

The goal of an adaptive real-time system is, to adjust the system behaviour dynamically to changes of the system environment. An adjustment of the functionality clearly requires detailed knowledge of the problem domain and thus support from the application. But there are still a lot of things that can be done on system level. The system can try to detect situations where an adaptation is needed. It can observe the behaviour of the application and trigger adaptations when application defined constraints are violated by the current execution. Of course, these activities can be coded into the application itself, but it can simplify the burden of writing a distributed real-time application a lot if these tasks are separated. The application provides the functionality to control the real-world system while the system checks concurrently on-line whether this execution is in accordance with its specification or not. The application can provide alternative control algorithms or emergency procedures that are activated, whenever a specification violation, and thus a reason for an adaptation is detected. The most simple system support for an on-line detection of a specification violation is a watch-dog timer. More sophisticated mechanisms for a specification of constraints in an object-oriented system are currently under development in the DIRECT project [4].

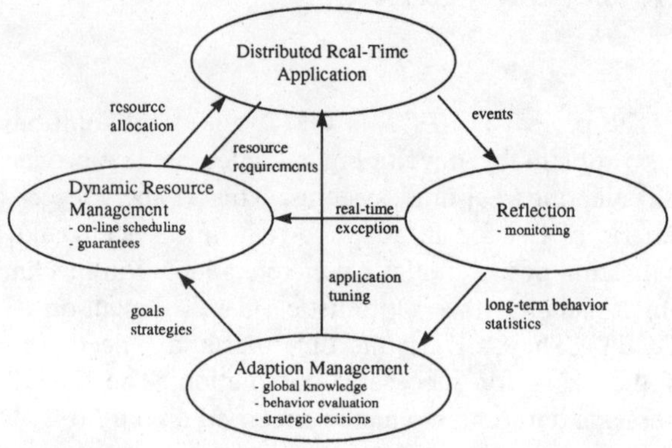

Figure 1. A generic model of an adaptive real-time system

Our generic model for a system architecture that supports on-line adaptation is depicted in figure 1. It consists of four components: the distributed real-time application, a monitoring component (reflection), that observes the behaviour of the application, a dynamic resource manager, that controls the execution of the application based on the data collected by the monitor, and finally the adaptation manager, that uses the input from the monitor to implement long-term adaptation strategies. Adaptations may be performed, either on the system or on application level. Due to the required flexibility of considered real-time systems, we design the resource scheduling scheme itself to be adaptable.

In the following sections these components are discussed in more detail:

2.1. Reflection

The reflection component observes the behaviour of the possibly distributed application. It collects events that report on the current status of the application. Events that are not important for checking the application defined constraints are filtered out at an early stage. The remaining events are either processed locally on the node where the events were produced or, combined to a global event stream, if it is necessary to assess global constraints. Such a global constraint is e.g. the condition, that all tasks, that belong to a joint activity are started within a small time window or that between a sending of a message and its reception on another node only a certain amount of time has passed.

Generating events is done by the application itself and/or by the operating system. Arbitrary events can be defined by the application programmer and some are predefined by the system.

2.2. Dynamic Resource Management

In the considered scenario multiple independent and dependent real-time application will exist, with different resource requirements. They may have different importance and different time criticalness (soft or hard deadlines). The resource manager has to know about these differences in order to „protect“ hard real-time tasks from disturbances by other ones.

A lot of work has been done for real-time scheduling and a lot of practical solution have been found, reaching from restricted scheduling algorithms suited for periodic independent tasks on a mono-processor considering only the single resource CPU (e.g. Rate Monotonic, Earliest Deadline First [9]) or multiple resources [12] up to scheduling of complex distributed applications (e.g. Mars [8], Alpha [1], DEDOS [15]). To ensure timeliness of all tasks always worst-case-execution times have to be used, often resulting in low resource utilization and a static structure of the whole system.

For applications with weaker timing constraints such solutions would be unsuitable or at least wasteful, e.g.

- worst-case execution times (WCET) cannot be computed, or
- changes in the environment are expected,
- if workload may vary dynamically.

This has led to dynamic real-time systems, e.g. the Spring kernel [12] or Chaos-Arc [11], and others. Such systems have in common, that tasks may be scheduled on-line. The main advantage of on-line scheduling is, that

- the current system's state can be taken into account and
- various scheduling strategies are implementable [10], reaching from simple Earliest Deadline First (EDF) to complex multiple resource scheduling, which unfortunately is mostly time-expensive
- adaptations on applications and on scheduling criteria can be done without stopping the system.

The DIRECT approach counts to the „complex reservation systems“ [13] in which resources can be reserved in a distributed system. These global resource management is implemented by communicating local resource manager (scheduler). A resource manager is responsible for „his“

local resources and trades with others about shared (e.g. network). So, local activities can be scheduled autonomously.

Resource requirements and current state of the system (including all running applications) build the base for scheduling decisions. The current state is defined by the set of already reserved resources. Resource requirements are initially defined by applications and may change over the time. Some types of changes are adaptations, performed by the application itself or the fourth component of our system, the adaptation manager.

2.3. Adaptation

Adaptation makes sense if something can be improved. Subject of improvement spans different levels of abstraction and is related to different strategies. These strategies may define optimisation of the overall system performance as the major goal or consider only the behaviour of single applications.

The strategy defined must be correlated to the scheduling scheme implemented by the resource manager, i.e. optimisation strategies cannot be realised by any scheduling scheme. So, the scheduling scheme itself must be subject for adaptation. We decide between system adaptations and application adaptations. System adaptations are adaptations of the behaviour of resource managers. These are strategic changes which may have immediate consequences on applications while application adaptations aim at changing the behaviour of a single application. The following should give some examples:

- Changing importance of tasks. The importance of tasks specifies an order on all considered tasks. In dynamic systems the set of considered tasks will vary over the time, making it necessary to recompute the importance relation. In addition the importance may change depending on the state of an application (e.g. a recovery point has been established).
- Adjustment of the frequency of periodic tasks. If e.g. the monitoring system detects declinements of a periodic task it could be useful to reduce the frequency of the task. This is different as to do it as fast as possible, because the task executes in a much more controlled way. Especially if other activations depend on such a task, a reduced but guaranteed quality should be better than a random, non-guaranteed quality.
- Changing the scheduling strategy. The scheduler itself consumes resources and this can lead to performance problems. Adaptation of the scheduler includes use of simpler algorithm, limiting the number of considered tasks or reserving resources only for tasks with a minimal importance.

3. An Example for an Adaptive Real-time System: Adjusting the OCET for TaskPair Scheduling

3.1. Worst Case Execution Time vs. Optimistic Case Execution Time

Nearly all existing hard real-time systems need to know about the worst case execution times (WCET) of the application tasks. Also, all of the well known scheduling algorithms are based on the knowledge of the WCETs. But in real life this imposes more and more problems and most programming environments do not provide tools for the determination of WCETs. This is because:

- A programming language that can be statically analysed for WCETs necessarily has to impose severe restrictions on the expressiveness of its statements. Loops must be bounded and recursion must be restricted in depth. Thus, also dynamic data-structures are restricted. Trees and lists must have a fixed depth and length and it is impossible to manage an object-space of previously unknown size. In dynamic systems, that schedules tasks dynamically at run-time, it is possible to give a function for a task that computes its WCET depending on the current system status (like e.g. the length of a list), but it usually requires additional effort from the programmer to provide the knowledge for this function.

- Another problem is, that WCETs are highly system specific. The WCET of a program does not only depend on its logical structure but also on the compiler, the processor, and the system environment. Thus, while there are some approaches to compute WCETs with only the source code and some system knowledge as input, most attempts to determine the WCET rely on the detailed knowledge of the processor, the source code, and the compiled object code. But this is still not sufficient. The characteristics of a single system, like its clock speed, its bus architecture, its memory access behaviour and its DMA controller have significant and often non-linear impact on the timing of a program. Thus, it is impossible to argue about WCET without a specific machine in mind.

- Even with the perfect knowledge about the executing machine, it becomes more and more difficult to give reasonable bounds for the WCET of a piece of code, because of the increasing complexity of today's hardware architectures. Most of the WCET analysis work in the past has been done on M68000 machines, which had nearly constant execution times per instruction. But all modern processors make intensive use of at least two levels of caches, internal pipelining and parallelism, branch prediction, and speculative execution. This leads to two observations, that are highly undesirable for a WCET analysis: 1) It is really hard to determine how long a single instruction takes in a certain execution context and 2) the difference in the execution time between the worst case and the best case may be in the order of 10. The first observation means, that it is even more difficult to build a WCET analyser than it was 5 years ago and the second observation tells, that even a good tool will provide WCET figures, that are far apart from the realistic execution times of a running application. The speedup of processors in the last years does not mainly result from faster transistors, but from a better usage of parallelism and locality. Modern processors are not built for deterministic execution times, but for a high probability of a maximum execution throughput. Drastically spoken, this means that real-time systems that rely only on WCETs will not benefit from the current development of new processors.

But if it so hard to get reasonable bounds for the WCET of a task, what can be done? One answer is, if all tasks in a hard real-time system must be executed all the time (exactly with their period or their maximum arrival time) and there is only one way of doing the job (like the task was coded): Nothing! The scheduler has to work with the WCETs, even if this causes an enormous underutilisation of the system during normal operation.

But the restrictions are not so hard. Thus, another answer to the question might be: Use probabilistic assumptions about the timing of your application and find a way to deal with the situation when a really bad case happens! Sometimes it can be more cost-effective to use a smaller sized system, that might fail in e.g. one percent of its task execution cycles to meet the deadline. This can be tolerable if the system still maintains a basic level of functionality and safety by guaranteeing a hard real-time core with known WCET, that is executed when this case happens.

A typical example for such a task is a sensor, that is polled with a certain period. The polling action might include distributed I/O and a complicated computation for preparing the value of interest form the input data (e.g. video image analysis). The WCET of this action might be unknown or at least so much longer than the average case execution time, that it might be worth thinking about dropping a measurement cycle if an extreme case happens. Especially, if it is acceptable to interpolate a missing value from the previous results with a simple computation. Thus, the resource requirement of this task can be redefined to be at least the resources necessary for the interpolation and enough to complete at least a sufficient percentage of the real measurements.

In order to handle these tasks the additional notion of the Optimistic Case Execution Time (OCET) of a task helps a lot. The $OCET_{t,x}$ of a task t defines the execution time (CPU time plus waiting time of a task) that is needed to get a probability of x that t is completed within this time. This probability is of course dependent on the system state and the environment. This means that the $OCET_{t,x}$ may change over the live-time of the system, but on the other hand the locality can be expected to be high. This means, in many cases the $OCET_{t,x}$ will change slowly over time. This is true in cases where the $OCET_{t,x}$ depends e.g. on the increasing size of an internal data-structure or on changing physical properties of sensors or actuators (e.g. caused by their increasing age). In case of transient errors (like e.g. message loss in the network and a repeated computation) this is surely not the case, but these cases can be regarded to be equally distributed over all executions. Thus, in order to predict the current $OCET_{t,x}$ of a cyclic task it should be a good heuristics to look at the last executions of the same task. This leads to the definition of $OCET_{t,k,n}$ of a task, which is the minimal execution time that was needed to successfully complete at least k out of the last n executions. If the assumption of locality is true, then $OCET_{t,k,n}$ is a fairly good approximation for $OCET_{t,x}$ with x=k/n. The case that the $OCET_{t,x}$ of t is monotonously increasing is pathological because either the task is unusable in a real-time system as it will exceed all bounds or the value will converge towards a fixed time.

Although many scheduling algorithms only consider CPU-usage, it is more realistic to take waiting time into account as well. The $OCET_{t,k,n}$ is defined to include both: *running* (CPU) time and *waiting* time, but in may cases it is useful to distinguish between both. Waiting-time is the time where a task could use the CPU, but it is waiting for something to happen, like. e.g. the arrival of a message from a decentralised I/O component. Usually this time is also hard to predict precisely and often it exceeds the running time by a magnitude. Thus we define also the $OCRT_{t,k,n}$ (the Optimistic Case Running Time) which describes the needed CPU time in those k out of n executions that defined the $OCET_{t,k,n}$.

3.2. Adaptation of the OCET

As stated above, the OCET (and the OCRT) of a task may change over time. As the OCET can serve as an input parameter for a dynamic scheduler, it is a good candidate for adaptation. Figure 2 depicts an instance of the generic adaptation model described above that does adaptation based on OCETs.

The first step for an adaptation is the reflection about the system. The monitoring component receives system generated events for measuring

- the *running* time (time, a thread uses the CPU),

- the *waiting* time (time, a thread waits by its own responsibility), and
- the *blocking* time (time, a thread wants to run, but is preempted) of a thread.

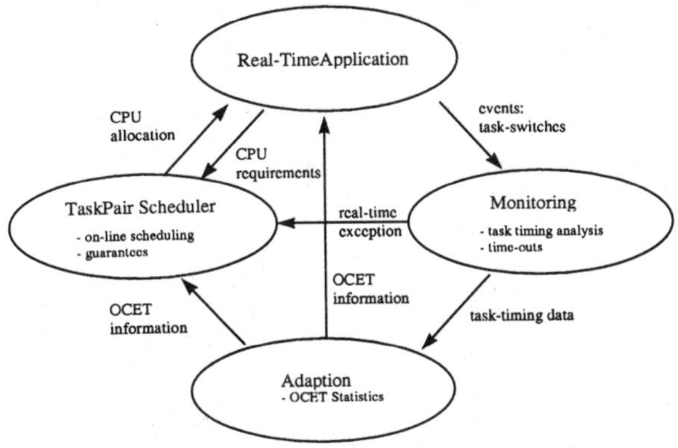

Figure 2. Adjusting the OCET for TaskPair Scheduling

What has to be observed are all state changes of threads. The required events are created by the low-level dispatcher of the system. If the dispatcher also informs the monitoring component about the deadlines of the task, the monitor can maintain a time-out queue and inform the scheduler if a deadline of a still executing task expires. These real-time exceptions are a direct feedback into the running system. If the scheduler requires information about the resource consumption of the currently running tasks, it also asks the monitor.

For a longer term adaptation the monitor sends the collected timing data about the (cyclic) tasks to the adaptation component. Here the statistics about the $OCET_{t,k,n}$ ($OCRT_{t,k,n}$) is maintained. The adaptation component stores the times of the last n executions of each task t and computes the required $OCET_{t,k,n}$ ($OCRT_{t,k,n}$) for a given k. If an execution of t fails to finish before its deadline is reached and it was terminated for that reason a „big" value is for its execution time and its running time is assumed. This value can be a fixed amount of time (e.g. if known, the WCET), or the maximum of the actual required time and a certain percentage of the OCET (OCRT).

Now, if a task changes its timing behaviour during the life-time of the system the dynamic scheduler as well as the application itself may ask for its current characteristics and react accordingly.

What is now needed is a scheduler that takes advantage from the knowledge about the OCETs and OCRTs. Within the DIRECT project we have developed TaskPair scheduling which addresses this requirement. The TaskPair-Scheduler runs tasks with an estimated OCET but an unknown WCET in a real-time environment. The scheduler starts a task t (SoftTask), if it can reserve the resources required for an execution within its $OCET_{t,k,n}$ and executes a short emergency procedure t´ (HardTask), if this task is going to violate its deadline.

If a single execution is going violate its deadline, it is the job of the emergency procedure t´ to clean up the system and preserve a consistent state before the deadline is really reached. It is assumed that the $WCET_{t'}$ of this emergency procedure is known, but this is not a severe restriction, as it is assumed to be very short. Thus, even a bad and very pessimistic execution time estimation should not lead to resource requirements that are comparable with those of the ordinary task ($OCET_{t,x} >> WCET_{t'}$). This approach is viable if $WCET_t > OCET_{t,x} + WCET_{t'}$. This means, the WCET of t is greater than its expected execution plus the time for a possible emergency procedure.

3.3. TaskPair-Scheduling

The TaskPair-Scheduling approach, preserves the advantages of dynamic scheduling while taking into account the requirements of hard real-time tasks. It is obvious, that in dynamic real-time systems deadline violations can occur, if no restrictions for creating new tasks exist and all dynamic arriving tasks are started. So the most stringent prediction, which can be made on-line is, to guarantee that an arrived task will meet its deadline or to decline this guarantee. This means deadline violations can be avoided for all tasks, in the sense, that tasks, which cannot be guaranteed are not started. If workload reduces, the guarantee may be given and the tasks are executed.

TPS is only briefly sketched here, a more detailed description is given in [14]. TPS can be viewed as a combination of the idea of on-line guarantees in the Spring kernel [12] and exception handling, known from programming languages.

The main idea is to split a real-time task into a pair of tasks, namely a SoftTask and a HardTask (In [14] SoftTask and HardTask were named MainTask and ExceptTask). The SoftTask is only executed if all required resources of the HardTask can be reserved by the resource manager (scheduler). The HardTask is only executed if the SoftTask could not succeed early enough. This guarantees, that either the SoftTask or the HardTask will succeed the operation before the deadline. The main difference to the guarantee concept of the Spring kernel is, that a task (SoftTask) can be started, even if its worst case execution time (WCET) is unknown. Only the WCET of the HardTask must be known. It is obvious, that such a construction makes sense, if the WCET of the HardTask can be computed, and is small in comparison to the OCET of the SoftTask. By construction the deadline of a Task-

Pair can be specified and termination before the deadline is guaranteed.

In the following we consider a task as the unit, which has to be planned by a scheduling algorithm. Tasks are pieces of code, which are executed by threads. Threads are the entities which are known to a dispatcher.

Similar to an exception construct with associated try- and exception handling block, a TaskPair is constructed as a pair of tasks, namely a SoftTask and a HardTask, the HardTask is syntactically placed as the exception handler.

```
TaskPair TP = (SoftTask,HardTask);
```

with the intention:

```
if( guarantee (TP,deadline)){
  try
      <SoftTask>
  except
      <HardTask>
  end
}
else // no guarantee
```

The TaskPair is described by a TaskPair descriptor, including a task descriptor for each task providing all necessary information about the TaskPair to the on-line scheduler, e.g. WCET (if possible) or a function to compute the WCET, earliest start-time, dependencies etc..

In contrast to currently used exception handling mechanisms in real-time programming languages (e.g. RTC [2, 16], RTC++ [5], Flex [6], RT-Euclid [7]), here the „exception handler (HardTask)“ is executed in order to *avoid* a deadline violation of the try-block and not after a deadline violation has happened.

In other words, it is guaranteed, that if the SoftTask is started either itself or the HardTask will terminate the TaskPair in a controlled way, leaving data in a defined state and without violating a deadline. The question is, what can or should effectively be done by a HardTask?

We distinguish between four possible functions:

- aborting the TaskPair,
- completion of what has been tried by the SoftTask,
- support of a *basic functionality*, and
- undoing, what the SoftTask has done.

In either case, predictability of the timing behaviour of the TaskPair is guaranteed.

While we are working on a distributed scheduler that is based on TaskPair scheduling we restrict this discussion of the scheduling algorithm to single processor machines. The simplest way to implement TaskPairs is, to dispatch SoftTasks round robin, all with the same (low) priority and preemptive, while executing HardTasks as late as possible but with maximum priority. This means that HardTasks are planned in a "time triggered" way and that they are non-preemptive.

The figure 3 shows two independent TaskPairs TP1 and TP2 running concurrently.

```
void TP1(deadline)
{
  if( guarantee(TP1,deadline) ) {
        try     ST1();
        except  HT1();
  } else ...; // no guarantee
}

void TP2(deadline)
{
  if( guarantee(TP2,deadline) ) {
        try     ST2();
        except  HT2();
  } else ...; // no guarantee
}
```

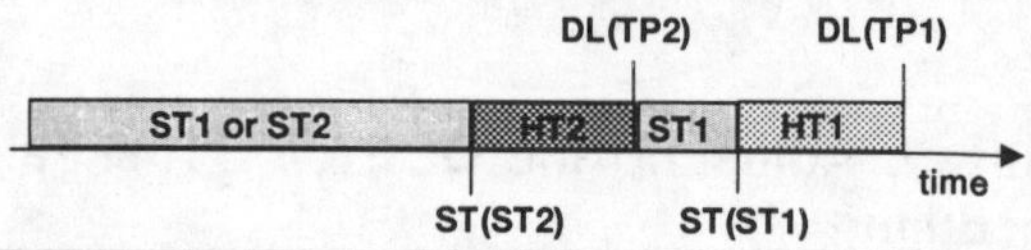

Figure 3. Simple strategy for scheduling TaskPairs

The scheduler places the HardTasks (HT1 and HT2) as late as possible with respect to their deadlines (DL) by reserving processor time on a time axis. For simplicity we do not consider time for run-time system activities like guaranteeing tasks, switching of threads etc.. All such activities must be bounded (see also [12]). In this example the deadline of TP2 is earlier than HT1's deadline. The SoftTasks (ST1 and ST2) start running. ST1 and ST2 may be preemptive or not and scheduled in an arbitrary way. If ST2 does not terminate before start time (ST) of HT2, HT2 starts execution. If ST2 succeeds, i.e. ST2 terminates before ST(HT2), HT2 is removed from the scheduler's plan. TaskPair T1 behaves analogously.

The base scheduling strategy considers only WCETs of HardTasks. This algorithm can be extended to reserve also resources for SoftTasks. Different alternatives seem to be interesting:

1. guaranteeing the WCET of the SoftTask,
2. guaranteeing a certain amount of time (e.g. the OCET) for the SoftTask, and
3. performing an acceptance test (e.g. considering the OCET), but without giving a guarantee,
4. no timing information considered, dispatching by EDF or importance based.
5. In the first case there is no need for a HardTask, since the guaranteed SoftTask always succeeds before the

deadline, but this still requires that the WCET of the SoftTask is known. That is what we want to avoid.

In the last case the SoftTask needs no timing specification other than the deadline. In this case in high load situations all resources will be spent for HardTasks and no SoftTask may succeed. This situation is acceptable, if HardTasks include a minimum functionality for the application (so we have a graceful degradation) but on the other hand it can be pathological, if all HardTasks only perform a „controlled abort" of a SoftTask, which in the end leads to full resource utilisation without any valuable computation.

Thus, in the scope of this paper cases 2 and 3 are interesting:

If the difference between waiting time and running time is not considered, it is known how to implement case 2, a guarantee for the OCET of the SoftTasks. One implementation is to check (as known from EDF scheduling) if enough CPU time (more than the $OCET_{t,k,n}$) is available before the deadline. For this computation the already consumed running time of the active tasks has to be taken into account as well as the time reserved for HardTasks. If this check succeeds, all current tasks can run with their guaranteed OCET. Thus, if $OCET_{t,k,n}$ is a correct approximation for $OCET_{t,x}$ the system can guarantee a certain percentage of successful executions of t. Unfortunately the assumption, that there is no waiting time, is unrealistic. If $OCET_{t,k,n} >> OCRT_{t,k,n}$ there is much more CPU time reserved for task t than needed and thus less time left for guaranteeing other tasks. Thus, case 3, that relieves this constraint, is interesting as well.

In this case no guarantee for the SoftTask is given, but the scheduler checks, if there is a good chance of completing the SoftTask before its deadline is reached. The check is done by testing if enough CPU time (more than the $OCRT_{t,k,n}$) is available before the deadline and if the deadline is far enough (current time + $OCET_{t,k,n}$). Unfortunately this check does not perform any guarantee, because it is not known, when the CPU can be used. This depends on the control flow of all SoftTasks and this is not known by the scheduler. This check can sort out those tasks that have a bad probability for success before the start of their HardTasks. This avoids wasting resources for SoftTasks that will probably fail anyway.

In both cases SoftTasks are sorted and executed by EDF or least laxity. But there are still different possibilities for the system, to react, if a task exceeds its OCET. If nothing is done, resource consumption of the actual task will be on the burden of other tasks. Another way to handle this situation is, to suspend the actual task, if other tasks are ready for execution, and have not exhausted their OCET. The latter seems to be reasonable for case 3, it is needed for case 2, in which we guarantee the OCET for the SoftTask, keeping in mind that this may be not enough for a successful execution. Here we cannot tolerate, that resources from already guaranteed tasks could be stolen.

So, from the application's point of view case 2 and case 3 differ in having a guaranteed execution time or not and the consequence in cases of exceeding the OCET.

From the scheduler's point of view, in both cases the available time, has to be computed, but obviously with a different quality. In general, the scheduler first decides whether it can guarantee the execution of the HardTask. This is done as described above. Then the scheduler has to determine whether the SoftTask has a chance to succeed before its deadline (case 2) or can be guaranteed (case 3).

```
if(available_time_for_SoftTask >
   OCET(SoftTask) + WCET(HardTask))
then start_TaskPair
else no_chance_for_success
end
```

As pointed out above, the guarantee can only be given, if we completely reserve the OCET, i.e. including the waiting time. In case of communication operations guaranteed resources are inherited by the called thread. Similar holds for nested TaskPairs. For more information see [14].

The question is how to compute the „available time". In case 2 we have to make pessimistic assumptions about all current guaranteed activities, for case 3 we define available_time as „seems to be available". So, in the latter case we can live with rough estimations to compute the available time, such that this can be done fast. For case 2 we have to do complete (pessimistic) scheduling. We believe, that both cases will have their place in the wide range of real-time applications.

4. Conclusion and Future Work

In this paper we have presented a generic model for an adaptive real-time system and a concrete instance of such a system: a system that does TaskPair scheduling based on dynamically computed OCETs and OCRTs. The potential of this approach is currently exploited in the prototype of the DIRECT system. We hope to be able to present performance figures for sample applications in the near future. We are currently working on a guarantee scheme for SoftTasks that takes the waiting time of tasks better into account than the current acceptance test does. We are also working on a more application oriented monitoring mechanism, that allows the system to know more about the progress of a single application, like e.g. the start and end of single object invocations or the number of iterations of a loop.

References

[1] K. Clark, E.D. Jensen, F.D.Reynolds, "An Architectural Overview Of The Alpha Real-Time Distributed Kernel", USENIX Association, "Micro-kernels and other Kernel Architectures", 1991

[2] N. Gehani, K. Ramamritham: "Real-Time Concurrent C: A Language for Programming Dynamic Real-Time Systems", July 1991

[3] M. Gergeleit, J. Kaiser, H. Streich: „Towards a Distributed Object-Oriented Real-Time Control System", Workshop on Concurrent Object-based Systems, in conjunction with the IEEE Symp. on Parallel and Distr. Processing, Dallas, TX, October 1994

[4] M. Gergeleit, J. Kaiser, H. Streich: „Checking Timing Constraints in Distributed Object-Oriented Programs", Workshop on Object-Oriented Real-Time Systems, in conjunction with the IEEE Symp. on Parallel and Distr. Processing, San Antonio, TX, October 1995

[5] Y. Ishikawa, H. Tokuda, C.W. Mercer: "Object-Oriented Real-Time Language Design: Constructs for Timing Constraints", ECOOP/OOPSLA '90 Proceedings, October 1990

[6] K.B. Kenny, K.-J. Lin: "Buiding Flexible Real-Time Systems Using the Flex Language", IEEE Computer 5/1991

[7] E. Klingerman, A. Stoyenko: "Real-time Euclid: A language for reliable real-time systems", IEEE Transactions on Software Engineering, Vol SE-12, September 1986

[8] H. Kopetz et. al., "Distributed Fault-Tolerant Real-Time Systems: The Mars Approach", IEEE Micro, Feb 1989

[9] C.L.Liu,MIT, J.W.Layland,Calif.Inst. of Technology, "Scheduling Algorithms for Multiprogramming in a Hard-Real-Time Environment", Journal of the ACM, Vol. 20, Jan 1973, No. 1, pp 46-61

[10] K. Ramamritham, J.A. Stankovic, "Scheduling Strategies Adopted in Spring: An Overview", Univ. of Massachusets at Amherst, COINS Technical Report 91-45

[11] K. Schwan, A. Gheith, H. Zhou, "From Chaos-min to chaos-arc: A family of real-time kernels", Proceedings of the Real-Time Systems Symposium, Orlando, Florida, pp. 82-92, IEEE, Dec 1990

[12] J.A. Stankovic, K. Ramamritham: "The Spring Kernel: A New Paradigm for Real-Time Operating Systems", ACM Operating Systems Review, Vol. 23 No.3, July 1989

[13] J.A. Stankovic, K. Ramamritham: "What is Predictability for Real-Time Systems", Journal of Real-Time Systems, No. 2, 1990

[14] H. Streich: "TaskPair-Scheduling: An Approach for Dynamic Real-Time Systems", Int. Journal of Mini & Microcomputers, Vol. 17, No. 2, pp 77-83, 1995

[15] J. Verhoosel, "A Formal Deterministic Scheduling Model for Hard Real-time Executings in DEDOS", Computing Science Notes, 93/08, Eindhoven University of Technology

[16] V. Wolfe, S. Davidson, I.Lee: "RTC: Language Support For Real-Time Concurrency", IEEE Computer 7/1991

Session 3

Real-Time Fault Tolerance

Session Chair:
Juergen Nehmer
University of Kaiserslautern, Germany

Adaptive Software Fault Tolerance Policies with Dynamic Real-Time Guarantees

Edgar Nett, Hermann Streich
GMD - SET.RS
D-53754 St. Augustin, Germany
{nett, streich}@gmd.de

Paolo Bizzarri, Andrea Bondavalli,
Fabio Tarini
CNUCE-CNR, I-56126 Pisa, Italy
A.Bondavalli@cnuce.cnr.it, tarini@ec.unipi.it

Abstract

Real-time applications with high dependability requirements demand for fault tolerance strategies. While for small systems with static behaviour policies based on worst case execution times can be used, this is not true for more complex systems, in which worst case execution times are partially unknown or differ drastically from their average execution time. In such cases often only a minimum of quality can be achieved. This paper proposes to combine fault-tolerant policies described by the FERT (Fault-tolerant Entity for Real Time) notation with the dynamic scheduling scheme TPS (TaskPair-Scheduling). TPS alleviates FERT's precondition of completely known WCETs and provides a flexible implementation base to enable an easy mapping of FERT strategies to a runtime system. In a first step, a significant subset of FERT is investigated, which implies: the Recovery Block Scheme, N-Version Programming, and Imprecise Computations. TPS is utilised to guarantee different levels of quality, tailored to the application and the required level of fault tolerance, while guaranteeing, that a common deadline is met.

1. Introduction

For Real-Time applications requiring high dependability, Fault Avoidance techniques, although necessary, do not allow to exclude the occurrence of faults at run-time. Then, Fault Tolerance (FT, in the following) policies must be employed. The necessary redundancy can be either statically sized on the worst case, or exploited by an adaptable behaviour of the application. In case of overload or lack of resources due to failures, adaptable use of redundancy can allow graceful degradation of the system; e.g. safety-critical tasks could be supported as long as possible, while non critical ones could receive fewer resources or even be omitted.

To reach an adaptive use of redundancy, the application designer and the run-time support must co-operate in such a way that the designer specifies needs and flexibility of the application, while the run-time support takes them into account in managing system resources. This specification could also allow to integrate Fault Tolerance policies in the design and implementation of the application, and in the validation of its Real-Time requirements. Special notations have been proposed to this purpose, both as programming-in-the-large notation [6] and as notations meant at an intermediate level [2] between programming-in-the-large and programming-in-the-small. These notations have to be general, powerful and expressive, in order to give the designer the maximum freedom. With the FERT notation, fault tolerance strategies can be defined for complex real-time systems and tailored to the existing environment.

One of the problems of providing fault tolerance in real-time applications is, that fault tolerance strategies are often very time-(resource-) intensive and that worst case execution times are often difficult to estimate. Imprecise estimations also add to more resource consumption which may lead in the end to unfeasible schedules. On the other hand, the effective resource consumption may often be much smaller than the estimated one, which means that the possibility for a successful computation before a given deadline is large. The problem for the run-time system is to guarantee, that given deadline is not violated, even if a computation does not succeed within its timing bounds.

The TaskPair-Scheduling (TPS) scheme [7] matches this requirement. It provides a method for flexible and predictable execution of tasks with hard timing constraints (no deadline violation allowed) in presence of „unsuccessful" computations as described above, by im-

0-8186-7570-5/96 $5.00 © 1996 IEEE

plicitly scheduling exceptional reactions if a task does not terminate within its time bound.

The main purpose of this paper is to discuss possible solutions how to realise FERT strategies by means of TPS. On the one hand TPS allows use of specifying average execution times while guaranteeing that deadlines are not violated, and on the other hand - if all worst case execution times are known (and usable) - it provides the full guarantee needed for the „original" FERT notation.

As a first step only a subset of FERT is considered. It consists of fault tolerance strategies which cover a large range of applications: the Recovery Block Scheme [1], N-Version Programming (NVP) [3], and Imprecise Computations [5]. In this paper the mapping of the Recovery Block Scheme and NVP to TPS is described as examples.

The rest of this paper is organised as follows. Section 2 outlines the FERT notation; section 3 outlines the TPS dynamic scheduling scheme; section 4 addresses some issues about usability of TPS to realise FERTs; section 5 describes two examples for possible implementations in terms of TPS.

2. The FERT notation

The FERT notation (Fault-tolerant Entity for Real Time), proposed in [2], allows the application designer both to define adaptable behaviour of the application and to specify different FT policies for each application component. This notation introduces an intermediate level in the design of the application, between the programming in-the-large and the programming in-the-small; at run-time, an application is composed of a set of functional modules, also called FERTs.

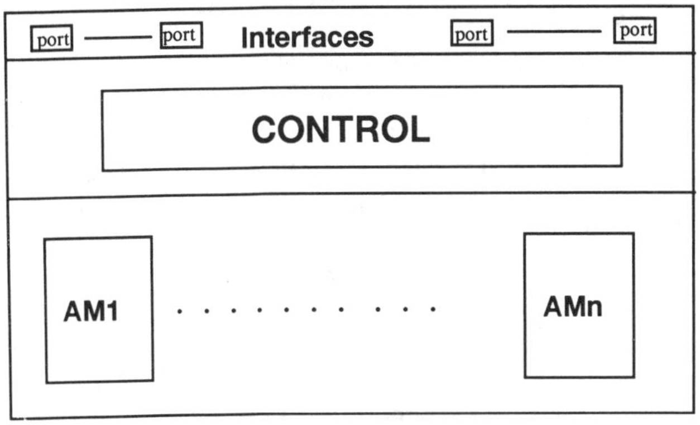

Figure 1. Structure of a FERT

A FERT includes three parts, each described below (see fig. 1):

1. an interface towards the other FERTs;
2. a set of Application Modules (AM in the following);
3. a control part which describes how to handle the execution of the AMs.

An interface is simply composed of a set of input and output ports.

The AMs are modules written in a language in-the-small, each one with its own worst case execution time and its input and output ports.

The control part controls the dynamic execution of the AMs in a FERT. In the control part the application designer can define an execution graph for the AMs through four primitives: POSSIBLE, EXEC, UNUSED and OUTPUT. The POSSIBLE primitive defines an alternative for the execution of the FERT, intended as a set of AMs with precedence constraints, a deadline for completing its execution and a value associated to it. With more than one POSSIBLE, the FERT designer can implement alternative computations (*''strategies''*) for the same FERT, with different values and costs. The remaining three primitives state how and under which conditions the execution of a single strategy must occur. The EXEC primitive specifies the inputs needed for the execution of each AM; such inputs may come from the output port of other AMs, from input ports of the FERT or may be given as constants. The OUTPUT primitive sends results out of the FERT, through the FERT interface. Lastly, the UNUSED primitive states that some AMs are no longer necessary to complete the strategy; such primitive is invoked after a run-time control on some condition of the output ports. For example, in a Recovery Block [Randell, 1976] the next alternative is not executed, if the results of previous one pass the acceptance test.

Externally, the FERT behaves also as a fault containment unit: the FERT resolves and masks all the faults admitted under the failure hypothesis made. Therefore the application designer can define the FERTs and their interfaces as parts of a functional design, without any explicit concern of faults and fault propagation between FERTs. Moreover for each *mode of functioning* of the system the FERTs belonging to an application can be divided in *fundamental* and *non fundamental* ones. The fundamental FERTs execute the computations that in the current mode of functioning are critical to the integrity of the system or for safety reasons: if the run-time support accepts to execute a given application, then it must guarantee that at least a *default strategy* (defined by the designer) will be executed within its deadline, of course under given fault and load hypotheses. The arrival time to the system of the fundamental FERTs of an application can be *periodic* (equally spaced, with a period characteristic for the FERT) or *sporadic* (arbitrarily spaced, but always with an interval longer than a fixed minimum, characteristic of the FERT). Non-fundamental FERTs can also be *aperiodic* (arrival time arbitrary spaced without a minimum).

In short, at run-time only one of the strategies of a FERT is chosen for execution, according to its „cost“ in terms of resources needed and its „value“ in terms of benefits gained by its execution. When accepting to execute a FERT, i.e., one of its strategies, the system is expected to guarantee deadline and internal precedence constraints of this strategy. The system must also guarantee that it will accept to execute possible safety-critical FERTs arriving with periodical rate or with a sporadic rate, although with an inter-arrival time not less than a given a minimum. Non critical FERTs can contend resources to each other, possibly utilising resources left because of unused redundancies and/or because of executions shorter than the specified worst case execution time (WCET).

3. The Task-Pair Scheduling scheme

The Task-Pair Scheduling scheme (TPS) is a central component of the DIRECT approach [9], aimed to dynamically guarantee to each application component accepted for execution, a minimum level of computation, and to execute the remaining part based on a best effort approach. To this purpose, the component is structured in a Task Pair (TP), made of a SoftTask (ST) and a HardTask (HT). A deadline is specified for the TP. The ST executes the normal computation, with a „best effort“ approach. With „best effort“ we mean, that we explicitly want to consider unknown timing behaviour of SoftTasks. The HT is executed if the SoftTask does not terminate within the specified time, in order to avoid a violation of the pair deadline. For a HT its WCET has to be specified. The scheduler reserves the CPU for this amount of time if the pair is accepted for execution. The scheduler is able to reserve resources also for STs. But in contrast to HTs the scheduler does not rely on timely correct behaviour of a ST.

The application programmer specifies the two parts (the *SoftTask* and *HardTask*, according to the TPS notation) while he's writing the code of the process itself.

```
TaskPair TP = (SoftTask,HardTask);
```

with the intention:

```
if( guarantee (TP,deadline)){
    try
            <SoftTask>
    except
            <HardTask>
    end
}
else // no guarantee
```

The TaskPair is described by a TaskPair descriptor, including a task descriptor for each task providing all necessary information about the TaskPair to the on-line scheduler, e.g. WCET (if possible) or a function to compute the WCET, earliest start-time, dependencies etc..

The basic idea was to execute the SoftTask only if all required resources of the HardTask can be reserved by the scheduler. This guarantees that either the SoftTask or the HardTask will succeed the operation before the deadline. Only the WCET of the HardTask must be known. It is obvious, that such a construction makes sense, if the WCET of the HardTask can be computed, and is small in comparison to the WCET of the SoftTask.

The smaller and simpler a HardTask is, the larger is the chance to guarantee it. The question is, what can or should effectively be done by a HardTask?

We decide between three main functions:

1. Undoing, what the SoftTask has done so far, which can be used for implementation of atomic behaviour of TaskPairs.
2. Completion of what has been tried by the SoftTask, e.g. for kinds of imprecise computation, the HardTask can utilise, what has been done so far by the SoftTask to compute a usable result.
3. Support of „basic functionality“, which is applicable to graceful degradation schemes. The HardTask performs a minimum quality of service, that is used if the desired quality cannot be achieved.

If the HardTask is empty (but exists), TPS guarantees the termination of the TaskPair before its deadline. No application relevant computation is performed, the only purpose of the HardTask is to avoid a deadline violation and to avoid the disturbance of other tasks. In either case, predictability of the timing behaviour of the TaskPair is guaranteed.

A Simple Guarantee Algorithm For TaskPairs

The simplest way to implement TaskPairs is, to dispatch SoftTasks with a simple dispatch strategy, preemptive and with low priority, while executing HardTasks as late as possible but with maximum priority. This means that HardTasks are planned in a „time triggered“ way and that they are non-preemptive. A simple dispatch strategy for SoftTasks can be e.g. EDF or least laxity first in combination with guaranteeing a minimum of execution time for SoftTasks.

The following picture shows two independent TaskPairs TP1 and TP2 running concurrently.

```
void TP1(deadline)
{
  if( guarantee(TP1,deadline) ) {
      try      ST1();
      except  HT1();
  } else ...; // no guarantee
}

void TP2(deadline)
{
  if( guarantee(TP2,deadline) ) {
      try      ST2();
      except  HT2();
  } else ...; // no guarantee
}
```

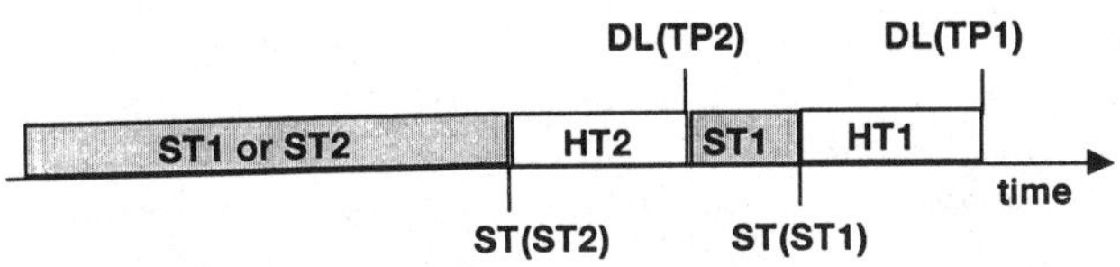

Figure 2: Simple strategy for scheduling TaskPairs

The scheduler places the HardTasks (HT1 and HT2) as late as possible with respect to their deadlines[1] (DL) by reserving processor time on a time axis. In this example the deadline of TP2 is earlier than HT1's deadline. The SoftTasks (ST1 and ST2) start running. ST1 and ST2 may be preemptive or not and scheduled in an arbitrary way. If ST2 does not terminate before start time (ST) of HT2, HT2 starts execution. If ST2 succeeds, i.e. ST2 terminates before ST(HT2), HT2 is removed from the scheduler's plan. TaskPair T1 behaves analogously.

TaskPairs may be nested, i.e. within SoftTasks and within HardTasks further TaskPairs may be entered. The scheduler knows about the dependency, i.e. an inner TaskPair has always to terminate before the outer one. E.g. for a TaskPair TP2 within a HardTask HT1 this implies that WCET(HT2) < WCET(HT1). In addition the time-difference is implicitly reserved for ST2. This construction is utilised when mapping FERT strategies to TaskPairs.

Currently distributed scheduling is under development. Communicating local schedulers guarantee TaskPairs with a common deadline executing on different nodes.

4. Issues on implementing FERTs on a TPS system

One of the basic assumptions of FERT is that at run-time system resources are allocated to the selected strategy before its execution, such that its deadline is guaranteed. In order to maintain this semantic for an implementation of FERTs on TPS, each strategy ought to be mapped on the HardTask of a pair, leaving empty the corresponding SoftTask. This method implements software fault tolerance covered by the applied FERT strategy but does not utilise the features of TPS. To take benefit of TPS for fault tolerance, the application should be implemented in SoftTasks, while fault management (e.g. recovery, acceptance tests, voting) should to be placed to HardTasks. This is reasonable because fault-management is assumed to be correct and has a known WCET, which is required for HardTasks. In contrast the application may fail and it is our aim to tolerate a faulty behaviour. Since resources can also be guaranteed for SoftTasks, there is no reason for putting the application into a HardTask. For the remainder of this paper, we assume, that the application is implemented in SoftTasks, while fault-management is implemented in HardTask.[2]

With this construction, two main benefits can be drawn from TPS for implementing fault-tolerance:

1. Applications, or parts of it, can be executed on a best effort basis.
2. Faults, which may result in a changed timing behaviour, can be tolerated (time faults).

Ad 1) This means, although if the system is not able to guarantee full availability of resources, the application can be started. If an application does not complete within time, the system will abort the SoftTask and start the HardTask. Now, since the HardTask contains the fault-management, fault-tolerance operations are activated.

Starting an application on a best effort basis is reasonable, only, if there is a realistic chance for a successful execution. If the system is already running on a high load, there will be not much resources available for an additional task. To deal with such situations, resources can also be reserved for SoftTasks. Here the question arises, how much resources shall be reserved. For this purpose we defined the „Optimistic Case Execution Time", OCET [4]. The $OCET_{,x}$ defines the time an application will need with a probability x of successful execution. If, e.g. 99% of all possible (correct) executions of an application will complete before $OCET_{99}$ time units and the WCET of the application is much larger than $OCET_{99}$, it seems to be reasonable to reserve $OCET_{99}$ time units instead of the WCET. Especially this is true, if the fault probability of the application has a comparable value.

Ad 2) Faults, which results in a changed timing behaviour are implicitly detected by TPS, if they appear in a SoftTask. TPS will automatically start the corresponding

1For simplicity we do not consider time for run-time system activities like guaranteeing tasks, switching of threads etc.. All such activities must be bounded (see also [8]).

2 For performance improvement it is sometimes reasonable to place parts of the fault-mangement additionally in a SoftTask.

HardTask as a first reaction on this fault. This mechanism can now be utilised by the fault-management executing in the HardTask. So, TPS enables the fault-management provided by FERT strategies to react on time-faults.

To implement a FERT strategy on TPS, the FERT components are mapped to a nested structure of TaskPairs.
A straightforward mapping exist, when *incremental* strategies are implemented by FERTs. A strategy Si is incremental with respect to a strategy Si-1, if Si is made by adding components and functionality over Si-1. For example (see figure 3), S2 could include an accurate but fragile algorithm V2 and, as a second chance, a robust but less accurate one V1. S1 could be made just of the latter algorithm. S0 contains the absolute minimum which has to be executed, i.e. in the FERTs notation an exception handler. Such strategies could be easily mapped on a TaskPairs, where strategy S2 is a TaskPair with V2 as SoftTask and the remainder (= S1) as the HardTask. S1 can now be implemented by another TaskPair, in which the SoftTask implements V1 and the HardTask V0.

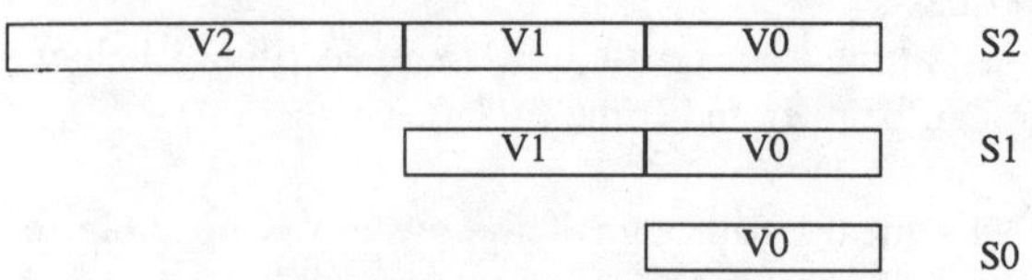

Figure 3: Incremental strategies

In the original idea of FERT the scheduler chooses a strategy for execution and reserves the requires resources. To achieve the maximum level of fault-tolerance, S2 must be guaranteed, S1 performs less fault-tolerance and S0 only an exception handler. So, the scheduler tries to guarantee S2. If this is not possible then S1, and if even this is not possible, S0.

By using best effort approaches for an algorithm, we get additional alternatives for execution. If, e.g. only S1 can be guaranteed it may be possible, that V2 has a reasonable chance to succeed, such that V2 should be tried. Or, if only S0 can be guaranteed, V1 can be tried. This means that beneath the guaranteed algorithms one more can be tried, enhancing the overall chance for a successful computation.

The flexibility of this approach is more enhanced if instead of WCETs OCETs are used. Since all algorithms of the example are implemented in a SoftTask (with respect to the related nesting level) the execution times specified for the algorithms need not be the WCET. This gives the designer additional possibilities for improving the fault-tolerance. If (in SoftTasks) WCETs can be replaced by OCETs with a similar success probability, this might enable the scheduler to guarantee a strategy with an additional algorithm, which in the end has a greater success probability than the strategy with WCETs.

To investigate concrete problems of implementing FERTs on a TPS system, the following set of fault tolerance policies can be taken into account, as representative ones: Recovery Block, N-Version Programming, and Imprecise Computations; these three FT policies seem to be sufficient for a programmer to specify a good variety of solutions covering almost every real-life scenarios. Recovery Blocks and Imprecise Computations have a similar dependency structure like TPS, so their mapping can take advantage of that. For N-Version Programming, a sequential execution of the version can be mapped, or a parallel version. If parallelism is required (e.g. to execute different versions on different processors at the same time or quasi-parallelism on one processor but with different threads of execution), a TaskPair have to be spawned, i.e. at least asynchronous activities has to be started, which have to inherit the initiating TaskPairs deadline; more precisely, it has to inherit the start time of the HardTask of the initiating TaskPair. In such cases, the HardTask of the TaskPair could be used for comparing the results of the N-Versions.

5. Implementation of some FT techniques by means of TPS.

In the following, two examples are proposed of implementation of FERTs on TaskPairs. The first one is based on a Recovery Block scheme, the second one on a N-Version scheme.

Before describing the implementation, we extend TPS by the primitive *execHardTask* due to the following problem:

TPS will terminate a TaskPair if a SoftTask completes timely. Since we place the fault-management in HardTasks (e.g. an acceptance test or voting) this will not be executed although it would be needed.

The new primitive execHardTask advises the scheduler not to terminate the TaskPair but to start the HardTask.

5.1. Recovery Block

The FERT is supposed to include following incremental strategies:

S_2 a Recovery Block with two versions;

S_1 the execution of a single variant with test on the output;

S_0 an exception handler.

The strategy S_2 has the following code:

```
begin {Strategy S2}
```

```
exec(V2(Fert.In), AT(V2.out));
if AT.out=success then
      begin
         output (V2.out)
         unused(RS,V1,AT,RS,EH)
      end
else  begin
        exec(RS,V2(Fert.In,
        AT(V1.out))
        if AT.out=success
          then begin
                 output (V1.out)
                 unused(RS, EH)
               end
          else exec(RS, EH)
      end
end {Strategy S2}
```

In the code, the following symbols have been used:

- V_i is the i-th variant of the RB;
- AT is the module implementing the acceptance test;
- RS is the component restoring the state after a failed execution of V_i;
- EH is the FERT exception handler.

The strategy S_1 has the following code:

```
begin {Strategy S1}
exec (V1(Fert.In), AT(V1.out))
      if AT.out=success
          then begin
                 output (V1.out)
                 unused(RS, EH)
               end
          else exec(RS, EH)
      end
end {Strategy S1}
```

Lastly, the strategy S0 is the following:

```
begin {Strategy S0}
      exec(EH)
end {Strategy S0}
```

According to the idea proposed in the previous section, these strategies can be implemented by means of a nested TPS, like the following one:

$TP_{S2} = \{ ST_{S2}, HT_{S2} \}$
$ST_{S2} = \{ V_2, AT,$
if AT.out=success then output(V_2.out)
else execHardTask }
$HT_{S2} = \{ RS, TP_{S1} \}$

$TP_{S1} = \{ ST_{S1}, HT_{S1} \}$
$ST_{S1} = \{ V_1, AT,$
if AT.out=success then output(V_1.out)
else execHardTask }
$HT_{S1} = \{ RS, TP_{S0} \}$

$TP_{S0} = \{ ST_{S0}, HT_{S0} \}$
$HT_{S0} = \{ EH \}$
$ST_{S0} = \{ execHardTask \}$

As described above „execHardTask" indicates, that the SoftTask must not return, becauses the corresponding HasdTask has to complete the TaskPair. The „unused" statement of FERTs is ignored, because resources are reclaimed by the scheduler implicitly. For all code within HardTasks we assume correct implementation and known WCETs. For SoftTasks arbitrary resources (WCETs, OCETs, or nothing) are guaranteed.

Assuming that the scheduler has guaranteed the strategy S2 (TP_{S2}) this implementation results in executing V_2 and then, if necessary, V_1 and then, if necessary, EH; V_1 is necessary in case of deadline-missing of V_2 (e.g. due to a time fault), or in case that V_2 gives a wrong result.

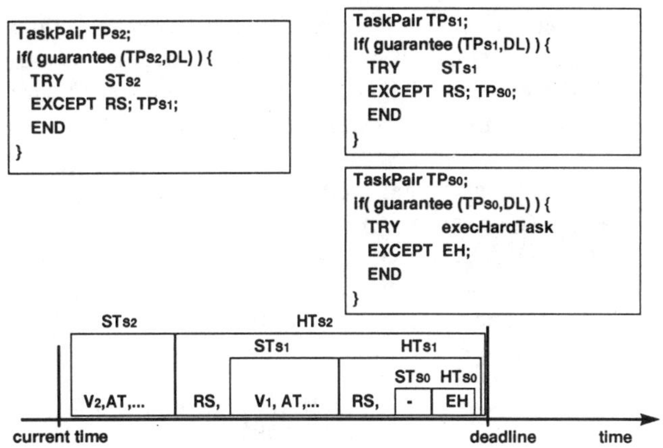

Figure 4: Implementation of Recovery Block Strategy

Figure 4 shows the scheduled Recovery Block Strategy S2 with TPS, like it is specified above. Since the HardTask HT_{S2} of TP_{S2} contains SoftTask ST_{S1} and HardTask HT_{S1} of TP_{S1}, resources for both tasks are reserved by construction. Since the Algorithms $V_1$and V_2 are running in the SoftTasks, the programmer has the freedom to specify the amount of execution time for reservation by maintaining the tolerance of timing faults. This is especially worthful, if the WCET differs extremely from the average (or „normal") execution time.

5.2. N-Version Programming

The following example does not claim to be a applicable solution for NVP and especially it cannot fully exploit the power and flexibility of FERT and TPS. The example is presented to give an idea about a possible implementation. Therefore, we consider a sequential execution of the different versions.

The FERT is supposed to include following incremental strategies:

- S_3: a N-Version scheme, with three versions;
- S_2: a N-Version scheme, with two versions;
- S_1: the execution of a single version plus an acceptance test;
- S_0: an exception handler.

In the code of the strategies the following symbols have been used:

- V_i is the i-th version of the NVP;
- EH is the FERT exception handler;
- Voter is a component which behaves according to the number of results on which it executes. It behaves as a voter if there are three results; it behaves as a compare test if it is called on two results; it behaves as an acceptance test, if it is called on a single result; last, it does nothing if it is called on no result at all.

The strategy S_1 has the following code:

```
begin {Strategy S3}
 exec((V3(Fert.In)||V2(Fert.In)||
       V1(Fert.In)),
       Voter(V3.out, V2.out,V1.out);
       if Voter.out=success
         then begin
                output (V.out)
                unused(EH)
              end
         else  exec(EH)
end {Strategy S3}
```

The strategy S_2 has the following code:

```
begin {Strategy S2}
 exec((V2(Fert.In)||V1(Fert.In)),
       Voter(V2.out, V1.out);
       if Voter.out=success
         then begin
                output (V.out)
                unused(EH)
              end
         else  exec(EH)
end {Strategy S2}
```

The strategy S_1 has the following code:

```
begin {Strategy S1}
 exec((V1(Fert.In)),
       Voter(V1.out);
       if Voter.out=success
         then begin
                output (V1.out)
                unused(EH)
              end
         else  exec(EH)
end {Strategy S1}
```

Lastly, the strategy S_0 has the following code:

```
begin {Strategy S0}
      exec(EH)
end {Strategy S0}
```

These strategies can be implemented by means of a nested TP, like the following one:

TP_{S3} = { ST_{S3}, HT_{S3} }
ST_{S3} = { V_3, execHardTask }
HT_{S3} = { TP $_{S2}$,
Voter(V_3.out, V_2.out, V_1.out);
if Voter.out = success
then output (V.out)
else EH }

TP_{S2} = { ST_{S2}, HT_{S2} }
ST_{S2} = { V_2, execHardTask }
HT_{S2} = { TP $_{S1}$,
Voter(V_2.out, V_1.out);
if Voter.out = success
then output (V.out)
else EH }

TP_{S1} = { ST_{S1}, HT_{S1} }
ST_{S1} = { V_1, execHardTask }
HT_{S1} = { TP_{S0},
Voter(V_1.out);
if Voter.out = success
then output (V1.out)
else EH }

TP_{S0} = { ST_{S1}, HT_{S1} }
ST_{S0} = { execHardTask }
HT_{S0} = { EH }

In normal conditions, this implementation is expected to execute V_3 , V_2 and V_1 plus the Voter plus, if necessary, EH. It executes only V_2, V_3 and the Voter (and then, if necessary, EH) in case of scarce resources. It executes only V_3 and the Voter (and again, if necessary, EH) in case of scarcer resources. In case of extremely scarce resources, only execution of EH is committed.

6. Conclusion

In this paper it is outlined how fault tolerance strategies for real-time applications can be implemented. Strategies denoted by the FERT notation are mapped to the Task-Pair-Scheduling scheme.

While for small systems with static behaviour fault tolerance policies based on worst case execution times can be used, this is not true for more complex systems, in which worst case execution times are partially unknown or differ drastically from their average execution time. Since FERTs originally depend on guaranteed execution times in such cases often guarantees cannot be given, although there may exist a good chance to succeed. Usage of the TPS scheme alleviates the precondition for executing FERT's strategies by mapping them to TaskPairs, which consists of a precisely (guaranteed) scheduled HardTask, based on WCETs and a SoftTask, which can be scheduled by best effort or for which arbitrary execution time can be specified. This enhances the flexibility of FERTs because additional strategies are definable by the specification of different resource requirements.

In all cases TPS guarantees a timely execution and termination of all TaskPairs. With this construction also time faults of applications can be tolerated.

A possible mapping to TaskPairs was proposed for the Recovery Block and for a sequential implementation of the N-Version scheme. Mappings for distributed strategies are currently under study.

References

[1] T. Anderson, P.A. Lee, „Fault Tolerance: Principles and Practice“, Prentice-Hall International, Inc. 1981

[2] A. Bondavalli, J. Stankovic and L. Strigini: "Adaptable Fault Tolerance for Real-Time Systems," in "Responsive Computer-Systems: Toward Integration of Fault-tolerance and Real-time", D. Fussell, Ed., Kluwer Academic Publishers, 1995, pp. 187-208.

[3] L. Chen, A. Avizienis, „N-Version Programming: A Fault-Tolerant Approach to Reliability of Software Operation“, Proc. 8 Int. Symp. on Fault Tolerant Computing Systems, Toulouse, June 1978

[4] M. Gergeleit, H. Streich: „TaskPair-Scheduling with Optimistic Execution Times - An Example of an Adaptive Real-Time System“, Proceedings of the WORDS '96, IEEE Comp. Soc. Second Int. Workshop on Object-oriented Real-time Dependaple Systems, Laguna Beach, California,

[5] J.W.S.Liu et al., "Algorithms for Scheduling Imprecise Computations", COMPUTER, May 1991, pp. 58-68

[6] D. Niehaus: Program representation and translation to support predictability of real-time systems. RTSS 1991

[7] H. Streich: "TaskPair-Scheduling: An Approach for Dynamic Real-Time Systems“, Int. Journal of Mini & Micro-computers, Vol. 17, No. 2, pp 77-83, 1995

[8] J.A. Stankovic, K. Ramamritham, "The Spring Kernel: A New Paradigm for Real-Time Operating Systems", ACM Operating Systems Review, Vol. 23 No.3, July 1989.

[9] M. Gergeleit, J. Kaiser, H. Streich: „Towards a Distributed Object-Oriented Real-Time Control System", Workshop on Concurrent Object-based Systems, in conjunction with the IEEE Symp. on Parallel and Distr. Processing, Dallas, TX, October 1994

Toward Dependable Safety-Critical Software*

Farokh Bastani, Bojan Cukic, Victoria Hilford, Anouar Jamoussi
Department of Computer Science
University of Houston
Houston, TX 77204-3475, USA
Email: FBastani@uh.edu

Abstract

The failure of safety-critical systems can result in catastrophic loss of life and property. Hence, it is necessary to assure the reliability of these systems to a high degree of confidence before they are put into operational use. However, at these extreme levels of ultra-high reliability requirements, typically failures rates of less than 10^{-7} *failures per hour, errors in the specification and in estimates of the operational profile become significant factors. An approach that has been suggested in practice is to use secondary and tertiary software that meet ultra-high reliability requirements but at a reduced functionality as compared with the primary software. Two major problems are (a) how to select appropriate functionality for the non-primary versions and (b) how to determine when to invoke these backup versions.*

In this paper, we present a unified approach for handling these two problems. It starts with a rigorous method for assessing ultra-high reliability requirements and then develops mechanisms for incorporating one or more backup versions. The reliability assessment procedure uses formal methods to amplify the effect of each test case and results in the construction of a reliability MAP (Measured Assurance Prediction system) for the software. This provides a confidence estimate for the correctness of the software for a given operational situation and serves as a trigger for switching to a backup version. The main requirement is that the MAP for the backup version must be known to a higher degree of confidence than that for the original version. The approach is illustrated using a simple process-control example.

1. Introduction

Embedded computer systems are used to control many safety-critical systems, such as aircraft control systems, chemical manufacturing plants, nuclear power plants, space vehicles, mass transit systems, and so on. Any failure of these systems can result in catastrophic loss of life and property. Hence, it is essential to ensure with a high degree of confidence that these systems meet their reliability requirements prior to deploying them in the field.

There are many sources of failures, such as physical wear and tear, design faults, faulty manufacturing, failures due to harsh environments (high temperature, radiation, noise), etc. These can be broadly classified as expected (or anticipated) failures and unanticipated failures. Anticipated failures include failures that have predictable effects, such as stuck-at-0 or stuck-at-1 type of faults, as well as unpredictable effects, such as single-event upsets that change code and data. In these cases, redundancy in the form of replication or parity codes, can be used to recover from the failure. Unanticipated failures, on the other hand, are usually due to design failures or specification faults that fail to anticipate some operating situations. All software faults are of the unanticipated category so that simple replication is not effective (all the versions will have the same fault). For hardware systems that are widely used in many applications, it is possible to ensure the absence of unanticipated faults with a high degree of confidence. However, as illustrated by the case of the Pentium^{TM} chip, unanticipated failures occur in practice even for hardware.

One way to deal with unanticipated failures is via independent specification and design of multiple versions of the system. Then, at run-time, the different versions can be executed in parallel with their outputs being compared by a voter [2] or they can be executed sequentially using the recovery block approach [14, 22]. The former approach requires specialized voters while the latter approach has to tolerate failures in the acceptance test. There have been some reports of correlated failures in spite of diverse ver-

*This work was supported in part by the U.S. Nuclear Regulatory Commission (NRC) under Award No. NRC-04-92-090 and in part by UH-ISSO. The opinions, findings, conclusions and recommendations expressed herein are those of the authors and do not necessarily reflect the views of the NRC.

sions [15]. Our preliminary experiments show that diversity is effective at reducing latent faults, but this is in the context of achieving *consensus designs*, i.e., using diversity within the software development process to achieve one ultra-high reliable version of the software. This still leaves open the possibility that the single consensus version will eventually fail, perhaps due to a fault in the specification.

A possible solution is to use "reduced capability" backup versions [17]. In this approach, the primary version provides full system functionality. When a failure is detected, the system switches to an alternative version that only provides critical functionality. There can be a tertiary version that provides even more limited capability that serves solely to prevent catastrophic failures. The expectation is that the reduced capability versions are simpler and, hence, can be designed and assessed to a higher degree of confidence than the primary version. This will be especially true if an evolutionary development effort is used in which the backup versions are evolved to the primary version [4]. The problem is how to determine when to switch from the primary version to the alternate version since there is the possibility of a failure in the acceptance test.

In this paper we show how the effort spent in assessing software reliability can be used to construct a run-time guide for determining when to invoke a backup version. The approach uses the assessment data to construct a reliability map of the program, i.e., record high confidence and low confidence regions in the input space. It requires the reliability map of the backup version to dominate (or be known to a higher degree of confidence than) that for the primary version. When the operating conditions move to a low confidence region, the backup version is invoked.

The raw assessment data is usually too sparse to provide sufficient confidence. For example, consider a program with 50 different input variables with each input having 10 different values. Then, testing the program at the rate of 1 test case every nanosecond for 100,000 years will directly validate only an insignificant fraction (less than 10^{-28}) of the input space. We briefly present formal methods that can be used to amplify the effect of each test data and to accelerate program testing so that an accurate map can be constructed in a reasonable amount of time.

The rest of the paper is organized as follows. Section 2 presents the background material, including the type of programs and the combination of formal methods with statistical testing. Section 3 presents the amplification method while Section 4 presents an outline of the MAP (Measured Assurance Prediction) system. Section 5 summarizes the paper and discusses some future research.

2. Background

The type of programs which we consider are process-control programs having the following characteristics. The program executes a series of cycles or frames. At the beginning of each frame, it reads inputs from the sensors, then it performs some computations, and at the end of the frame it sends commands to the actuators. The inputs are numerical quantities, such as temperature, pressure, velocity, and so on, rather than symbolic strings. Similarly, the computations are numerical computations. The computation in each frame depends on the inputs read at the beginning of the frame as well as the values of some internal variables called history or state variables. These are necessary for computing the rate of change of an input quantity, the current state of actuators (such as the state of valves), etc. These state variables are updated at the end of every frame.

Process-control computations can be viewed as functions that map points in $I \times S$ to points in O, where I denotes the input space, S denotes the state space, and O denotes the output space. Further, because of the limited accuracy of sensors and actuators, these functions are often piecewise continuous. Similarly, the fault regions are also clustered rather than being randomly scattered over the input/state space [3]. If the control function is smooth, then interpolation can be used to obtain an estimate of the ouput value from nearby points that have already been tested. The accuracy of the estimate improves as the distance between the tested points decreases.

These characteristics have been used to develop models for measuring the correctness probability [3, 21] (and correctness possibility [5, 6]) of a process-control program based on the set of test cases used. In these models, each test case provides some probabilistic (or fuzzy) information that the program is correct for a test case x given that it has been tested and found to be correct for test case y. The models then proceed in two steps to

- (a) combine the information from the entire set of test cases to estimate the correctness probability (or correctness possibility) of the program for a given untested point and

- (b) combine the information for all the untested points to obtain an estimate of the probability (or possibility) that there is no fault in the program.

The MAP system is related to our earlier work but with two differences. First, it only proceeds to step (a) to provide an estimate, based on interpolation, of the confidence in the correctness of the program for a particular input [11]. Second, it uses a combination of formal methods with statistical testing to eliminate any assumptions regarding the distribution of fault regions in the input domain. Formal methods are used to amplify the effect of testing so that each test case provides information about the behavior of the program for a large number of points in $I \times S$.

Ideally, one would prefer to formally show that the program meets its specification in which case there is no need for any backup versions. The advantage of formal verification is that it is independent of the distribution of inputs to the program. A major disadvantage is that it cannot cope with specification errors. Also, current proof methods assume ideal machine conditions (infinite precision, and no compiler, operating system, or CPU faults, etc.) and are susceptible to errors in the proof process unless the verification process is fully automated and the verifier itself has been completely verified.

An alternative approach is to use statistical testing to obtain an estimate of the reliability of the program. Its main advantage is that it directly yields a measure of the system reliability [13]. That is, it accounts for numerical round-off errors and timing errors as well as faults in the compiler, operating system, CPU, etc. However, there is no single widely accepted approach for obtaining the reliability estimate. Three major classes of software reliability models that have been proposed over the past 25 years are fault-seeding models, reliability growth models, and sampling models [21]. Fault-seeding models make assumptions regarding the distribution of the remaining faults in the program which are difficult to justify rigorously. Reliability growth models make difficult-to-justify assumptions about the size of faults that are removed at each step. The sampling model [24] is theoretically sound, but it has several practical problems. The reliability estimate is sensitive to variations in the operational profile (this is true for any statistical approach). Also, it needs a large number of test cases [9] and an accurate test oracle [1].

A natural way of overcoming these problems is to combine formal verification with statistical sampling. This has been done to some extent in the input-domain based approach [19] [20] which divides the input space into a number of partitions. Only those partitions that cannot be formally verified are subjected to statistical testing. The advantage of this method is that verification of partitions that correspond to normal operating situations (i.e., have a high probability of occurrence) results in a significant reduction in the testing requirement. Also, since normal operating regimes are usually well understood, one can have more confidence in the correctness of their specification so that verification is more acceptable. However, this approach requires the verification of complete program paths which is difficult to perform in practice. Other proof techniques, such as hierarchical proof methods [23] and proof of general properties (such as showing the absence of deadlocks, absence of violation of mutual exclusion, satisfaction of real-time constraints, etc.), do not reduce the amount of testing that is needed to attain a certain level of confidence in the reliability of the program! This is an apparent paradox since one would expect that the effort expended in proving some portion of the code would result in a corresponding decrease in the amount of testing needed. However, with the exception of the proof of complete paths, no other formal method appears to directly decrease the testing requirements.

The novel aspect of the approach proposed for developing the MAP system is to use formal methods, not for the purpose of program verification which can fail to uncover specification errors, but instead to dramatically increase the effectiveness of testing. This marriage of formal methods with statistical testing combines the strengths of both approaches without incurring their weaknesses. In particular, formal methods are used to enhance program testing in the following ways:

1. *Test Case Amplification* by using partial proofs and source code transformation rules to infer the behavior of the program for many inputs based on its behavior for one input.

2. *Execution Acceleration* by transforming a program P into a version Q such that the correctness of Q implies the correctness of P and Q runs faster than P.

3. *Efficient Test Data Generation* by automatically generating the code for an optimized test data generator for each partition in the input space.

Based on the set of directly or indirectly validated points in $I \times S$, the MAP system will construct an assurance profile

$$f_P :: I \times S \rightarrow 0 \cdot\cdot 1$$

for the program P. This profile is a function of the density of validated points in the neighborhood of each point as well as the shape of the computed function. If P is the primary program and Q is the backup program, then the following condition must be satisfied:

$$\forall x : x \in I \times S :: f_Q(x) \geq f_P(x),$$

i.e., the confidence in the backup version must be uniformly greater or equal to that in the primary version.

In the next two sections, we present more details of the approach. The following control problem is used to illustrate the method [16]. Consider a spacecraft, C, that has six engines, two of which cause it to move linearly in the forward or reverse directions (L_f and L_r, respectively), two of which cause it to rotate along one plane (S_{1f} and S_{1r}, respectively), and two cause it to rotate along an orthogonal plane (S_{2f} and S_{2r}, respectively). (Pair of thrusters are needed to be able to invoke starting and braking actions, respectively.) The goal of the control program is to move the spacecraft from a starting point toward the space station, S, and then move around it at a specified radius r. The spacecraft should avoid collision with all obstacles and should conserve fuel. The inputs to the program are as follows:

Coordinates of S: (x_S, y_S, z_S);
Coordinates of C: (x_C, y_C, z_C);
Coordinates of nearby objects, O_i: (x_i, y_i, z_i);
Velocity of S: (v_{xS}, v_{yS}, v_{zS});
Velocity of C: (v_{xC}, v_{yC}, v_{zC});
Velocity of O_i: (v_{xi}, v_{yi}, v_{zi});
Desired radius: r;
Setting of engines: $l_f, l_r, s_{1f}, s_{1r}, s_{2f}, s_{2r}$.

A possible sequence of evolutionary versions [4] is as follows:

1. Version 1. Avoid collisions with nearby objects and the space-station.
2. Version 2. Avoid collisions and approach the station.
3. Version 3. Avoid collisions and approach the station and then move around the station.
4. Version 4. Enhance Version 3 to minimize fuel consumption.
5. Version 5. Enhance Version 4 to provide a smooth control, i.e., bounded linear and angular accelerations and decelerations.

In this sequence, the recovery block can be structured in order to have Version 5 as the primary version with Version 3 as the secondary version and Version 1 as the tertiary (safety-related) version.

3. Test Case Amplification

To see how formal methods are used to amplify test cases, consider a program *classify_triangle(a,b,c)* that reads in the three sides of a triangle and classifies it. If we can prove that the final portion of the program operates on x, y, and z which are permutations of a, b, and c such that $x \leq y \leq z$, then each test case with this proof corresponds to 6 test cases without this proof. This type of partial proofs can also collapse complete dimensions of the input space, with the result that fewer test cases are needed and the reliability estimate becomes less sensitive to the operational profile.

In the MAP system, test cases are amplified using the following steps.

1. *Domain decomposition.* The first step forms vertical slices of the program [25] that decompose the input domain into smaller portions. Typically, each slice will target the computations related to a distinct output or state variable. This has a significant benefit if each portion has fewer dimensions than the overall input domain. For example, consider an input space with 5 dimensions where each dimension has 10 different values for a total of 10^5 input points. Suppose that it is possible to slice the program to obtain two portions where each portion has 3 dimensions. Then, the total number of input points in this case is 2×10^3 which is just 2% of the original 5-dimensional input domain. Alternatively, every test case in the partitioned domain corresponds to 100 points in the original domain, which is a substantial amplification. Vertical program slicing is shown in Figure 1.

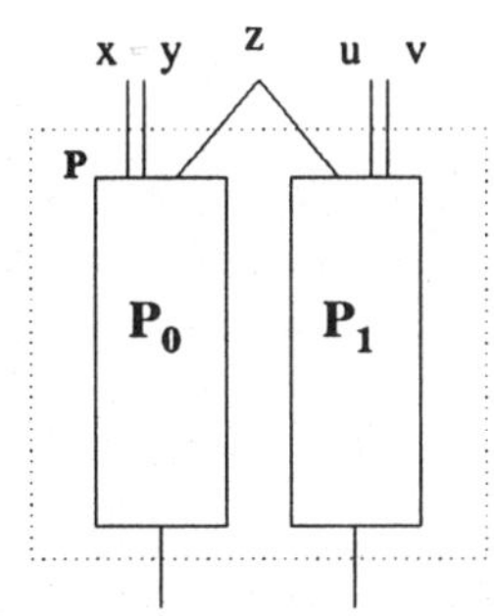

Figure 1. Vertical slicing program decomposition.

2. *Dimension elimination.* Dimensions in the input domain can also be eliminated by horizontal slices of the program. One typical example occurs when the same physical quantity can be passed to the program in many different ways, as illustrated in the *classify_triangle* program. The verification of the preprocessing stage can often result in the elimination of some dimensions. For example, if the pertinent physical input is the distance between the spacecraft and the space-station while the inputs to the program are the coordinates of each one relative to some reference frame, then verification will reduce the 6-dimensional input space to a 1-dimensional space. Horizontal slicing is schematically represented in Figure 2. The number of input dimensions can also be reduced by eliminating nondeterministic data structures and control structures.
3. *Dimension reduction.* The third step is to use partial proofs to reduce the number of points along each dimension. This can be achieved in several ways, including modeling of stress testing and bounding the numerical precision. Since the sensors and actuators in process-control systems are electromechanical devices, they have bounded accuracy, e.g. able to

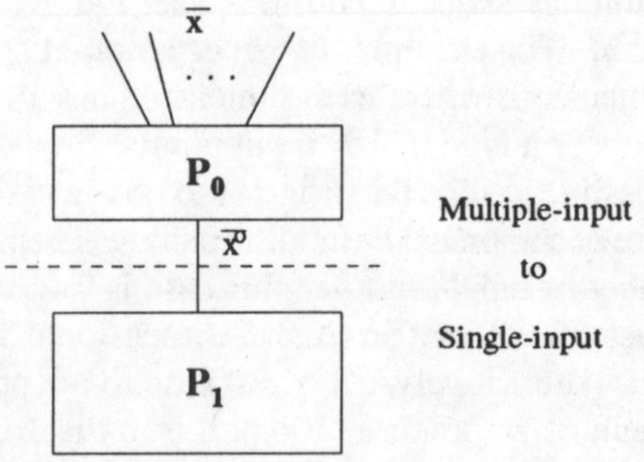

Figure 2. Horizontal slicing program decomposition.

measure speed accurately to only within 0.1 miles/hr. (say). Hence, by incorporating this, the number of points along a dimension can be reduced substantially, say from 2^{32} for a 32-bit machine to only 200,000 if the maximum specified speed is 20,000 miles/hr.

These effects can also be achieved by using source code transformation rules [8] [10] that add some functionality to the software, such as graphical user interfaces, fault tolerance, and safety features, and design rules that eliminate dependence of test outcomes on the internal states of the program [7].

Once each test case has been sufficiently amplified, the next step is to speed up the generation and execution of test cases in order to be able to perform a large amount of testing within a given validation time. For the MAP system, we are developing an automated test data generation system that efficiently generates a large number of random test cases. It takes the predicate that characterizes a test region and ranks the variables in a sequence that increases the probability that a randomly generated set of input values will satisfy the condition [12]. In addition, the region is enclosed within a tight "box" to achieve an even higher hit ratio with random sampling [12]. Initial experiments show that these refinements yield a speedup of 300%. Combined with an algorithm that automatically customizes the test data generator for each predicate, the speedup is substantially higher (over 100 times).

To speedup the execution of test cases, we have developed source code transformation rules that use table lookup, lower precision numbers, slower sampling frequency, coarser inputs, etc., to achieve transformed versions that satisfy the monotonic equivalence condition and are faster by several orders of magnitude [12]. For example, consider an image processing program that operates on 1000×1000 pixels and a coarser version that operates on 100×100 pixels. If it can be proved that for all inputs, the correctness of the coarser version implies correctness of the original program, then we have a 100-folds speedup.

3.1. Example

The input space of a simplified version of the spacecraft/space-station problem outlined in Section 2, i.e. without any other nearby objects, consists of 19 dimensions which is a very huge input space. Now, clearly the same physical configuration can be presented to the program in many different ways by adjusting the coordinates and velocities of S and C. Hence, we first take a horizontal slice to eliminate this 1-to-m mapping between the application level inputs and the inputs to the program. Suppose that in addition it can be shown that the coordinate system can be rotated so that the position of C is within the xy plane. Then, the reduced set of parameters are:

Coordinates of C: (x_C, y_C);
Velocity of C: (v_{xC}, v_{yC}, v_{zC});
Desired radius: r;
Setting of engines: $l_f, l_r, s_{1f}, s_{1r}, s_{2f}, s_{2r}$.

Thus the number of dimensions has been reduced to 12. Suppose it can then be shown that only the distance of C from S is relevant. This replaces (x_C, y_C) by d which reduces the number of parameters to 11. There is still another 1-to-m mapping between the application level input space and the program level input space. This is due to the possibility that only the difference between d and r matters and not their actual values. If this can be verified, then d and r are replaced by $d - r$. The new set of parameters, using polar coordinates, are:

Difference between d and r: $d - r$;
Velocity of C: $(v_r, v_\theta, v_\alpha)$;
Setting of engines: $l_f, l_r, s_{1f}, s_{1r}, s_{2f}, s_{2r}$.

The number of dimensions has been reduced to 10 from the original 19. Now, let us apply another horizontal slice to remove the portion that actually fires the thrusters. Let the first portion of the program computes set_v_r, set_v_θ, and set_v_α based on the desired change in the position. Let the second portion of the program control the engines based on the values of set_v_r, set_v_θ, set_v_α, l_f, l_r, s_{1f}, s_{1r}, s_{2f}, and s_{2r}. If this portion can be formally verified, then the number of dimensions for testing the first portion is reduced to 4. Further reductions can be achieved by taking vertical slices for the computations of set_v_r, set_v_θ, and set_v_α. If these computations use disjoint input parameters, i.e., if set_v_r uses only $d - r$ and v_r, set_v_θ uses v_θ, and set_v_α uses v_α, then the state spaces consist of only 2 dimensions, 1 dimension, and 1 dimension, respectively. In this case, it is practical to test the angular controls exhaustively.

4. Measured Assurance Prediction (MAP) System

Several sources of validation data are needed to construct the confidence profile for a given program. Because of the large number of input parameters for process-control programs, the first step is to use the partial proof and source code transformation methods discussed in the previous section to reduce the number of dimensions in $I \times S$. To factor in the shape of the control function, the system requires symbolic expressions for the output functions. The set of test cases used during the reliability assessment phase are then used to construct the confidence profile.

The profile is first computed exactly based on the amplification factor, i.e., for each $x \in I \times S$, $f(x) = 0$ if it is not directly covered by any test cases or $f(x) = 1$ otherwise. Since the reduced set of dimensions is used, the only coverage that must be computed are those that reduce the number of points along a dimension. Afte: this step, the confidence for uncovered points is computed by using sampling without replacement in the neighborhood of those points.

The coverage information provided by partial proofs can be grouped into three categories.

- *Equivalence Poofs*: Let $x \Rightarrow y$ denote that if the program is correct for point x then it is also correct for point y. The best situation is when equivalence can be established, i.e., $x \Rightarrow y \wedge y \Rightarrow x$. This allows either x or y to be tested. It occurs whenever a proof eliminates a dimension or when nondeterministic data or control structures are replaced by deterministic structures or when a horizontal slice is taken to achieve an m-to-1 reduction [7]. Several examples of this scenario appeared in the previous section, including the 6-to-1 reduction for the triangle classification program.

- *Implication Poofs*:The next case occurs when it can be established that $x \Rightarrow y$. In this case, it is necessary to test x which will then cover both x and y. This situation occurs often in recursive or iterative programs. Examples include the well-known GCD program, array operations such as sorting and searching, and matrix operations. If a linear sequence can be established, such as

$$x \Rightarrow y \Rightarrow \cdots \Rightarrow z,$$

then only 1 test case is needed. More often, the set of points will be divided into two portions, say U and V, such that

$$\forall v \in V :: (\exists u \in U :: u \Rightarrow v),$$

in which case it is necessary to sample points from U using sampling without replacement.

- *High Stress Benefit Poofs*: The third type of coverage is similar to the second type except that it is less direct. It usually arises when the set of possible correct outputs for point x is a subset of that for point y or when the available execution time for point x is less than that for point y. In both these cases, point x is a high stress input for the program as compared with point y. It needs to be formally established that if the program works for high stress points then it will also work for lower stress points.

To illustrate these coverages, consider the obstacle avoidance portion of the spacecraft/space-station problem. A trivial example of an equivalence proof is to show that n objects that are nonintersecting with the paths of the spacecraft and the space-station is equivalent to 0 objects. An example of implied coverage is to prove that a sequence of n objects along the path of the spacecraft $\Rightarrow$ a sequence of $n - 1$ objects. That is, it must be shown that the program deals with the objects in the sequence in which it encounters them. There are several possible stress coverages, including showing that an input with moving objects (relative speed) covers an input with stationary objects, an input with large objects covers one with small objects, an input with near objects covers one with far away objects, and so on. It may then be possible to simply test the program for the highest stress inputs (maximum number of objects, highest speed, nearest distance) and prove that the program should work for lower stress inputs.

The MAP system is being developed using Mathematica^{TM} for the symbolic manipulation portions. A first version was developed for the NeXT^{TM} workstation to graphically display the partitions in the input space and the shape of the outputs [18]. It did not use any dimension reduction steps; instead, the user could select two dimensions for display with all other coordinates being fixed. A second version is being developed for the PC environment. It uses color graphics to display more information. Also, methods are being developed to efficiently store and access the validation information. After this step, the system will directly incorporate the dimension reduction proofs as well as the three coverage proofs.

5. Summary

In this paper, we presented an approach for using the reliability assessment data to replace the acceptance test in a restricted form of recovery block program in which the backup version provides reduced capability but to a higher degree of confidence. The confidence profile is computed using a combination of formal methods and statistical testing. The approach does not make any assumptions about the correlation between points in the input space other than those for which formal proofs can be provided. It also uses statistical sampling (without replacement) to provide nonzero confidence values for points that are not covered by the amplified set of test cases. Some future research areas include identifying specific proof procedures and applying the approach to complex process-control programs.

References

[1] P.E. Amman, S.S. Brilliant, and J.C. Knight, "The effect of imperfect error detection on reliability assessment via life testing," *IEEE Trans. on Softw. Eng.*, Vol. 20, No. 2, Feb. 1994, pp. 142-148.

[2] A. Avizienis, "The n-version approach to fault-tolerant software," *IEEE Trans. Softw. Eng.*, Dec. 1985, pp. 1491-1501.

[3] F.B. Bastani, *An Input-Domain Based Theory of Software Reliability and its Application*, Ph.D. Dissertation, Dept. of EECS, Univ. of Calif., Berkeley, Aug. 1980.

[4] F.B. Bastani, "Experiences with a feedback version development methodology," *IEEE Trans. Soft. Eng.*, Vol. SE-11, No. 8, Aug. 1985, pp. 718-723.

[5] F.B. Bastani, "On the uncertainty in the correctness of computer programs," *IEEE Trans. Soft. Eng.*, Vol. SE-11, No. 9, Sep. 1985, pp. 857-864.

[6] F.B. Bastani, G. DiMarco, A. Pasquini, "Experimental evaluation of a fuzzy-set based measure of software correctness using program mutation," *15th Intl. Conf. on Softw. Eng.*, Baltimore, MD, May 1993, pp. 45-54.

[7] F. B. Bastani, B. Cukic, "A transformational approach for measuring software reliability," 4th *IEEE Int. Work. on Eval. Tech. for Dep. Sys.*, San Antonio, TX, Oct. 1995.

[8] Bauer, F.L., Moller, B., Partsch, H., Pepper, P., "Formal program construction by transformations - computer-aided, intuition-guided programming," *IEEE Trans. Soft. Eng.*, Vol. 15, No. 2, Feb. 1989, pp. 165-180.

[9] R.W. Butler and G.B. Finelli, "The infeasibility of quantifying the reliability of life-critical real-time software," *IEEE Trans. Soft. Eng.*, Vol. 19, No. 1, Jan. 1993, pp. 3-12.

[10] B. Cukic and F.B. Bastani, "Automatic array alignment as a step in hierarchical program transformation," *Intl. Par. Proc. Symp.*, Cancun, Mexico, April 1994, pp. 578-582.

[11] D. Hamlet, "Connecting test coverage to software dependability," *Proc. 5th Intl. Symp. on Soft. Rel. Eng.*, Monterey, CA, Nov. 1994, pp. 158-165.

[12] A. Jamoussi and F.B. Bastani, "An automated test data generation system for massive program testing," *Tech. Rep.*, Dept. of Comp. Sc., Univ. of Houston, April 1995.

[13] K. Kanoun, M. Kaaniche, and J.C. Laprie, "Experience in software reliability: From data collection to quantitative evaluation," *Proc. 4th Intl. Symp. Soft. Rel. Eng.*, Denver, CO, Nov. 1993, pp. 234-246.

[14] K.H. Kim, "Distributed execution of recovery blocks: An approach to uniform treatment of hardware and software faults," *Proc. 4th Intl. Conf. on Dist. Comp. Sys.*, May 1984, pp. 526-532.

[15] J.C. Knight and N.G. Leveson, "An experimental evaluation of the assumptions of independence in multi-version programming," *IEEE Trans. Softw. Eng.*, Vol. SE-12, Jan. 1986, pp. 96-109.

[16] R. Lea, Private communication, 1992.

[17] B. Littlewood and L. Strigini, "The risks of software," *Scientific American*, Nov. 1992, pp. 62-75.

[18] N. Puttagunta, *VISUALPATHS: A Program Visualization Tool to Aid in Software Reliability Analysis*, M.S. Thesis, Dept. of Comp. Sc., Univ. of Houston, August 1994.

[19] C.V. Ramamoorthy, F.B. Bastani, Y.R. Mok, and C. Nam, "A systematic approach to the development and validation of critical software for nuclear power plants," *4th Int. Conf. on Soft. Eng.*, Munich, 1979.

[20] C.V. Ramamoorthy, Y.R. Mok, F.B. Bastani, G.H. Chin, and K. Suzuki, "Application of a methodology for the development and validation of reliable process-control software," *IEEE Trans. Soft. Eng.*, Vol. SE-7, No. 6, Nov. 1981, pp. 354-371.

[21] C.V. Ramamoorthy and F.B. Bastani, "Software reliability - status and perspectives," *IEEE Trans. Soft. Eng.*, Vol. SE-8, July 1982, pp. 354-370.

[22] B. Randell, "System structure for software fault tolerance," *IEEE Trans. on Softw. Eng.*, Vol. SE-1, June 1975, pp. 220-232.

[23] Robinson, L., Levitt, K.N., "Proof techniques for hierarchically structured programs," *Comm. of the ACM*, Vol. 20, No. 4, April 1977, pp. 271-283.

[24] T. A. Thayer, M. Lipow, and E.C. Nelson, *Software Reliability*, North-Holland, 1978.

[25] M. Weiser, "Program slicing," *IEEE Trans. Soft. Eng.*, Vol. SE-10, No. 4, 1984, pp. 352-357.

Ada95 Object-Oriented and Real-Time Support for Development of Software Fault Tolerance Reusable Components

Eltefaat H. Shokri   Kam S. Tso
SoHaR Incorporated
Beverly Hills, CA 90211
{shokri,tso}@sohar.com

Abstract

This paper reports our experience on exploiting the object-oriented and real-time features of Ada95 to support the development of a reusable software fault tolerance testbed. The testbed is a hardware and software platform for the creation of software fault tolerance systems from reusable components and provides a fault-injection environment for evaluating their effectiveness. The reusable components were identified from an in-depth analysis of the software fault tolerance domain and designed using an object-oriented approach based on the Booch Method. The result of the analysis and design is a set of objects and their communication patterns. The identified objects were classified as active, passive, and shared-data objects and implemented as reusable components by mapping them into the corresponding Ada95 object-oriented constructs. A distributed recovery block system with a simplified air traffic control application were developed from the reusable components to demonstrate effective reuse and meeting soft real-time requirements.

1. Introduction

This paper reports our experience on exploiting the object-oriented and real-time features of Ada95 [5] to support the development of a software fault tolerance testbed based on reusable components. Ada95 is the revised standard of the original Ada language (Ada83) [1]. The revised language increases the flexibility and applicability of Ada by introducing new features supporting object-oriented programming, hierarchical libraries, and development of real-time systems. Since the ISO approval of Ada95 in early 1995, only a handful of pre-release Ada95 compilers are available and limited experience has been reported on its use. This research represents an early effort of using the new features of Ada95 to validate its effectiveness in supporting software reuse and real-time applications.

The Reusable Software Fault-tolerance Testbed (ReSoFT) [13] is a hardware and software platform developed under the sponsorship of the U.S. Air Force for experimenting with software fault tolerance. Software fault tolerance (SWFT) has been proposed as an additional measure to achieve ultra-high dependability required for large-scale safety-critical applications [11, 2]. Although several SWFT techniques have been proposed, they are not widely used partly because (1) their effectiveness in tolerating design faults is yet to be validated, and (2) they are difficult to implement in practice. Therefore, an effort to develop a flexible and reusable testbed to allow the creation of SWFT systems from reusable components and the evaluation of their effectiveness through experimentation is a critical research and development task. In response to this need, the ReSoFT project was initiated. Ada95 was chosen for implementing ReSoFT mainly because it provides object-oriented and soft real-time facilities while retaining its main reliability goal.

The rest of the paper is organized as follows: Section 2 gives an overview of the main characteristics as well as the overall structure of the ReSoFT testbed. In Section 3, we describe Ada95 support for implementing the SWFT reusable components. Section 4 discusses the implementation of fault-injection facility and the support by Ada95. Finally, Section 5 provides a conclusion to the paper.

2. ReSoFT: A Testbed for Software Fault Tolerance

ReSoFT (Reusable Software Fault-tolerance Testbed) [13] is an integrated software fault tolerance testbed developed on an open architecture and uses standard off-the-shelf hardware and software components. The main goals for developing ReSoFT are satisfying (i) the need to experiment with various well-established SWFT schemes for investigating its effectiveness and (ii) the desire to reduce repetitive effort in implementing those SWFT systems. Re-

SoFT consists of a number of computer nodes (workstations) communicating over a redundant network. Software components of ReSoFT are developed and executed on top of the Sun Solaris operating system. It also makes use of available POSIX 1003.1b thread libraries for improved response times. Figure 1 depicts the system architecture for ReSoFT.

The software architecture of ReSoFT, which consists of the reusable components identified during an in-depth domain analysis [14], is an integration of the following layers:

- *Reusable SWFT Components*:

  To identify reusable SWFT components, an object-oriented analysis of well-established SWFT schemes was conducted [13]. In this process, maximum care was taken to make sure that (i) the components capture the common functionality of a wide range of SWFT schemes, and (ii) application-dependent components are identified and separated from the generic (application-independent) components. The Reusable SWFT layer include full implementations of application-independent components as well as templates for application-dependent components.

- *Network Communication Components*:

  Network communication components provide application transparent reliable inter-node communications among the ReSoFT nodes. Moreover, a diagnosis facility for communication faults is provided for monitoring.

- *Fault Injection Components*:

  Fault injection components are designed to support testing and evaluation of the dependability of the SWFT components and systems. They can inject a vast variety of faults into other layers such as SWFT components and application modules.

- *Graphical User Interfaces*:

  Graphical user interfaces have been implemented:

  1. to automate integration of SWFT schemes with application modules, such that the application developer is relieved from the complexity of the SWFT schemes (Graphical Application Integrator).
  2. to accept fault injection commands from the user and direct the fault injection experiment (Graphical Fault-Injection Manager).
  3. to monitor the execution of SWFT schemes and to supervise them to a safe state when an unexpected situation arises (System Status Monitor).

- *Application Modules*

  The reusable SWFT components are designed in a way that can be integrated by various applications with a minimal change. The application developer should implement the application preferably in the form of reusable components and then complete the templates of application-dependent SWFT components according to the application characteristics using the Graphical Application Integrator. The Graphical Application Integrator will then automatically build the program(s) supporting fault-tolerant execution of the application.

In this paper, our focus is on the first three layers (i.e., the Reusable SWFT layer, Network Communication layer, and Fault Injection layer). Reusable components in these three layers are classified into three general categories (i) *active components* which have their own execution threads, (ii) *passive components* which do not possess any execution thread but rather provide services to other components, and (iii) *shared-data components* which store shared data items. The mapping of reusable components into Ada95 tasks and/or packages will be discussed in details in the following sections.

3. Object-Oriented Support of Reuse in ReSoFT

There have been several efforts to develop application-specific frameworks of reusable components. One of the major reuse efforts to date is the Common Ada Missile Packages (CAMP) Project [8] which has the objective of identifying common missile functions and the design of reusable Ada components. The CAMP experience has identified four major technical issues that have impeded meaningful levels of software reuse. These problems, along with our solutions, are:

1. *Inappropriateness to the domain*

   Domain analysis has been found to be the key factor in the success of reusability [10]. Components that result from domain analysis are better suited to reuse because they capture the essential functionality of the domain. In the ReSoFT project [13], we conducted an in-depth domain analysis of software fault tolerance to identify reusable components.

2. *Difficulty in integration of reusable and custom code*

   A typical concern about reusability is that software developers prefer to create their own modules because available components are difficult to understand and adapt to new applications. This is the case when components designed for reuse lack a rigorous specification for their functionalities and interfaces. Object-

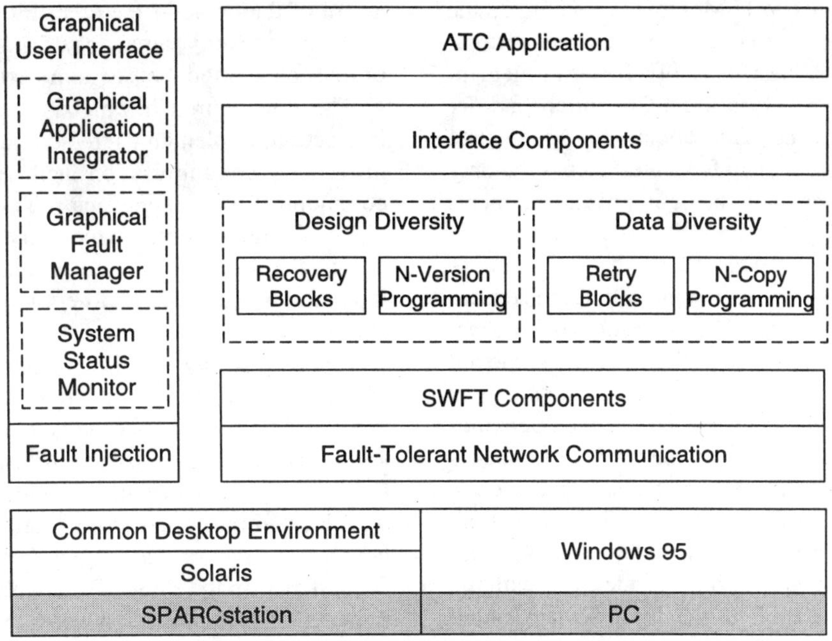

Figure 1. System Architecture of ReSoFT

oriented design has been found to be the most promising technique for attaining the goals of extensibility and reusability [9, 12]. In the ReSoFT project, an object-oriented methodology, based on the Booch method [3] and Rational Rose CASE tool [15], has been used for analysis and design of the reusable SWFT components.

3. *Skepticism about the quality of reusable components*

   Reusable components should be efficient, robust, and most importantly reliable. Rigorous testing of reusable software components is the only conceivable way of assessing their reliability. Therefore, it is important that these components support testing and data collection. Our approach to reliability assessment and assurance is to provide supports for fault injection and data collection for evaluating the fault detection and recovery capabilities of SWFT components.

4. *Lack of support tools for reuse*

   Although identification and implementation of reusable components are by their nature the most important tasks facing the developer of these components, another key issue is an efficient integration of system reusable software components into application modules to create a robust application. We implemented a graphical application integrator which builds the SWFT scheme chosen by the user and integrates it with the application.

3.1. Mapping Reusable Components into Ada95 Constructs

The result of object-oriented design is a set of objects, their interfaces as well as communication patterns. Identified objects are classified into three types:

- *Active objects*
  An active object is an object which possesses its own execution thread(s). Additionally, an active object may also provide services to other objects.

- *Passive objects*
  Passive objects do not own any execution thread and merely act as service providers to other objects. Passive objects are stateless, meaning that they do not maintain any long-life data used by successive services.

- *Shared-data objects*
  Shared-data objects maintain data stores to which multiple objects may have mutually-exclusive access.

To illustrate this classification, Figure 2 depicts implementation of the Distributed Recovery Block (DRB) scheme [7] using reusable components. The DRB scheme is essentially an approach for structuring a duplex redundant computer nodes dedicated to execute different versions of the applications. Each version of the application is structured in the form of the recovery block abstraction [11]. The acceptability of the result of the application execution is validated by an acceptance test. Moreover, to facilitate the detection of node-crash failures, each node sends periodical status messages called heartbeats [4]. In our implementation, the DRB Executive, Heartbeat-Generator, and Try Block are active components, Checkpoint and Acceptance Test are passive components, and RB Data-Store is the only shared-data object. As shown in the figure, five components share the data stored in the RB Data-Store component.

The role of the Application Execution Interface component needs some clarification since it interrelates the execution of the application and the DRB protocol. The Application Execution Interface component is the only interaction component used in the implementation of DRB and acts as a communication interface between the Try Block component and the application in the following way. It first makes a rendezvous with the Try Block component to receive the command for initiation of a new cycle of the application. Upon completion of the rendezvous with Try Block, Application Execution Interface makes another rendezvous with the application to start a new cycle of the application. Using a similar approach, Application Execution Interface passes the result of the completed cycle to Try Block.

To implement reusable components in Ada95, the following mapping was adopted:

- A passive object is implemented as an Ada package which does not contain any task. Care was taken to make sure the package for a passive object does not possess any persistent global data item. In other words, all data communication between a server package and its client are realized by passing parameters.
- An active object is implemented as an Ada package with a task for each of its execution threads.
- A shared-data object is implemented using the protected record type provided in Ada95, thereby the Ada runtime system will guarantee mutually exclusive accesses to the shared-data.

3.2. Tightly-Coupled Objects vs. Loosely-Coupled Objects

Two approaches to the implementation of object interactions in Ada95 may be used. The first, called *tightly-coupled* interactions, employs the rendezvous concept for the interaction a server and its clients. In this approach, all objects residing in a processor share a single address space and should be compiled into a single process consisting of concurrent tasks interacting with synchronous rendezvous. This method of packaging all objects as a single program is easy to implement. However, it has a major drawback that any change in an object leads to the rebuilding of the entire program. Moreover, one of the interacting objects should be aware of the name of the other interacting object (because of the rendezvous semantics). This reduces the autonomy of reusable components to some extent.

The other approach, which is denoted as *loosely-coupled* interaction, implements interactions between objects using lower-level communication facilities such as socket-based communication mechanisms. This approach provides a higher degree of object autonomy. However, we learned during the ReSoFT implementation that loosely-coupled interactions introduce non-negligible overhead if not used carefully.

It is hard to say whether one approach has to be generally preferred to the other. Our experiments suggested that one practical hybrid solution is to partition the system into several subsystems based on criteria such as communication pattern, and use tightly-coupled interactions among objects located in the same subsystem while employ loosely-coupled interactions among objects residing in different subsystems. In the ReSoFT implementation, reusable components and the application were considered as two separate subsystems and their interactions were implemented using loosely-coupled interactions.

3.3. Use of Ada95 in the DRB Implementation

In Section 3.1, a component-based implementation of the DRB scheme was shown to illustrate the use of reusable components. In this section, we describe how features of Ada95 were employed in this implementation.

3.3.1. Tagged Types for Inheritance

The ultimate goal in designing SWFT components is their maximum reusability. These components should be designed in the way that they can be reused by any application with no or minimal changes. However, there are functionalities which may vary from one application to another. As an example, the Checkpoint component, which is responsible for establishing and restoring the application checkpoints, is highly dependent on the application. We used the inheritance mechanism to separate application-specific functionality from application-independent features.

Ada95 introduced tagged types for implementing inheritance. A tagged type can be extended by adding new items as well new operations to the derived types. The `CHECKPOINT` record type is designed as a tagged record type on which the `GET_CURRENT_CHECKPOINT_NUMBER` procedure can be applied. The application designer can then introduce a new type `APPLICATION_CHECKPOINT` type which is defined based on the `CHECKPOINT` type. As shown in the following piece of code, two new procedures, namely

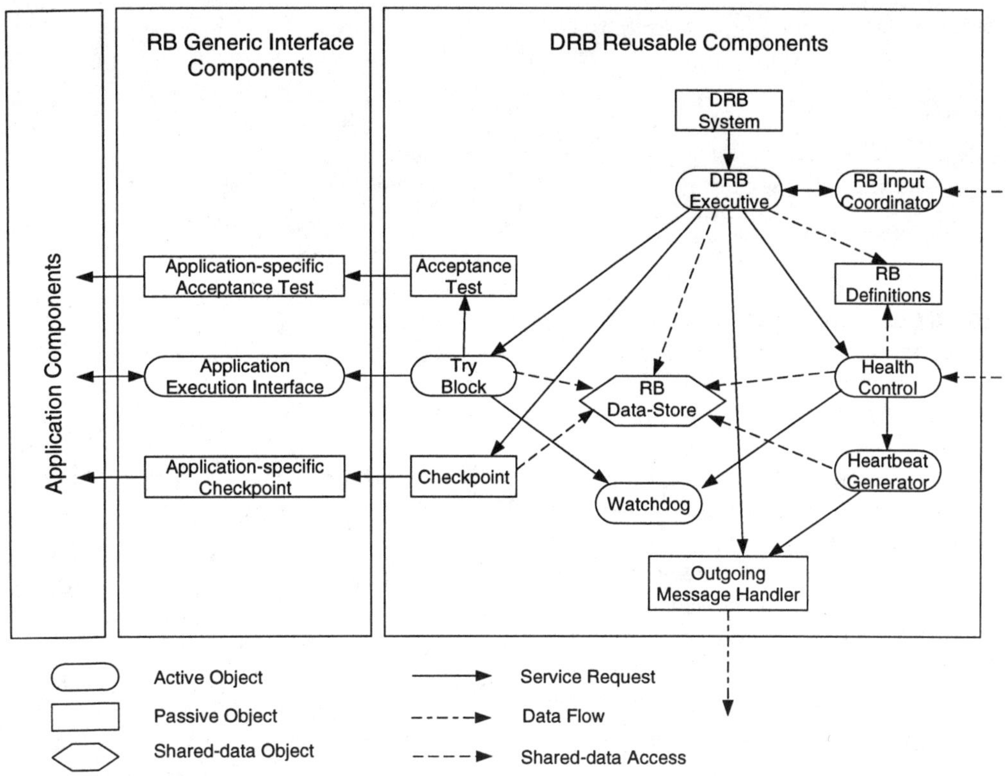

Figure 2. Component-based Implementation of the Distributed Recovery Block

ESTABLISH and RESTORE, can be additionally applied to APPLICATION_CHECKPOINT type.

```
-- Specification for general checkpoint package.
package CHECKPOINT_PACKAGE is
 type CHECKPOINT is tagged
    record
       CHECKPOINT_NUMBER : integer;
    end record;
 function GET_CURRENT_CHECKPOINT_NUMBER
         return integer;
 -- ...
 -- other application-independent procedures
 -- ...
end CHECKPOINT-PACKAGE;

-- Specification for application-specific
-- checkpoint package.
with CHECKPOINT_PACKAGE;
use  CHECKPOINT_PACKAGE;
package APPLICATION_CHECKPOINT_PACKAGE is
 type APPLICATION_CHECKPOINT is
   new CHECKPOINT with
     record
       -- Application-specific checkpoint fields
    end record;
  procedure ESTABLISH
  (CURRENT_CHECK : out APPLICATION_CHECKPOINT);
  procedure RESTORE
  (CHECK_TOBE_RESTORED:
          in APPLICATION_CHECKPOINT);
  -- ...
  -- Other application specific procedures
  -- ...
end APPLICATION_CHECKPOINT_PACKAGE;
```

We adopted the same technique for other application dependent features such as the Acceptance Test component.

3.3.2. Asynchronous Transfer of Control for Time-Bounded Computations

There are occasions during the execution of the SWFT schemes in which an activity should be abandoned if certain condition (such as when the execution time of an activity exceeds a predetermined limit) arises and an alternative action should be taken. As an example, when DRB Executive orders the Try Block to executive a new application cycle, the DRB Executive component should receive the result of the execution by a predefined time-limit, say T. If T time-units elapses and DRB Executive does not receive the result, it should abandon normal execution of the scheme and take a recovery action. Thereby, waiting for the result of application execution should be terminated before T time-units elapsed, otherwise the action (i.e., waiting for the application result) should be abandoned and it should then be concluded that the application has either experienced a timing fault or died. This set of actions can be taken using "select-then-abort" [5] statement introduced in Ada95 as shown below.

```
SEND_ORDER_TO_TRYBLOCK;
select
   delay MAXIMUM_APPL_EXEC_TIME;
   TAKE_RECOVERY_ACTION;
then abort
   WAITING_FOR_APPL_RESULT(APPL_RESULT);
end select;
```

SEND_ORDER_TO_TRYBLOCK orders the Try Block component to manage execution of a new application cycle. Then the DRB Executive component executes the select statement which executes the WAITING_FOR_APPL_RESULT procedure until it is completed or the MAXIMUM_APPL_EXEC-TIME timeout expires. If the control does not return from the procedure by this time-period, then the execution of WAITING_FOR_APPL_RESULT will be abandoned and the TAKE_RECOVERY_ACTION procedure will be activated. This cannot be easily implemented in Ada83. The only solution in Ada83 is dedicating a task for WAITING_FOR_APPL_RESULT so that the task can be aborted as the time limit expires. Of course, dynamic creation and abort of tasks are not recommended in critical applications.

3.3.3. Protected Types for Shared Data

In Ada83, the rendezvous model is used for both task synchronization and data-sharing purposes. Unfortunately, use of rendezvous has not been entirely satisfactory specially for data-sharing. It requires additional tasks to manage shared data and this often leads to poor performance. Ada95 introduced the concept of protected types which provides orderly and synchronous access to shared data without creating additional task.

The following piece of code illustrates the use of protected type in DRB implementation. RB_DATA_STORE maintains DRB configuration status such as the role of each node, the versions of application software being executed, and the health status of participant nodes. MY_ROLE, for example, is an item of shared data and can only be accessed through procedures GET_NODE_ROLE and SET_NODE_ROLE. For each shared data item RB_DATA_STORE, possible types of access were studies and implemented as procedures.

```
protected RB_DATA_STORE is
    function GET_NODE_ROLE return ROLES;
    procedure SET_NODE_ROLE(NEW_ROLE : ROLES);
    -- ...
    -- Other procedures to manipulate other
    -- protected data items
    -- ...
private
   MY_ROLE : ROLES;
   PEER_ROLE : ROLES;
   -- ....
   -- Declaration of other protected data items
   -- ...
end RB_DATA_STORE;

protected body RB_DATA_STORE is
  function GET_NODE_ROLE return ROLE is
  begin
     return MY_ROLE;
  end GET_NODE_ROLE;
  procedure SET_NODE_ROLE(NEW_ROLE : ROLES) is
  begin
    MY_ROLE := NEW_ROLE;
  end SET_NODE_ROLE;
  -- ...
  -- bodies of other procedures
  -- ...
end  RB_DATA_STORE;
```

4. Object-Oriented Support of Fault-Injection Testing in ReSoFT

To ensure that ReSoFT meets its reliability as well as timing requirements, a fault-injection subsystem was implemented. In order to preserve timely as well as functional characteristics of the target system, existing fault-injection approaches [6] inject faults by altering object codes or data items using underlying hardware and/or operating system facilities, making the fault-injection subsystem platform-dependent. To reduce such an undesired dependency, we decided to implement the fault-injection subsystem as a collection of reusable components which are compiled along with operational SWFT components. To be more specific, each SWFT component is paired with a fault-injection component responsible for injecting faults in the paired SWFT component. Using this component-based approach, robustness of each individual SWFT component as well as correctness of SWFT schemes can be evaluated in a platform-independent manner.

The fault-injection subsystem, as illustrated in Figure 3, has the following characteristics:

- *Centralized fault-injection manager*:

  The global fault-injection manager (FI Manager) accepts attributes of the faults to be injected from the user via a graphical user interface. The FI Manager then directs the fault-injection experiment and generates a fault-injection schedule. It also sends timely orders to the corresponding nodes to emulate faults. Moreover, the FI Manager monitors any changes occurring in the system in response to the injection of a fault. Finally, it analyzes system behavior for each fault injected and logs the necessary data for off-line reliability assessment. Whenever an undesired change occurs (e.g., a nodes crashes) in response to a fault, the FI Manager takes appropriate actions (e.g., restarting the crashed node) to reset the fault tolerant system to a normal state so that the FI experiment can be continued.

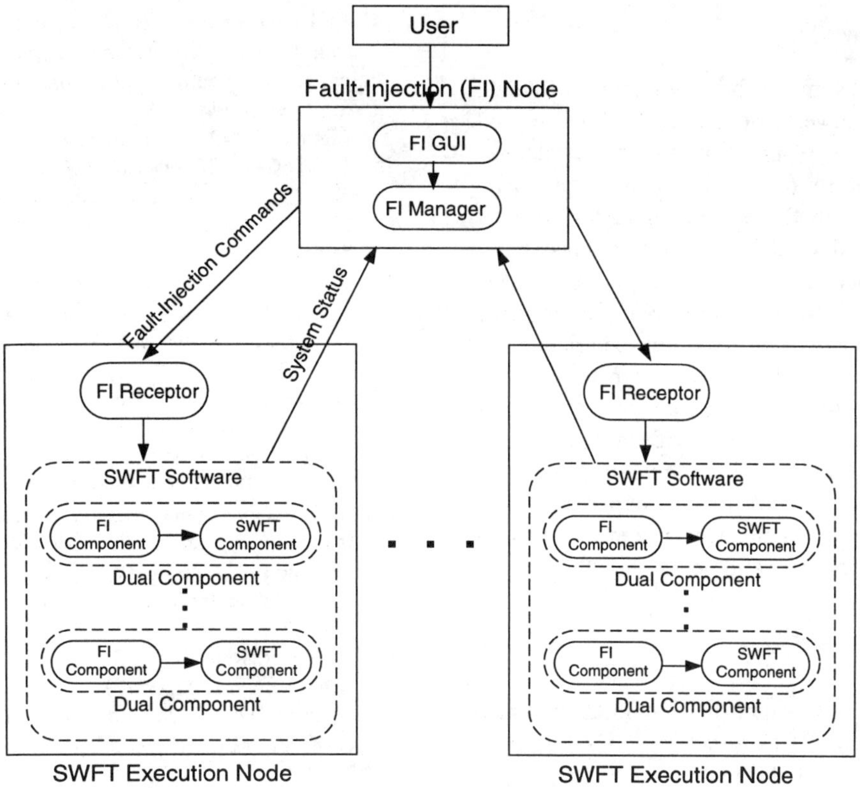

Figure 3. Overall Architecture of the Fault-injection in ReSoFT

- *Fault-injection receptors residing in each SWFT execution node*:

  Each of the nodes executing the SWFT scheme is designed to receive fault-injection commands issued by the FI Manager. After the fault-injection receptor (FI Receptor) receives a fault-injection command, it invokes the necessary procedures of the corresponding fault-injection component (FI component) to initiate the emulation of the given fault scenario.

- *Child-parent packaging to avoid recompilation*:

  As shown in Figure 3, each SWFT component is paired with a FI component creating a virtual component which is called here a *dual component*. Each FI component is designed for injecting faults in its peer SWFT component (i.e., the SWFT component located in the same dual component). Isolating fault-injection functionality from the SWFT components and locating them in separate components guarantees the minimal code alteration in SWFT components. In our Ada95 implementation, each FI component is implemented as a child package of its peer SWFT component. By doing so, SWFT components (specifications as well as bodies) do not need to be recompiled for fault-injection.

- *Minimal change in timing behaviors of the target SWFT scheme*:

  An important concern in using software fault-injection for evaluation and validation of a software system is that the fault-injection components might alter the timing as well as logical behavior of the target system. In ReSoFT, we attempt to minimize such undesired behavioral changes by the following techniques:

  1. Only one additional task (the fault-injection task) was introduced for fault-injection. Since this task will be in the blocked state most of the time waiting for a fault-injection command, the timing overhead is expected to be very small. Early fault-injection experiments on ReSoFT indicated that the timing overhead introduced by fault-injection components remains tolerable.

  2. Since injecting faults is mostly done by FI components (i.e., child packages), the code changes on SWFT components are kept minimal, thereby reducing the possibility of any behavioral changes in the SWFT target components.

5. Conclusions

The paper discussed some of the new features of Ada95 which were used in the development of the ReSoFT testbed. The development effort reported in the paper demonstrated that the new features (especially the ones supporting object-oriented development) are well-suited for developing reusable software. We used features such as class-based construction, inheritance, and child-parent hierarchical library structuring in developing reusable SWFT components and schemes. We are convinced from our experience that although Ada95 compilers and tools are not mature enough yet, but if used conservatively (i.e., using new features with extensive care and tests), they will provide significant cost and reliability benefits. We did not use any annexes mainly because none was available. We plan to experiment with features of the Real-Time System Annex to evaluate its effectiveness, when it becomes available.

Acknowledgments

This work was partially supported by Small Business Innovation Research (SBIR) Contract F30602-94-C-0013 from Rome Laboratory, U. S. Air Force.

References

[1] ANSI/MIL-STD-1815A. *Reference Manual for the Ada Programming Language*. American National Standards Institute, Feb. 1983.

[2] A. Avižienis and L. Chen. On the implementation of N-Version Programming for software fault-tolerance during program execution. In *Proceedings of COMPSAC-77*, pages 149–155, 1977.

[3] G. Booch. *Object Oriented Analysis and Design with Applications*. Benjamin/Cummings, Redwood City, CA, 2nd edition, 1994.

[4] M. Hecht, J. Agron, H. Hecht, and K. H. Kim. A distributed fault tolerant architecture for nuclear reactor and other critical process control applications. In *Digest of the 21st Annual International Symposium on Fault-Tolerant Computing*, pages 3–9, Montreal, Canada, June 1991.

[5] ISO/IEC-8652:1995. *Ada Reference Manual: Language and Standard Libraries*. International Organization for Standardization and International Electrotechnical Commission, Jan. 1995.

[6] R. K. Iyer and D. Tang. Experimental analysis of computer system dependability. In D. Pradhan, editor, *Fault-Tolerant Computing: Theory and Techniques*. Prentice Hall, 2nd edition, 1995.

[7] K. H. Kim and H. O. Welch. Distributed execution of recovery blocks: An approach for uniform treatment of hardware and software faults in real-time applications. *IEEE Trans. Comput.*, 38(5):626–636, May 1989.

[8] McDonnell Douglas Missile Systems Company. Developing and using Ada parts in real-time embedded applications. Common Ada Missile Packages — Phase 3 (CAMP-3), McDonnell Douglas Missile Systems Company, St. Louis, MO, Apr. 1990.

[9] B. Meyer. Reusability: The case for object-oriented design. *IEEE Software*, pages 201–215, Mar. 1987.

[10] R. Prieto-Diaz. Domain analysis for reusability. In *Proceedings of COMPSAC'87*, pages 23–29, Tokyo, Japan, Oct. 1987.

[11] B. Randell. System structure for software fault tolerance. *IEEE Trans. Softw. Eng.*, SE-1(2):220–232, June 1975.

[12] S. Shlaer and S. J. Mellor. An object-oriented approach to domain analysis. *ACM SIGSOFT Software Engineering Notes*, 14(5), July 1989.

[13] K. S. Tso and E. H. Shokri. ReSoFT: A reusable software fault tolerance testbed. In *Proceedings of Pacific Rim International Symposium of Fault-Tolerant Systems*, pages 98–103, Newport Beach, CA, Dec. 1995.

[14] K. S. Tso, E. H. Shokri, A. T. Tai, and R. J. Dziegiel, Jr. A reuse framework for software fault tolerance. In *Proceedings of AIAA Computing in Aerospace 10 Conference*, pages 490–500, San Antonio, TX, Mar. 1995.

[15] I. White. *Rational Rose Essentials — Using the Booch Method*. Benjamin/Cummings, Redwood City, CA, 1994.

Multiple Invariant System Design for Fault-Tolerant Real-Time Applications *

I-Ling Yen
Department of Computer Science
A-714 Wells Hall
Michigan State University
East Lansing, MI 48824-1027

Phone: (517) 353-3542
Fax: (517) 432-1061
Email: yen@cps.msu.edu

Abstract

Multiple invariant design technique can be used for writing parallel programs to enhance the performance and achieve fault tolerance. Essentially, a system can have multiple invariants, ranging from strong ones to weak ones. After a change in the system state (due to client requests, environmental changes, processor failures, and so on), the system is allowed to converge to a state that satisfies a weaker invariant with the convergence time being minimal. When the strong invariant of the system is satisfied, the system operates at peak performance. If the system is in a state that satisfies a weaker invariant, then it will operate correctly but perhaps at a reduced performance. The performance is restored by establishing the strong invariant in various ways, such as using a background process for periodic or on-demand maintenance.

We apply this model to fault-tolerant real-time transaction processing systems. The system consists of multiple servers accessing a shared state space and should satisfy the following properties: (1) Fail-stop processor failures at any point of execution should be tolerated, and (2) Whenever a high priority task interrupts the current execution, the system should switch immediately to the new task without affecting data integrity in the shared state space. These goals are achieved by using multiple-invariant data structures. In this paper, we discuss the multiple-invariant system design concept and multiple-invariant data structures. A set of rules that facilitate the design of the multiple-invariant data structures are derived. Some example data structures are used to illustrate this design technique.

Index Terms: Multiple-invariant systems, Performance and reliability improvement, Real-time response, Systematic design strategies, Data structure graphs.

1 Introduction

Multiprocessor systems have been used increasingly in various application areas, such as banking systems, stock-exchange systems, airline reservation systems, intelligent-vehicle highway systems [?], air-craft control systems, medical systems, and manufacturing systems. There is the potential of enhancing the reliability, availability, and performance of these systems by using multiprocessor systems. However, it is also a very challenging task to design a system that exploits the potential performance and reliability gains from multiple processors. The multiple invariant system is one approach for multiprocessor system design for enhancing the reliability and availability without sacrificing the performance.

A multiple invariant system is designed so that changes in the system state (due to client requests, environmental changes, failure occurrences, and so on) can be dealt with by the most efficient algorithm to achieve a rapid response. In this case, the system may converge to a state that satisfies a weaker invariant so that subsequent system operations may run less efficiently though they will still be performed correctly. The degraded system state will be overcome by restoring the system state to satisfy its strong invariant. This can be done in many ways, such as using a background process for periodic or on-demand maintenance or restoring the system state at the most convenient time during subsequent operations.

*This material is based in part upon work supported by NSF under Grant No. CCR-9521419.

Consider multiple processors that concurrently access a shared search table. If a linked list is used as the data structure to implement the search table, then the insertion operation can be implemented very efficiently while a search operation may suffer from performance problems. A tree structure can greatly improve the average search time while yielding a more expensive insert operation. It is possible to design a multiple invariant data structure to optimize the performance. For the above example, we can have a modified tree structure whose root node also serves as the header of a linked list. For an insert operation, the system simply inserts the node at the tail of the linked list. For a search operation, the system searches through the tree and then the linked list. Idle processors will move the nodes in the linked list to an appropriate place in the tree. By doing so, the average response time can be reduced when the system load is below a certain threshold.

Just as the multiple invariant approach can be used to improve the performance of application systems, it can also be used to achieve fault tolerance. Since failures can occur at any point of the execution, a fault-tolerant algorithm should guarantee that after the execution of each instruction, the system invariant will be satisfied. However, if a weak system invariant is chosen, (i.e., allowing the system to degrade to a less efficient structure) then the system performance may degrade after several failures. If a strong system invariant is chosen (i.e., the data structure requirement should always be satisfied) then some costly algorithm may have to be used in order to maintain system integrity. Using the multiple invariant system approach allows the system to selectively satisfy the best possible invariant depending on the current operating situation. During normal execution, the strong invariant of the system is guaranteed and the system operates at peak performance. Upon failure, the system may degrade and only satisfy a weak invariant and, hence, may operate at a reduced performance level. The strong invariant can be restored to recoup the performance by using idle processors.

The most significant effect of the multiple invariant system design is for real-time systems where a critical change in the system state (for example, due to environmental changes that invoke a high priority task) can be handled in real-time even if some regular tasks are in execution. Since a higher priority task can interrupt at any point of the execution, the case is almost the same as the case of achieving fault tolerance. If the system is designed as discussed above, then a current task with lower priority can be aborted immediately and the system will be left in a state satisfying only the weak invariant. Under this circumstance, the system will still operate correctly and the high priority task can be run immediately without violating the consistency (weak invariant) of the system state.

In this paper, we discuss the design of multiple invariant systems for real-time applications and for fault-tolerance purposes. In the next section, we define our system model. In Section 3, we discuss the general design rules. The paper is summerized in Section 4.

2 System Model

We consider a system consisting of multiple processors. The processors accesses the system state concurrently. We model the system state as a shared data structure, such as a tuple space [?] or a blackboard [?]. The goal is to maintain the data structure integrity considering interruption of any concurrent process at any point of its execution.

Our fault model is primarily concerned with processor failures. We consider clean processor failures (fail-stop behavior). Robust data structures [?], [?] and other resilient storage strategies [?] [?] can be integrated into the system to detect and correct storage failures.

3 Design Strategies for Multiple Invariant Data Structures

To provide a uniform framework for designing multiple-invariant data structures, we map data structures to corresponding graphs. A data structure DS can be represented by a data structure graph (DSG) $G = (N, E)$, where $N = \{n_1, \cdots, n_h\}$ corresponds to the set of h nodes in DS and E corresponds to the set of relations among the nodes in N. Each node may have some information associated with it. Let $n_i.info$ denote the information contained in node n_i. Each edge in G is directed, starting from n_i and pointing to n_j, for some i and j. We represent the directed edges in E in terms of pointers instead of conventional graph representations. We use a list associated with n_i, namely, $n_i.edges$ to store the edges outgoing from n_i. In particular, $n_i.edges_j$ denotes the jth outgoing edge of n_i.

Let B ($Before$) denote the *prior* function, where $B(n_i)$ is the set of nodes that have outgoing edges directly to n_i, i.e., $B(n_i) = \{n_j | n_j.edges_k = n_i$, for some j and $k\}$. Also, let A ($After$) denote the *posterior* function, where $A(n_i)$ is the set of nodes that n_i's outgoing edges point to, i.e., $A(n_i) = \{n_j | n_i.edges_k = n_j$, for some j and k}. A node n_{i_k} is said to be reachable from n_{i_1} if there exists a sequence of nodes $(n_{i_1}, n_{i_2}, \ldots\ldots, n_{i_k})$ such that for all j, $1 \leq j < k$, $n_{i_{j+1}} \in A(n_{i_j})$.

The type of update operations considered here are those which modify either the information or the edges associated with a node. Some basic update operations for a data abstraction are the *delete* operation, which removes a node and related edges from DS; the *insert* operation, which inserts a new node and related edges into DS; and the *update* operation, which updates the information content of a node. Also, for some data structures, it is necessary to restructure the edges after the *insert* and *delete* operations. In this section, we consider methods for transforming fault-intolerant algorithms into fault-tolerant ones considering the operations identified above. Similar design strategies can be used to develop methods for transforming the code for other types of operations.

For each operation considered, we first start with the fault-intolerant code. Then, the fault-tolerant version is developed by (1) identifying a critical statement which commits the operation, (2) selecting a detector pointer and the corresponding detection condition to determine whether a critical statement has been executed (the detection condition generally involves the information that is changed by the critical statement); and (3) discussing the detection of the violation of the strong invariant and subsequent actions that restore the strong invariant.

3.1 Simple Node Removal and Insertion

Removing a node from a data structure graph also involves updating edges of the prior nodes. If more than one edge has to be updated, then a partially performed *delete* operation can cause violation of the data-structure invariant since the node may be reachable when traversed via one path but not reachable via the other. It can also cause the loss of a subgraph because of the disconnection of an edge. To resolve these problems, we mark the node to be deleted as invalid without removing it, so that the structure of the DSG will be maintained until all the involved edges have been updated. A *status* bit is associated with each node for this purpose. Let $n_i.status$ denote the status field of n_i. To delete node n_i, we first set $n_i.status = invalid$. Traversal through n_i along any path will recognize the status of n_i and consider n_i as being deleted.

The strong invariant and weak invariant cannot be formulated in general, since they depend on the relations among nodes in the DSG G. However, concerning only the nodes, the weak invariant allows the status of each node n_i, $n_i.status$, to be either *valid* or *invalid* while the strong invariant requires that for all i, $n_i.status = valid$. Consider deleting node p. The general fault-tolerant code should first set $p.status$ to *invalid* and then update the involved edges. The critical statement is $p.status := invalid$. Before all edges are updated, a processor failure can cause violation of the strong invariant. It can be detected by inspecting the status bit of the node since the node will not be fully disconnected until all the edges that need to be updated have been updated.

```
if p.status = invalid then
    update involved edges;
end if;
```

Updating the involved edges for the *delete* operation can be complicated. There are three possible cases.

1. The only changes are adjustments of the prior nodes to point to appropriate posterior nodes.
2. The node cannot be deleted; instead, the information in some other node must be transferred to this node followed by subsequent removal of that (or some other) node.
3. More extensive restructuring is needed to restore the strong invariant after the node is deleted.

Here, we only consider the first case. When more sophisticated restructuring is required for node removal, we can first set $n_i.status = invalid$ and then use the general restructuring techniques discussed in Section 3.2.2 for updating edges. The code for deleting a node by simply readjusting the edges of the prior nodes is given in the following. First, we consider the fault-intolerant code.

```
delete(x):
    traverse G and locate p according to x;
    lock nodes in B(p);
    signal(tcb.lock);
    for each n_i in B(p) do
        update n_i.edges_j, where n_i.edges_j points to p;
        unlock(n_i);
    end;
```

To obtain the fault-tolerant code from its fault-intolerant version, we need to add the statement which updates the detector pointer. Also, the fault-tolerant version should use the *status* field as discussed above. The following is the fault-tolerant code for the *delete* operation.

```
delete(x):
   traverse G and locate p according to x;
   lock nodes in B(p);
   tcb.detector_ptr := p;
   p.status := invalid;
   signal(tcb.lock);
   for each n_i in B(p) do
      update n_i.edges_j, where n_i.edges_j points to p;
      unlock(n_i);
   end;
```

A recovery server can examine *detector_ptr.status* to determine whether the critical statement has been executed. If *detector_ptr.status* = *valid* then the server had failed before executing the critical statement, otherwise, after. The overhead in the fault-tolerant code is the assignment of *detector_ptr* and *p.status*, for a total of two extra assignments.

If a failure occurs before updating all the edges for the *delete* operation, then the strong invariant may be violated. During traversal, this violation of the strong invariant can be determined by examining the *status* field. If a node with *invalid* status is encountered, then we need to update the edges to restore the strong invariant. The code for the traversal is given in the following. For simplicity, we assume that the beginning node for traversal is not the node to be located.

```
p := beginning node for traversal; lock(p);
x := the edge to be traversed following p;
q := p.edges_x;
while not (DSG-dependent termination
   conditions) loop
   if q.status = invalid then
      update p.edges_x;
      perform other necessary updates;
   elsif q is the node to be located then
      perform the action (e.g., delete); exit loop;
   else lock(q); unlock(p); p := q;
   end if;
   x := the edge to be traversed following p;
   q := p.edges_x;
end loop;
```

During traversal, an overhead is incurred because of the extra check of the *status* field for each node encountered. This overhead is significant since it occurs in every iteration. One way of reducing this overhead is to activate the checks for integrity violation whenever a processor failure is detected and to deactivate the checks whenever it is known that the strong invariant has been reestablished. This can be achieved by introducing appropriate compiler directives and using additional support from the operating system.

Similar to the *delete* operation, for the *insert* operation we need to consider multiple updates to the edges to connect the new node. To cope with failures, we perform a sequence of steps that is the reverse of the sequence for the *delete* operation. Consider inserting node p in the general case. First, we create p and assign *p.info* accordingly. Fields *p.edges* will be filled during the traversal when their values can be determined. Field *p.status* should be set to *invalid* first. Then, the involved edges will be assigned to point to p or other appropriate nodes. Finally, we set *p.status* to *valid*. The critical statement is *p.status* = *valid*. Thus, the node will not be considered inserted until all the involved edges are updated. The *detector_ptr* can be assigned to the newly inserted node and whether the critical statement has been performed or not can be determined by examining the *status* of the node pointed to by the *detector_ptr*. Partially performed edge updates will cause violation of the strong invariant. As with deletion, violation of the strong invariant can be detected by examining the *status* field of each node during the traversal. To restore the strong invariant in this case, we adjust the pointers to the state before insertion. The possible cases for edge updates include

1. The new node can be inserted by updating the pointers of nodes that will become its prior nodes.
2. Some information has to be moved to create space for the new node.
3. The data structure needs more extensive restructuring.

The outline of the fault-tolerant code considering only the first case is given in the following.

```
insert(x):
   p := new'node;
   p.info := x; p.status := invalid;
   traverse G and locate nodes in A(p);
   for each n_i in A(p) do
      p.edges_i := n_i;
   end loop;
   lock nodes in B(p);
   tcb.detector_ptr := p;
   for each n_i in B(p) do n_i.edges_{j_i} := p; end;
   unlock nodes in B(p);
   p.status := valid; signal(tcb.lock);
```

If a failure occurs before the *status* bit is set to *valid*, then the recovery server will find that *detector_ptr.status* = *invalid* and decide that the critical statement has not been executed; otherwise, it has been executed. The traversal code is the same as that in the *delete* operation. The overhead of the fault-tolerant program for the *insert* operation includes one assignment of the *detector_ptr* and two assignments for *p.status*.

The code developed in this section can be applied to several data structures, such as single and double linked data structures and sorted lists.

3.2 Graph Restructuring

The *insert* and *delete* operations may sometime require simultaneous changes in the edge information of several nodes and incomplete adjustments may result in invalid structures. The shadow-page technique can be used to achieve a one-step change in the structure of the information [?]. Under certain conditions, which are often satisfied in practice, it is possible to have a more efficient approach using multiple-invariant data structures. We discuss two strategies in the following subsections.

3.2.1 Information Transfer

Sometimes, while updating the edges for the *delete* and *insert* operations, it is necessary to transfer the information of a valid node to an invalid node to accomplish the restructuring. In this case also, efficient fault-tolerant code can be developed to achieve the goal. Consider transferring information from a source node s to a destination node d. A precondition for the transfer operation is that $status(s) = valid \wedge status(d) = invalid$ and postcondition $status(d') = valid \wedge status(s') = invalid \wedge d'.info = s.info$, where d' and s' are the new contents of d and s, respectively. In order to make the code fault-tolerant, it is necessary to ensure that $s.info$ will not be duplicated or lost during the execution even if failures occur. Consequently, a single-step operation that makes node s *invalid* and node d *valid* is required. The strategy we use first moves the information record out of the node. For node p, $p.info\_ptr$ is now used, which is a pointer that points to p's information record. The key idea is to add a field, *back_ptr*, in the information record, instead of using a *status* field to mark the status of a node. A node p is considered to be invalid if

$$p.info\_ptr = \mathbf{nil} \vee p.info\_ptr.back\_ptr \neq p.$$

Using this design, a single-step swap of the status of two nodes can be achieved by redirecting the *back_ptr* properly.

As we can see from the design, the critical statement should be the redirection of the *back_ptr* from s to d to make s invalid and d valid. The *detector_ptr* can be d or s and whether the critical statement has been executed can be determined by checking whether $detector\_ptr.info\_ptr.back\_ptr = detector\_ptr$. The fault-tolerant code for information transfer is given in the following. Note that in a fault-intolerant situation, we do not have an invalid node and, hence, we do not give the fault-intolerant code for this case.

```
{precondition: d.info_ptr.back_ptr ≠ d∧
   s.info_ptr.back_ptr = s}
transfer_info(s, d):
   lock(s); lock(d);
   d.info_ptr := s.info_ptr;
   tcb.detector_ptr := d;
   s.info_ptr.back_ptr := d;
   signal(tcb.lock);
   unlock(s); unlock(d);
```

In the first step, we set $d.info\_ptr$ to point to the same information record as $s.info\_ptr$. In the third step, setting d to be valid and s to be invalid is done in a single update step by simply updating the $info\_ptr.back\_ptr$ to point to d. The critical statement is $s.info\_ptr.back\_ptr := d$ and whether it has been executed can be detected by examining whether $detector\_ptr.info\_ptr.back\_ptr = d$.

Note that if *transfer_info* is used as part of another operation, such as a *delete* or an *insert* operation, then the *detector_ptr* should not be set and the locking and unlocking actions are not necessary. These operations should be performed in the main procedure.

3.2.2 General Restructuring Strategy

To design a data structure DS that facilitates general restructuring, we start with the normal representation graph $G = (N, E)$ and construct an alternate graph $G' = (N, E')$, where E and E' are fully disjoint sets. In addition, for all i and j, such that $n_i, n_j \in N$ and n_j is reachable from n_i in G, n_j should be reachable from n_i in G'. A simple design of E' can meet these requirements. Let the fields *next* and *prev* be associated with each node n_i, where $n_i.next = n_{(i+1) \bmod h}$ and $n_i.prev = n_{(i-1) \bmod h}$, for all i. The set of pointers $n_i.next$ and $n_i.prev$, for all i, consists of all the edges in E'. Essentially, the edges in E' link all nodes together as a circular double-linked list. Other designs of E' can be considered for various data structures.

During restructuring, the edges associated with several nodes may have to be updated atomically to ensure data integrity. When these edges are partially updated, the traversal along these nodes may be incorrect. To resolve this problem, we use an extra boolean field $n_i.edge\_state$ for node n_i to indicate whether the edges in $n_i.edges$ are fully correct. When $n_i.edge\_state = correct$, all the edge information in $n_i.edges$ is fully correct; and when $n_i.edge\_state = uncertain$, the edge information in $n_i.edges$ may have been partially updated. To update the edges of a group of nodes in DS, we perform the following sequence of steps: (1) update the edges of the nodes that belong to E'; (2) for all nodes n in the group, $n.edge\_state = uncertain$; (3) update the edges of the nodes that belong to E; (4) for all nodes n in the group, $n.edge\_state = correct$. During traversal of DS, if a node with an *uncertain* edge status is encountered, the traversal should continue along edges in E' instead of E since the edges in E may not be correct.

Generally, the original data structure is designed such that traversal along E will be very efficient. The design for E' should ensure that the correct edge information can be maintained or recovered in the presence of failures (like the double-linked-list example given previously). Thus, when the edge information in E cannot be trusted, the data structure accesses still can be processed correctly by traversing along E', though, generally, the traversal may be much less efficient. The strong invariant requires that for all i, $n_i.edge\_state = correct$ while under the weak invariant, $n_i.edge\_state$ can be *uncertain* because of partial updates. When the strong invariant is true, the traversal through the data structure can be very efficient along the edges in E. When the strong invariant is violated because of failures, the traversal will be done along edges in E' which will be less efficient but correct.

For example, consider a binary-search-tree data structure, T, where each node in T has *left* and *right* pointers to its left and right subtrees, respectively. Assume that node q is to be deleted and $q.left \neq nil \wedge q.right \neq nil$. Let p denote the parent node of q where $p.left = q$. If we set $p.left$ to $q.left$ after deleting node q, then the subtree pointed to by $q.right$ is lost. To handle this problem, we design the tree data structure so that it contains two sets of edges, E and E', where E contains the *left* and *right* pointers and E' contains the *next* pointer. Essentially, E organizes T as a tree structure while E' organizes T into a linearly linked list. More precisely, we want E' to organize T into a sorted linearly linked list. When a failure occurs, the edges of a node may be partially updated. In this case, some nodes may have their *edge_state* set to *uncertain*. During traversal, instead of continuing to traverse the tree when a node with value *uncertain* for its *edge_state* is encountered, we switch and traverse via the linearly linked list. The traversal time is logarithmic for traversing along E while it is linear along E'.

3.3 Update Node Information

Updating the node information does not involve any modification to the field *edges* and, thus, it is generally simpler than other operations. Let p denote the node for which the associated information $p.info$ is to be updated. The fault-intolerant code is

```
update(x):
    traverse and locate node p according to x;
    lock(p);
    p.info := f(···)
    signal(tcb.lock);
    unlock(p);
```

Here, f is the function that computes the final value that replaces x for the update operation.

Consider the fault-tolerant code. To ensure that the information update is done in one step, we use the information pointer $p.info\_ptr$ to point to the information record. The update can be done by simply redirecting $p.info\_ptr$ to point to the new information record. Thus, the critical statement is the redirection of *info_ptr*. However, to allow the detection of whether the critical statement has been executed, we still need to use the *back_ptr* field in the information record as discussed in the previous subsection. In this case, *detector_ptr* should be the original *info_ptr* of p. The use of *tcb.detector_ptr* will be discussed after the fault-tolerant code is given in the following.

```
update(x):
    traverse and locate node p according to x;
    lock(p);
    q := new'info_node;
    q.info := f(···);
    q.back_ptr := p;
    tcb.detector_ptr := p.info_ptr;
    p.info_ptr := q;
    signal(tcb.lock);
    unlock(p);
```

In this code, the new *info_node* q is created to store the new information and $q.back\_ptr$ is set to point to p. Thus, when $p.info\_ptr$ is set to q, the update is done. The statement $p.info\_ptr := q$ is the critical statement. Whether the critical statement has been executed can be detected by checking *detector_ptr*. If $detector\_ptr \neq nil$ or $detector\_ptr = detector\_ptr.back\_ptr.info\_ptr$, then the critical statement has not been executed. Since there is only a single statement (the critical statement) which updates the shared data structure, the strong invariant will always be true. Thus, the data-structure invariant can be used as the strong invariant and there is no need for the weak invariant. The overhead in the fault-tolerant code includes the creation of node q, assigning $q.back\_ptr$, assigning $p.info\_ptr$, and the assignment of $detector\_ptr$.

The code given above is designed for general circumstances. If some special conditions are satisfied for a specific application program, then the code can be simplified to further reduce the overhead. First, if function f is idempotent then there is no need to use any detector node since the code can be reexecuted without problems. If f is scalar and idempotent, then the information pointer is not needed and the code is identical to the fault-intolerant version.

4 Summary

We have developed a multiple invariant system design strategy for real-time transaction processing system. The strategies can be used for designing fault-tolerant data structures. With this design, a high priority task can preempt a lower priority task at any point of the execution without violating the system state integrity. The approach gives better performance than the shadow page technique since it is less costly (in terms of the execution time overhead) and does not require any global recovery. Our approach has some resemblance to self-stabilizing systems in the sense that upon failure, the system may converge to a state that violates the regular system invariant and the system invariant can be restored by subsequent execution steps. However, self-stabilizing systems allow the system to converge to illegitimate states upon a failure which, therefor, results in vulnerable periods. This problem is avoided in our approach by imposing the constraint that the system should converge to a state that satisfies a weak invariant upon fail-stop processor failures at any time during the execution.

We also discussed various design rules in the paper, including information transfer, general restructuring, and so on. These strategies can be used to transform ordinary data structures into semantically equivalent multiple-invariant data structures.

References

[1] S. Ahuja, N.J. Cerriero, and D. Gelernter, "Linda and friends," *IEEE Computer*, Vol. 19, No. 8, Aug. 1986.

[2] R.K. Jurgen, "Smart cars and highways go global," *IEEE Spectrum*, Vol. 28, No. 5, May 1991, pp. 26-36.

[3] B.W. Lampson, "Atomic transactions," in B.W. Lampson, M. Paul, and H.J. Siegert (Eds.), *Distributed Systems - Architecture and Implementation*, Springer-Verlag, New York, 1981, pp. 246-265.

[4] E.K. Lee and R.H. Katz, "The performance of parity placements in disk arrays," *IEEE Trans. on Computers*, Vol. 42, No. 6, June 1993, pp. 651-664.

[5] C.-C.J. Li, P.P. Chen, and W.K. Fuchs, "Local concurrent error detection and correction in data structures using virtual backpointers," *IEEE Trans. on Computers*, Vol. 38, No. 11, Nov. 1989, pp. 1481-1492.

[6] P. Mazumder, "Design of a fault-tolerant three-dimensional dynamic random-access memory with on-chip error-correcting circuit," *IEEE Trans. on Computers*, Vol. 42, No. 12, Dec. 1993, pp. 1453-1468.

[7] D.J. Taylor and C.-J.H. Seger, "Robust storage structures for crash recovery," *IEEE Trans. Computers*, Vol. C-35, No. 4, April 1986, pp. 288-295.

[8] P.H. Winston, *Artificial Intelligence*, 2nd ed., Addison-Wesley, 1984.

Panel Session 1

Major Research Issues in ORDS

Moderator:
Tom Lawrence, USAF Rome Laboratory

Panelists:
David Bakken, BBN Systems and Technologies
Michael Davis, SRI International
Michel Gien, Chorus Systems, France
Phillip Q. Hwang, Defense Mapping Agency
Hiroaki Nakanishi, Hitachi Ltd., Japan

Commentators:
Luise Moser, University of California, Santa Barbara
Vicraj Thomas, Honeywell Technology Center

QoS Issues for Wide-Area CORBA-Based Object Systems[1]

David E. Bakken Richard E. Schantz John A. Zinky

BBN Systems and Technologies
10 Moulton Street
Cambridge, Mass. 02138 USA
{dbakken,schantz,jzinky}@bbn.com

Abstract

CORBA is a distributed object technology which shows great promise in simplifying the task of programming applications which must operate over a wide area network (WAN). However, WAN environments are dynamic and hostile, making applications which are distributed over a WAN much more difficult to develop and maintain than those only distributed across a local area network (LAN). Quality of service (QoS) is a technology emerging in a number of communications substrates which needs to be provided at the object level to help distributed applications be easier to develop and be more adaptive at runtime. This position paper described some of the issues which CORBA-based midleware must address to provide QoS for objects distributed across a WAN and overviews QuO, our middleware framework for wide-area QoS.

Introduction

Distributed computing has become mainstream in recent years. Distributed applications are increasingly being deployed across wide area networks (WANs), and the growth (and benefits) of such wide-area applications would seem poised for explosive growth as fiber optics and internet technologies become deployed ever more widely.

However, writing and maintaining distributed applications which must operate across a WAN is difficult. Unlike a LAN, WANs are less predictable and have much more operational variance — they are a much more dynamic and hostile environment than a LAN environment — and more likely to change from configuration to configuration. Worse, performance and other non-functional issues (which we call *system properties*) are buried beneath an object's functional interface, so an object can't easily help an application adapt to a changing environment or reuse its code for a new environment.

A compounding factor in writing and maintaining distributed applications is that few application programmers are either trained or experienced in dealing with system properties, because the system properties of the local objects with which they are used to programming are ideal. For example, local objects do not fail independently of a client, for all practical purposes, and the delay is minimal and the bandwidth infinite when invoking a method. However, the system properties of distributed objects are far from ideal: they can fail unexpectedly, and the delay for a method invocation to return may be long and have high variance. As a result of this, most invocations to remote objects in a real-world distributed application are bracketed with extra code to handle errors and performance anomalies. Another difficulty is that the knowledge of system properties is available at different times, in different locations, and from different sources, and it is very difficult for the programmer to synthesize them.

The above difficulties can be greatly reduced by the judicious use of middleware, a layer of software above the communications substrate which offers a consistent, higher-level abstraction throughout the network. This paper describes some of the issues involved with developing CORBA middleware to help simplify the development of wide-area distributed applications by providing application-level quality of service (QoS).

The remainder of this position paper is organized as follows. We first offer opinions on developing such middleware, to stimulate discussion and frame our research. We then discuss our project to provide such middleware to the application layer, and conclude with a brief discussion of the open issues for such middleware.

1 This research was funded by Contract No. F30602-94-C-0188 from USAF Rome Laboratory.

0-8186-7570-5/96 $5.00 © 1996 IEEE

Comments on Supporting Wide-Area Distributed Objects

Our group at BBN has been developing distributed object middleware and using it to deploy wide-area applications with it for fifteen years [3,1,9,8]. Based on this experience, we offer the following comments:

1. CORBA is an excellent vehicle for wide-area distributed R&D. Distributed object technology not only encapsulates but also integrates the three kinds of resources which operating systems typically encapsulate separately — processing, storage, and communications — and is thus quite general. CORBA is by far the best distributed technology available, in our opinion, with its technology arguably 3 years ahead of that of network OLE. The research issues involved in supporting CORBA's functional interoperability have largely been resolved, with only engineering issues remaining to be resolved.
2. To be most useful in the wide area, an object's functional interface (e.g., CORBA IDL) must be augmented to deal with non-functional aspects such as delay and throughput [2,7,6]. Bad performance cannot be hidden beneath a functional interface: it will always show through.
   Often these non-functional issues are ignored during development, often due to deadline pressure or to ignorance of some of the system properties the application will encounter. However, when the application is deployed over the wide area logic to handle these non-functional issues has to be added as anomalies are encountered, i.e. in an *ad hoc* manner. After just a few rounds of such changes the code becomes expensive to maintain and evolve.
3. Client-object QoS contracts which span wide areas *will* be broken, because WAN environments are so dynamic and hostile. To deal with this, we must have compound contracts with fallback positions which are specified and automatically deployed when necessary.
4. Adaptivity implies multiple implementations. A given implementation of an object makes design tradeoffs *viz.* resources required: processing versus storage (including memory) versus communications. If an application is to be successfully fielded in multiple environments, including WANS, a number of implementations with different tradeoffs should be available and deployed as conditions merit. A client should be able to access one implementation if the object is local to its desktop, another implementation if it is across a WAN, and yet another if the bandwidth across that WAN becomes unusually low or high.
5. Adaptivity and multiple implementations imply late binding. A program will be more adaptive if it can delay some binding decisions and even rebind them at runtime as conditions warrant. Different binding decisions should be able to be made at design time, install time, program startup, when a connection/binding is set up, and when a method is invoked using this connection/binding. If a programmer is forced to make commitments unnecessarily early, and later the environment does not match what the programmer assumed, then the program will perform poorly.

QoS for CORBA Objects (QuO)

We have developed an architecture, QuO, for providing QoS for CORBA objects in a manner which addresses the above concerns [10]. QuO provides Quality Description Language(QDL) with three sublanguages:

- Contract Description Language (CDL). CDL captures the behavior between a client and an object; it is described further below.
- Object Description Language (ODL). ODL describes for each implementation of an object how appropriate it is in given conditions and what abstract resources it requires. ODL can be used to dispatch the most appropriate implementation for the current conditions.
- Resource Description Language (RDL). RDL maps from an implementation's abstract resource requirements onto physical resources.

CDL captures the behavior of the client and the object viz. their connection. It allows the specification of two different kinds of regions, which are specified with predicates defined in terms of both the client's usage patterns and the QoS the object will provide. Negotiated regions specify the expected (agreed) behavior of the client and object, while reality regions specify the measured or observed behavior. The contract is adjudicated by middleware residing in the client's address space. When a transition occurs between negotiated or reality region, CDL allows handlers in either the client or the object or both to be invoked to help them adjust their behavior or expectations or both to the changing conditions.

QuO provides multiple binding times for system properties. At QDL definition time the structure of the regions (e.g., a region depends on the maximum invocatiosn/second) and their handler routines are specified. When a connection is set up parameters can be passed which shape the regions (e.g., the value for the maximum invocations/second for a region). At negotiation time the client and object set their expected behavior. At invocation time, the QuO predicates defining the reality regions are evaluated and their changes propagated.

Finally, we hope to support a broader notion of QoS in future QuO work. The notion of QoS generally held by researchers and practitioners involves mainly communications and performance issues and assumes a multimedia setting across only a LAN (or at most a MAN) and without any adaptivity. We presently are supporting communications and performance issues with adaptivity. In the future we hope to also support availability, security, and non-functional other aspects which a client may wish to specify and regulate when connecting to a remote object across a wide-area environment. Eventually we hope provide higher levels of abstractions for the programmer to trade off such as accuracy versus timeliness versus precision.

We thus view QoS as the architectural umbrella for integrating those system properties needed to better develop, deploy, and maintain wide-area object-oriented distributed application. We feel that the application programmer can benefit from this QoS framework which allows the specification and trading off of these different aspects. Much can be gained by the implementing of such middleware in terms of understanding the fundamental tradeoffs between these aspects.

Open Issues

We consider the following to be open issues in how to best have middleware deal with system properties in the wide area:

1. What is the best way to control an entire stack of the resources (below the middleware) necessary to provide end-to-end client-object reservations in the wide area? Should the OS provide for processing and storage levels of reservation similar to which the integrated services (INTSERV) working group [4] of the IETF [5] is providing for communications:
    - Guaranteed
    - Predictive
    - Controlled
    - Best-Effort

   With these, it would be possible to provide end-to-end reservations of all resources needed for a client-object connection yet with fallback positions and priorities.
2. How can we best layer contracts (in the middleware)?
3. What kind of contract abstractions (above the middleware) are useful to application programmers?

Acknowledgments

We thank Jerry Dussault, Richard Soley, and Tom Wilkes for their feedback on this research.

References

[1] Bruce M. Anderson and John P. Flynn, CASES: A System for Assessing Naval Warfighting Capability, Proceedings of the 1990 Symposium on Command and Control Research, SAIC Report 90/1508, June 1990.

[2] Shigeru Chiba, Open Implementation Home Page, Internet Publication (URL http://www.xerox.com/PARC/spl/eca/oi.html).

[3] Programming Support in the Cronus Distributed Operating System, Proceedings of the Sixth International Conference on Distributed Computing Systems, May, 1986.

[4] IETF Home Page, Internet Publication (URL http://www.ietf.cnri.reston.va.us/home.html).

[5] Integrated Services (intserv) Charter, Internet Publication (URL http://www.ietf.cnri.reston.va.us/html.charters/intserv-charter.html).

[6] Kiczales, G. and J. des Riveres and D. G. Bobrow. The Art of the Metaobject Protocol, The MIT Press, 1991.

[7] Kiczales, G. ed. Workshop on Open Implementation '94, Internet Publication (URL http://www.xerox.com/PARC/spl/eca/oi/workshop-94).

[8] Edward F. Walker and Michael A. Dean, The Migration of Cronus to CORBA, Proceedings of the Fifth IEEE Dual Use Technologies and Applications Conference, May, 1995.

[9] Edward F. Walker and Richard Floyd and Paul Neves, Asynchronous Remote Operation Execution in Distributed Systems, Proceedings of the Tenth International Conference on Distributed Computing Systems, June, 1990.

[10] John A. Zinky and David E. Bakken and Richard E. Schantz, Supporting Quality of Service for CORBA Objects, BBN Report No. 8119, November 1995. Submitted for publication to Theory and Practice of Object Systems, Special Issue on CORBA and the OMG.

Position Paper: Resource Management for Complex Distributed Systems

Michael B. Davis and Jaroslaw J. Sydir
SRI International
Information, Telecommunications, and Automation Division
Menlo Park, California 94025
mdavis@erg.sri.com, sydir@erg.sri.com

Abstract*

Complex distributed systems require resource management capabilities that can allocate resources to applications, monitor and control the use of resources, and reallocate resources in response to anomalies. To utilize resources efficiently and meet application and system-wide requirements, the resource management system must consider resource availability, control policies, and application's quality-of-service (QoS) requirements when making allocation decisions. We identify several open research areas in QoS-based system resource management and briefly describe the ongoing Global Resource Management (GRM) project, which will address some of these areas.

Keywords: quality of service, resource management, distributed operating systems

1. Background

Commercial and military computing enterprises are becoming increasingly large and complex. To provide high availability and good performance, applications and system resources are typically replicated and distributed. The distributed nature of these enterprises, together with dynamic usage patterns, the existence of many failure modes, and contention for limited resources, leads to challenges in managing the enterprises' computing resources. A resource management system is required to allocate resources to applications, monitor and control the use of resources, and reallocate resources in response to anomalies. The resource management system should strive to continuously provide the appropriate quality of service (QoS) for each application and user, based on the characteristics of the application, the needs of the user, the operational mode of the system (e.g., peacetime or crisis), the available resources, and so forth, during changing network and battle scenarios.

*The work described in this paper was funded under Rome Laboratory Contract Number F30602-95-C-0299.

Figure 1 shows a high-level concept of resource management. The management system monitors the QoS requirements of applications and the resources that are available to meet the QoS requirements, makes allocation decisions based on this information and on policies determined elsewhere, and sends control directives to applications and resources. The resource management system attempts to satisfy the QoS requirements of each application by allocating to it the required resources. When the QoS requirements cannot be fully met, the application has several choices: (1) don't execute; (2) execute, using alternative implementations or algorithms; or (3) execute with varying QoS parameters. The resource management concept can be applied recursively at different levels of the system being managed.

Traditionally, the component parts of distributed systems (e.g., applications, systems, and networks) have been managed by separate, independent management elements that are only slightly integrated, if at all. Commercial-off-the-shelf (COTS) products have been designed to manage networks rather than resources in general, and have been designed to work in commercial rather than dynamic military and crisis-response environments. Also, most resource management has attempted to use resources efficiently without taking into account the QoS requirements of applications. Research is

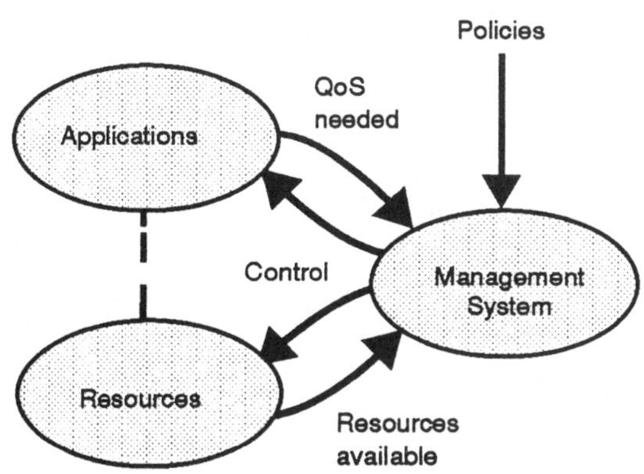

Figure 1. Resource management concept

required to develop new techniques that will manage system resources in a uniform and coordinated way in a dynamic environment, in accordance with application QoS requirements; and the validity of the techniques must be demonstrated.

2. Open research areas

To support the integrated resource management that will be required for the efficient and dependable operation of future distributed systems, research is needed in the following areas.

1. *Comprehensive understanding of QoS.* There is no common definition of QoS. Research is needed to develop a definition of QoS in terms of a basic set of primitives and their relationships. Current candidates for these primitives include timeliness (when); precision (how much), and accuracy (how good); their first derivatives; and possibly their second derivatives. (An example of a first derivative is jitter, which expresses variations in delay.) In addition to the instantaneous values of QoS, aggregations (e.g., system-wide QoS values) and statistical distributions (e.g., distributions of packet delays) must also be considered. Research is needed to define the mapping between QoS values and computing resources such as processing, communications, and storage, and to identify basic trade-offs among the resources relative to QoS. For example, one can satisfy the same timeliness criteria either by compressing data before sending it (more processing, less communication) or by not compressing it (more communication, less processing). There may have to be qualitative mappings and relationships that must be quantified for specific systems or types of systems.

2. *Application-level QoS framework.* There is no systematic way of categorizing the QoS requirements of applications. Research is needed to classify applications in terms of their QoS requirements, including the classification of data into groups for which certain sets of QoS requirements make sense (e.g., various kinds of video streams, images, or database records). The resulting framework would allow an application domain to define its QoS requirements and map them to the basic QoS primitives described above in connection with Area 1.

3. *Architectural framework for QoS-driven resource management.* Current resource management systems use QoS in an ad hoc way. Research is needed to develop an architectural framework in which QoS specifications resulting from work in Areas 1 and 2 above can be used for managing the resources of a distributed system. This research should include the defining and specifying of (1) system layers and major services; (2) the overall management structure (e.g., domains and agents); (3) high-level interfaces to applications; (4) the types of management information to collect; and (5) requirements for low-level system components (e.g., the operating system, the communication system, and network management components).

4. *Application-specific QoS trade-offs.* The trade-offs that can be made among various QoS parameters for particular application domains have been incompletely studied. Research is needed in specific application domains (e.g., computer-supported collaborative work) to find the relevant QoS trade-offs that can be made, and to develop application-specific, discrete QoS levels.

5. *Additional QoS primitives.* The QoS primitives discussed earlier (timeliness, precision, and accuracy) relate mainly to performance. Other categories of QoS primitives should also be considered, such as cost. Research is needed to identify these additional QoS primitives and incorporate them into the framework of Areas 1 through 4 above.

6. *Resource management algorithms.* Resource allocation and control decisions must be made at several levels and degrees of locality within the distributed system, using information about resource status and application requirements that may be incomplete, incorrect, or out of date. Because of the complexity of the problem, it will be necessary to use heuristic approaches. Research is needed to develop efficient, distributed, resource allocation algorithms that make trade-offs among different resource types to achieve given QoS levels. The algorithms must work despite inexact status information.

7. *Distributed operating systems.* The QoS and resource management frameworks developed above will be useful only if they can be incorporated into operating systems. Research is needed in mapping the QoS primitives, resource management architecture, and resource management architecture into the framework of soft real-time distributed operating systems.

3. Global resource management project

In an ongoing project sponsored by Rome Laboratory, entitled Global Resource Management (GRM), SRI and Stanford Telecommunications are addressing several of the research issues described above: in particular, those in Areas 1; 2 (for certain classes of applications); 3; and parts of 6 and 7. We are developing a concept of adaptive, integrated enterprise management that will manage system resources for the next generation of large, distributed command and control systems. GRM will allocate and control system resources in accordance with user QoS requirements, despite changes in operational modes, variations in the availability of resources, mobility, and anomalies. The development will be done in the context of resource-intensive multimedia command, control, communications, computers, and intelligence (C4I) applications. The products of the effort will be (1) a reference model and an abstract model that express the pertinent aspects of the global resource management problem domain in an implementation-independent way, using object oriented analysis methodologies; (2) a detailed functional architecture that describes a reference implementation of the global resource management system; and (3) a set of recommendations and designs for further investigations, the development of a prototype implementation, and the integration of the prototype into a testbed environment. We intend to begin this research with the frameworks provided by the Reference Model of Open Distributed Processing (RM-ODP) [1]; the Telecommunications Information Network Architecture Consortium (TINA-C) Network Resource Information Model [2]; and the Common Object Resource Broker Architecture (CORBA) [3]. We will extend these frameworks to include QoS and management functions.

GRM is a continuation of work begun on another Rome Laboratory-sponsored project, System Resource Management (SRM) [4]. In the SRM project, we developed a method for expressing quality of service requirements using *benefit functions*. A benefit function specifies the benefit to the user or to the system as a whole of attaining a given QoS level. Figure 2 shows an example benefit function, in which the relative benefit to the user of a video conferencing application is specified as a function of the frame rate attained.

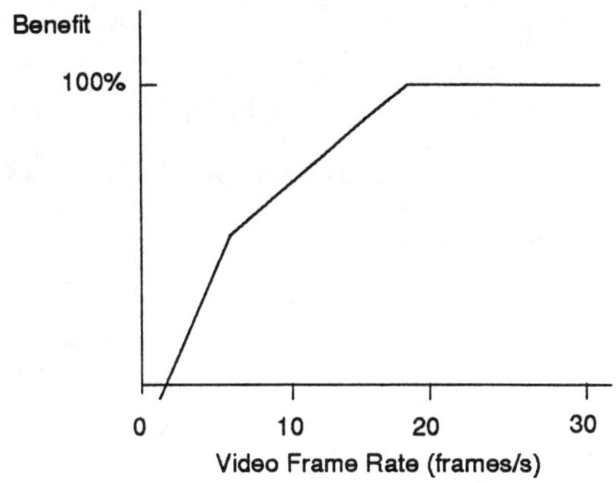

Figure 2. Example benefit function

References

[1] ISO/IEC. *ITU-T Recommendation X.901, ISO/IEC 10746-1, Reference Model of Open Distributed Processing—Part 1: Overview*, International Organization for Standardization and International Electrotechnical Committee, May 1995.

[2] TINA-C. *Network Resource Information Model Specification*, Telecommunications Information Networking Architecture Consortium, December 1994.

[3] OMG. *CORBA: Architecture and Specification*, Object Management Group, August 1995.

[4] M. Davis, A. Downing, and T. Lawrence. "Adaptable System Resource Management for Soft Real-time Systems," Symposium on Command and Control Research and Decision Aids, Monterey, California, June 1994.

Object-Oriented Programming of Complex Fault-Tolerant Real-Time Systems

L. E. Moser, P. Narasimhan, P. M. Melliar-Smith
Department of Electrical and Computer Engineering
University of California, Santa Barbara, CA 93106
moser@ece.ucsb.edu, priya@alpha.ece.ucsb.edu, pmms@ece.uscb.edu

Abstract

The challenge is to build, on top of standard operating systems, complex fault-tolerant real-time systems that operate fast enough to meet real-time deadlines even under fault conditions, using an approach that simplifies the application programming. The Realize system, being developed at the University of California, Santa Barbara, aims to meet that challenge.

1 Introduction

Research and design paradigms for real-time systems have focused primarily on hard real-time systems. For relatively simple preplanned synchronous systems, effective strategies have been devised that aim to guarantee real-time deadlines using worst-case workloads and worst-case performance measures.

Unfortunately, complex real-time systems involve considerable uncertainty. There is variation in processor performance resulting from caches, cycle stealing, and interrupt handling; there is variation in the execution of the programs caused by many special cases and heuristic algorithms; there is variation in message transmission and delivery time; there is variation resulting from fault recovery; and there is variation in the workload. A preplanned synchronous design for a complex real-time system is necessarily a conservative design, with adverse effects on performance. Consequently, large complex systems are typically designed as asynchronous event-driven distributed systems, running on top of standard operating systems.

Fault tolerance is often regarded as an afterthought, rather than an integral part of a system design. Such an approach greatly increases the probability that fault recovery will itself introduce additional errors; this is particularly unfortunate since fault recovery is difficult to test. Sound fault-tolerance depends on replication of information and on the ability to distinguish data that are known to be consistent from data that are possibly inconsistent after a fault. In a real-time system, there is little enough time for regular processing, let alone for recovery from faults. Thus, replication of data and its certification as stable must be a continuous on-going activity involving all of the operations on the data.

Programming complex fault-tolerant real-time systems is difficult, particularly for programmers who are skilled in the application domain but unskilled in operating system implementation. Ensuring consistency in the presence of asynchrony, concurrency, replication and faults is not easy, and meeting a deadline achieves little if the answers are wrong. To address these issues, researchers have adopted an object-oriented approach to the development of fault-tolerant real-time applications [3, 6].

2 Object-Oriented Programming

The advantages of an object-oriented approach to the development of complex fault-tolerant real-time systems are evident:

- Object-orientation provides a framework that can reduce the complexity of programming a complex application.
- The correspondence between the objects of the application program and the objects of the physical world simplifies programming, particularly for programmers whose skill and experience is in the application domain.
- Object-orientation abstracts the implementation details and enables the programmers to concentrate on the behavior required for the application.

0-8186-7570-5/96 $5.00 © 1996 IEEE

- An object-oriented approach provides powerful constructs for the composition of software components and allows the contributions of different programmers to be coordinated more easily.
- Object-oriented programming facilitates interoperability across the distributed and heterogeneous platforms that complex systems may require.
- Programming with objects can contain the side-effects of faults and can aid in detection of, and recovery from, faults.

The challenges of object-oriented programming for the development of complex fault-tolerant real-time systems are also evident:

- The local focus of object-oriented programming, which is valuable for controlling complexity, masks global properties such as real-time deadlines and performance.
- The interface definitions of objects currently have no notion of real-time or performance.
- The interface definitions of objects currently make no attempt to specify operations invoked on other objects so that, even if performance characteristics were available in the object interface definitions, it would be difficult to analyze system performance without analyzing the detailed implementation code.
- Support facilities are needed to allocate the objects across a distributed system and schedule the tasks so that real-time performance requirements are met.
- Inheritance of real-time attributes of objects is not yet fully understood.
- The interface definitions of objects provide no fault detection criteria.
- No mechanisms exist for defining the type or degree of replication that is needed to achieve fault-tolerance.
- Maintaining consistency between the replicas of an object across a large distributed system may be difficult.
- Creating a replica of an object during recovery from a fault requires the current state of the object to be copied. For some objects the state may be too large to be copied in one atomic operation within the real-time performance constraints of the system.

Nevertheless, the benefits of object-oriented programming encourage us to explore ways of using objects in the programming of complex fault-tolerant real-time applications.

3 Providing Fault Tolerance

Fault detection and recovery is a significant problem for real-time systems. Clearly, the quickest possible detection of a processor or communication fault is essential. The system is then reconfigured, and the fault and reconfiguration are reported so that all of the processors in the system have a consistent view of the new configuration. If there are new processors in the configuration, a set of pending operations is transferred to them. All processors then reschedule the tasks of the failed processors and reallocate those tasks to the survivors.

To provide fault tolerance and to improve performance, objects are replicated at different sites within the distributed system. Two methods are available for handling replicated objects: active replication by operation distribution and passive replication by update distribution.

In *active replication by operation distribution*, the operation is communicated to all of the replicas of the object and is performed at the site of each object. This method of replication is advantageous when the computational cost of the operation is low, compared to the cost of communicating the updated state of the object. If the operation itself invokes a nested operation on an object, the nested operation must, of course, also be performed on each replica, but only once on each replica.

In *passive replication by update distribution*, the operation is performed on only one of the replicas and the updated state of the object is communicated to all of the other replicas. This method of replication is appropriate when the computational cost of performing the operation is higher than the cost of communicating the updated state of the object at the end of the operation. If the operation invokes a nested operation on another object within the same domain, the identity of that object is recorded, so that its updated state can also be distributed at the end of the operation.

Active replication by operation distribution yields substantially more rapid recovery from faults than passive replication by update distribution and is the method of choice for systems that require rapid recovery within real-time deadlines. Provision of both means of replication allows the application programmer to opt for either, depending on the real-time requirements of the task in question.

4 Scheduling Operations in Real Time

There is no single perfect scheduling strategy for all multiprocessor real-time systems. A simple least-laxity/importance algorithm [2] is appropriate for a wide range of soft real-time applications, but other strategies may be more effective for specific applications. In theory, least-laxity scheduling is preferable to priority scheduling (such as rate-monotonic scheduling) if the mean duration of an operation (including context switching and communication) exceeds about twice the mean additional cost of heap-sorting the laxities within the scheduler. In our opinion, priority scheduling is unlikely to be preferable to least-laxity scheduling for most complex real-time applications.

Least-laxity scheduling is effective provided that the system is not overloaded. With least-laxity scheduling, many operations may miss their deadlines if an overload occurs. An *importance* parameter allows the system to abandon less important operations to ensure that more important operations meet their deadlines. An operation is executed only if the scheduler estimates that all of the more important operations will meet their deadlines. Use of the importance parameter, however, increases the scheduler overhead.

The ability to schedule tasks in real-time depends on the granularity of the operations and the tightness of the deadlines of the application, on the overheads within the object-oriented mechanisms themselves, and on the characteristics of the operating system on which the program is run.

5 The Realize System

The Realize system, being developed at the University of California, Santa Barbara, is middleware for use on top of a standard operating system. It provides an object-oriented platform for development of fault-tolerant soft real-time applications. Our primary objective for Realize is a simple programming paradigm that is easily understood by typical programmers and that helps them to maintain consistency and correctness despite the difficulties introduced by distributed processing and faults.

In Realize, a programmer creates a single object-oriented program, defining objects and operations on them and hierarchically constructing complex objects from simpler ones, exactly as though only one processor were being used. The programmer also specifies the degree of fault-tolerance required and real-time performance constraints. The Realize system assumes responsibility for distributing the objects across the system, for replicating objects, for fault detection and recovery, and for scheduling operations to meet real-time constraints.

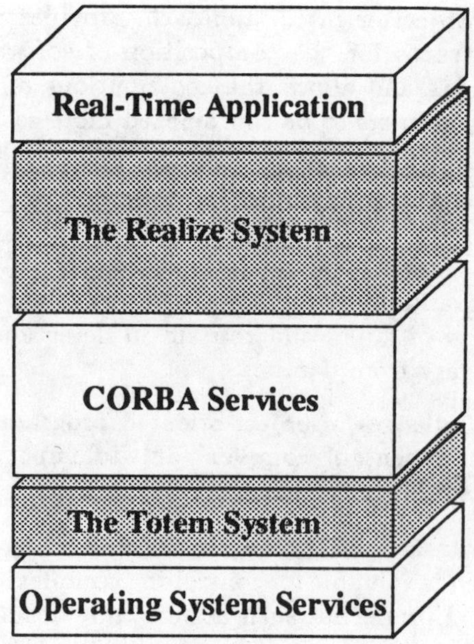

Figure 1. The Realize system provides object replication, fault recovery and real-time scheduling, using the reliable ordered multicasts of the Totem system.

The Realize system uses a CORBA-based approach [1, 5] in which calls on objects are replaced by calls on stubs and skeletons. In the terminology of CORBA, Realize is an ORB of a specialized nature that must accommodate object replication, fault detection and recovery, and real-time scheduling. Realize services include the ability to encapsulate calls in multicast messages and communicate them to other processors. As shown in Figure 1, Realize is layered on top of the Totem system [4], a multicast group communication system that provides reliable ordered delivery of messages, as well as membership and topology maintenance services, within a distributed system.

6 Conclusion

The utility of object-oriented platforms, such as Realize, for the development of complex fault-tolerant real-time applications will ultimately depend on their ability to reduce development costs, and development risks, which are increasing as real-time systems become more complex. With faster processors and communication media becoming readily available, we can incur overheads and still meet real-time deadlines. The benefits of object-oriented middleware, such as Realize, in reducing the costs and the timescale of software development will greatly exceed the additional costs of faster processors and communication media, unless the system is very simple or many copies of the system are to be deployed.

Acknowledgments

This research has been supported in part by ARPA Grant No. N00174-95-K-0083.

References

[1] R. M. Adler, "Emerging standards for component software," *Computer*, vol. 28, no. 3, March 1995, pp. 68-77.

[2] M. L. Dertouzos and A. K. Mok, "Multiprocessor online scheduling of hard-real-time tasks," *IEEE Transactions on Software Engineering*, vol. 15, no. 12, December 1989, pp. 1497-1506.

[3] K. H. Kim, K. Mori and H. Nakanishi, "Realization of autonomous decentralized computing with the RTO.k object structuring scheme and the HU-DF inter-process-group communication scheme," *Proceedings of the 1995 IEEE International Symposium on Autonomous Decentralized Systems*, April 1995, Phoenix, AZ, pp. 305-312.

[4] L. E. Moser, P. M. Melliar-Smith, D. A. Agarwal, R. K. Budhia, C. A. Lingley-Papadopoulos and T. P. Archambault, "The Totem system," *Proceedings of the 25th Annual International Symposium on Fault-Tolerant Computing*, Pasadena, CA, June 1995, pp. 61-66.

[5] *The Common Object Request Broker: Architecture and Specification*, Object Management Group, Inc, Framingham, MA, 1995.

[6] S. K. Shrivastava and A. Waterworth, "Using objects and actions to provide fault tolerance in distributed real-time applications," *Proceedings of the Twelfth Real-Time Systems Symposium*, San Antonio, TX, December 1991, pp. 276-285.

Operating System Structure Model for Real-Time Systems

Toshiro Kamiuchi, Hiroaki Nakanishi and Keijiro Hayashi
Hitachi Ltd.
6 Kanda-Surugadai 4-chome, Chiyoda-ku, Tokyo, Japan
Email: kamiuchi@perl.hitachi.co.jp
nakanish@cm.head.hitachi.co.jp
hayashik@omika.hitachi.co.jp

Abstract

For the real-time computing, an operating system is one of the most important component, to realize the required computing functions with time critical conditions. In this paper, OS Route MIX is proposed for evaluating OS performance. By applying this performance measure, the practical OS designs are shown. To improve OS performance, "Three layers structure model" is defined, to add an OS process as one of OS function implementation, that is a way of simplifying OS design and making performance improved. The practical results and data are shown.

1. Introduction

The architecture for real-time system depends on its application. If the application does not have to handle large volume of data, simple scheduler which initiates tasks at each control cycle is enough. However, as the real-time system is becoming popular with expansion of its application fields[1][2], such simple scheduling scheme is not sufficient. And the demand for integrating total system forces real-time system to have the capability of multi-tasking and interrupt handling and various functions of operating system which are important features for operating system design[3][4][5]. Adding these features may suffer response time of the system which is one of the most important factors of the real-time system.

To develop a new real-time operating system which has multi-tasking functions, we analyzed detail behavior of the existing multi-task real-time operating systems and introduced OS Route Mix to measure its overhead which may be one of key data of Operating System for real-time system designers.

We have developed three layers model of the operating system. By implementing the middle layer as an OS Process and applying light weight scheduling service to it, the overhead of the operating system becomes one half of that of the conventional operating system with firmware implementation. The OS Process is designed as a service object of the operating system. This implementation helps to add portability and expandability of its functions to the operating system.

2. OS Route Mix

To meet the needs for integrating total system, including real-time control system and information processing system, much more functions have to be added to the operating system. This kind of functional enhancement often leads to large overhead of the operating system. Sometimes such increase makes real-time operating system unsuitable for its target applications. Especially, an object-oriented approach requires a clear function definition for the system objects. As for the function itself, there might be powerful tools to define them. However, for timing issues, a very traditional calculation of dynamic steps is only one method for evaluation. So we really need to have such measuring tools. From the viewpoint of operating

0-8186-7570-5/96 $5.00 © 1996 IEEE

system designer, we have to resolve the problems of the performance, especially response, to develop a new real-time operating system.

There are many measures for evaluating the performance of a computer system. However, there is not a good measure for evaluating performance of a real-time operating system.[6] To analyze and evaluate the operating system performance, we have introduced OS Route Mix (OSRX). To analyze which portion of the operating system's functions is consuming the overhead time, all the routes in the operating system are picked up and for each route processing time Ti is calculated.

OSRX is defined as follows,

$$OSRX = \Sigma\ Wi*Ti$$

Where Ti is the overhead of the route i and Wi is occurrence of the route i measured from the typical 30 application systems such as control systems of an electric power generating plant, a steel manufacturing plant, factory automation, a train traffic control system and so on. As for measuring environment, several computer systems, which are relatively traditional, are sampled. That means they have not virtual storage functions but fully supporting of executing non-residence programs and data, that are stored in secondly storage devices like magnetic discs.

Table 1. Occurrence of routes

Route	Occurrence(%)
Task scheduler	47
I/O Management	16
Task Management	15
Timer Control	6
Memory Management	4
Shared Resource Management	12
Total	100

OSRX is equivalent to the mean overhead time when application program or interrupt calls for the operating system's assistance. Table. 1 shows Wis. Task scheduler, I/O processing and Task management are three major Wis. These three makes 78% of total Wis. I/O processing is 36%, Task scheduler is 16% and Task management is 14% of the total OSRX respectively. This means these three functional modules have to be improved to get better performance of the real-time operating system.

Those data need to be re-calculated through the real application samples because the real-time computing environment changes dynamically. Especially, Memory management complexity, such as virtual storage mechanism, has affected the basic operating system steps. On the other hand, from the viewpoint of practical real-time system design, there might be an opinion that those occurrence weights are still effective because those items are really essential operations. So we are continuing to apply those OSRX to measure the basic performance evaluation of our new generation Operating Systems at first.

3. Three layers model and OS Process

I/O processing (I/O Driver) has sometimes big issue for implementation. Because I/O processing comes with the problem of how to make synchronization with input and output operation. A typical implementation of the I/O processing is using a task level processing such as "sync demon" in UNIX operating system.[7] By using "process" or "task" we can make operating system's kernel structure relatively simple. This is because we do not have to support recursive call of the kernel itself by making "process" or "task" take care of the "wait for I/O completion".

However, this strategy brings large overhead of swapping operation for non-resident application programs or data. Because it means that memory management service which perform disc I/O for swap in and out operation has to be implemented in "process" or "task" level, resulting a large overhead of task switching including task scheduling overhead. To avoid such overhead, we have introduced OS Process. OS Process is placed in the middle layer of the kernel and the tasks.

Fig. 1 shows the difference between the conventional structure and the new one developed. Functions implemented as OS

Processes are selected from the viewpoint of avoiding the above overhead and minimizing developing cost of wide variety of the each functions.

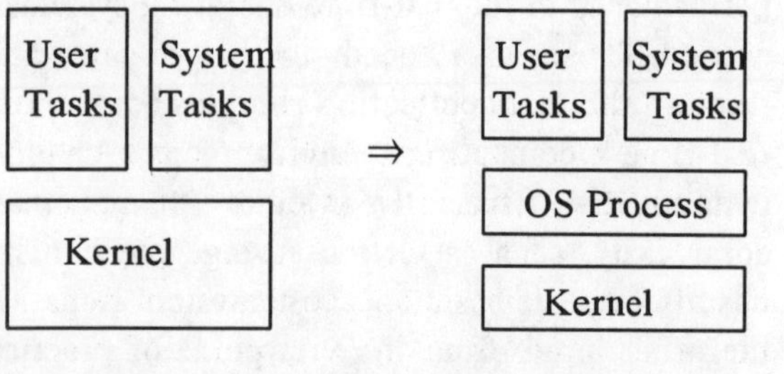

Fig.1 Operating System Structure

For example, in case of memory management, fixed memory allocation is preferable to realize deterministic scheduling than virtual memory type dynamic allocation. However, because fixed memory allocation is simple and impose some limitation on the application scheduling, one allocation strategy is not enough to support wide variety of the application's scheduling needs. This means we have to develop several types of memory management programs which are also very much dependent on the memory management hardware.

To minimize development cost of these memory management programs, object oriented approach is taken. By selecting memory management function as a service object of the operating system, operating system kernel may just send a message for requesting memory allocation, including swap operation. And this implementation makes operating system kernel independent form a specific memory management hardware and memory allocation strategies.

For the similar reason above, we have additionally picked up "Multi-computer Management", "Network Management" and "Error Data Gathering" which are to be implemented as OS Processes. Fig. 2 shows the detail structure of these layers.

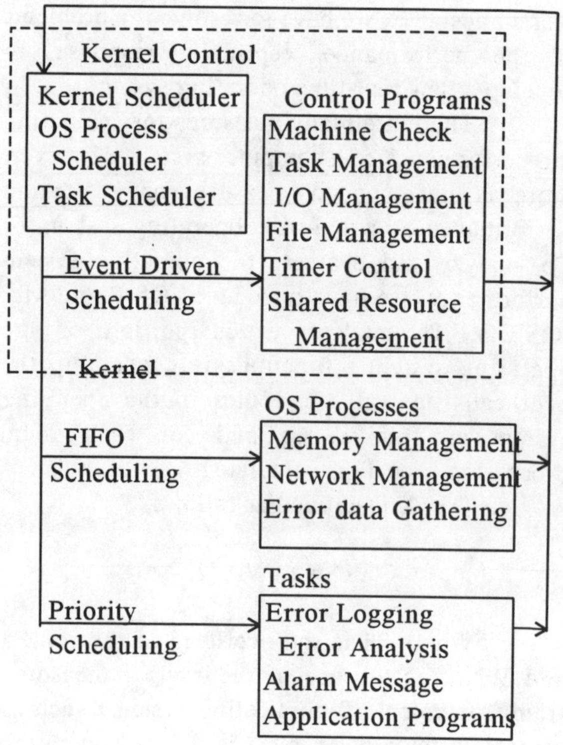

Fig.2 Three layers model

4. Controlling three layers model

Since OS Process is the part of the operating system and implemented as a service object, it can take advantage of having the privilege for accessing system table and saving/resuming the minimum set of information when switching occurs. Although several schemes for scheduling real-time objects are applicable[8], we have selected simple message-triggered FIFO scheduling, because those objects shall process requested service in the shortest time upon request.

By applying these scheduling and control mechanism and implementing major scheduling modules by firmware, overhead of initiating OS Process has become only 17 μ s, while conventional task initiation overhead was 271 μ s in case of 4MIPS class micro-processor. Tasks are given priority scheduling to allow the user, system designer, to take the full control over tasks. Table 2 summarizes the characteristics of the OS Process compared with the kernel and the task.

Table 2. Summary of three layers Program

—	Control Programs	OS Processes	Tasks
Priority	High	⇔	Low
I/O System Call	NA	A	A
Access to System tables	Direct	Direct	via Sys-Calls
Privileged Instruction	A	A	NA
Scheduling	Event Driven	FIFO	Priority
Error Handling	Shutdown	Abort OS Process	Abort Task
Use of Control Programs	A	A	NA

NA : Not Available, A : Available

5. Real implementation examples

As for a real design work of operating systems, there might be various types of implementations. in hardware and software environments of computers. For examples, The following items might be very important factors, that is, as for the hardware, whether a commercially standard micro-processor is applied for the main engine or not and as for the operating system functions, whether a general purpose environment is required or not

In a case of our original hardware, utilizing firmware technique can reduce all the overhead of operating system, specially when it is applied to common routines which serve to kernel, scheduler and OS Process. In a case of a high end of mini-computer we have chosen about 25% of the basic operating system steps and implemented them as micro-programs. Those micro-programs are designed to fetch necessary data for processing into a local storage in the CPU board. They help to reduce the overhead of the OS Processes implemented as service object of the operating system.

To use the micro-program to wider range of processing, we have implemented them in the way that software modules such as user own coding routines can be linked from them. User routines can control micro-programs by returning specific values to the caller micro-program. By this implementation, firmware can be applied to the modules where minimum overhead is required and at the same time flexibility to link to user own coding routine are needed.

Introducing suitable scheduling and control mechanisms for each layer and utilizing firmware technique in major routines, the developed operating system shows 2.4 times faster OSRX than that of conventional model. In Fig. 3, the result is shown with the OS Route Mix of other implementation of real-time operating systems. Hitachi 80E is a traditional 16bits mini-computer of which operating system is a proprietary real-time multi-tasking operating system. Hitachi V90/50 is a 32bits relatively large scale mini-computer of which OS is a multi-tasking / multi-user complexes OS. And Hitachi V90/x5 series is based on standard micro-processors such as MC68030 and MC68040 which do not allow the firmware implementation.

Hitachi V90/45's operating system is based on UNIX system V, and its kernel is modified to support real-time features such as fixed priority control, fixed memory allocation and so on. Although Hitachi V90/45 has two times faster processing power than Hitachi V90/50, the modified UNIX operating system which utilizes conventional swapper in process level instead of taking the three layers model structure shows bigger overhead of Memory Management and the total OSRX is about two times bigger than the three layers structured Hitachi V90/50.

For practical real-time system design, such basic data of operating system performance is essential. However, for a dead line timing control, we still have more enhanced control schemes for response evaluation and design tools.

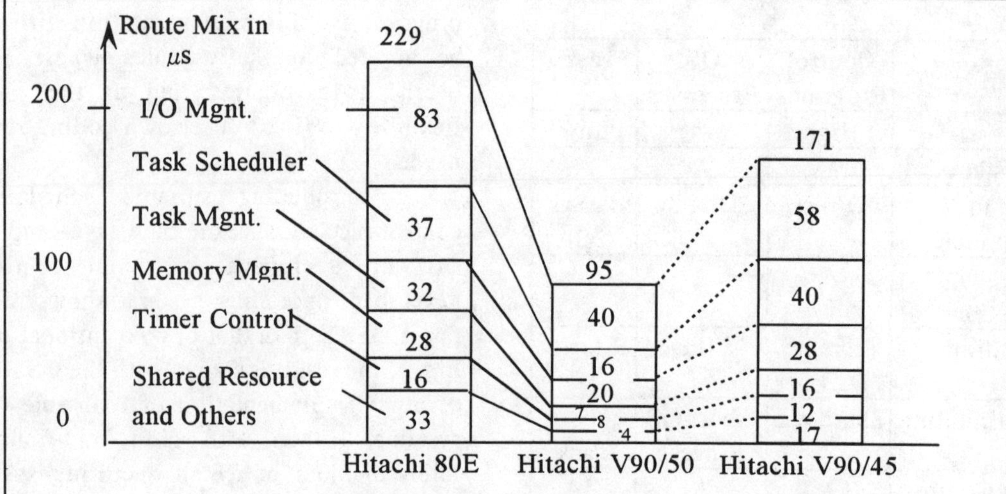

Fig. 3 OS Route Mix(OSRX) of Hitachi Control

6. Concluding remarks

Adding useful functions to the real-time operating system may suffer from increase of the overhead. Since the real-time system has to be designed to give minimum overhead of operating system, to make our enhancing the operating system functions easier and to provide good real-time system design environments to real-time system designers.

For those targets, we proposed a measure for evaluating the operating system performance. And we have developed three layers model multi-tasking operating systems.

In addition to that implementing the middle layer, OS Processes, as service object, makes operating system easy to be ported to various hardware platform and to support different types of services.

The trend of operating system is to go to an open domain, such as UNIX, at first, Windows and some compact operating systems of which functions are compatible with standard OSs. Micro-Kernel, OS-7 and so on may follow such direction. However, practical time critical software base must be guaranteed. We would like to pursue for those technologies.

References

[1] __: Real-Time Systems", Computer Vol.24, No.5, May 1991

[2]H. Wataya et al.: The Cooperating Autonomous Decentralized System Architecture, Proc. ISADS95, p40 1995

[3]J A Stankovie et al., Implications of Classical Scheduling Results for Real-Time Systems, Computer Vol.28, No.6 p16, Jun. 1995

[4]B. Furth et al..: Real-Time UNIX Systems, Norwel, MA:Kluwer, 1991

[5]K. Ramamritham, et al.: Effective Scheduling Algorithms for Real-Time Multiprocessor Systems, IEEE Transactions on Parallel and Distributed Systems, Vol.1, No.2, Apr. 1990

[6]___: Misconceptions about real-time computing, IEEE Computer., Vol.21, no.10, Oct. 1988

[7]m.J.Bach:The Design of The UNIX Operating System, Prentice-Hall, 1986

[8]Kim. K.H. et al.: A Real-Time Object Model RTO.k and an Experimental Investigation of Its Potentials, Proc. 1994 CS Computer Software and Applications Conf.(COMPSAC)

[9]Kim K.H. et al : Realization of Autonomous Decentralized Computing with the RTO.k Object Structuring Schemes and the HU-DF Inter-Process-Group Communication Scheme, Proc. ISADS95, p305, 1995

Failure Notifications: A Useful Extension to the Object Programming Model

Vicraj Thomas
Honeywell Technology Center
Minneapolis, MN 55418.
thomas@htc.honeywell.com

Abstract

This position paper argues for extending the object programming model with failure notifications that are generated when an object fails to meet its specification.

1. Introduction

Object-oriented system design and development methods have proven useful in building large and complex systems. Since real-time, dependable systems tend to be large and complex systems it is natural to use object-oriented methods for building such systems. Unfortunately, the object-oriented programming paradigm and the languages that support it were not designed with real-time or fault-tolerant system in mind.

It is therefore important to look at ways in which the object model can be extended to simplify the construction of real-time, fault-tolerant systems. This paper proposes extending the object model to include failure notifications. A case is made for the need for such failure notifications and the advantages of having such notifications.

2. Failure Notifications

Designers of real-time dependable systems have to make assumptions about the number and kinds of failures the system can suffer. These assumptions determine the techniques used to build the system. For example, if a component can suffer at most N failures, a fault-tolerance technique such as the replicated state machine approach may be used with $N+1$ replicas of the component. This assumption may however not hold during system operation, resulting in the failure of the component. If other components in the system are not prepared to deal with this failure, the entire system is liable to fail.

A failure notification is generated if a component or object fails, i.e., it is unable to meet it specification because assumptions made during system design no longer hold. For example, the system may have suffered unanticipated failures or may be overloaded due to excessive demands for service. Other objects in the system may ask to be notified if an object fails and may take corrective action on receipt of the notification.

3. Usefulness of Failure Notifications

The ability to detect the failure of a object can greatly simplify the construction of a real-time dependable system. Ideally, a real-time dependable system as a whole behaves as an object that executes commands correctly and in a timely fashion, or a failure-notification is generated. This assures users that, unless a failure notification is generated, their commands have been correctly processed and any results produced can be relied upon. Such a system is much easier to design and implement if each of its component objects in turn produce correct results or a failure notification is generated. Since the failure of any object is detectable, other object do not have to implement complicated failure detection schemes. These objects may in turn be implemented by other such objects, and this process continued until the simplest objects of the system are reached. At each level, the guarantees made by these objects simplify their composition to form more complex objects.

Failure notifications is even more important in systems built out of commercial-off-the-shelf (COTS) components since any assumptions regarding the run time behavior of the system are more easily violated.

A programming model that provides for such failure notification is described in [1]. This model was originally developed for building fault-tolerant software and is currently being extended to handle timing failures.

References

[1] R. D. Schlichting and V. T. Thomas. Programming langauge support for writing fault-tolerant distributed systems. *IEEE Transactions on Computers*, 44(2):203–212, February 1995.

0-8186-7570-5/96 $5.00 © 1996 IEEE

Session 4

Real-Time Software Engineering

Session Co-Chairs:
Ray Paul
Office of the Undersecretary of Defense, USA
Kinji Mori
Hitachi SDL, Japan

Integration of Formal Verification with Real-Time Design

George Krasovec
Advanced System Technologies
Englewood, CO USA
gkrasovec@advsystech.com

Natarajan Shankar
SRI International
Menlo Park, CA USA
shankar@csl.sri.com

Paul Ward
Paul Ward Associates
New York, NY USA
ward@acm.org

Abstract

As computers play greater roles in critical functions of complex systems, an increased reliance on formal methods for verification of critical components will be required. However, formal methods are often criticized regarding their intractability to large problems domains and lack of mechanical support. This paper reports on preliminary research[1] aimed at integrating formal verification techniques with an OO, real-time CASE tool. In this work we established the feasibility of translating a system implementation model expressed the Real-Time Object-Oriented Modeling (ROOM) notation into the Murphi model checker notation. Fragments of the translation algorithm were implemented in a proof-of-concept prototype; two formal analysis benchmarks were manually replicated using the translator definition. Methods for graphically expressing safety and timing constraints were demonstrated.

Introduction

Formal methods are mathematically based techniques for describing computer system properties. They provide frameworks within which computer systems can be specified and verified in a systematic and mathematically sound manner. Although the complete formal verification of large complicated systems is currently impractical, the application of formal methods to critical subsystems can establish that a particular design meets requirements.

The reliability requirements of modern life-critical computer systems are extremely high, typically specified at 10-7 (or less) failures per hour. Assessing the reliability of hardware components can be relatively easily accomplished using Markov modeling techniques, but assessing the reliability of designs is a much more difficult problem. It has been shown that testing and reliability growth modeling techniques are not feasible for demonstrating the reliability of ultra reliable computer system designs [1]. Moreover, fault tolerance mechanisms, such as N-version programming, are not necessarily a solution to the ultra reliability requirement since key assumptions of failure independence generally cannot be verified. The only defensible strategy for ensuring the reliability of ultra reliable system designs is through design fault avoidance. A promising way to accomplish design fault avoidance is through the application of formal methods.

A long-standing criticism of formal methods is their intractability to large problems domains and lack of mechanical support. However, much has been accomplished in recent years regarding the application of these methods to the critical components of systems, such as flight control logic, communication protocols, and various fault tolerance mechanisms. In addition, a continuum of automated reasoning tools, based on increasingly rigorous notations from process algebras to formal theorem languages, have been developed and applied with notable success. For example, the Reliable Computing Platform (RCP) project sponsored by NASA Langley Research Center has successfully specified and verified a multi-level design using the Ehdm verification system for a generic, ultra reliable digital flight control system [2].

As computer systems are used to support future life-critical functions in transportation, industrial, medical, and weapon control system applications, increased reliance on formal methods for specification and verification of critical system components will be required. Formal methods must be made available to the designers of these critical systems to ensure that they meet safety, reliability and liveness requirements.

Formal Specifications and Tools

A practical taxonomy of formal methods distinguishes between model-oriented and property-oriented formal methods [3]. Using the former method, a specifier directly defines a system's behavior by building an abstract model of the system using mathematical constructs such as relations, transformations, sets, functions, and sequences. Model-oriented methods applicable to sequential programs and abstract data types include state machines, VDM, and Z, whereas Petri nets, CCS, CSP, Unity and TSL support the specification of concurrent and distributed system behavior.

In the property-oriented method, a specifier defines a system's behavior indirectly by enumerating a set of properties the system must satisfy. Property-oriented methods can be broken further into two categories, sometimes referred to as axiomatic and algebraic.

[1]Work sponsored under SBIR Phase I grant from the Office of Naval Research, contract N00014-94-C-0036.

0-8186-7570-5/96 $5.00 © 1996 IEEE

Axiomatic methods are based on proofs of correctness of implementations of abstract data types which use first order predicate logic to specify operations for each type. Hoare, Iota, OBJ, Anna, and Larch are examples of formal specification methods which fall in this category. In the algebraic method, data types and processes are defined to be heterogeneous algebras, and axioms (restricted to equations) are used to specify system properties. Property-oriented methods for specifying the behavior of concurrent and distributed systems include extensions to the Hoare-axiom method, temporal logic, and Lamport's transition axiom method. The strong typing mechanism of Z allows it to be used for property-oriented specifications as well.

A tool-oriented taxonomy of formal methods and their supporting tools separates formal reasoning tools into the classes of process algebras, model checkers, and general-purpose theorem provers. Process algebra based tools supporting CSR and CSP, as well as model checkers such as Murphi, CTL and the Modechart verifier, are highly specialized and require little human intervention during verification. In contrast, theorem provers, such as Boyer-Moore, EVES, Ehdm, PVS, and HOL, usually need considerable human interaction. The latter three tools support a more expressive high order specification language (at some burden to the verification process) which allow variables to range over functions, arguments of functions to be functions, functions to be written as lambda abstractions, and terms to have variable types. The Boyer-Moore prover provides effective verification of the specification, but the specification must be expressed in terms of raw logic.

Formal method system specifications are often based upon a state machine paradigm in which all system states (or modes) are identified in terms of the transitions among them, the events and conditions under which they occur (i.e., a control law), and the constraints or invariants which they must satisfy. Although the syntax of system specifications will vary among different notations, the semantics will be quite similar. Most notations surveyed in recent literature are constrained to a finite state space. In some applications, such as the NASA reliable computing platform project, methods are required to support an infinite state space specification. Specification techniques have been developed to minimize state explosion problems in large problem domains, including Harel's higraphs and Mok's Modecharts, a higraph variant.

With the variety of formal methods and reasoning tools currently in use, formal methodologists must be proficient not only in the underlying fundamentals of formal methods but also in the language and environment of particular reasoning tools. Moreover, in many general purpose reasoning tools the user is burdened with the encoding of the target system specification as well as the construction of a verification framework (such as the representation of the underlying state machine, notation/method primitives and domain-specific language operators). Users of the PVS verification system, for example, have partitioned the verification framework into a reusable component called the real-time specification/verification method (RSVM), and the target system specification and verification logic have been partitioned into another application specific component [4]. Experimental versions of RSVM have been developed and used to check proofs of the Boyer-Moore majority algorithm, ordered binary tree insertion, Byzantine agreement, a railroad crossing gate system, and others. More recently, an extension to temporal logic theory called durational calculus has been encoded into an RSVM for application to hard real time system verification [5]. New verification techniques emerging from PVS are being used to enhance Ehdm, a mature verification system developed by SRI in the early 1980's under NASA sponsorship [6]. Although PVS was developed initially in 1990 by SRI as a prototype for R&D purposes, in recent years it has gained acceptance as a robust verification system. HOL-based general purpose theorem provers, particularly those with a module library capability (e.g., Ehdm), are also capable of partitioning tool inputs into reusable and system specific components to facilitate proof construction.

The partitioned data architecture of HOL-based verification systems and the commonality of state machine based verification frameworks create an opportunity for practicing engineers and analysts to have access to formal verification techniques for a wide variety of applications. In contrast to the traditional use of formal methods and reasoning tools strictly by formal method experts, these methods can extend to the adequately trained (but not necessarily formal methods expert) design analyst the capability to use conventional real-time CASE tools to specify the required behavior of critical system components. In theory, a CASE tool - formal methods translator can subsequently be used to map the real-time behavior model onto the semantics of the target automated reasoning tool. Accordingly, the ultimate goal of this research is to develop commercially viable translators to link compatible sets of CASE and formal verification tools in order to establish the correctness of critical system components and algorithms.

CASE technology and real-time design

The basic data flow diagram notation and terminology, on which modern CASE technology was founded, was first proposed by DeMarco in 1978 [7]. Using this notation, the system description is defined by a set of diagrams called data flow diagrams. These diagrams each can contain a number of elements including: 1) oval or rounded rectangle elements called process bubbles or transformations representing data transformations (i.e., they transform input data into output data), 2) boxes called external entities representing elements which are outside of the system being described but which need some interface to the system, 3) parallel lines or a three sided boxes called data stores which represent a repository of data, and 4) data flows connecting the other elements.

The data flows represent data flowing between the elements and the names and descriptions of the data flows are collected in a data dictionary. The diagrams are hierarchical with the top level diagram containing a single process but showing all the external entities which interface to the system being called the context diagram. Each process on a diagram can either be expanded to the next level (i.e., the higher level process is an abstraction of the processes shown at the lowest level) or described in a mini-specification. The mini-specification describes the processing that is to occur for a particular process. This description may take many forms including English, tables, or trees. A set of data flow diagrams which is completely consistent (i.e., all inputs and output on one level match up with corresponding inputs and outputs on the next level) is called a leveled set.

Extensions to the basic data flow notation were developed by Ward-Mellor [8] and Hatley-Pirbhai [9] to accommodate real-time system description. The extensions added control processes or transformations, depicted as dotted ovals, to the data flow diagrams. A second addition was the control flow which is shown with a dotted line and which represents flow of control information (versus data flow). Finally, to provide a description of the control processes, the state transition diagram was added. The state transition diagram depicts the states the control process can be in, the events which cause a transition out of any given state, and the result of the transition (e.g., triggering, signaling, activating, or deactivating another process). The control processes can also be hierarchical in nature, although a higher level control process is not necessarily abstraction of the lower level process, but can represent control processes that go on in addition to any lower level control processes.

Another real-time description modification to the basic data flow diagram concepts is the separation of data flows into discrete and continuous data flows. Discrete data flows are the data flows described by DeMarco. Continuous data flows are required for some real time systems where an input is provided continuously (e.g., some sensor reading which is an input into the system—the value is always available although the process may not use it continuously).

Early proponents of real-time design methodologies soon discovered that their notations were not scalable to large problem domains. Work principally by Harel introduced a more concise but abstract notation for state representation using higraphs, later formalized in a real-time notation called statecharts [10, 11]. In the statechart notation, states can be nested within states and transition edges are allowed between sets of states, thus providing a significant reduction in behavior model volume and considerably greater expressive power to deal with the state space explosion problem. It is important to note that the higraph-based Statechart notation does not reduce the number of states of a system; rather, it provides much more economical method for expressing them. Analysis and verification tools, particularly those based on finite state process algebras and model checking, still have to deal with the entirety of the system state space.

The most recent new direction in CASE technology involves the replacement of the fundamental process model with the object oriented design paradigm [12,13]. This change has been motivated by many of the strengths inherent in an object oriented design methodology, such as reusability, information hiding, and inheritance, as well as by a shift in the development community toward object based development techniques and tool sets. Examples of object oriented design methodologies include the object modeling techniques of Rumbaugh [14] and Booch [15], and the real-time object oriented modeling method of Selic, Gullekson, and Ward [16]. These object oriented methodologies have been incorporated into commercial CASE tools (such as Rational's Rose, Cadre's ObjectTeam and ObjecTime Limited's ObjecTime). The ObjecTime CASE tool, in particular, has been specifically designed to support the development of real-time distributed systems and has the expressive power of the statecharts semantics for modeling system behavior.

Research objectives

The purpose of our research was to develop and demonstrate an interactive software tool that supports formal specification and reasoning about temporal, liveness, and safety properties of hard real-time systems. The benefit of the resulting technology is to extend the availability of formal methods capabilities to trained system and software engineers who are not necessarily experts in the formal methods domain. This is to be accomplished by providing a mechanical link between the familiar real-time design domain of the engineer and the complex mathematical notations of the formal methodologist. With this goal in mind, we identified technical objectives to establish the feasibility of our approach:

Technical Objective 1: Show that a non-trivial set of extensions can be made to real-time design notations which express semantics of the formal methods domain.

Initially, our research considered traditional real-time design notations based on the real-time/structured analysis model, such as is implemented in Cadre's Teamwork and Intersolve's Accelerator real-time CASE tools. Similar RT/SA models are defined within the Electronic Industries Institute's CASE data interchange format (CDIF) standards. However, the motivation for defining extensions to this type of design notation was minimized when we considered the ROOM notation of the ObjecTime CASE tool, whose semantics proved to be advantageous for this type of research. Nonetheless, it was important to establish during our research that a real-time design notation existed which could support extensions for additional semantics of the formal methods domain.

Technical Objective 2: Show that a mapping can be defined between relevant semantics of the extended RT/SA notation and the input space of a general purpose formal verification system.

A critical capability enabling our approach is the mechanical, automated translation of models expressed in a suitable real-time design notation to the input space of the target formal verification system. An assessment was made of the syntax and semantic content of each candidate input statement type of the formal verification system to ensure that it can be adequately and unambiguously composed by the translator. It was essential that our work demonstrate that the forward transformation exists; the reverse transformation would be of obvious value, but its existence is not critical to our approach and was thus not addressed. The satisfaction of this objective required a limited prototype demonstration of a transformation algorithm, and the comparison of mechanically generated formal specifications with controlled ("truth") specifications.

Work performed

We prepared and conducted a concrete demonstration of the feasibility of linking CASE technology with formal methods, and have experimented with translating between specific CASE and FM notations using AST's meta-tool capabilities.

Tool integration strategy

An overview of CASE technology was conducted and trade-offs were considered among specific real time design notations and associated CASE tools which are candidates for our research. Important properties relative to our task included executability, scalability, and abstraction (e.g., object-orientation, or OO). SAS's ObjectBench supports an OO framework, but is not scalable to large problems (it has a flat design model with no hierarchies). The StateMate CASE tool supports very robust state modeling, but has no OO capability. Cadre's Teamwork Sim CASE tool has a relatively weak state modeling capability, no OO capability, and is marginally executable.

ObjecTime, a new generation CASE tool, is based the ROOM notation supports execution of model components (individually or as a system) at all levels of design abstraction. An additional, important feature of ObjecTime is that its ROOM compliant models are directly translatable into both analysis scripts and target system code. Consequently, analyses which establish properties of models developed in ROOM notation imply the existence of the same properties in the mechanically generated target system source code. This is an important feature for our work, since there may be little or no burden in proving that the implementation implies the design (i.e., design and implementation are isomorphic). This assumes, however, that the code generator is reliable. Moreover, the ROOM notation is used for model building throughout the system development life cycle, thereby avoiding the discontinuities inherent the traditional "waterfall" based structured analysis/structured design methodologies.

ObjecTime, since it was originally developed by a telecommunications project team as a tool for building large switching systems, has a number of features to enhance scalability. The incorporation of the Statechart notation within an object-oriented framework permits representing the state of a large system in a very succinct manner, as a collection of cooperating concurrent hierarchical state machines. Furthermore, ObjecTime permits system components to contain other components, to any desired depth of nesting. This hierarchical containment structure, together with total encapsulation of the contents of a model component, permits system components of arbitrary scope to be packaged and inserted within other components. Finally, a contained system component or communication channel may be statically or dynamically replicated, with built-in tracking of the identities of the replicated components and a simple syntax for designating one of a set of replicated components as the target of a message. These scalability features have been demonstrated in practice. ObjecTime has been used to create production systems in the range of a million lines of code. These systems include packet networks and voice services frameworks. Thus, ObjecTime appears to be an excellent prospect for use as a front-end for formal verification tools.

For evaluation purposes, PVS [17] and Murphi [18] were exercised on several formal specification scripts. One of the examples, Fischer's mutual exclusion algorithm, was developed in detail and selected as a candidate for encoding in the ROOM notation via the ObjecTime CASE tool. The logic of this algorithm was encoded in ROOM using ObjecTime and checked via a functional simulation (the purpose of the simulation was to ensure that the model of the Fischer algorithm behaved in a reasonable manner). After the model was completed and checked, it was exported in an ASCII linear form for comparison with the formal specification used by PVS and Murphi. After manually comparing the contents of these proof scripts, we convinced ourselves that at least for this class of problem (control/state intensive with temporal duration constraints), ObjecTime could be used to capture the formal specification within the real time design domain by a non-formal methods person.

Our evaluation revealed that the formal specification type declarations and state transition functions of the Murphi syntax can be derived from the ObjecTime export. However, the expression and capture of the system invariants (the properties to be proved) is not evident within the ROOM notation and graphic user interface. The invariants in the Fischer case study all involved limits on various state durations. Extensions to the ROOM notation and/or a new model presentation (design view) must be developed to facilitate the natural and convenient expression of these invariants. Invariants involving durations in and among states, as well as mutual state inclusion/exclusion, are good places to start.

Second, concurrence was reached as to the use of Murphi as the canonical form for the output of the ROOM translator. This would leverage our value by allowing users to choose between proof checkers (such as Unity or Murphi) for smaller scale designs which can be verified through exhaustive proof checking techniques and theorem provers (such as PVS or Ehdm) for more complex applications. The selection of Murphi as the output format is further indicated by the recent development of a Murphi-PVS translator by Shankar and Park. Thus the Murphi notation provides a multi-use gateway to both model verification and theorem proving tools.

Third, it is apparent that more case studies are needed in different problem domains to establish the general applicability of ROOM/ObjecTime as a formal methods front end. Other case studies for further analysis were discussed, including problems in fault tolerance and communication protocols.

Suitability of ROOMcharts

A system, or a component of a system, is considered to have a state when the effect of an input depends on the history of previous inputs. Certain system components have no state - a square root routine, for example, always returns the same root for a given argument independent of the arguments that have been used in previous invocations. However, ROOM Actors, which can represent arbitrarily large, complex system components must be capable of having state.

We refer to the internal operation of an Actor over time as its behavior. When an Actor is required to operate differently during different time periods, we represent the behavior changes as caused by changes in state. The high-level behavior of an Actor is represented by an extended state machine called a ROOMchart. ROOMcharts are based on the state-chart formalism of Harel [10][2]. At any point in time, an Actor has a state that determines how it will react when it receives a message through one of its ports. The set of such states, and the possible sequences in which the states can be visited, are described by a ROOMchart that is part of the class definition for the Actor.

Figure 1 is a ROOMchart for the processor Actor of our Fischer mutual exclusion model. The states are shown as rounded rectangles, and the transitions are shown as arrows with states at their heads and tails. Choice points, a ROOM extension to the statechart notation which provides greater detail for specifying transitions, are shown as triangles.

[2] This formalism has been implemented in a number of commerical CASE tools and successfully applied to a variety of real-time hardware and software problems.

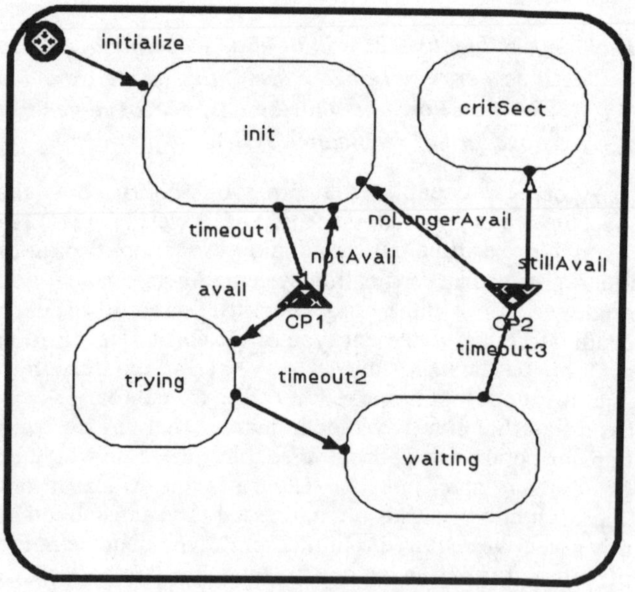

Figure 1. Sample ROOMchart for a processor

A state represents a period of time during which an Actor is capable of exhibiting a particular kind of behavior. A transition represents an allowed path from one state to another. For example, the state labeled *trying* represents a period of time during which the processor makes a reservation for a semaphore, and the transition from *trying* to *waiting* places the processor in the *waiting* state to allow the semaphore reservation to age. After the *timeout3* transition occurs, choice point *CP2* tests whether the reservation persisted during the waiting period or whether it was revoked. In the former case, a transition is made to the *critSect* state[3]; in the latter case, a transition is made to the *init* state to begin the process again. The arc from the circle in the upper left-hand corner represents the transition into the initial state *init* (that is, the state in which a processor Actor finds itself when first created). The states are enclosed within a single, all-encompassing state called top.

Based on our previous experience with ROOMCharts, and in particular with our construction of ROOM models for the Fischer mutual exclusion and railroad crossing problems, we found that the ROOM notation has excellent expressivity for capturing detailed model behavior.

Preliminary translator algorithm definition

The process of translator development occurred in two phases. First, we manually constructed several models in the Murphi notation and then established mappings to the equivalent semantics in the ROOM notation. Not unexpectedly, this involved a relatively small subset of the ObjecTime semantics, and totally ignored the message passing formalisms in ROOM (since Murphi has no

[3] A transition out of the *critSect* state was omitted from the model since it was irrelevant to the proof of the algorithm.

inherent concept of communications or their attendant delays, protocols, etc.). It was apparent that any translation based on this "reverse-engineering" approach would constrain the system designer by restriction to limited semantics of the ROOM notation, clearly an undesirable consequence. Moreover, it became apparent that modeling of the communication process is fundamental to ROOM-based design methodology, so that explicit support for these semantics needed to be added to the Murphi grammar. For the near term, however, we added explicit mechanisms to Murphi scripts for asynchronous and synchronous communications among Actors.

During the second phase of this task, we analyzed the semantics of ROOM models and mapped their semantics into equivalent Murphi statements. In this "forward-engineering" phase, we concluded that a faithful translation of a ROOM model requires the interpretation of not only model objects, attributes and relationships, but also the parsing of embedded code segments associated with transition and action objects. The entire code segment grammar was not relevant; only those statements dealing with the transmission, receipt, and synchronization of message traffic were needed.

Proof-of-concept prototype development

During this task, we built, tested, and experimented with a limited capability, proof-of-concept translator. This preliminary translator algorithm was implemented in the C programming language and integrated into AST's meta-tool framework on a Sun Sparcstation. The translator consisted of three components. The first component provided for the parsing of the ObjecTime export file and building of an equivalent object data base in the translator's object management system. The Lex and Yacc utility programs were used to tokenize the linear form input stream and execute object production rules to populate the translator's object data base. The second component consisted of an adaptation of AST's meta tool framework. The AST framework is a highly adaptable, development resource which provides fundamental tool services such as data base management, graphical data interface, and tool integration services. The bulk of the adaptation of the tool framework is accomplished through data tailoring. Additional code modification was required to complete the adaptation of the graphical user interface and add special drawing support for rendering new constrain relationships between separate Actor state machines.

Figure 2 summarizes our experimentation with the prototype CASE/FM translator. Starting from manually prepared Murphi formal specifications, we built equivalent ROOM models. The ROOM models were then mechanically translated into the Murphi notation as defined by the translator mappings. Fragments of the mechanically generated specifications were successfully compared with the original Murphi models as a demonstration of translator development feasibility.

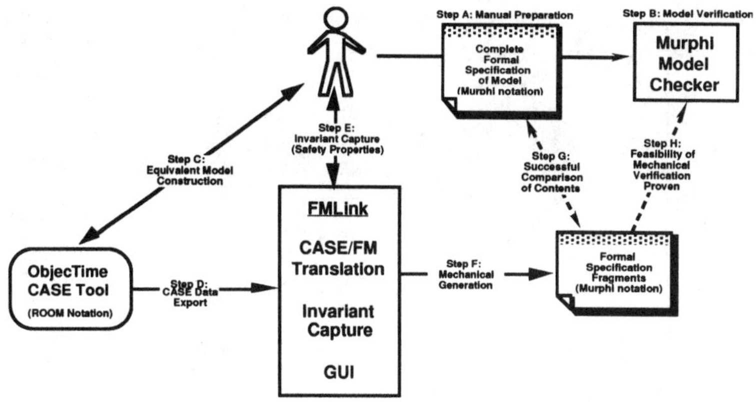

Figure 2. Demonstration of CASE/FM integrability

In related research completed in 1993 at the University of British Columbia, the HOL-Voss verification tool was integrated with the Statemate CASE tool [19]. Sample problems of limited size were translated to the HOL-Voss tool set; BDDs were used to reduce the problem state space and model checker run time. Problems with excessive run times and high memory utilization were manifesting themselves even for relatively small problem domains and despite the use of BDDs (since the models used in the UBC study were not optimized in any way, obvious improvements to model structure could be made manually to yield better results).

Research results

Our initial work has yielded several results with implications on verification practice, translator design, and capture of invariants.

Verification practice implications

As summarized in Figure 3, system verification techniques evolve through the system development life cycle from relatively simple activities, such as manual code inspection and static analysis on one end, to abstract theorem proving techniques on the other. In between these two extremes are several verification techniques with increasing technical complexity. In normal practice, each verification method is applied until some high percentage the errors discoverable by that method are removed. As more powerful and complex verification methods are applied, latent errors become not only more scarce but also more difficult (and costly) to detect. Ultimately, theorem proving is applied to critical system components to ensure that design errors related to specific assertions of safety or liveness, for example, are non-existent.

The process of manual proof construction for non-trivial applications is complex and time consuming. It has

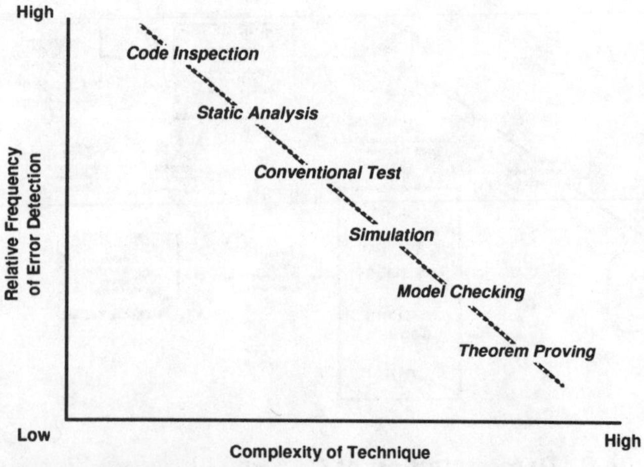

Figure 3 The continuum of system verification techniques

been the practice of SRI formal methodologists to attack such proofs in stages, initially with model checking techniques followed by the application of theorem proving. It has been SRI's experience that (a) most common design errors can be found quickly and more easily using model checking techniques, and (b) that proof construction invariably proceeds more smoothly for models which have passed through a model checker. Consequently, it became apparent that a model translation capability which omitted support for model checkers would be lacking when used in conjunction with the common approach for formal verification.

This realization led us to consider an translator architecture as illustrated in Figure 4. The key feature of this architecture is that the FMLink translator has the ability to emit verification scripts for either a model checker or a theorem prover. The feasibility of a dual-notation script emitter was established by the demonstration of a limited capability emitter for the Murphi notation and the applicability of a recently developed technique for converting Murphi scripts into PVS notation. Whether FMLink implements this converter or emits PVS scripts directly is a future design decision; it is sufficient to recognize for now that both notations can be supported.

Essentials of the ROOM/Murphi translation

Murphi is a model-checker based on the Unity formalism for state transitions. A Murphi program consists of a section each for the constants, the types, the program variables, the procedures and functions, the rules/rule-sets, and the program invariants. The types must be finite and include scalars, subranges, records, and arrays. The expressions of the language include the various arithmetic and logical operations and bounded quantification. The statements of the language include assignment, if-then-else, for, while, clear, and error statements, and procedure calls. A rule consists of a guard expression and a transition statement. It is also possible to define rule-sets indexed by a parameter. A start state section initializes the program variables. A valid program trace is one that starts from the start state so that each execution step corresponds to one of the rules. A program is considered correct if no error or deadlock state occurs, and every state satisfies the given set of invariants.

The key difference between the two is that ROOM uses a highly structured description of the state space and the behavior, whereas Murphi uses a global state space with a flat collection of transitions. The ROOM notation is convenient for a well-structured specification whereas the Murphi model is better suited for the verification of global safety and liveness assertions. The translation from ROOM to Murphi must therefore flatten the Actor hierarchy and conflate the state space. The steps in this translation are summarized below.

1. To simplify our task, we ignore any hierarchical nesting of ROOM Actors and assume that we are dealing with a flat collection of Actor instances with bindings between them.

2. A replicated Actor instance is associated with array variables in Murphi corresponding to the Actor variables and Murphi rule sets corresponding to the Actor transitions.

3. The Murphi variables encoding the state of each Actor instance include:

- a program counter variable _PC ranging over the substate labels of the Actor instance. If the state space is nested, there can be multiple such _PC variables to reflect

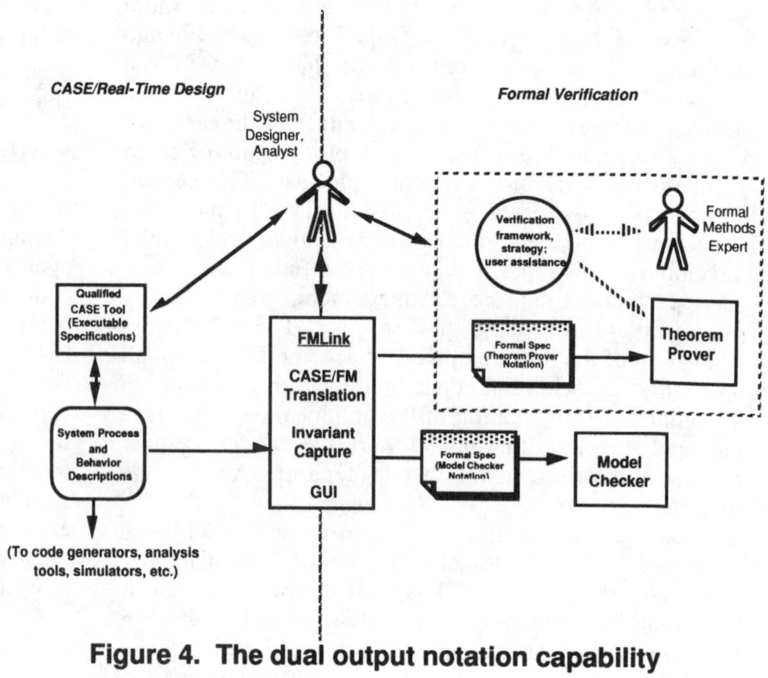

Figure 4. The dual output notation capability

the nesting.

- a Phase variable reflecting whether the Actor instance is Ready, Enabled, Blocked, or Unblocked. This will be explained below.

- a Boolean variable for each input port instance indicating whether there is a message on that port. - a data variable for data to be sent or received on a given (input or output) port instance

- a timer variable for each timer employed in an Actor instance. - any other data variables that are used by an Actor instance.

4. Once we have the global state derived from a ROOM description, the resulting constants, types, and variables can be declared in the CONST, TYPE, and VAR section of a Murphi program.

5. The ROOM initializations for each Actor instance are translated to Murphi using the STARTSTATE command in Murphi. Each Actor instance phase is initialized to Ready, all the port variables are set to be empty, and only the timer variables corresponding to initialized timers in the Actor instance are activated. All the other variables are initialized to default values in Murphi though it would be more accurate to allow these to be chosen non-deterministically.

6. A ROOM timer has an associated (optional) lower and upper bound and it can trigger an event when it is within these bounds. is modeled as a subrange variable in Murphi whose value is 0 when the timer is deactivated. A timer is activated by being set to 1. In order to increment time, there is a tick action in Murphi that increments all active timers by one provided no active timer is already at its maximal value. In other words, time is not allowed to progress beyond the upper bound of a timer.

7. A transition is triggered by either a timer or a communication along an input port. Our translation separates the triggering of a transition from its actual execution in order to reflect the asynchrony of the ROOM model and to ensure that no invariants are violated during a completion of any internally triggered transitions. When an Actor instance is in its Ready phase, a transition can be triggered by setting its phase to Enabled. A triggered Actor instance can then execute the transition by returning to its Ready phase and turning off the trigger. The case of transitions triggered by synchronous communication is more complicated since the sending Actor instance must block until a reply has been received.

Translator design implications

The ROOM notation is quite rigid but very amenable as modeling language for our two benchmark examples, Fischer's real-time mutual exclusion protocol and a railroad crossing controller. The graphical simulator of the ObjecTime CASE tool was useful in observing and understanding the behavior of the model.

Second, the translation between ROOM and Murphi is not straightforward. ROOM is a graphical notation that is intended for design and simulation. It is based on a radical encapsulation of state where the state in one Actor instance is invisible to any other Actor instance. The only interaction is through explicit communication. Transitions must be triggered either by communication or time-out events. The Murphi language is intended for verification and debugging. It is based on a model of asynchronous computation with a global state (i.e., no encapsulation). Transitions are specified as a collection of rules with guards (not triggers) and (deterministic) actions. Murphi does not formalize any notion of real time. Despite this semantic gap between the two formalisms, we have been able to accurately translate a subset of ROOM into PVS. This fragment includes guards, assignments, choice points, timers/time-outs, synchronous and asynchronous communication

One challenge in the translation was the modeling of real time, timers, and time-outs. These were modeled in Murphi through the use of timer variables that were incremented in a global tick event. This "old-fashioned" approach is described, for example, by Abadi and Lamport [20].

A second challenge was to accurately model the enabling and execution of transitions. The difficulty here is that one transition can trigger several other transitions which must subsequently be taken. This was modeled in Murphi by introducing a state component called the Phase which could either be Ready, to indicate that state was ready to be triggered, or Enabled, to indicate that a transition (indicated by a signal variable) has been triggered. Additional phases were used to model synchronous communication where a transition is blocked until a response to a query has been received. In addition, we have defined translations for several ROOM constructs and data types, including Actor replication which is modeled by arrays and rule sets in Murphi.

Third, we were able to debug/verify these benchmark examples using Murphi. To our knowledge, this is the first time the Murphi has been applied to the verification of real-time systems. Even with the overhead of our translation, these examples did not tax the Murphi verifier. We were able to find significant errors and correct our misunderstanding of our ROOM models using Murphi. For example, Murphi detected that we had overlooked communication delays in our ROOM simulations since they were run on a single workstation. These delays had to be bounded by means of timers in order to achieve a correct design. We also used Murphi to experiment with the relative bounds on the timer values in order to seed errors to confirm our understanding of the model.

Finally, we have used AST's meta-tool to encode some of the translation rules and emit fragments of the target Murphi description. In so doing, we have developed some preliminary ideas on using graphical notation for capturing correctness assertions. This is difficult in ROOM since the typical assertions violate the restrictions of radical encapsulation.

Our research has exposed several issues that need to be addressed in future R&D:

• Completeness - Our translation between ROOM and Murphi ignores several key aspects of the ROOM notation.

• Scalability - Murphi has been used to verify and debug hardware and communication protocols of commercial interest. Even so, Murphi is a finite-state machine verifier and can suffer from the state-explosion problem when given descriptions that can reach a large or even an unbounded number of states. There are a number of approaches that are being explored to minimize the state explosion, such as symmetry and BDDs.

• Generality - The version of Murphi used in our work is limited to verification of safety properties. There are some experimental extensions to Murphi that can cope with a limited but common class of liveness properties.

• Graphical and textual capture of assertions - CASE tools have historically had no motivation to capture the kinds of constrains, invariants, and temporal assertions that are necessary to mechanically produce executable formal specifications. There are a few classes of invariants which have obvious applicability to CASE models in the ROOM notation, particularly those involving relationships among states and transitions.

• Animating error traces - Formal methods, particularly mechanized ones, are most useful when used as debugging aids. For example, in our experiments the Murphi error traces were crucial aids in understanding both the ROOM design and the sources of any errors. These error traces could be made more accessible if they can be animated by means of ObjecTime's graphical simulator.

• Graphical aids for directing PVS proofs - The translation from ROOM to Murphi and eventually to PVS may result in a loss of information. In particular, the structural information present in the ROOM notation will not be as salient in the translated result. This structural information can often be very useful in organizing a formal proof.

Conclusion

Our research has demonstrated that the semantics of the ROOM notation and the Murphi model checker are compatible. This suggests that a rigorous, statemate-based CASE tool, like ObjecTime, is capable of serving as a front end for formal analysis tools. We have also demonstrated that semantics for specification of common constraint types can be asserted graphically, and have discovered a technique for feedback of Murphi error traces to the ObjecTime simulator. These traces can be of considerable value for analysis and test of faulty component behavior. Although our research did not address the generation of formal proof scripts, such as for PVS, we suspect that the automatic generation of these scripts is feasible as well, and that for given classes of systems, re-usable proof strategies can be developed.

References

1. Butler, R. W., and G. B. Finelli, "The Infeasibility of Quantifying the Reliability of Life-Critical Real-Time Software" IEEE Transactions on Software Engineering, January 1993, pp 3 - 12.
2. Butler, R. W., and B.L. Di Vito, "Formal Design and Verification of a Reliable Computing Platform For Real-Time Control (Phase 2 Results)", NASA Technical Memorandum 104196, January. 1992.
3. Heitmeyer, C. L. and Labaw, B. G., "A Benchmark for Comparing Different Approaches for Specifying and Verifying Real-Time Systems," Workshop on Real-Time Operating Systems and Software. May, 1993. IEEE, New York.
4. Owre, Sam, Rushby, John M., and Shankar, Natarajan., "A Prototype Verification System," Automated Deduction - CADE-11, 11th International Conference on Automated Deduction, Lecture Notes in Artificial Intelligence, June, 1992, pp 748-752, Springer Verlag.
5. Ravn, A., Rischel, H. and Hansen, K., "Specifying and Verifying Requirements of Real-Time Systems" IEEE Transactions on Software Engineering, January 1993, pp 41 - 55.
6. Rushby, J. M., von Henke, F., "An Introduction to Formal Specification and Verification using Ehdm", SRI Intl., Menlo Park, CA, 1991.
7. DeMarco, T. *Structured Analysis and System Specification*. Englewood Cliffs, NJ: Prentice Hall, 1978.
8. Ward, P. and S. Mellor. *Structured Development for Real-Time Systems*. Englewood Cliffs, NJ: Prentice Hall, 1985.
9. Hatley, D. and I. Pirbhai. *Strategies for Real-Time System Specification*. New York: Dorset House Publishing, 1988.
10. Harel, David, "On Visual Formalisms," Communications of the ACM. 1994. v31, n5, pp 514-530.
11. Harel, D., Pnueli, A., Schmidt, J. P., and Sherman, R. "On the formal semantics of statecharts," Proc. 2nd IEEE Symp. on Logic in Computer Science. 1987. pp 54-64.
12. Ward, P. and L. Williams. *The CASE/Real-Time Method: An Object-Oriented Approach to Systems Engineering*. New York: Dorset House, 1992.
13. Shlaer, S. and S. Mellor. *Object Lifecycles: Modeling the World in States*. Englewood Cliffs, NJ: Prentice Hall, 1992.
14 Rumbaugh, J., M. Blaha, W. Premerlani, F. Eddy, and W. Lorensen . *Object-Oriented Modeling and Design*. Englewood Cliffs, NJ: Benjamin/Cummings, 1991.
15. Booch, G. *Object-Oriented Design with Applications*. Menlo Park, CA: Prentice Hall, 1992.
16. Ward, P., Selic, B. and Gullekson, G. *Real-Time Object Oriented Modeling*. Englewood Cliffs, NJ: Prentice Hall, 1994.
17. Shankar, N., S. Owre, J. M. Rushby, "The PVS Proof Checker: A Reference Manual", SRI Intl., Menlo Park CA, Feb. 1993 .
18. Dill, David L. and Melton, Ralph.. "Annotated Murphi Reference Manual," March, 1993, Stanford University.
19. Day, Nancy. "A Model Checker for Statecharts (Linking CASE tools with Formal Methods)," Masters Thesis. 1993. University of British Columbia.
20. Abadi, M. and Lamport, L. *An Old-fashioned Recipe for Real Time*. Pages 1-27.

Specification and Analysis of Real-Time Systems in Statecharts

Sung Deok Cha and Hyoung Seok Hong
Department of Computer Science
Korea Advanced Institute of Science and Technology (KAIST)
373-1, Kusong-dong, Yusong-gu, Taejon, Korea
{cha,hshong}@salmosa.kaist.ac.kr

Abstract

Increased use of software in controlling safety-critical systems produced an urgent need to specify and analyze behavior of these systems systematically and rigorously. Statecharts formalism, a popular extension of conventional finite state machines, has been successfully used for specifying requirements of many reactive systems including the TCAS II, an aircraft collision avoidance system.

However, little has been published on specific guidelines on how one can best specify and analyze requirements in Statecharts. In this paper, we present a framework for specifying requirements of real-time systems in Statecharts and analyzing them for completeness, consistency, and safety. We use the requirements taken from an emergency shutdown system for a Korean nuclear power plant, called Wolsung SDS2, as an example.

1 Introduction

Software controlling safety-critical real-time systems has already become a part of our society. There are many examples of such systems that impact daily lives of general public. For example, commercial fly-by-wire jetliners (e.g., Airbus A320 and Boeing 777) are currently in service. Many countries operate nuclear power plants, and some, including Korea, are developing software-based emergency shutdown systems. Should emergency situations such as reactor overheating occur, software is required to initiate emergency shutdown procedures while maintaining the plant in a safe state. Some medical devices depend on software to control the amount of radiation to deliver for cancer therapy and to monitor various vital status of patients in an intensive care unit.

When software is used as a control agent in such systems, safety becomes a paramount concern. In the worst case, computer malfunctions, especially unsafe software control outputs, could result in serious and unacceptable consequences such as death, injury, or environmental damage. Software development process is known to be a costly and often error-prone activity, and software quality assurance becomes an essential concern. Of the several phases involved in typical software development, requirements specification and analysis phase is empirically known to play the most crucial role in determining the overall and final software quality.

There are several attributes desired of high quality requirements specification including understandability, maintainability, analyzability, scalability, unambiguity, etc. Many approaches have been suggested on developing high-quality requirements specification. Examples include development of formal specification languages, visual formalisms such as Statecharts[1] or Modecharts[6], and completeness and consistency criteria[5]. These approaches are closely related and tend to be complementary to each other. Formal specification languages enable automated analysis on the requirements but might be difficult to understand. On the other hand, requirements specification in visual languages, while easily understood, might lack formal and unambiguous semantics, thereby making the task of detecting flaws, especially subtle ones, difficult.

Statecharts formalism is a popular language to specify behavior of reactive systems. RSML, an extended version of Statecharts, has been successfully used to specify the behavior of the TCAS II whose logic is very complex[7]. However, there have been few research efforts on methodologies on how software requirements should be organized and specified in Statecharts. Note that a specification language in itself does not provide any methodologies or guidelines. For example, in Statecharts a state can communicate with any of the rest states using events, that is, an event is assumed to be broadcasted to all the states. For another example, the scope of a variable is assumed to be global in that it can be used and defined in the guard and action of any transitions. We have found that undisciplined and unrestricted uses of these features may result in software requirements of poor quality and erroneous.

In this paper, we propose a framework for developing and analyzing real-time safety-critical systems in Statecharts. Usefulness of our approach is demonstrated using an real-world industrial system, called Wolsung SDS2, a software-based emergency shutdown system whose functionality is similar to that of the Darlington plant in Canada[10].

Our paper is organized as follows: Section 2 briefly reviews Statecharts formalism. Section 3 proposes a framework for specifying and analyzing real-time systems using Statecharts. Additionally, a set of criteria to statically analyze Statecharts is proposed. In Section 4, si afety analysis method for Statecharts, based on forward and backward simulation, is presented. Finally, Section 5 concludes the paper.

2 Statecharts

Statecharts, proposed by D. Harel for specifying complex reactive systems[1], consists of $<S, T, E, V>$ where S is a set of *states*, T is a set of *transitions*, E is a set of *events*, and V is a set of *variables*.

States are either *BASIC*, *OR*, or *AND* states: *BASIC* states have no substates. *OR* states have substates that are related to each other by an *exclusive-or* relation. Being in an *OR* state implies being in only one of its substates. *AND* states have substates, called *orthogonal components*, that are related by an *and* relation. Being in an *AND* state implies that being in all of its orthogonal components.

Changes among states are represented by a transition whose format is

event [*condition*] / *action*

An *event* is an instantaneous occurrence of a stimulus (trigger), a *condition* is a predicate which should be satisfied for a transition to occur and an *action* may generate other events or perform computations.

In summary, we can say[1]:

Statecharts = finite state machine + depth
+ orthogonality + broadcast

The concept of *depth* is achieved by *OR* states, and *orthogonality* is achieved by *AND* states. *Broadcast* are used for communications between states. The communication is achieved by an action of a transition. That is, when a transition triggered, an action generates an event and this event is assumed to be globally broadcasted.

3 A Framework for Real-Time Systems

A real-time system consists of the four basic components: plant, sensors, actuators, and controller (Fig 1). The plant is a physical, mechanical, or electrical process. As far as the controller is concerned, the plant represents the external environment with which the controller interacts. Sensors measure the plant's current state and provide inputs to the controller. Based on the sensor's inputs, the controller performs a set of computations to determine how the current state of the plant should be changed. The controller's decisions are then passed back to the plant by actuators via controlled variables.

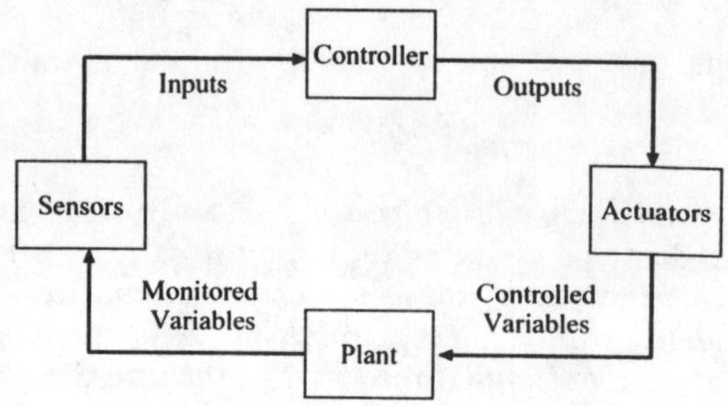

Figure 1: The structure of real-time systems

There is an important distinction to be made between the plant and the controller. The plant is usually "given" in that the developer of the controller may not change characteristics of the plant. In contrast, there is usually more freedom in the development of the controller. In fact, the controller is often implemented by real-time software so that the controller logic can be changed easily if desired. Furthermore, the behavior of the controller cannot be developed unless that of the the plant is first specified.

The controller must regulate the plant's behavior with respect to a set of pre-defined physical laws, objectives, and constraints required of the system. Therefore, requirements of real-time systems should be expressed in terms of not only the controller's functionality but also the plant's behavior. Several methodologies that explicitly take into account these characteristics of real-time systems have been proposed. Examples include the SCR method[11], Heitmeyer and Labaw's method[2]. and TTL/RTTL[9].

However, none of those methodologies supports the use of Statecharts. Several features of Statecharts formalism, when used in unrestricted and ill-disciplined manners, can result in erroneous or in the worst cases, unrealistic requirements specifications. Especially, we have found that the following features of Statecharts can potentially cause ill-disciplined requirements specifications:

- Every event is assumed to be broadcasted to all the states. Therefore, every state can communicate with each other using broadcasted events.

- Every variable is assumed to be global in that it can be used and defined in the guard and action of any transitions.

It should be emphasized that Statecharts in itself do not enforce restrictions or provide guidelines on how various states and transitions should be organized. While one may argue that such complete freedom provides maximum flexibility to the developers, it is often the case that subtle errors caused by the ill-disciplined uses can be quite difficult to detect.

The main purpose of the framework is to properly specify the requirements of the software in real-time systems in a systematic and disciplined manner using Statecharts. The software in real-time systems interacts with the plant, i.e, its environment via input and output variables. The requirements for the software are, therefore, described by specifying the required characteritics of output variables as a response to the observed input variables. In our framework, the software is represented as an *AND* state which has three orthogonal components: *Inputs*, *Functions*, *Outputs* (Fig 2).

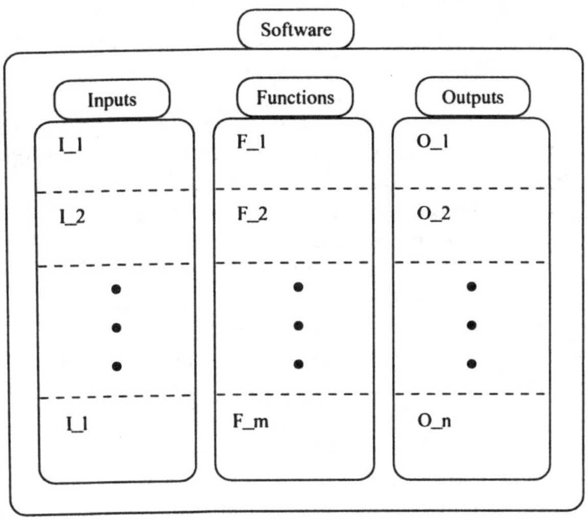

Figure 2: A framework for real-time systems

The first step in our framework is to identify all relevant input and output variables to the software. The behavior of input and output variables is specified by the orthogonal components of *Inputs* and *Outputs*, respectively. Then, the required behavior of the software is identified as a set of functions which are specified by the orthogonal components of *Functions*. We use the prefix "i_", "f_" and "o_" for the orthogonal components states, triggering events, and variables in *Inputs*, *Functions* and *Outputs*, respectively.

We propose several criteria on how to use Statecharts constructs and semantics for specifying the behavior of input, output variables and functions in a disciplined way. After the behavior of all input, output variables and functions are specified, the next step is to statically analyze Statecharts using the following criteria.

C.1 The orthogonal components of *Inputs* and *Outputs* should have a default state specified with a conditional connective. In contrast, the orthogonal components of *Functions* should have a deterministic default state.

Default states of the orthogonal components of *Inputs*, and *Outputs* represent the state of the plant at the time of software initialization. No assumptions can be made about the default state of the plant, because the changes of states of the plant may occur independently on the software. Therefore, when the software is initialized, every state in the orthogonal components of *Inputs* and *Outputs* may be a possible default state. In contrast, default states in *Functions*, should be designed to have a pre-defined deterministic value, because they represent the initial configuration of the real-time control software.

C.2 The format of transitions in *Inputs* must be

$$i\_e\ [true]\ /\ f\_e$$

Transitions in *Inputs* represent a change of values of input variables. The triggering events in *Inputs* are external events which are generated outside the software, i.e., in the plant. Therefore they should not directly trigger other transitions in *Functions* or *Outputs*. Note that it is enforced by the naming conventions because only the events whose prefix is "i_" can trigger transitions in *Inputs*. The guard of transitions should be *true* because Transitions in *Inputs* is triggered unconditionally when an external event occurs. The action should generate one and only one event which triggers transitions in *Functions*. In our framework, the behavior of input variables should be specified without any considerations of how they will be used in *Functions*. That is, the logic of the software should be confined in *Functions* for the modularity of the entire specifications.

C.3 The format of transitions in *Functions* must be

$$[i\_e \mid f\_e \mid timeout \mid \lambda]\ [condition]\ /\ [\lambda \mid f\_e \mid o\_e]$$

where λ is a null event.

A transition in *Functions* represents certain computation taking place by the software. Therefore, the triggering event must have been generated externally by *Inputs* (i_e), internally by other parts of the *Functions* (f_e), or implicitly by the *timeout* event and λ event. When the computation is completed, it can optionally trigger transitions in *Functions* or *Outputs.* The latter models the generation of external output to the plant. However, generated events may not trigger a transitions in *Inputs.*

C.4 The format of transitions in *Outputs* must be

$$o\_e \ [true] \ / \ \lambda$$

Transitions in *Outputs* represent a change of values of output variables. The triggering events for transitions in *Outputs* should be generated only by transitions in *Functions.* Because the actuators are passive in nature, transitions in *Outputs* should be triggered unconditionally (i.e., the gaurd is *true*) and should not perform any computation or generate any events on its own.

The criteria **C.2, C.3, C.4** can be summarized as follows: only three types of communications, i.e., broadcasting of events, between states are allowed:

- from *Inputs* to *Functions*,
- from *Functions* to *Functions*, and
- from *Functions* to *Outputs.*

All other types of communication are to be prohibited.

C.5 Each input variable should be used at least once in the software and each output variable should be defined at least once by the software.

If an input variable is not used at all in *Functions*, then it is an indication of a missing functionality in the software. Similarly, there is a missing functionality in the software if an output variable is not defined at all in *Functions.*

C.6 Every state must have a transition defined for every possible event, i.e., conditions of outgoing transitions should be complete.

C.7 Every state should have one and only one corresponding transition for every possible event. That is, conditions of outgoing transitions should be deterministic to be considered consistent.

Jaffe et al. discussed completeness and consistency equivalent to the criteria **C.8** and **C.9**[5] in terms of conventional finite state machines. Recently, Heimdahl and Leveson proposed similar concepts for RSML[3] and developed an analysis method for these properties.

3.1 A Case Study: Wolsung SDS2

To demonstrate our framework and analysis methods, we use a real-world application, Shutdown System Number 2 (SDS2) of Wolsung Nuclear Power Plant. as an example. The SDS2 monitors the state of the nuclear reactor states such as pressure and power and generates a trip signal if the nuclear reactor becomes unsafe, i.e., the pressure is too high or too low. There are several trip parameters in SDS2 and we describe only the *PDL delayed trip parameter*[1] as follows:

- If *i_Pressure* $\leq$ 950 kPa (delayed trip setpoint) and *i_Power* $\geq$ 80%FP, then continue normal operation without opening *o_PDLTrip* for three seconds.
- After three seconds has expired, the value of *o_PDLTrip* becomes *Open*, if *i_Power* $\geq$ 80%FP. Otherwise, the value of *o_PDLTrip* becomes *Close.*
- Once *o_PDLTrip* has been opened, keep it *Open* for 1 seconds.
- Then the value of *o_PDLTrip* becomes *Close* once *i_Pressure* $>$ 950 kPa or *i_Power* is $<$ 80%FP.

In this example, there are two input variables: *i_Pressure*, *i_Power* and one output variable: *o_PDLTrip* and one function: *f_PDLDelayedTrip.* Fig 3 shows a Statechart for the *PDL delayed trip parameter* part of the SDS2.

Transitions in *Inputs* are triggered by the external events and generate events which are broadcasted into only *Functions.* The default state of i_PressureState is determined by a conditional connective which contains predicates over the corresponding input variable, *i_Pressure.* The *Outputs* consists of *o_PDLTrip* which represents whether the trip parameter is open or close. Note that the events *o_open*, *o_close* should be generated only by transitions in *Functions.* The behavior of *f_PDLDelayedTrip* is as follows: the default state is S_1 which represent that the controller is idle and in waiting state. Once the state S_2 is entered, the software waits for three seconds. After three seconds the software checks the current values of *i_Pressure* and *i_Power* and determines whether or not to generate a trip signal.

4 Safety Analysis of Statecharts

Safety analysis techniques often employ backward analysis approach where a hazardous system state

[1]For simplicity, we have slightly simplified the actual requirement.

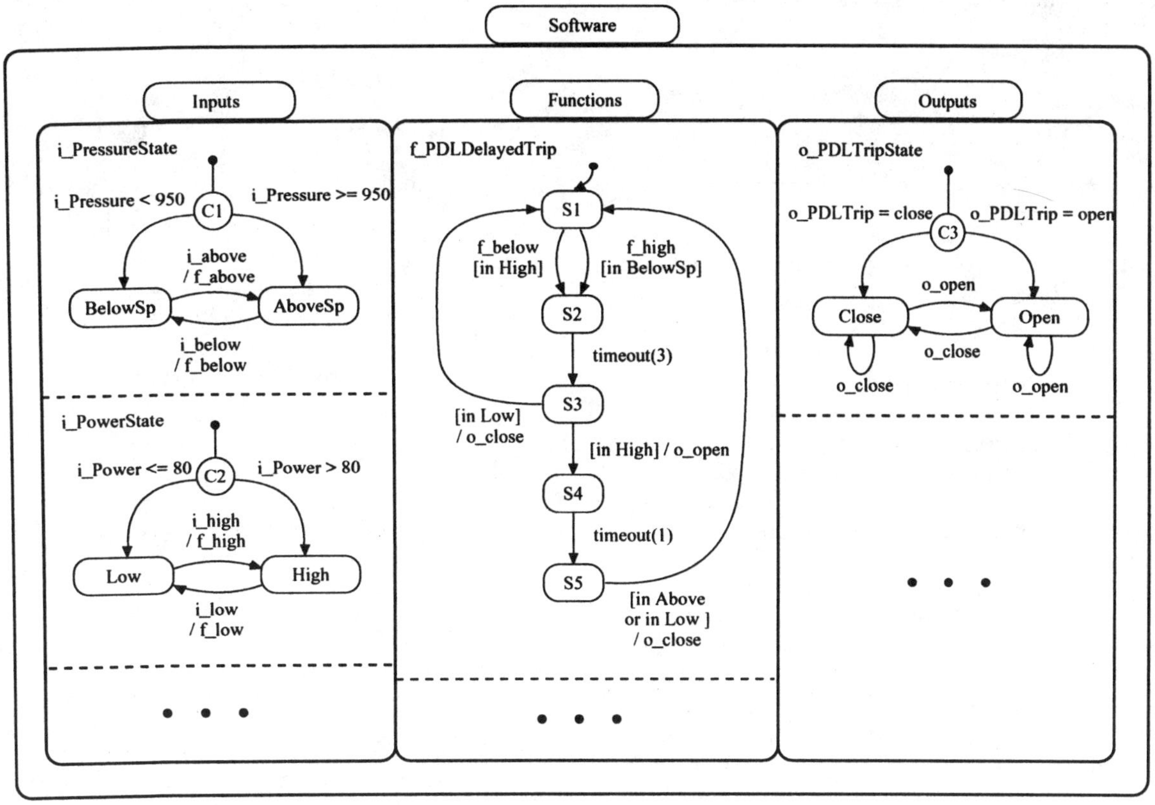

Figure 3: A Statechart for Wolsung SDS2

is assumed to have occurred and where the credible causes and their relationships (when applicable) are explored. When performing fault tree analysis, backward analysis is usually performed manually without any assistance of automated tools. Therefore, the effectiveness of the FTA method depends heavily on the capability of system safety analysts.

An alternative is a model-based backward analysis approach using Petri nets[8] or other state-based models. Backward step-by-step simulation can provide analysts with deeper understanding of if and how the hazardous system states might occur. Once event sequences leading to hazards are identified, they can be eliminated or risks minimized by introducing additional safety requirements.

4.1 Forward Simulation

The semantics of Statecharts [4, 12] are based on the *synchrony hypothesis* which assumes that the system is infinitely faster than the environment. Hence, the response to an external events is always generated in the same time when the event is introduced. The semantic description of Statecharts is based on the notion of *micro-step*. A micro-step is initiated when an external event arrives at the system, causing a cascade of subsequent internal events. A micro-step is completed when no more internal events are generated or there are no more transitions triggered by the events that were previously generated.

Unfortunately, when applying forward simulation to verify safety property of complex systems, *state explosion problem* is encountered. While our framework does not completely solve the state explosion problem, various restrictions on the transition formats may reduce the complexity of forward simulation considerably. In order to further simplify the forward simulation process, we distinguish the two types of events:

- *Spontaneous* events refer to the events "i_e" in *Inputs* and *timeout*, λ in *Functions*. The occurrence of these events does not depend on any other component in Statecharts.
- *Non-Spontaneous* events refer to the events "f_e" in *Functions* and "o_e" in *Outputs* These events may be generated only as a consequence of actions associated with events triggered by spontaneous events.

For example, in Fig. 4, *i_below* is a spontaneous

event and *f_below* is a non-spontaneous event.

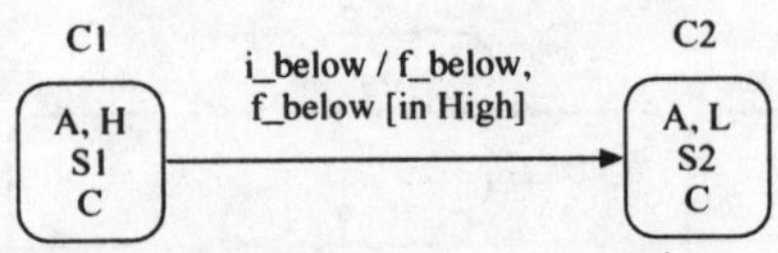

Figure 4: An example of spontaneous and non-spontaneous events

In summary, every event in *Inputs* is spontaneous and every event in *Outputs* is non-spontaneous. Events in *Functions* can be either spontaneous or not. Transitions in *Functions* whose triggering event is *timeout* or λ represent a spontaneous function. Transitions whose triggering event is *f_e* represent a non-spontaneous computation initiated by the sensor or by other function of the controller.

We note that any changes in the Statecharts configuration can be initiated only by a spontaneous event and forward simulation can be conducted by identifying all spontaneous events in the current configuration. Note that this forward simulation procedure is different from that of pure Statecharts where there is no notion of spontaneous or non-spontaneous events.

An example of forward simulation is shown in Fig. 5. Assume that the current configuration is c_0 = $(A,\ L,\ S_1,\ C)$[2] There are two spontaneous events which trigger outgoing transitions from c_0: *i_below* and *i_high*. Secondly, we generate the next configurations based on the micro-step semantics for each event.

4.2 Backward Simulation

In contrast to forward simulation that usually starts from a initial system configuration, backward analysis typically starts from the system configuration that is assumed to be hazardous. For example, a configuration $(B, H, *, C)$ means that whenever the system is in the states B, H, and C, it is hazardous regardless of the states of the *f_PDLDelayedTrip*. Note that the states in *Functions* need not necessarily be included in the definition of the hazardous system state because the system could be in the hazardous state regardless of the software's state. In this case, the configuration should be regarded as follows:

$$(B, H, *, C) = \{(B, H, S_i, C) \mid 1 \leq i \leq 5\}$$

Once the hazardous system configuration is defined, backward analysis can proceed by considering

[2]For notational convenience, we denote states in Fig 3, *AboveSp*, *BelowSp*, *Low*, *High*, *Close*, *Open* with A, B, L, H, C, O, respectively.

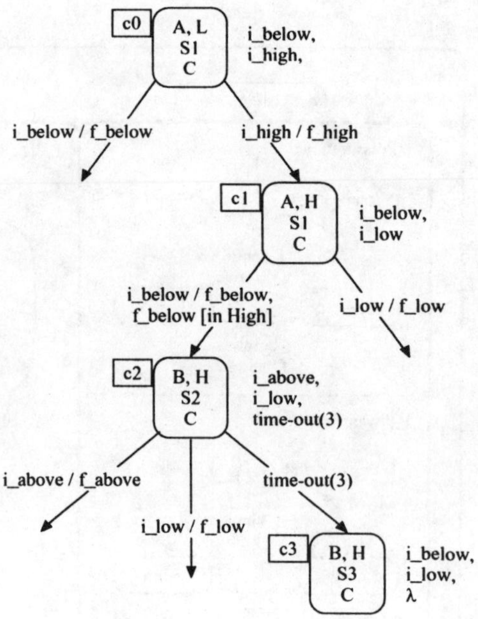

Figure 5: An example of forward simulation

only spontaneous events. As an example, let's assume that the hazardous configuration from which backward analysis begins as $c_0 = (B,\ H,\ S_5,\ C)$ which means that "trip signal is not generated when the value of *i_Pressure* $\leq$ 950 kPa, *i_Power* $\geq$ 80 %FP and the software is in the state S_5" (Fig. 6). There are three spontaneous events which trigger incoming transitions to c_0: *m_below*, *m_high* and *timeout*(1). For each event, the preceding configuration can be generated. For example, taking a step backwards following the event *i_below*, $c_1 = (A,\ H,\ S_5,\ C)$ is obtained.

To demonstrate how the λ event can be properly handled, let's take a step backward from the configuration c_2. Brute-force and reverse application of the micro-step semantics would result the configuration c_3 = $(B,\ H,\ S_3,\ C)$. However, this path is infeasible because the transition [*in High*] / *o_open* would result in a configuration $(B,\ H,\ S_4,\ O)$, not c_2, due to the action of the transition, *o_open*. Hence, we need not further continue the backward analysis from $(B,\ H,\ S_3,\ C)$. In Fig 6, infeasible paths are shown in dashed line.

5 Conclusions

In this paper, we have proposed a framework to specify and analyze the behavior of real-time systems in Statecharts and demonstrated its effectiveness using Wolsung SDS2 shutdown system. It should be

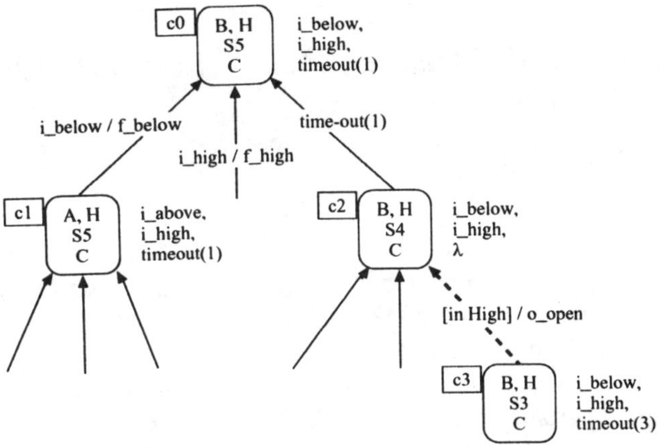

Figure 6: An example of backward simulation

emphasized again that Statecharts formalism does not provide any guidelines or restrictions on how one can properly specify and analyze the behavior of real-time systems. In our framework, we provide several guidelines and enforce several restrictions on how the features of Statecharts are used. It is important to note that such restrictions have been introduced for systematic specifications of software requirements based on the inherent characteristics of real-time systems.

In our framework, software requirements are described by specifying the behavior of output variables in terms of input variables using three orthogonal components: *Inputs*, *Functions*, and *Outputs*. Briefly reviewed, Statecharts are developed and analyzed as follows:

1. Identify all the relevant input and output variables of the plant. The behavior of each input and output variable is specified by the orthogonal components of *Inputs* or *Outputs*, respectively. A set of functions representing the required behavior of the controller is specified next as the orthogonal components of *Functions*.

2. Statecharts are statically analyzed for several criteria such as consistency and completeness criteria. Restrictions enforced on states and transitions reduce the possibility of introducing flaws in the model.

3. Statecharts that have passed the static analysis are then dynamically analyzed. We have demonstrated how forward and backward simulations can be performed in Statecharts. Specific guidelines on formulating the definition of hazardous system states are also suggested.

References

[1] D. Harel, "Statecharts: A Visual Formalism for Complex Systems," *Science of Computer Programming*, 8, pp. 231-274, 1987.

[2] C. Heitmeyer and B. Labaw, "Requirements Specification of Hard Real-Time Systems: Experience with a Language and a Verifier," In *Foundations of Real-Time Computing: Formal Specification and Methods*, Kluwer Academic Publishers, pp. 291-313, 1991.

[3] M.P.E. Heimdahl and N.G. Leveson, "Completeness and Consistency Analysis of State-Based Requirements," in *Proceedings of the 17th International Conference on Software Engineering*, Seattle, pp. 3-14, Apr. 1995.

[4] D. Harel, A. Pnueli, J.P. Schmidt and R. Sherman, "On the Formal Semantics of Statecharts," in *Proceedings of Symposium on Logic in Computer Science*, pp. 54-64, 1987.

[5] M.S. Jaffe, N.G. Leveson, M.P.E. Heimdahl, and B.E. Melhart, "Software Requirements Analysis for Real-Time Process-Control Systems," *IEEE Transactions on Software Engineering*, Vol. 17, No. 2, pp. 241-258, Mar. 1991.

[6] F. Jahanian, and A.K. Mok, "Modechart: A Specification Language for Real-Time Systems," *IEEE Transactions on Software Engineering*, Vol. 20, No. 12, pp. 933-947, Dec. 1994.

[7] N.G. Leveson, M.P.E. Heimdahl, H. Hildreth, and J.D. Reese, "Requirements Specification for Process-Control Systems," *IEEE Transactions on Software Engineering*, Vol. 30, No. 9, pp. 684-707, Sept. 1994.

[8] N.G. Leveson and J.L. Stolzy, "Safety Analysis Using Petri Nets," *IEEE Transactions on Software Engineering*, Vol. 13, No. 6, pp. 386-397, Mar. 1987.

[9] J.S. Ostroff, "Verification of Safety Critical Systems Using TTL/RTTL," *Proceedings of the REX Workshop on Real-Time: Theory in Practice*, LNCS 600, Springer-Verlag, 1991.

[10] D.L. Parnas, G.J.K. Asmis, and J. Madey, "Assessment of Safety-Critical Software in Nuclear Power Plants," *Nuclear Safety*, Vol. 32, No. 2, pp. 189-198, April-June, 1991.

[11] D.L. Parnas and J. Madey, "Functional Documentation for Computer Systems Engineering," CRL Report 237, Faculty of Eng. McMaster University, Hamilton, Ontario, Sept. 1991.

[12] A. Pnueli and M. Shalev, "What is in a Step: On the Semantics of Statecharts," *Theoretical Aspects of Computer Science*, LNCS 298, Springer-Verlag, pp. 244-264, 1991.

An Analytical Approach to Change for the Design of Reusable Real-Time Software

Carol L. Hoover and Pradeep K. Khosla
Carnegie Mellon University

Abstract

In this paper, we present an analytical method for incorporating knowledge about change into the design of reusable real-time software components. We apply this method to the construction of algorithmic software solutions that minimize the effect of anticipated changes in the solution. The motivation for our research is based on two premises: (1) software solutions that can easily be adapted to changes in the problem and solution domains are more readily reused and (2) the reuse of reliable software promotes the construction of high quality software systems and reduces development costs. This paper also briefly overviews our current research goals involving analysis and design methods to enable software engineers to rapidly adapt their real-time software solutions to changing application requirements.

Keywords: *Real-time software design, reusable software, change analysis.*

1 Background

It is well-known that real-time systems often have stringent timing and safety requirements [1]. The *ad hoc* way in which software engineers develop and maintain real-time software systems is often difficult, time-consuming, and costly [12]. To make matters worse, as in the example of the software problem that delayed the first NASA Space Shuttle orbital flight, the software process is sometimes error-prone. In discussing the Shuttle incident, Garman notes a key idea: though software is easy to change, it is also via change that software is the easiest part of a real-time system to compromise [3].

Software reuse, broadly defined as the use of engineering knowledge or artifacts from existing systems to build new ones, is considered to be an important technology for improving software quality, reducing development cost, and reducing the time-to-market [2].

Some researchers argue that reuse does not guarantee the safety of a software solution. For instance, Leveson states that safety is not just a property of the software itself but also a property of the software design and environment in which it is used [10]. It appears to be an open question as to whether or not it is easier to construct a safe software system totally from scratch or from parts of a "safe" existing software solution. As evidenced by the comprehensive literature dedicated to the topic of software reuse, it is clear that software experts consider software reuse to be an important technology [6, 8, 11].

The Chimera software infrastructure prescribes a standard format for building reusable software objects. These port-based objects automatically map to executable processes. The format includes operations for initializing, executing, and deactivating these dynamic objects as well as for error handling and inter-object communications [16]. The Chimera infrastructure includes a real-time kernel designed to ensure a performance-efficient and reliable real-time operating environment for robotics and factory applications [17].

Chimera objects are software packages that become processes activated by the real-time kernel. The concern of the Chimera infrastructure is with the *dynamic* and real-time world of the computer and its operating environment. In the *static* world of the human, software engineers design software solutions for real-time applications and partition these solutions into logical and eventually physical software components.

Static software components are mapped into dynamic processes in multiple ways and can ideally be executed in more than one operating environment. While Chimera addresses the reuse of dynamic process packages, we are concerned here with the reuse of static software components.

We would like to partition algorithms, such as those used to provide advanced motor control for robotic and factory automation applications, into software components that can be reused across various software solutions and if possible across different operating environments. A reusable software component is, for the purposes of our study, an executable or compilable module carefully designed to be useful in several programs, including unanticipated ones [18].

Before presenting our approach, we should mention that other researchers have also studied ways to decompose mathematical software solutions into software components. For instance, Westerberg's work reflects the common approach of intuitively decomposing algorithms into functional modules. Westerberg's modules can be used in various permutations to solve systems of linear equations [19].

We think our approach differs because we *systematically* decompose a software solution into small-effect operations. More importantly, we *analytically* recombine these small-effect operations into software components that localize the expected changes to the software solution.

2 Problem of Change

Concerned with the safety implications and the high cost of maintaining systems with stringent timing and reliability requirements, our goal is to study the nature of changes to software (*software metamorphosis*) and to develop analytical methods for designing and building real-time software solutions that can reliably and easily be changed.

From here on, we use the terminology "reduce the impact of change" to mean that a software solution can be modified in a reliable, timely, and cost-efficient manner. Our goal is to design reusable software components that *reduce the impact of making changes* to real-time software solutions.

From our experience in developing (and maintaining) industrial control software, we have found that changing a real-time software system can be tedious and time-consuming. Depending upon the type of change and the organization of the software, the software engineer may need to examine and modify multiple parts of the software system [5].

During the design of a new application, we would like to be able to experiment with alternative software solutions to achieve real-time performance or reliability. Sometimes we can achieve our goals via tunable parameters. Other times, we need to tune the solution by replacing poor-performing parts of the system [15].

We would like to localize the parts of a software system affected by a change because it is easier to modify software when the related changes are close together. Also, we are less likely to introduce errors into unrelated parts of the system if the unrelated part is logically or physically separate from the part affected by the change. In essence, we would like to examine and modify as few software components as possible.

We think it is reasonable that software engineers may decide not to reuse a whole or partial software solution if they perceive or discover that it would take more effort to modify an existing software solution than to create a new one. Therefore, we would like to find a systematic way to design reusable software components that localize change to as few software components as possible.

3 Analytical Approach

In this section, we propose target software properties and attributes that will serve as evaluation criteria for software resulting from our case study. We discuss the fundamental rationales for our analytical approach to building reusable software components that minimize the impact of change. Lastly, we outline the basic steps in our analytical approach and show how it applies to the construction of software components to implement optimization algorithms.

3.1 Target Software Properties and Attributes

Ideally, we would like to maximize a real-time software solution's capacity for change and to minimize the required engineering effort. So, our goal is to design software solutions that are *changeable* with respect to changing application requirements. But change to a software solution should preserve the predictability and reliability of the system.

In addition to the design of software components that enable flexible software solutions, we would like software solutions which are *partitionable* into concurrent processes. We would like to be able to easily distribute or replicate these processes across multiple processors if necessary to satisfy reliability and performance requirements.

Most importantly, we want to build software solutions whose components are *reusable* across various levels of granularity. This means that partial as well as whole solutions (software components as well as systems) can be used in alternative applications. We think the key is for software components to be used intact or replaced entirely if possible: modifying the internals of a software component requires substantial human effort and risks the introduction of error.

3.2 Fundamental Rationales

Two fundamental rationales direct our approach to designing reusable software components. One rationale is based on the idea of recasting large-effect algorithms into small-effect or partial algorithms. The idea is that software components that encapsulate whole algorithms more likely require internal modification in order to be reused; whereas, components that contain parts of an algorithm can be alternated to achieve variations in the algorithm's logic or implementation features [18].

Our interpretation of this rationale is that software solutions implemented as large-effect components are not likely to be as reusable as those implemented as small-effect components. One research question is how to recast a monolithic software solution into parts whose implementations will facilitate changes to the software solution via replaceable and reusable components.

The other rationale behind our approach is our idea that those parts of a software solution that are affected by the same expected changes should be located in the same software component. If we must replace or modify components to alter a software solution, we would like to minimize the number of affected components. Therefore, a second research question is how to design software components that minimize the impact of change.

3.3 Detailed Approach

Our approach involves decomposition of the large-effect operations of a software solution and recomposition into software components based on an analysis of expected changes. Since many of our target software solutions involve experimental algorithms, we focus on decomposition at the algorithmic level and conduct a case study of three optimization algorithms.

Our approach has the following five basic steps.

1. Identify the algorithmic features.
2. Decompose large-effect operations.
3. Determine the expected changes.
4. Componentize the operations using change analysis.
5. Add other necessary components.

In *step one*, we analyze the characteristic features of the algorithmic solution. For example, the characteristic features of a genetic algorithm (a type of optimization algorithm with which we are experimenting to select feasible control configurations) are an initial state of a population *P* of samples, a potential goal state of the population, and a transition function *T*. Note that samples x_i are members of a set *S* of all possible samples.

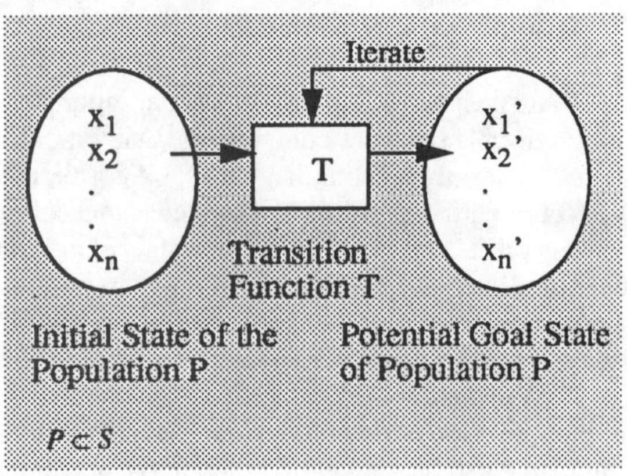

Step two is the decomposition of the large-effect transition function T into smaller-effect operations (a sample evolver and a sample evaluator). The interested reader should note that this version of a genetic algorithm always maintains a fixed-size population; therefore, a randomly generated new sample x' replaces a sample x_i to become x_i' only if it is more fit than x_i.

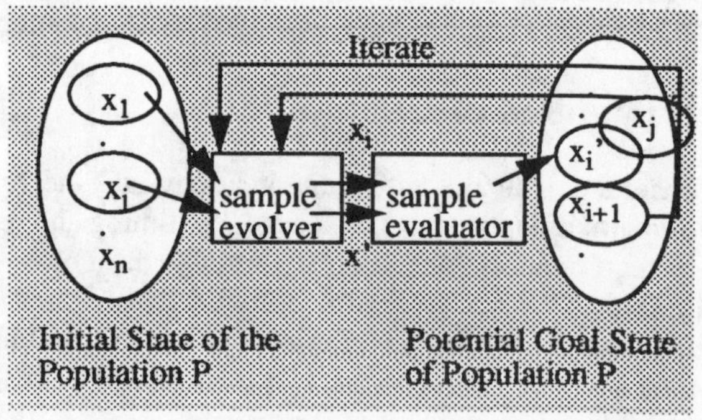

We continue the decomposition of the sample evolver into a randomized genetic operation selector and various genetic operations used to generate new samples. Likewise, we decompose the sample evaluator into a function to retrieve the saved fitness value of a particular sample, an objective function to calculate the fitness of a sample, and a comparator function to select the most fit of two samples. Further decomposition of these suboperations would result in smaller-effect operations which are trivial and therefore not significantly reusable.

We can now identify a set *O* of small-effect operations.

$$O = \{o \Diamond operator\}$$
$$= \{RandOperSel, GenOper, ...\}$$

$x \Diamond xtype$ means that *x* is of type *xtype*.

Step three is to enumerate the types of changes that the researcher would expect to make to the software solution. These could be logical or algorithmic changes such as using a different genetic operator to generate a new sample or implementation changes such as changing the representation of a population member. We create a set C of expected changes.

$$C = \{c \Diamond change\}$$
$$= \{ChangeGenOper, ChangeSearchSpaceRep, ...\}$$

Step four involves the recomposition of the small-effect operations into software components based on the analysis of anticipated changes to the software solution. The goal is to group operations affected by the same changes into the same components. Then changes to the software solution or its implementation can be made simply by replacing whole components or by modifying a minimal number of components whose contents are affected by the changes.

Step four is a combination of several substeps. We

relate each anticipated change to the software solution to the operations whose implementations would be affected by this change via a change impact relation *CI*.

$$CI \lozenge C \leftrightarrow O = \{(c,o) \lozenge C \times O \mid (c \rightarrow impact(o))\}$$

Applying the relation *CI*, we obtain a set of operations for each expected change. We will call these *change sets*. In totality, we now have a set of change sets *CS*.

$$CS = \{cs \subseteq O \mid {}^{(\exists c \lozenge C \mid (cs = CI(c))) \wedge}_{(\forall (o \in cs), \exists ((c,o) \in CI))}\}$$

The size of a change set indicates the magnitude of the impact of the change associated with the set. For instance, modifying a particular genetic operator affects only the software implementation of this operator. The cardinality of the change set *CGO* associated with this change is one.

$$CGO = \{GenOper\}$$

Alternatively, changing the search space representation impacts the genetic operators, objective function, initialization of the population, and all other operations on population members. As we would expect, the cardinality of the change set *CSSR* associated with modifying the search space representation is larger than one.

$$CSSR = \{GenOper, ObjFunc, InitPop, \ldots\}$$

We might think that we are done in that each set of operations affected by a change could be the contents of a software component. Then in the future, all we would have to do to make the expected change is to exchange or modify the related software component. But such compositions could result in duplicate copies of operations. In the above example, both *CGO* and *CSSR* contain the genetic operations *GenOper*.

What is needed is a way to combine change sets that intersect one another. We define a relation called *Overlap* that associates two change sets that have a non-empty intersection.

$$Overlap \lozenge CS \leftrightarrow CS$$

$$= \{(cs_1, cs_2) \lozenge CS \times CS \mid (cs_1 \cap cs_2 \neq \emptyset)\}$$

We must also consider the case when change sets are related transitively through repeated application of the *Overlap* relation. For instance, in the diagram which follows, the change set *CGO* (operations affected by changing a genetic operation) is related to the change set *COF* (operations affected by changing the objective function) because they both intersect *CSSR* (operations affected by changing the search space representation). The two-directional arrows indicate overlapping (intersecting) change sets.

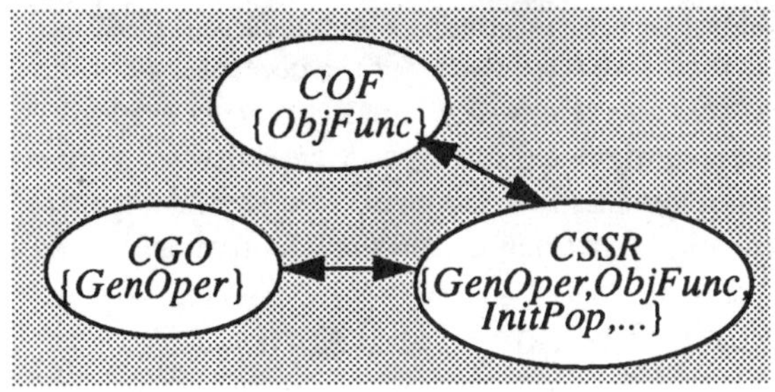

To relate those change sets which overlap directly or transitively, we define an *Affinity* relation.

$$Affinity \lozenge CS \leftrightarrow CS = \{(cs_1, cs_2) \lozenge CS \leftrightarrow CS \mid$$

$$\left.\begin{matrix} ((cs_1, cs_2) \in Overlap) \vee (\exists (cs_3 \lozenge CS) \\ ;((cs_1, cs_3) \in Afffinity \wedge (cs_3, cs_2) \in Affinity)) \end{matrix}\right\}$$

By applying the *Affinity* relation, we analytically group change sets *CGO*, *COF*, and *CSSR*. The union of these sets ($\{GenOper, ObjFunc, InitPop, \ldots\}$) guarantees that we will have no duplicate copies of the resulting operations. Now we encapsulate these operations into a software component. The resulting software component contains operations on population members. More will be said later about the interesting aspects of the results of this analysis.

By applying the *Affinity* relation (an equivalence relation) across all of our change sets, we effectively partition the change sets. The equivalence class described above contains *CGO*, *COF*, and *CSSR*. Each equivalence class becomes a software component. The formal definition for this partition of change sets into software components follows.

$$SameComponent = \{(ec \subseteq CS) \mid (\forall cs_1, cs_2 \in CS$$

$$;\left(\begin{matrix} ((cs_1 \in ec) \wedge \\ (cs_2 \in ec)) \end{matrix} \Leftrightarrow ((cs_1, cs_2) \in Affinity)\right))\}$$

Lastly in *step five*, we add software components to encapsulate those parts of the software solution not delineated in steps two through four. In the case of our genetic algorithm, the iterator logic that directs the detailed steps (e.g. sequence of function calls) of the algorithm can be encapsulated in a controller component.

Good design principles dictate that a controller component not be cognizant of the internal details of the operations that it activates. In addition, though the control logic is impacted by changes in the order by which small-effect operations are activated, an operation's implementation should not be affected by these changes.

It is therefore reasonable for the control logic to be logically separate from the small-effect operations with respect to anticipated changes in the software solution or in its implementation. Likewise, we should organize the control logic as a separate software component.

4 Application and Results

Our analysis of a genetic algorithm resulted in the six basic software components shown below. Each component has a boldface *change signature* which represents the anticipated changes associated with that component. The operations contained within the component are shown in italics. We were able to make the expected changes either by replacing or by modifying only the software component associated with a particular change.

CROS (Change Randomized Operator Selector)
Randomized Operator Selector

CCP (Change Control Parameters)
Input/Output Control Parameters

CVBIS (Change Value-Based Item Selector)
Value-Based Item Selector

CMC (Change Master Control Logic)
Genetic Master Control

CGO *U* CSSR *U* COF (Change Genetic Operator, Change Search Space Representation, Change Objective Function)
Representation of Population
Genetic Operators *Objective Function* *Initialize Population* *Output Population* *Retrieve Population Member* *Update Population Member*

CFVR (Change Fitness Value Representation)
Representation of Fitness Values
Initialize Fitness Values *Output Fitness Values* *Retrieve Fitness Value* *Update Fitness Value*

We also applied our analysis technique to two other optimization algorithms: a simulated annealing algorithm and a genetic annealing algorithm which is a hybrid algorithm having characteristics of both a simulated annealing algorithm and a genetic algorithm.

Interestingly, we found that we could reuse the software components across the three different optimization algorithms. The primary difference occurred in the data format used to represent the solution space being optimized. The genetic and hybrid genetic annealing algorithms operate on populations of samples and can share the population software component. The simulated annealing algorithm optimizes a configuration.

As we would expect, each optimization algorithm has different control logic and therefore requires its own controller software component. Shown below are the software components resulting from our analysis. We indicate the extent to which each component could be reused across the three algorithms.

GA GAA	SAA
Population	Configuration

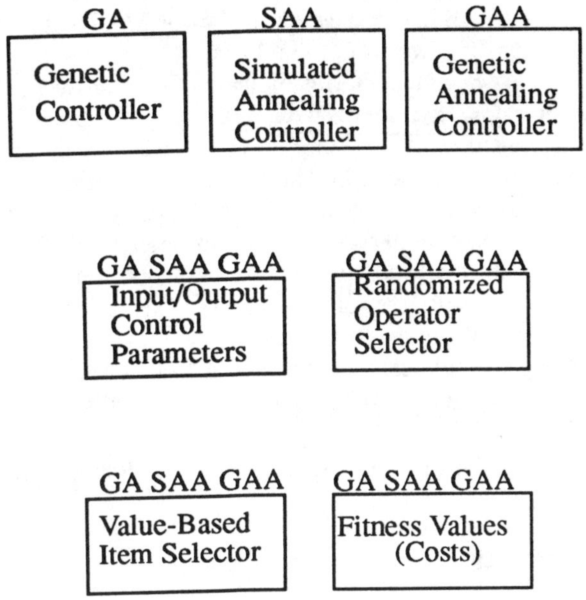

GA - Genetic Algorithm
SAA - Simulated Annealing Algorithm
GAA - Genetic Annealing Algorithm

5 Conclusions

In this section, we experientially and rationally explain how the software components resulting from the application of our analytical approach satisfy the software evaluation criteria discussed earlier. We relate the results to the design of objects and also discuss the potential for automating the technique.

We restate our goal to build software components that reduce the impact of making changes to software solutions, that promote software reuse, and that further the potential for concurrency and parallelization. Equally important is the constraint that we must be able to make changes to software solutions while preserving the predictability and reliability of the existing solutions.

By localizing the parts of a software solution affected by a particular change, we found it easier to make experimental changes to a componentized software solution than to a monolithic software solution (all parts of the solution in one component). For instance, we could easily change the representation of the population by modifying or replacing only the population component.

In the monolithic solution, we had to locate and modify those sections that operated directly on population members. We could more quickly and easily make the necessary changes in a smaller, more functionally coherent population component because there was less distance between related changes than in the monolithic program component.

By reusing those parts of the solution not impacted by a change, we could guarantee that the unaffected parts perform as predictably and reliably as before the change was made. But in the monolithic program component, we had to be careful not to introduce errors into unrelated sections of the monolithic solution.

From a real-time point-of-view, the ability to make localized changes by replacing components is valuable [15]. To improve the performance of a genetic operator, we modify or replace only the population component. As expected, we found that our decomposed software solution (in contrast to our monolithic solution) could be executed as concurrent processes. We found that it would take substantial manual labor to separate the monolithic solution so that it could be executed in multiple processes: the only feasible scenario was to execute it as a single process.

We were easily able to execute our small-effect, reusable components as concurrent processes. In one mapping from reusable software components to processes, we packaged the input/output component in one Chimera process and the other components together in another Chimera process. Other mappings to Chimera processes are certainly possible.

The interesting point is that had we followed an intuitional partitioning with sole attention to abstract data types (ADT's), we may not have organized the input and output of the control parameters as a separate component. Though the control parameters could have been encapsulated as an ADT, actual input and output of these parameters may have been intertwined with the control logic. Experience or engineering guidelines may have directed us to separate the input/output logic from the master control logic [13], but even well-known design principles are not always applied systematically.

In our monolithic optimization programs the input/output operations and control logic were integrated and could not easily be mapped into separate processes. But by analyzing the expected changes, we thought of changing the facility for the input and output of control parameters. Our analysis *systematically* directed us to create a separate component for these operations.

Our analytical approach supports the fundamental design principles of decomposition, partitioning, and encapsulation [14]. The population and cost components encapsulate data objects similar to those that would result from an object-oriented analysis. Our approach also gave us direction about how to organize non-data parts of the software solution such as the randomized operator selector and the master control logic. While the definition of objects still depends largely on the creative thinking and intuition of the human, our components originate from a systematic application of specific changes that we expect to make to software solutions.

We have outlined in detail the mathematical expressions for step four of our analytical approach to demonstrate its potential for automation. Identifying the algorithmic features, decomposizing the large-effect operations, determining the expected changes and their impact on small-effect operations, as well as adding control logic components currently require significant human thought.

6 Future Research

In this case study, the results of analytically applying knowledge about change to the construction of software components was valuable: our change analytic approach systematically guided us to partition our three optimization algorithms into software components that minimized the impact of change. We think further application to software solutions expressed as concise algorithms would yield similar results. We plan to test this hypothesis using real-time applications such as feed-back and control and digital filters.

We also note that in our study change analysis helped us to partition the software solution into reusable entities. Further study is needed to determine the precise relationship between change and the granularity level of reuse. Clearly, software solutions which are difficult to change are less likely to be reused (unless they can be used intact). But do software components organized on the basis of localizing change necessarily lead to the construction of reusable components and software solutions?

Jackson and Jackson suggest that hierarchical decompositions yield reusable software modules in numerical applications but not necessary so in other domains [7]. Would our approach yield reusable components in less mathematically-oriented real-time applications such as multi-media and real-time financial data reporting?

Another problem that we need to research is the possibility that components organized according to anticipated changes may exhibit undesirable software qualities and discourage reuse. For instance, is it possible that a software component could become too large or complex because many change sets of operations are transitively related (the *Affinity* relation in step four)? Could this size or complexity discourage reuse of the component?

Decomposing software solutions and relating anticipated changes to the decomposition solution is still vastly a human effort. We plan to study the types of changes that software engineers make to real-time software systems and to develop a model for categorizing change and relating change to software solutions during the design phase.

Our overall goal is to guide the software engineer in the design of reusable software components that facilitate cost-effective and risk-reduced changes to real-time software systems. We need models for building knowledge bases of change and for incorporating information about change into the design process. Similar to the type of analysis used to estimate the dependability of MARS application designs [9], we would like to be able to estimate the cost of software change. Halang states that real-time software systems are more likely to be predictable and dependable if they are simple [4]. Our analytical techniques should seek to reduce software complexity.

We think there is a need to formally study the way in which software engineers think about change during the design of new software components and about reuse when building new software solutions. Some research questions to be answered are the following.

- How should software engineers think when designing software solutions from reusable components?
- How can software engineers better anticipate and plan for change?

References

1. A. Burns and A. Wellings, *Real-Time Systems and Their Programming Languages*, Addison-Wesley Publ., Wokingham, England, 1990, pp. 2-10.
2. W. B. Frakes and S. Isoda, "Success Factors for Systematic Reuse," *IEEE Software*, Sept. 1994, pp. 14-19.
3. J. R. Garman, "The 'Bug' Heard 'Round the World," *ACM SIGSOFT: Software Engineering Notes*, Vol. 6, No. 5, Oct. 1981.
4. W. A. Halang, "Real-Time Systems: Another Perspective," K.M. Kavi, ed., *Real-Time Systems: Abstractions, Languages, and Design Methodologies*, IEEE Computer Society Press, Los Alamitos, Ca., 1992, pp. 11-18.x
5. C. L. Hoover, "The Role of the Real-Time Software Engineer: An Introductory Course," in *Proc. of the Eighth SEI Conference on Software Engineering Education*, New Orleans, La., Mar./Apr. 1995, Springer-Verlag, Berlin, pp. 167-186.
6. *IEEE Software* (issue featuring systematic reuse), Sept. 1994.
7. D. Jackson and M. Jackson, "Problem Decomposition for Reuse," Technical Report: CMU-CS-95-108, School of Computer Science, Carnegie Mellon University, Pittsburgh, Pa., 1995.
8. T. Capers Jones, "Reusability in Programming: A Survey of the State of the Art," *IEEE Trans. on Software Engineering*, Vol. 10, No. 5, Sept. 1984, pp. 488-494.
9. H. Kopetz et. al., "Distributed Fault-Tolerant Real-Time Systems: The Mars Approach," *IEEE Micro*, Vol. 9, No. 1, Feb. 1989, pp. 25-40.
10. N. G. Leveson, *Safeware: System Safety and Computers*, Addison-Wesley Publ., Reading, Mass., 1995, pp. 30-31.

11. H. Mili, F. Mili, and A. Mili, "Reusing Software: Issues and Research Directions," *IEEE Trans. on Software Engineering*, Vol. 21, No. 6, June 1995, pp. 528-562.
12. A. K. Mok, "Towards Mechanization of Real-Time System Design," A.M. van Tilborg and G.M. Koob, eds., *Foundations of Real-Time Computing: Formal Specifications and Methods*, Kluwer Academic Publ., Boston, 1991, pp. 1-37.
13. D. L. Parnas, "On the criteria to be used in decomposing systems into modules," *Communications of the ACM*, Vol. 15, No. 12, Dec. 1972, pp. 1053-1058.
14. D. L. Parnas, P.C. Clements, and D. M. Weiss, "The Modular Structure of Complex Systems," *IEEE Trans. on Software Engineering*, Vol. SE-11, No. 3, Mar. 1985, pp. 259-266.
15. M. Sitaraman, "Performance-Parameterized Reusable Software Components," *Int'l Journal of Software Engineering and Knowledge Engineering*, Vol. 2, No. 4, 1992, pp. 567-587.
16. D. B. Stewart and P. K. Khosla, "The Chimera Methodology: Design of Dynamically Reconfigurable Real-Time Software Using Port-Based Objects," in *Proc. of the IEEE Workshop on Object-Oriented Real-Time Dependable Systems* (WORDS'94), Dana Point, Ca., Oct. 1994.
17. D. B. Stewart, D. E. Schmitz, and P. K. Khosla, "The Chimera II Real-Time Operating System for Advanced Sensor-Based Control Applications," *IEEE Trans. on Systems, Man, and Cybernetics*, Nov./Dec. 1992, pp. 1282-1295.
18. B. W. Weide, W. F. Ogden, and M. Sitaraman, "Recasting Algorithms to Encourage Reuse," *IEEE Software*, Sept. 1994, pp. 80-88.
19. K. Westerberg, "Development of Software for Solving Systems of Linear Equations," Technical Report: EDRC-05-35-89, Engineering and Design Research Center, Carnegie Mellon University, Pittsburgh, Pa., 1989.

System Development based on a Real-Time Object Model

Seung-Min Yang
Tae-Myung Yoon
Soongsil University
Seoul 156-743, Korea
e-mail: yang@se.soongsil.ac.kr

Moon Hae Kim
Kunkuk University
Seoul 133-701, Korea
e-mail:mhkim@galaxy.konkuk.ac.kr

Byoung-Joon Min
Inchon University
Inchon, Korea

Jung-Guk Kim
HUFS
Seoul, Korea

Shin Heu
Hanyang University
Seoul, Korea

Abstract

This paper presents an approach to developing real-time dependable systems (RDSs) based on a real-time object model. A process to develop an RDS is divided into two phases (specification phase and construction phase), and a real-time object model, named RTO.e, plays the role of a connector as well as that of a fire wall between two phases. RTO.e, an extension of the RTO.k model proposed earlier by Kim and Kopetz, embeds the Z specification language and the exception section. Discussions are focused on the concept of the two-phased development process and a real-time object identification procedure using the RTO.e specification. An IVS (Intelligent Vehicle System) prototype is being under development to confirm the ideas and techniques investigated.

1. Introduction

Traditional development techniques are not suitable for development of a real-time dependable system (RDS). Recently, attempts to apply the object-oriented technology to development of an RDS are being made. The object-oriented technology [Boo91, Rum91] has potential advantages, but it also has difficulties in specifying and verifying temporal aspects of an RDS. To overcome these difficulties, research on real-time object models is actively on progress [Ell94, Kim93, Kim94, Sel94]. The real-time object models proposed in [Ell94] and [Sel94] extend traditional object model with temporal aspects. For example, the object model is extended with a deadline for each execution of a method. However, such structural extension is not sufficient in developing an RDS, especially in a distributed computing environment. A more practical and general real-time object model is proposed by Kim and Kopetz [Kim93, Kim94], named RTO.k. Unlike other real-time object models, RTO.k includes time-triggered objects which can be used to model real-time computing environments, and provides a specification language with clear semantics. With RTO.k, many different kinds of objects can be specified. RTO.k provides a basis for our research described in this paper.

Since 1994, the DRAGON (Distributed Real-time Availability Guaranteed Operation and Networking) group, formerly called CREG (Collaborative Real-time computing Environment Group), has been conducting the IDEEA (Integrated Development Environment for Embedded Applications) project. The purpose of the IDEEA project is to build an integrated environment for development and operation of a large complex RDS. Currently, IDEEA is based on RTO.e, which is an extension of RTO.k, and the scope of works involved in IDEEA is two-fold: (1) development support (such as modeling, specification, development process, testbed, and tools), and (2) operation support (such as DREAM kernel [Kim96], DRAGON extension [Kim95], and various runtime servers) for an RDS.

In this paper, we focus on the concept of two-phased development process and a real-time object identification procedure using the RTO.e specification. The next section gives a brief overview of the IDEEA project, and Section 3 presents the two-phased development process (specification phase and construction phase) and RTO.e. Detailed description on a real-time object identification procedure is given in Section 4. Section 5 concludes by giving current status of the IDEEA project.

0-8186-7570-5/96 $5.00 © 1996 IEEE

2. Overview of the IDEEA Project

The objective of the IDEEA project is to provide a unified frame and an integrated method along with supporting tools for developing large complex RDSs. Several issues must be addressed when developing embedded RDSs [Coo91, Kar92, Kim92, She92].

(1) These systems interact heavily with their environments. We assume that the environments are dynamic, large, complex, and evolving.

(2) Parallel software consists of concurrent threads for processing inputs and effecting the outputs (i.e., sensor-input and actuator-output).

(3) Reliability depends not only on logical correctness but also on temporal constraints (e.g., response deadlines, data validity time, periodic execution, and synchronization among threads) which underlie predictable and timely behaviors.

(4) Distribution of embedded systems provides a more natural, and often a more efficient processing configuration. In turn, distribution can lead to more robust and reliable functionality.

(5) The mission criticality of many embedded systems is characterized by life-safety and very high cost risk. This can demand ultra high reliability.

(6) Embedded systems constantly evolve due to an increasing need for greater functionality and responsiveness. Thus, routine support, upgrade and maintenance are commonplace.

The major difficulty in developing embedded RDSs stems from the fact that the system requires both logical and temporal correctness. Moreover, due to the criticalness of the mission that the system carries out, high reliability including fault tolerance capability must be provided. Since these requirements are directly affected by the runtime kernel and the target architecture, it is desirable to have a unified and integrated system engineering approach.

Three major focal points of the IDEEA project to handle these difficulties are real-time object (RTO) modeling, integrated system-level diagnostics, and time-guarantee kernel. These reflect various concepts and techniques which are considered as cores of real-time system development as depicted in Figure 1. The core techniques shown in the left side of the figure are accommodated into three focal points of IDEEA. RTO modeling acts as a frame with which system designers can specify hardware and software objects in a unified manner. Integrated diagnostics process [Web92] encompasses mechanisms for testing, on-line/off-line monitoring, and fault tolerance of the system. Time-guarantee kernel supports scheduling of real-time objects and provides primitives and infrastructure for fault tolerance. In the IDEEA project these are realized via the RTO.e model based specification and construction, three-level monitors, and the DRAGON extension with the DREAM kernel [Kim96], respectively.

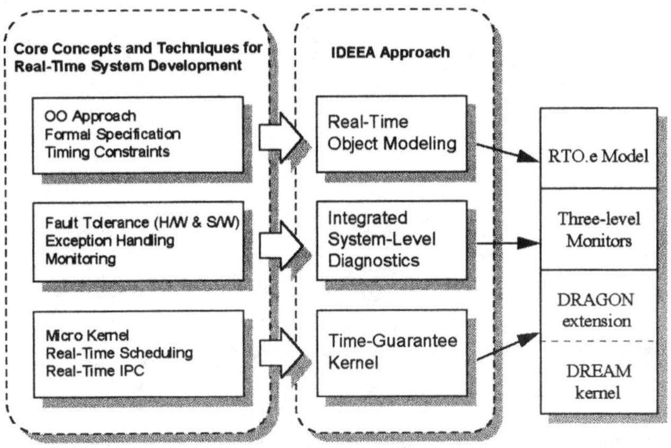

Figure 1. IDEEA Approach for Real-Time System Development

In the next sections, we discuss an RTO.e model based specification and construction in some detail along with a brief introduction of three-level monitors and the DRAGON extension.

3. Object Model based Development with RTO.e

3.1 Development Process based on RTO Model

The key idea of the RTO model based development is, as shown in Figure 2, that the system development is done in two phases and RTO.e specification is in between. The first phase is the Specification Phase where the system requirements are decomposed into RTOs and the identified objects are written in RTO.e form. (RTO.e is an RTO.k extension and will be discussed in detail in Section 3.2.) The second phase is the Construction Phase where the identified objects, hardware as well as software objects, are instantiated and combined. Therefore, RTO.e plays the role of a connector and also that of a fire wall between two phases. In next two subsections, we discuss some details of each phase.

3.1.1 Specification Phase. The Specification Phase consists of two stages: Problem Conception and Decomposition.

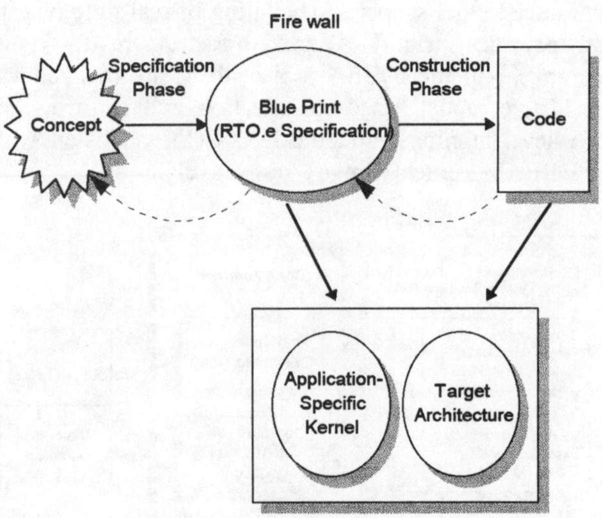

Figure 2. Two Phases of the RTO Model based Development

Stage I: Problem Conception. The target application is described. Its functionality (i.e., user needs) and the operational environment (i.e., constraints and tolerances) are formalized. Functional and performance requirements for three types of components described below are derived, i.e., requirement translation is done.

Type-1 Components consist of mechanical and electrical hardware (e.g., turbines, wing-flaps).

Type-2 Components consist of actuators and sensors needed to monitor and affect the behaviors of Type-1 components.

Type-3 Components perform control and operation of the system. These components can be either hardware units (e.g., CPU, coprocessor, memory) or software units (i.e., functional programs).

Stage II: Decomposition. An OO approach is natural for large complex RDSs: sensor and actuator objects are defined with control objects in between. While this stage focuses on logical correctness, timing and reliability requirements must be considered so that objects are allocated in such a way that these performance requirements are achievable. Object attributes such as size range, relative complexity, execution time range, and communication needs should be specified. Timing criteria, inherited from requirement specifications are carried forward into this decomposition framework. Real-time monitor objects (RTMOs) are also identified in this stage, which will be discussed in the next section.

The outcome of Stage II consists of a set of communicating RTO.e objects. Temporal consistency is checked . Some objects may need to be clustered to minimize the overhead due to communications, duplicated functionality, coupling and locality of resources.

3.1.2 Construction Phase. The Construction Phase consists of three stages: Instantiation, Evaluation, and Integration.

Stage III: Instantiation. RTOs are instantiated and tested. Objects are combined based on their connectivity (i.e., coupling or data sharing) and functionality if necessary. This reduces the complexity and overhead incurred by the excessive number of objects. Domain specific kernel objects are framed uniquely for a given application and instantiated for this specific use based on a system specification. The frame must be optimized by selecting specific functions, services, and policies (e.g., scheduling and communication protocols) prior to instantiation. Thus, the final outcome of this stage provides total system functionality including hardware. At this stage, clustering and framing are done in a coarse grain fashion. Finer grain optimizations, e.g., load balancing, placement of data, and communication optimization, are performed in the next stage.

Stage IV: Design Evaluation. This stage is to determine the "optimal" candidate target architecture which satisfies the system's performance requirements such as timeliness and reliability. This "mapping" of tasks is critical with respect to performance because of two major factors: (1) configuration, and (2) scheduling of tasks and messages. These factors should be all considered simultaneously (i.e., many different combinations must be evaluated). This approach advocates three mutually validating analyses, namely, mathematical analysis, simulation-based analysis [She92], and testbed-based analysis. These analyses are not exclusive, and are often supplementary.

(1) Mathematical Analysis: Mathematical proof, although restricted, is needed for obtaining high confidence from the design specification with respect to reliability prediction and timeliness guarantee. The timing requirements of very critical tasks must be guaranteed to meet execution deadlines. Reliability modeling with the consideration of both hardware and software is yet another area to be studied.

(2) Simulation-based Analysis: The simulation-based technique can be used to analyze the performance whenever the mathematical approach is impossible. Many mathematical models are not applicable for analyzing complex RDSs due to approximation or simplification of many parameters. The simulation model should realistically reflect the actual physical

environment or system.

(3) Testbed-based Analysis: A testbed is also utilized to evaluate the candidate configuration. The major focus here is to confirm the simulation result. Software (including the kernel) is executed on a hardware emulator. Actual timing results are measured. Faults are injected into software system. Features that depend heavily upon the hardware configuration and networking benefit substantially. For example, synchronization and communication methods are refined depending on whether shared memory or message passing mechanism is used.

Stage V: System Integration. The target configuration is developed while the software can continue to be refined using the testbed. Integration testing is performed once the target hardware has been "essentially" debugged and is ready to host the software. The system is subject to a "Post Integration Optimization". Decisions such as scheduling, load balancing, placement of data, context switching overhead, and interprocess communication are refined and/or validated. Figure 3 shows five stages of the proposed model.

3.1.3 Discussion. The Specification Phase defines system's functionality, reliability and timeliness, and decomposes these requirements into RTOs. The Construction Phase defines stages toward realization of the system, i.e., how the requirements are satisfied. In a traditional view, the model has two design steps: The first step is completed during the decomposition (Stage II) and the second step is completed during the evaluation (Stage IV).

In the proposed model, decomposition focuses on feasibility meeting both functional and nonfunctional requirements. The analysis in this phase includes factors such as workloads, memory and communication capacity, as well as reliability of hardware and software components. Evaluation, on the other hand, focuses on assessing software components, implemented in the prior stage (i.e., Instantiation), along with virtual hardware components as a simulated hardware configuration. Various configurations of hardware and software components such as the number of processors, allocation of software components into hardware components, and connectivity of hardware components are examined to determine if all performance and reliability (including fault-tolerance) requirements are satisfied.

Contrary to the "build-the-hardware-first" approach, we advocate that a more effective solution is to first meet system's functional and nonfunctional requirements through analysis and refinement. Then, based on the best design solution, the best hardware solution can be obtained.

In summary, the model has two potential advantages for developing large complex RDSs. First, embedded software can be designed with fewer restrictions imposed by hardware organization and operating system. This could reduce software development cost. Also, we believe that reuse based on a DSSA (Domain Specific Software Architecture) becomes realistically more achievable. Second, the performance and reliability of the system is examined in two stages which not only improves the analyzability but also possibly reduces the need to "undo" and "redo" the higher level design specification (i.e., going back to Stage II from Stage IV for redesign of software).

3.2 RTO.e

The RTO.k model is being extended, named RTO.e, to provide the capability for modeling all components of an RDS in a unified manner as shown in Figure 4.

Currently, works on RTO.e focus on two aspects as mentioned below so that RTMOs can be extracted from the RTO.e specification, and the way to include RTKO and RTHO in a unified RTO.e frame is being investigated.

There are three types of RTMOs: system-level, node-level, and branch-level. The roles of these monitor objects are to supervise the status, diagnose it, recover from faults, reconfigure, and handle exceptions in system-level, node-level, and task-level, respectively. In the following, we briefly describe two extensions made to RTO.k.

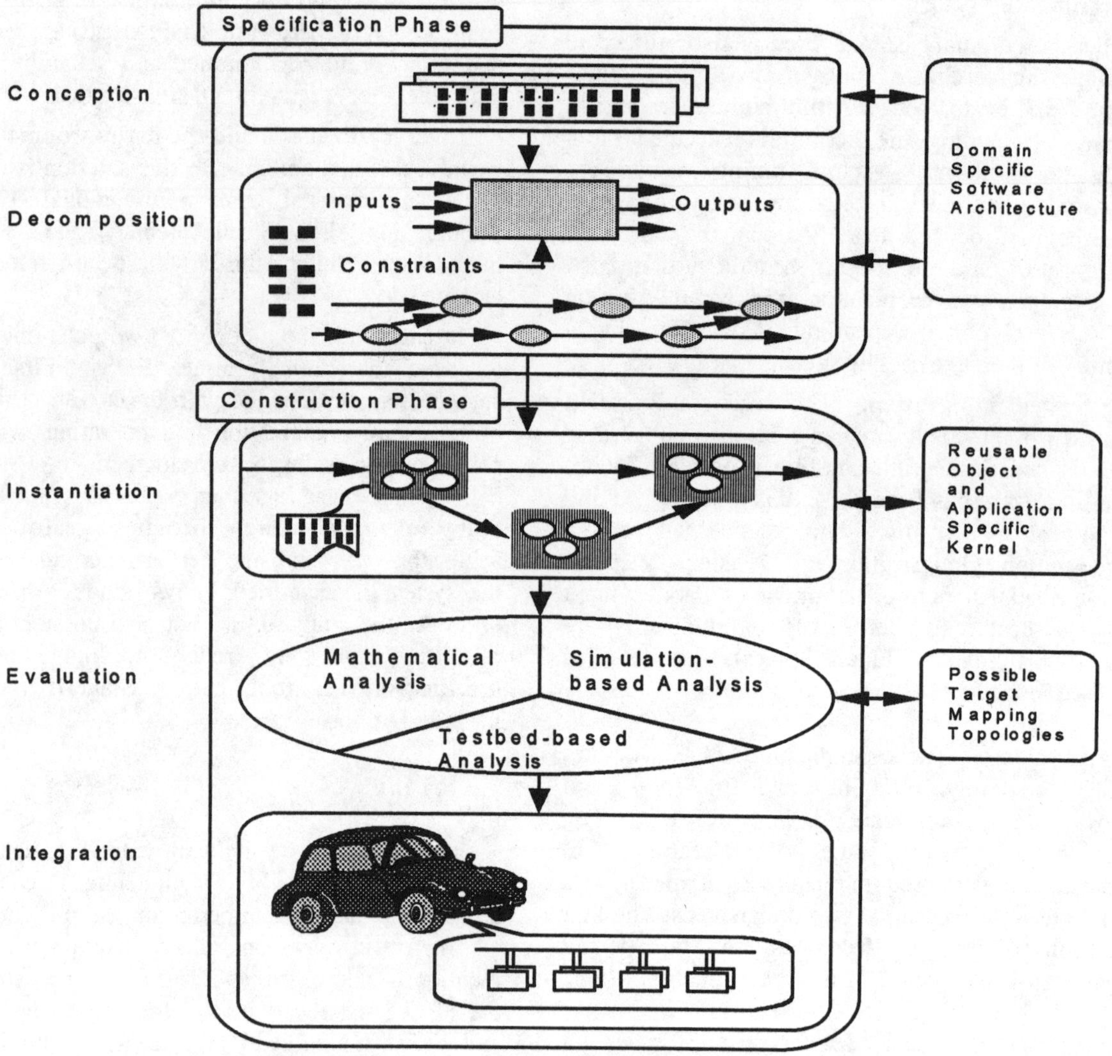

Figure 3. Five Stages of the RTO.e Model based Development

Application Layer	Real-Time Object (RTO.k)
Monitor Layer	Real-Time Monitor Object (RTMO)
Kernel Layer	Real-Time Kernel Object (RTKO)
Hardware Layer	Real-Time Hardware Object (RTHO)

Figure 4. RTO.e

(1) Addition of exception section

The exception section is added to specify the exception handling capability of monitor objects. There are two kinds of exceptions occurred during the execution of an RTO, that is, ones which can be handled within the object and ones which can not be handled within the object.

The former ones should be handled within the RTO in the form of branch-level RTMOs, and the latter ones should be handled by node-level or system-level RTMOs.

Therefore, it is desirable to specify handling procedures within corresponding RTOs. Figure 5 shows the specifications for exception handling. The notation "[e:h]" shown in Figure 5 is adopted from [Cri95], and the keyword "DEFAULT" refers occurrences of any undeclared exceptions in a method, that is, if any undeclared exception is occurred, the handler indicated by "default_handler_name" is activated.

system_exception <Name>
begin
    <system exception handling>
end.

(a) System exception

[exception_name or condition : handler_name] or
[DEFAULT : default_handler_name]

(b) User exception declaration in a method body

handler <handler_name>
begin
    <user exception handling>
end.

(c) Definition of an exception handler

Figure 5. Specification in Exception Section

(2) Formal specification of condition section

The condition section in RTO.e specifies requirements in the object level. Since this section can be utilized in testing, object-level fault-tolerance and design of RTMOs, it is desirable to use a formal specification. For this purpose, we are currently investigating Z notations in order to identify whether requirements for RTMOs can be extracted from the Z specification. Figure 6 shows an example of using Z notations in the condition section.

[IVS, SENSOR, ACTUATOR]
CS : SPEED
ACC : $\mathbb{P}$THROTTLE
BRAKE : $\mathbb{P}$BRAKEPRESS

$CS = \{\exists s : SPEED \mid s = ran(ACC' \mapsto (\exists ss{:}SPEED))\}$
$ACC \neq \{0\} \Rightarrow BRAKE = \{0\}$

The meaning of above specifications is that the values of accelerator and brake output from actuators in the IVS system should not be more than zero simultaneously.

Figure 6. Example of Z specification

With the above extensions, specification of an RTO.e model is shown in Figure 7.

RTO.e_class <class_name>
begin
    use_section : a list of names of other RTO.e's which provide services to this object;
    object_data_space_section : definition of data storages used by methods defined in this object;
    spontaneous_method_section : definition of time-triggered methods;
    service_method_section : definition of message-triggered methods;
    exceptioin_section : definition for handling exceptions occurring in the object;
    condition_section : formal specifications for important constraints, fault-tolerance, etc.;
end

Figure 7. RTO.e specification

4. Object Identification Procedure

4.1 Multi-Dimensional Decomposition (MDD)

In developing complex RDSs, system modeling scheme and decomposition scheme are inseparably related to each other. System model should be effective in representing the system characteristics, and the decomposition needs to be carried out in such a way that it complies with the modeling concept. In this section, an RTO.e model based system decomposition scheme is presented.

Design technologies for real-time systems may be classified into two categories. One is the decomposition based on functional structure and the other is the decomposition based on object-orientation. The structural design scheme has been used and some extensions of the scheme for real-time systems have been suggested in the literature [Gom89, San89, War85]. The decomposition based on functional structure, however, can not properly represent the parallel execution of modules and data flow among modules. Especially, in distributed environments, it is difficult to represent temporal constraints spread among several modules. The object-oriented approach, on the other hand, conveys several advantages. One of them is that software objects coincide with the system objects. It suits well especially for the decomposition of large-scale concurrent processing systems. However, the traditional object-oriented approach reveals the following drawbacks:

- Timing prediction which is essential for the design of a real-time system is difficult.
- Thorough identification of objects is not easy.
- Conversion of identified objects to tasks/processes is not straightforward

In order to overcome the above limitations, we propose a scheme called Multi-Dimensional Decomposition (MDD). The scheme goes through several steps and a different object identification criterion is applied to each step. The criteria are orthogonal.

Before we describe the details of the scheme we define physical objects and logical objects as follows. Physical objects are RTOs which are identified as the final product of the MDD, whereas logical objects are object groups which are identified in the course of the MDD. Therefore, a physical object is a unit for instantiation while a logical object is not. The concept of logical objects is useful in three aspects. First, identification of physical objects can be done in a top-down manner. Second, it provides a means to inheritance. Third, diagnostic and monitoring requirements can be derived in a systematic way. That is, by virtue of the logical objects, we can specify functional and temporal requirements imposed on a group of objects in the course of MDD and extract diagnostic and monitoring requirements. These are used to design real-time monitor objects (RTMO). RTMOs are special kinds of RTOs, which monitor the system status, i.e., behavior of ordinary RTOs. RTMOs perform the functions such as supervision, diagnosis, recovery, recording, and emergency handling. RTMOs are classified into three levels as discussed in Section 3.2. Branch-level RTMOs are associated with physical objects while node-level and system-level RTMOs are associated with logical objects.

4.2 Five Steps in MDD

The object identification procedure of MDD scheme consists of five steps including RTMO identification step as shown in Figure 8.

(1) 1st Step : by Mode

The status of a system under operation generally has several operation modes. For instance, IVS (Intelligent Vehicle System) could be in an automatic operation mode, a manual operation mode, a semi-automatic operation mode, an inspection mode, and an off mode. In this step operation mode objects are identified and the conditions for the state transition among modes are described in the mode transition diagram. A mode may have submodes.

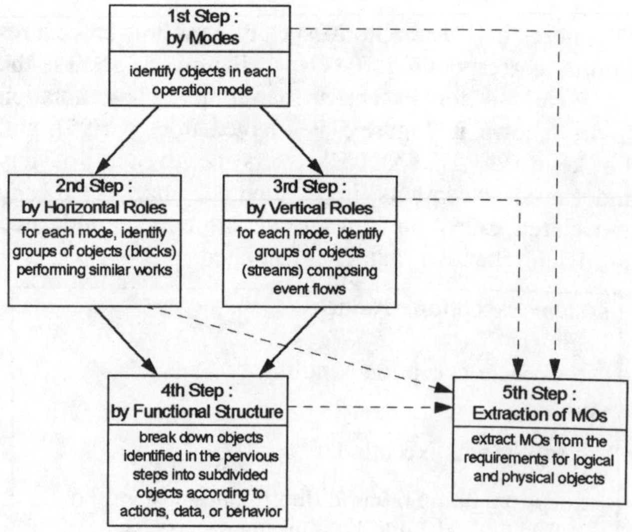

Figure 8. MDD Object Identification Procedure

(2) 2nd Step : by Horizontal Roles

In this step a group of logical objects are identified for each mode based on horizontal roles. By the horizontal roles we mean object groups which perform similar works. These logical object groups are called blocks. For instance, block objects in the automatic operation mode of IVS include input and output block objects, analysis block object, decision block object and management block object.

(3) 3rd Step : by Vertical Roles

In this step, logical objects are identified based on event flows or streams among block objects. Examples in IVS are streams for controlling speed and streams for controlling direction. We can derive a stream either starting from source points (i.e., input devices) to destinations (i.e., output devices) or starting from destinations back to source points. In MDD, the latter approach is used. That is, logical objects are identified for each output object or actuator.

(4) 4th Step : by Functional Structure

Objects identified in the previous steps are subdivided into smaller objects or methods if necessary. Identification can be done based on actions, data structures, or behaviors. Relationship among physical objects are also defined and presented using graphic notation.

(5) 5th Step : Extraction of RTMOs

Monitor objects are derived from the requirements of logical and physical objects. The object identification procedure described above has the following advantages in comparison with the traditional object-oriented technology.

1) Systematic identification of objects can be achieved and the relationship among objects can be easily formalized.
2) Multiple inheritance can be obtained.
3) It is easy to extract RTMOs by defining logical objects which are essential for development of a system with high reliability.
4) It is easy to convert identified objects to executable models.

5. Current Status

This paper has discussed an RTO.e model based development approach for RDSs. The two-phased development process has been proposed with the multilevel RTO identification technique. RTO.e takes a major role in the two phased development as a connector as well as a fire wall between two phases. RTO.e extends RTO.k with (1) addition of the exception section and (2) use of formal specification in the condition section. We currently develop an RTO.e editor and various tools for analyzing the RTO.e specification as discussed in Section 3.2. RTO.e to C++ translator is also under study.

Another major task of the IDEEA project is to develop a time guarantee kernel, named the DRAGON kernel. The kernel is designed to handle RTO.e methods and real-time data objects. The kernel also provides primitives and infrastructure for fault (or any abnormal behavior of the system) detection and recovery, which include timeout, dynamic scheduling, data replication, state save, hardware and software component shutdown, and initiation of redundant components.

An IVS (Intelligent Vehicle System) prototype is under development to examine the ideas and techniques investigated in the project. The first version of the IVS prototype has sonic sensors, a lane sensor, and a marker reader. The prototype handles direction and speed of the vehicle. Vehicle's body size is $40cm \times 40cm$ with an Intel 80486 main CPU and two 80196s. IVS software is being developed based on the RTO.e model and specification.

References

[Boo91] G. Booch, *Object Oriented Design*, Benjamin/Cummings, 1991.

[Coo91] J. E. Cooling, *Software Design for Real-Time Systems*, Chapman and Hall, 1991.

[Cri95] F. Cristian, "Software Exception Handling and Fault-Tolerance," *Software Fault-Tolerance*, ed. by M. Lyu, Chapter 4, Wiley, 1995.

[Ell94] J. R. Ellis, *Objectifying Real-Time Systems*, SIGS Books, 1994.

[Gom89] H. Gomaam "A Software Design Method for Distributed Real-Time Applications," *Journal of Systems and Software*, Sep. 1989, pp. 81-94.

[Kav92] K. M. Kavi and S. M. Yang, "Real-time systems design methodologies: an introduction and a survey," *Journal of Systems and Software*, April, 1992, pp. 85-99.

[Kim92] K. H. Kim, "Design of Real-Time Fault-Tolerant Computing Stations," *Lecture in NATO Advanced Science Inst. on Real-Time Computing*, Oct. 1992.

[Kim93] K.H. Kim and L.F. Bacellar, "A Real-Time Object Model: A Step toward an Integrated Methodology for Engineering Complex Dependable Systems," *Proc. CSESAW '93*, July 1993, pp. 56-64.

[Kim94] K. H. Kim and H. Kopetz, "A Real-Time Object Model RTO.k and an Experimental Investigation of It's Potential," *Proc. IEEE COMPSAC*, Nov. 1994.

[Kim95] M.H. Kim and S.M. Yang, "Environment for Development of Reliable Real-Time Distributed Applications," *Proc. FTDCS'95*, Aug. 1995, pp.120-126.

[Kim96] K.H. Kim, C. Subbaraman, and Y. Kim, "Imitation of RTO.k Objects Using PCD Components in C++," *Proc. WORDS'96*, Feb. 1996.

[Rum91] J. Rumbaugh, et. al., *Object-Oriented Modeling and Design*, Prentice Hall, 1991.

[San89] B. Sanden, "Entity-life modeling and structured and analysis in real-time software design -A comparison," *CACM*, Dec. 1989, pp. 1458-1466.

[Sel94] B. Selic, G. Gullekson, and P. T. Ward, *Real-Time Object-Oriented Modeling*, Wiley, 1994.

[She92] F. T. Sheldon, S. M. Yang, and T. Bornejko, "Simulation-Based Analysis for Real-Time Systems Development," *IEEE AutoTestCon '92*, Sep. 1992, pp. 361-366.

[War85] P. T. Ward and S. J. Mellor, *Structured Development for Real-Time Systems*, Yourdon Press, 1985.

[Web92] N. J. Weber, "Integrated Diagnostics for Software," *Proc. NAECON '92*, 1992, pp. 565-571.

Object-Oriented Software Development with Fault Tolerance for Distributed Real-Time Systems *

Stephen S. Yau and Jun Wang

Computer Science and Engineering Department
Arizona State University
Tempe, AZ 85287-5406, USA

and

Jai-Woo Kim

Computer and Information Science Department
University of Florida
Gainesville, FL 32611-6120, USA

Abstract

An object-oriented application software development approach with fault tolerance for distributed real-time systems is presented. This approach is based on the PaRallel Object-Oriented Functional computation model (PROOF) with real-time and fault tolerance features. The real-time constraints are satisfied by using mutli-version methods, which are based on virtual method definitions with different underlying implementations, and encapsulation of time constraints in objects. The fault tolerance in application software layer is supported by monitoring, checkpoints and recovery.

Keywords: object-oriented software development, real-time system, fault tolerance.

1 Introduction

As the requirements for application software of distributed real-time computing systems (DRCS) become larger and more complex, such as those for telecommunications systems, transportation systems, factory automation, and command, control, communication, and intelligence (C3I) systems, which have many interacting components, shared resources, computation intensive tasks, time critical events, and constraints for fault prevention and tolerance, effective software development approaches are needed to efficiently develop high quality application software. Currently, lack of proper software development approaches is the major obstacle for the wide use of the DRCS in various areas. Although many traditional approaches, such as structured analysis/structured design (SA/SD) and functional decomposition [1, 2], have been extended with distributed real-time and fault-tolerant features, the absence of strong human cognition reduces its effectiveness in developing large-scale applications. Most of software development approaches with fault tolerance for DRCS emphasize software development at the system level, and are not suitable for solving general applications with great complexity in real-time and fault recovery.

In this paper, we will present an object-oriented application software development approach with fault tolerance for DRCS. It is based on the PaRallel Object-Oriented Functional (PROOF) computation model which incorporates the object-oriented paradigm with functional features. To support real-time aspects, *multi-version* methods will provide time transparency vs. states of objects, and time constraints will be encapsulated at either method-level or object-level of aggregation and inheritance hierarchy for the application software [3]. Resistance to faults will be realized by applying an extended method based on *primary-site* and *process-pair* approaches at the application software layer, whose purposes are to reduce the degrees of replication of processes and computing resources and to ensure certain reliability.

2 Overview of Our Approach

Our approach is based on the PROOF computation model [4], which characterizes application software into a set of objects which encapsulate data and implementations of related operations along with time constraints. Object interfaces for prototypes of public functions are provided. States of an object are defined by the internal data of objects. Besides quantifying real-time constraints within methods of an object, optional precondition *guard* constructs can be defined to achieve synchronization among objects. Checkpoint and recovery facilities based on user-defined critical data of applications and monitor daemon processes will provide fault tolerance features. Development guidelines for our approach are also expanded from those in PROOF software development framework to accommodate real-time and fault-tolerant features.

*This work is partially supported by Hitachi, Ltd.

0-8186-7570-5/96 $5.00 © 1996 IEEE

Although many models, such as finite state machines, Petri nets, and temporal logic, have been extended to capture different aspects of a complex real-time software, but none of them can individually represent the whole scheme. Extended approaches for SA/SD have also been developed, such as DARTS (Design Approach for Real-Time Systems) [2], for functional decomposition of software modules, which shows much complexity. The feasibility of object-orientation for real-time software development has been examined through its suitability to represent and encapsulate temporal characteristics of objects in the real-time software [5]. Our object-oriented approach for supporting real-time features in software will emphasize the following aspects:

- Object adaptability will be realized through *multi-version* methods, in which each virtual method will have different implementations depended on object states, to provide time transparency at the method level;
- Specific combinations of timing attributes will be defined at the method and object levels to quantify real-time constraints.

Therefore, real-time requirements will be satisfied in object structures as shown in Figure 1.

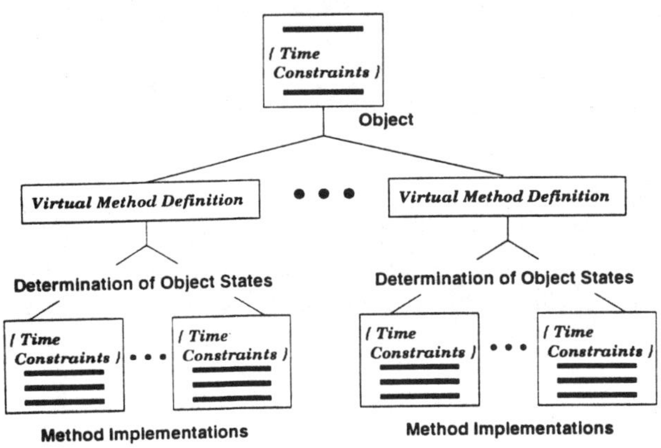

Figure 1: Real-time constraints in the structure of an object.

Many fault-tolerant issues have been realized at the hardware and operating system layers, which needs many resources to prevent, detect, and recover hardware and software faults with great complexity and low efficiency. Although common but misleading arguments are against embedding fault-tolerant schemes inside or around application software, the well-tested high abstractions offered by fault-tolerant schemes can provide efficient and portable software across different DRCS. Various approaches for fault tolerance at the application software layer have been proposed for the DRCS, such as voting (e.g., N-versions), primary-site, process-pair, and distribute recover block. Our fault tolerance approach is based on *primary-site* and *process-pair* approaches. Advantages are reducing the number of processes and computing resources based on the *process-pair* approach with great reliability offered by the primary-site approach. Our fault-tolerant environment built on the application software layer will contain library routines of high abstractions for checkpoints and recovery, and monitor daemons, which is shown in Figure 2.

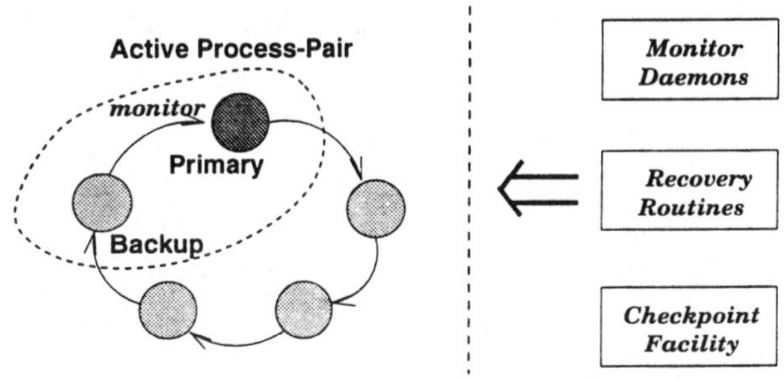

Figure 2: The application software layer fault-tolerant environment.

3 Support for Real Time

For distributed real-time applications, besides exploiting inherent concurrency and communications among objects, specific timing constraints are needed to specify time critical events in applications. *Multi-version* methods will support object adaptability based on object states. To support large-scale application software with great scalability through aggregation and inheritance hierarchy, time constraints need to be specified at either object-level or method-level. To increase timing transparency vs. states of objects, *multi-version* methods are defined for specifying virtual methods in objects with different implementation details depended on object states [3].

3.1 Object adaptability through multi-version methods

Function overloading and overriding in object-oriented development provide a powerful scheme to relieve users from many housekeeping activities, which shows a strong advantage over any other paradigm. But function overloading and overriding, such as *virtual* methods, have the limitation to effectively handle timing-related events throughout lives of objects. Normally virtual methods are defined along the inheritance hierarchy for a set of close-related objects. Each virtual method at a certain level of object hierarchy optionally makes aggregated refinements based on specifications of the upper level and explicitly associates with objects belonging to this level, not object states at certain moments. One-to-one mapping between a virtual method to an object can not satisfy real-time requirements, which is shown in Figure 3.

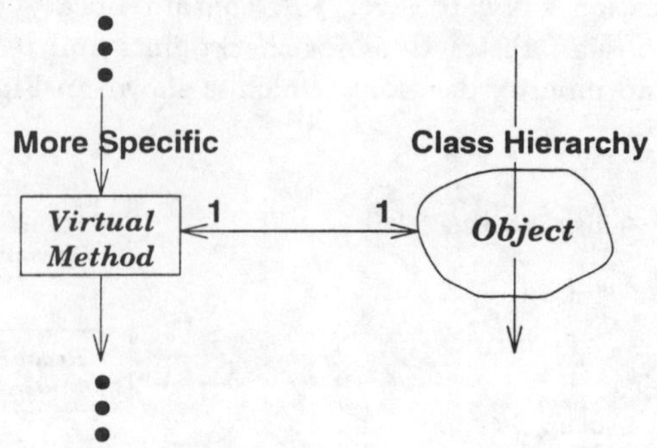

Figure 3: The limitation of the virtual method.

Control structures available in high-level languages, such as CASE and IF-THEN-ELSE, can be used manually through the software to state explicitly about which methods of objects to be invoked, which is similar to the situation for other programming languages before virtual methods have been introduced in object-oriented languages. Several strong drawbacks can affects qualities of software:

- object states for any moment need to be explicitly maintained, which generates many overheads,
- modularities among objects can not be kept, which increases efforts for understanding and maintaining the software,
- and individual object methods can lock object states from modification by others, which significantly limits concurrent invocations of methods of the same objects and reduces the inherent parallelism.

To extend the concepts of function overloading and overriding, a set of object-state-dependent method implementations associated with a virtual method definition for an object are needed. Actions for different object states will be encapsulated into a single method called a *virtual* method, which can be used as the external interface to the object. Internally, each action associated with certain object state will be implemented as a single method called an *actual method.* Comparing to Figure 3, A *virtual* method associated with any object mapping to a set of *actual* objects is called a *multi-version* method as shown in

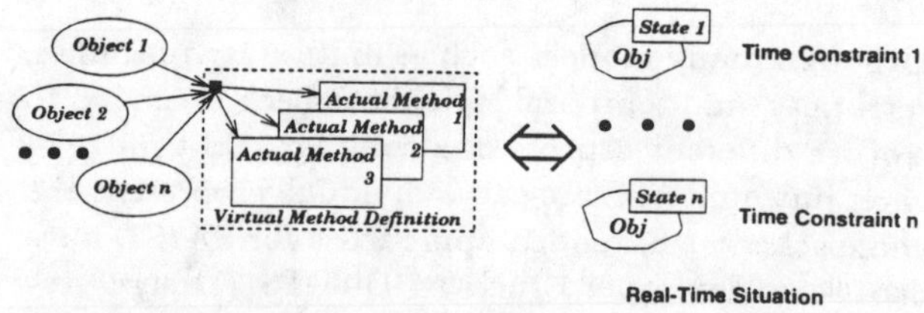

Figure 4: Multi-version methods.

Figure 4. Object states, which are declaratively specified through data attributes of an object at different moments during the life of the object (not different objects), will be used to determine which method implementation should be chosen when the corresponding virtual method is invoked.

Multi-version methods provide a flexible and easily maintainable mechanism to satisfy special properties of certain DRCS, such as autonomous decentralized systems (ADS) [6, 7]. Since other method implementations which are suitable to the modified problem specifications can be dynamically inserted to virtual methods without significant changes in applications themselves during the maintenance stage, special properties of ADS systems, such as on-line expansion, maintenance, and testing based on autonomous controllability and coordinability, can be easily supported.

3.2 Encapsulation of time constraints in objects

Real-time requirements should be localized to objects through encapsulating time constraints inside methods or bodies of objects. In our approach, combination of different timing attributes can be used to specify time constraints as follows:

- *Start-time* determines starting execution time.
- *End-time* determines ending execution time.
- *Duration-time* determines time interval during which execution should be performed.
- *Period* determines a time interval between consecutive executions of a periodic method. When a method is not periodic, this attribute can be used to specify minimal time interval between consecutive activation of the aperiodic method.

Encapsulation helps to build object hierarchy through inheritance. To encapsulate time constraints in objects will be able to make time constraints local to objects, which is suitable for the abstraction and inheritance purposes. Through the timing inheritance in the object hierarchy, the range of problem statements can be significantly scaled up in order to apply to large-scale real-time application software. Therefore, besides that the optional *guard* can be specified inside the method, time constraints expressed in terms of timing attributes will be optionally quantified too. Encapsulation of time constraints in each method definition can allow early determination of schedulability at the design phase, improve predictability of the software system by specifying any specific action in which time constraints may be violated, and prevent the problem of priority inversion in the case of concurrency and synchronization controls of *shared* objects which are accessed for referring or updating.

When there are no explicit time constraints specified in a DRCS, such as the ADS system, time attributes for time critical events described above can be flexibly incorporated to realize real-time requirements.

4 Support for Fault Tolerance

Fault tolerance has been realized at three different layers of DRCS: hardware, operating systems, and application software. Although many studies have been conducted at hardware and operating system layers to cover fault recovery from general hardware and software failures occurred in the application software, fault tolerants provided by these two layers can be inefficient. For instance, that a checkpoint scheme in an operating system checks entire in-memory data for an application can be inefficient whereas application-based one checks user-defined critical data.

In order to keep software reliability, the *process pair* approach [8] reduces degrees for replication of processes and wastes of computing resources, while the *primary site* approach [9] provides strong reliability. Our approach for fault tolerance at the application software layer is based on both approaches, which is shown in Figure 5.

The fault-tolerant facility provided by our approach includes process/node replications, process/node monitoring, and process critical data checkpoints. The tasks shown in Figure 5 to provide fault recovery are listed as follows:

- Each node in a circle has its neighbor (left or right) nodes as its main backup node;
- The application process will be executed on the primary node, while the application process will be inactive on backup nodes.
- A *process check* daemon is running on primary node to monitor crashes or hangs of application processes;
- A *processor check* daemon is running on each backup node to monitor for the crash of its own primary node;
- The backup node will periodically check user-defined critical data on the node where the applicatin process is running for fault recovery.

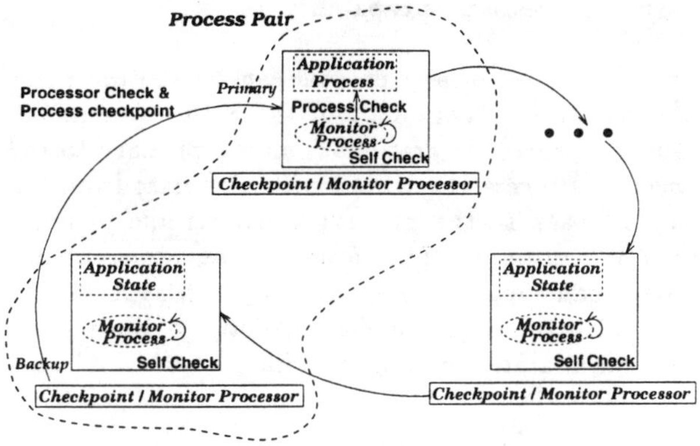

Figure 5: The fault-tolerant approach.

Any faults happen to processes or nodes, the situations will be handled as follows:

- When an application process on the primary node crashes or hangs, it is restarted on the primary node if appropriate;
- When a node crashs, all the application processes on the crashed node will be shifted to its backup node;
- An application process will be restored to its latest recoverable internal state using critical data before its last failure;
- Processes for application software will be migrated back to their primary nodes whenever the primary nodes are up and running.

Since our approach only allows the primary node to actually execute application software, and backup nodes records states of the corresponding process, no concerns are needed to keep state consistency and memory coherence among primary and backup nodes. Besides our approach, other software fault tolerance techniques, such as N-version programming or recovery blocks, can be used inside application software, such as object bodies or object methods, to provide more fault resistance capability.

Primary nodes of any process can be shutdown for maintenance without halt of the entire system because of process migrations from the primary to the backup. Process migrations will take place from the backup back to the primary whenever the primary node is recovered. Therefore, our fault-tolerant approach can support certain DRCS, such as ADS systems with these special characteristics of on-line expansion, maintenance, and testing.

4.1 Monitor process/node

In order to detect faults or failures of processes for application software and nodes on which processes are executed, daemons are needed to monitor processes and nodes, and when any exception happens, proper operations will be invoked by daemons for recovery. Criteria for monitoring and recovery should be simple, fair, flexible, and distributed controls. An adaptive distributed fault monitoring and recovering protocol is developed for the daemons.

The initial configuration includes: participated nodes and their circular monitoring arrangements; application software, primary node for execution, period for monitoring and checking, communication ports, and critical data to check. Daemon on each node will hold a unique status file which records: for each process, process ID p_i, node ID n_i, primary node ID pn_i, number of consecutive failures f_i, and maximum number of retries $retrymax_i$; local host name *local_host*; name of the node that is being monitored *cur_monitor_node*; name of the last node that was being monitored *pre_monitor_node*. The algorithm for the fault monitoring and recovering protocol is shown as follows:

1. /* Initialize the status file */
   (a) assign *local_list*, *cur_monitor_node*, *pre_monitor_node* based on the configuration.
   (b) get the status table from other working node (randomly chosen);
   initialize the content for each process in my own status file;
   (c) for each process i,
   if (($n_i \neq pn_i$) and (pn_i = *local_host*)) then
   get the most recent state of the process from node n_i;
   set n_i to pn_i and restart process i;
   broadcast changes around the network;
   end if
   (d) end for
2. /* Processing */
   Loop forever:
   (a) for each process i:
   i. if n_i = *local_host* then
   A. if process i is not crash or hung,
   $f_i = 0$;
   continue;
   B. end if
   C. increment f_i;
   D. if $f_i <$ $retrymax_i$,
   restart process i;
   else if f_i = $retrymax_i$,
   /* guess that the application inherently has problems */
   print out an urgent warning message;
   end if
   ii. else if node *cur_monitor_node* is failed,
   A. set *pre_monitor_node* to *cur_monitor_node*;
   B. set *cur_monitor_node* to a new node;
   C. if n_i = *cur_monitor_node* then
   /* backup upgrades to primary */
   check critical data of process i from *cur_monitor_node*;
   D. else if n_i = *pre_monitor_node* then
   /* primary just failed */
   restart process i and set n_i to *local_host*;
   broadcast changes around the network;
   check critical data to backup of *local_host*;
   E. end if
   iii. else if ($n_i \neq$ *local_host*) then
   stop running the process;
   iv. end if
   (b) process queueing events (updates/time out/process crash);
   end loop;

Besides monitoring processes/nodes, daemons will check periodically to record critical data in backup nodes for the corresponding primary nodes of all the participated nodes.

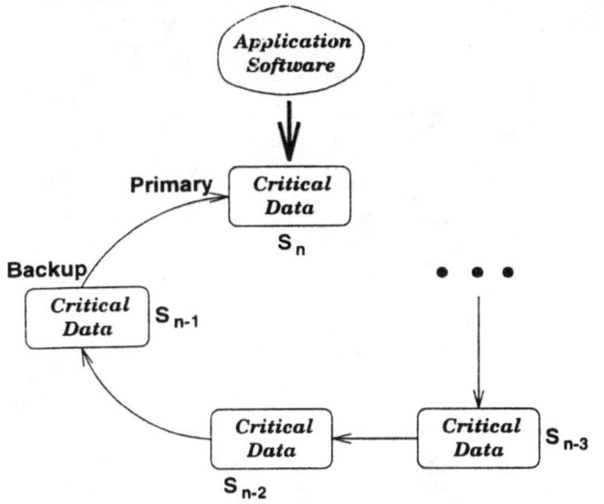

Figure 6: Our checkpoint scheme.

4.2 Check the critical data

Critical data will help to restore a failed process to the state before the last checkpoint is performed. Since nodes will be monitored in pairs, critical data will be available with different states in all the participated nodes (checkpoints will be performed at different moments at different nodes), as shown in Figure 6, except that the primary node for any process does not need to check the critical data on the backup node because the primary node holds the genuine critical data for the process.

Critical data for application software can be specified as follows: statically specify critical data file of processes for application software during the initialization before execution; dynamically register/remove critical data buffer through specification inside application software for store important variables. The data file will store the information about specified variables at the execution time. Checkpoints can be performed periodically or specified explicitly in the software.

4.3 Process recovery

In order for daemons to recover any failed process to the last recorded states, the system-level recovery routine will invoke user-defined routines implicitly to utilize the corresponding critical data for recovery. Since our approach for fault tolerance is at the application software level, no general recovery mechanism can be used to universally recover failed processes without specifications from users to properly utilize critical data. Unlike approaches for fault tolerance at the operating system level where execution image (e.g., core image) will be dumped periodically during the checkpoint instead of partial data, where much unnecessary space will be wasted for simplicity in recovery without alternative specifications, where user-define routines can provide not only efficient recovery but also much more flexibility in providing alternative decisions through using other fault-tolerant constructs, such as recovery blocks or N-version.

5 Software Development Framework

Our software development framework for real-time application software with fault tolerance is based on PROOF software development framework. It consists of development guidelines for different phases in the software development life cycle [3, 4]: object decomposition, object design, object verification, code implementation, and testing.

Development guidelines for object decomposition includes the following steps for the analysis of problem statements:

1. identify classes (input, output, and process) and objects,
2. determine class interface through public functions,
3. specify interactions (dependency, communications, and synchronization) among objects,
4. identify shared objects which heavily interact with other others,
5. identify periodic and aperiodic control threads for objects,
6. determine the priority among control threads,
7. build class hierarchy,
8. and identify critical data in terms of object data attributes.

Following results of the object decomposition (such as class/object diagrams and state transition diagrams), object design is done as follows:

1. determine bodies of objects for control threads,
2. design methods of each object,
3. and define recovery routines for using critical data in case of faults.

Object verification will check the completeness, consistency, and schedulability of the object design in terms of object algebra and Petri nets. After analysis and design, code implementation can be implemented in PROOF/L, a C++-alike language with real-time

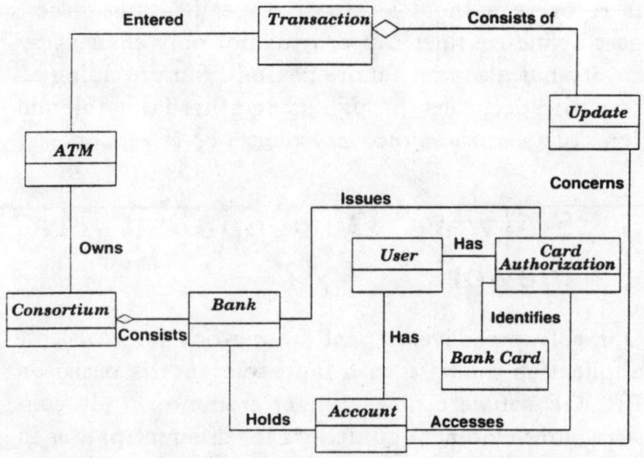

Figure 7: The object diagram for the simplified ATM system.

and fault-tolerance extensions, which is followed by testing of units and the entire application.

Currently, Constructs for satisfying real-time constraints, library routines of checkpoints and recovery, and monitoring daemons for fault tolerance are being implemented. CASE tools for assisting users to realize our software development framework are also under development.

6 An Example

In this section, we will illustrate our approach by applying it to a simplified Automated Teller Machine (ATM) system. The object diagram for static relationships among objects is shown in Figure 7. The state transition diagram for describing dynamic behavior of ATM is shown in Figure 8.

To demonstrate how to realized real-time and fault-tolerance features in the application software, pseudo code for the class *ATM* is shown as follows:

```
class ATM
        composition
# Define critical data for checkpoints
        critical
          id : int X
          location : string X
          owner : Bank X
          transaction: Transaction X
        ...
        end critical
        transaction_account : Account X
        transaction_card : Card X
        transaction_passwd : string
        end composition
```

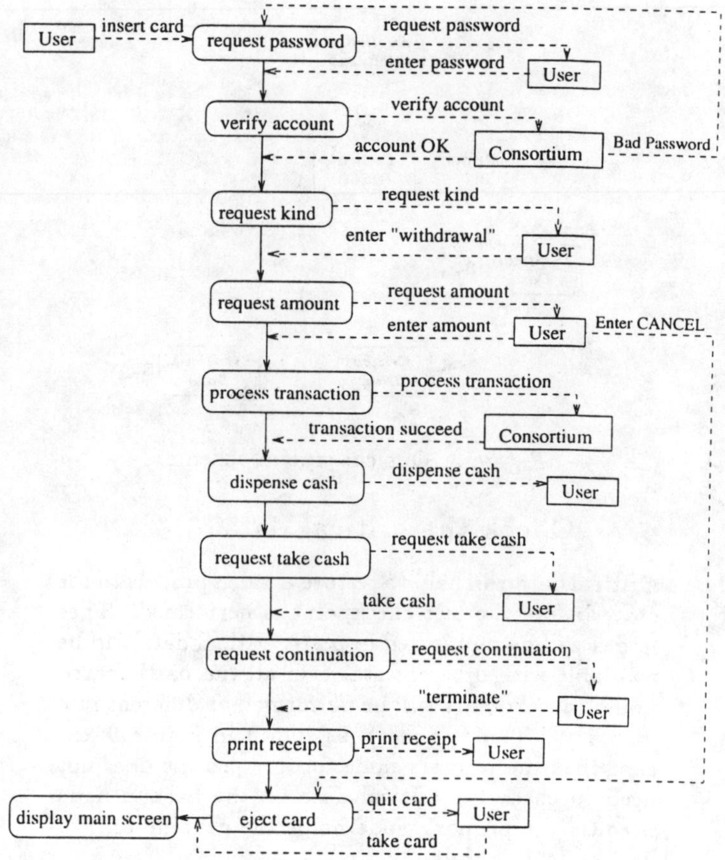

Figure 8: The state transition diagram for any instance of class *ATM*.

```
        method request_password( ->
                   verified : boolean)
          guard: none
# Time constraints
          duration-time: 5 sec
          version 1:
# Condition
          (transaction_passwd == NULL)
          expression
              transaction_card.eject();
          ...
        version 2:
          (transaction_passwd <> NULL)
          expression
              transaction_account.verify(
                    transaction_passwd, verified)
          ...
        ...
        end method

        ...

        method recover()
```

```
# Use critical data to restart process
# at the last checkpoints
                ...
        end method
end class
```

Variable *id*, *location*, *owner*, and *transaction* of the class *ATM* will be critical data to be used by the method *recovery* for the recovery purpose of any ATM. The method *request_password* will process within certain duration (10 seconds) to decide whether to permit the transaction based on the partial object state, the value of *transaction_passwd*.

During the initialization of DRCS for the ATM system, nodes and their circular assignments for monitoring will be determined. Initial configuration will specify primary nodes on which the ATM system will be run, critical data file, period for checkpoints, and communication ports. Built on the top of ADS system software, on-line expansion, maintenance, and testing can be realized through injection of different method implementations and process migrations between primary and backup nodes.

7 Discussion

In this paper, an object-oriented approach with real-time and fault-tolerant capabilities is presented for developing distributed real-time application software with fault tolerance. Flexible object adaptability with the support from *multi-verion* methods provides a predictable change environment for application software, simplifies object interface, and makes application software more reliable and easier to modify. To encapsulate time constraints within methods or bodies of objects will localize complexity of time constraints within objects themselves, which can make aggregation and inheritance hierarchy easy to build. Fault tolerance based on *primary node* and *process pair* approaches provides capabilities of monitoring, checkpoint, and recovery at the application software layer. It not only reduces the duplication of processes and computing resources, but also ensures the reliability of the entire real-time software systems.

Besides the implementation, some aspects of an approach can be improved. Current monitoring capability detects process failures and hang-ups, and processor failures. It should be extended with detection of errors in application software. How to define critical data which can be useful for process recovery and how to assist users to define recovery routines need to be explored further. We will define graphical notations for representing timing constraints which can provide concrete specifications to assist users to easily understand schemes of real-time application software. We will also investigate how verification can be more effective to check schedulability of object specifications with time constraints.

References

[1] P. T. Ward and S. J. Mellor, *Structured Development of Real-Time Systems*, Yourdon Press, Englewood Cliffs, NJ, 1985.

[2] H. Gomma, "Software Development of Real-Time Systems," *Comm. ACM*, Vol. 29, No. 7, July 1986, pp. 657-668.

[3] S. S. Yau, D.-H. Bae, "Object-Oriented and Function Software Design for Distributed Real-Time Systems," *Computer Communications*, Vol. 17, No. 10, October 1994, pp. 691-698.

[4] S. S. Yau, D.-H. Bae, and M. Chidambaram, "Object-Oriented Development of Architecture Transparent Software for Distributed Parallel Systems," *Computer Communications*, Vol. 16, No. 5, May 1993, pp. 317-327.

[5] T. Bihari, P. Gopinath, and K. Schwan, "Object-Oriented Design of Real-Time Software," *Proc. Real-Time Systems Symposium*, December 1989, pp. 194-201.

[6] K. Mori, "Autonomous Decentralized Systems: Concept, Data Field Architecture and Future Trends," *Proc. 1st Int'l Symposium on Autonomous Decentralized Systems (ISADS'93)*, 1993, pp. 28-34.

[7] S. S. Yau, et al., "An Object-Oriented Software Development Framework for Autonomous Decentralized Systems," *Proc. 2nd Int'l Symposium on Autonomous Decentralized Systems (ISADS'95)*, 1995, pp. 405-411.

[8] J. Gray and D. P. Siewiorek, "High-Availability Computer Systems," *IEEE Computer*, Vol. 24, No. 9, September 1991, pp. 39-48.

[9] Y. Huang and P. Jalote, "Effect of Fault Tolerance on Response Time – Analysis of the Primary Site Approach," *IEEE Trans. on Computers*, Vol. 41, No. 4, April 1992, pp. 420-428.

Panel Session 2

Real-Time Demands on CORBA and Other Established Object-Oriented Systems

---

Moderator:
Richard Mark Soley, Object Management Group, Inc.

Panelists:
Dock Allen, Computing Devices International
S.E. Mitchell, The University of York, UK
Bhavani Thuraisingham, MITRE Corporation
Douglas Wells, Open Systems Foundation Research Institute

Position Paper: Real-Time is a Critical Feature for CORBA, Just Like Everything Else

Richard Mark Soley, Ph.D.
Object Management Group, Inc.
492 Old Connecticut Path
Framingham, MA 01701
U.S.A.
soley@omg.org

The Roots of Middleware

What is middleware? As the technical leader of a consortium dedicated to promoting middleware, the Object Management Group[1], I am asked this question on a daily basis. My usual answer is, "Middleware is the software no-one will pay for." After all, ubiquity is the prerequisite for successful middleware; else the portability and interoperability that middleware users seek is simply not to be found.

A more important (and technically accurate) characterization of middleware might be, "Middleware is a collection of widely portable API's supporting interoperability between applications." This isn't such a new idea, applications have been interoperating via closed static protocols (i.e., *file formats*) for decades. The success of the networking movement, outside electronic mail, has primarily been in the area of extending this interoperability metaphor (file-based interoperability) across network boundaries.

But as one explores more deeply the middleware phenomenon, one finds an even more deeply rooted need that portability that middleware attempts to exploit. Middleware packages, whether based on remote procedure call (RPC), pipes, messaging or some other metaphor, attempt to support single-thread model of computing in a multiple-thread world. In essence, they instill a programmer's illusion that he can continue to operate the same way in a multiple-processor world as he did in a single-processor world. Perhaps even the same languages, compilers and applications can gain efficient new form in that world.

Orthogonality and Transparency

In fact, even the earliest RPC schemes emphasized a model of computing in which networked computers were simply linear extensions of the existing single-processor computing model. The Open Software Foundation's Distributed Computing Environment (DCE)[2] and other RPC mechanisms are essentially presented to the programmer as precompilers which allow automatic insertion of network code across specific procedure call boundaries. Such calls are in fact even synchronous (as they would be in the single-processor, single-threaded world) across the naturally asynchronous network boundary.

This is not intended as a critique of such an approach. In fact, this

0-8186-7570-5/96 $5.00 © 1996 IEEE

"network extension" model is a particularly powerful one for programmers who wish to ignore the complexities of network programming, programming in the face of communication failure, etc.

Later technologies such as the Object Management Group's Common Object Request Broker Architecture (CORBA)[3] have similarly concentrated on extending a programming model across networks. In the case of CORBA, that model is an object messaging scheme, in which messages represent synchronous or asynchronous requests on resources ("objects" or "applications"), with the usual object-orientation concepts in place. That is, method selection (in CORBA parlance, selection of the "implementation") is disambiguated at run-time; interfaces are described in a multiple-inheritance, object-oriented declaration language (the Interface Definition Language, or IDL[CORBA]); and object (application) state is encapsulated by the interface mechanism.

Though CORBA and DCE represent somewhat different computing models, both are attempting to provide the illusion of extending a simple model across network boundaries. In fact, CORBA is a part of a wider Object Management Architecture (OMA)[1] which embraces further network model extension. OMA already includes facilities ranging from event notification and naming to concurrency control, transaction management and query service. All seek to confer the blessings of a complex systems model on new and existing programs without the overhead of understanding those models in detail.

The Object Transaction CORBAservice (OTS)[4] is an excellent case in point. OTS allows users of CORBA-based systems to include series of CORBA requests within a transaction, which can later be committed or rolled back. In fact, such transactions may be nested; and not only may many different computers take part in any given transaction, but many different *transaction monitors* may share in a transaction. This is a very complex task indeed, and OTS masks all of the details of that task's implementation.

A particularly good model of such "boundary extensions" is presented in the new ISO Reference Model for Open Distributed Processing (RM-ODP)[5]. Growing primarily out of the ANSA distributed computing model,[6] RM-ODP among other things seeks to outline the *transparencies* that a distributed computing model presents to programmers and users. This powerful abstract framework for representing interoperability support is destined to supplant the OSI seven-layer model, which had the same impact on the lower echelons of the networking world.

Let's revisit that claim above, that OTS masks all of the details of transaction management, allowing programmers to concentrate on the specific application task. Could this possibly be true? After all, someone must implement rollback handlers, someone with application knowledge. That's the long answer. The short answer is: no.

The Illusion Breaks Down

Unfortunately, the illusions of transparencies that middleware systems attempt to invoke always break down somewhere. Sure you might have default handlers for various transaction failures and communications problems. But unfortunately, transaction management is never totally orthogonal to application behavior.

A superb example of this situation is the support of real-time behaviors. It would be delightful if a middleware platform like CORBA could hide and automate all of the support mechanisms for presentation of real-time behavior: fault tolerance, priority-managed queues, watchdog timers, automatic program and data replication, voting, etc. And in fact, various groups within the OMG process are working on all of these areas.

The main point of this position paper is, however, that *they will not completely succeed.* There will always be pieces of the solution that are not totally orthogonal, that programmers will have to understand. And in the real-time area, this sea of complexity will be an ocean.

Where Do We Go From Here?

The key must be to minimize such effects, and encapsulate wherever possible the extended demands on programmers caused by not-totally-transparent infrastructure features like those required by real-time systems. What is most important is that we do not shrink from including such features simply because of the potential complexity. CORBA (and other middleware systems, for that matter) represent a great hope to simplify software integration and maintenance, the two highest-cost and lowest-interest parts of the software lifecycle. Therefore research should continue, but with an eye to the truth that total orthogonality and transparency are myths.

References

[1] Richard M. Soley (ed.), *Object Management Architecture Guide, Third Edition,* John Wiley & Sons, New York, 1995.

[2] See multiple articles at http://www.osf.org/comm/lit/lit-dce.html

[3] *The Common Object Request Broker: Architecture and Specification*, Object Management Group, Inc., Framingham, MA, U.S.A., 1995.

[4] *CORBAservices: The Common Object Services,* Object Management Group, Inc., Framingham, MA, U.S.A., 1995.

[5] ISO/IEC standard 10746 (RM-ODP, see ftp://ftp.omg.org/pub/docs/tc/94-12-31.ps)

[6] See multiple articles at http://www.ansa.co.uk/ANSA/index.html

CORBA Technology for Cross-Domain Interoperability in Embedded Military Systems, and Issues in Its Use

Ms. Dock Allen
Computing Devices International,
Chair of the OMG Realtime SIG

Abstract

Object Technology, particularly the Common Object Request Broker (CORBA), offers several benefits for embedded military systems including support for application diversity, interface management, technology insertion, system evolution, distribution, reconfiguration, and open standards, as well as mainstream commercial availability. These systems place requirements on the technology including support for embedding, hard and soft realtime, and fault tolerance, as well as a high level of assurance of correct operation. To serve this market, the existing technology should be extended to allow for and encourage increased configurability, predictable timing and resource utilization, performance characterization, the ability to configure static implementations, application management of time and system resources, asynchronous interactions and/or time-outs, full prioritization, implementations free of priority inversion, fault tolerance, and application fault and error handling.

1. Introduction

CORBA [1] offers a potentially important technology for bringing the benefits of Object Technology into embedded military systems, especially for integrating diverse applications and subsystems. This paper briefly explores the benefits and issues this technology. Although the emphasis is on airborne systems, the discussion is applicable to ground-based, shipboard, and spaceborne systems. This paper also presents some of the likely impacts of these systems on the requirements for CORBA implementations and the CORBA Application Program Interface (API).

2. Background

Two developmental characteristics of avionics platforms make Object Technology particularly attractive: the diversity of applications, and the distributed nature of system development both geographically and over time. On the technical side, the distributed nature of the architectures and the need for reconfiguration, even in the context of Integrated Core Processing, also make these systems a good candidate for Object Technology. Finally, the trends towards using commercial components and open systems standards make CORBA particularly attractive. However, the benefits of commercial open technology are accompanied by concerns about its suitablility for these systems.

Avionics platforms are characterized by having multiple subsystems. The diversity among the applications is wide, including safety-critical applications, hard realtime control applications, high performance / high data volume signal processing applications, general purpose management and planning applications, graphical applications for pilot / crew interactions, and low level onboard diagnostic applications. **Support for application diversity** is a key requirement.

A list of example applications for a typical avionics platform [2] is:

- sensor signal processing for several different sensors
- multi-sensor fusion
- vehicle management
- pilot-vehicle interface
- mission management
- stores management
- navigation
- communications

These applications have a diverse set of requirements. For example, signal processing requires high performance processing and high data rates, but tends to be done with a small kernel application; the operating system support required is minimal. Mission management requires moderate processing and data rates, but involves large programs; the abstraction provided by a high order language and a feature-rich operating system is important. Weapons control is safety-critical, and requires an extremely high level of assurance of correct operation. Vehicle management involves "hard" real time

0-8186-7570-5/96 $5.00 © 1996 IEEE

applications, which have stringent requirements for predictable timing behavior.

The requirements of these subsystems in terms of performance, dependability, security, predictability, and support for high levels of abstraction vary widely, as do the ways in which these subsystems stress the available technology. A system architecture that can present different support environments to different applications has an increased chance of meeting the full range of system requirements.

Avionics systems usually have more that one mode of operation, and system **reconfiguration** is needed during made changes. In a small system, this can be accomplished by a complete reload of some software. However, in an integrated system, the ability to selectively reconfigure portions of the currently operating components is useful. This minimizes time required to change modes, and can allow onboard software define a new configuration dynamically, instead of being limited to a fixed set of possible configurations. In addition to mode changes, most systems have fault tolerance requirements, and it is common for systems to use graceful degradation, operating in a degraded mode, when there are onboard failures. For these reasons, support for reconfiguration is a key requirement.

The subsystems are developed by a team of contractors, dispersed over a wide geographical area. Rigorous **interface management** is required for success. This is typically accomplished by Interface Control Documents, which are developed early in the design and maintained, with varying degrees of automation, over the life of the system. In addition, the platform life cycles are measured in decades. Several major block upgrades occurring during this time, so that the system development occurs over a wide span of time. This makes the careful documentation and management of interfaces ever more critical.

As mentioned about, platform life cycles cover decades. System upgrades include both technology insertion and the addition of new functions. The ability to perform near-**seamless technology insertion** over several generations of technology, and to support graceful **system evolution** through the incorporation of new and modified system functions, are key requirements.

The resulting avionics platforms require processing performance beyond the capability of a single processor. For example, it is estimated that the F-22 required 400 Million Instructions Per Second (MIPS) of data processing and 8000 Million Operations Per Second (MOPS) of signal processing [3]. Estimates for a Joint Advanced Strike Technology (JAST)-class aircraft are 1000 MIPS of data processing and 15,500 MOPS of signal processing [3]. For this reason, the processing architectures are characterized by **distribution**. The approaches to distribution have ranged from federated architectures to varying degrees and types of integrated processing.

In the past, avionics platforms were designed as federated (very loosely coupled) collections of subsystems. Interfaces between subsystems were documents and managed using Interface Control Documents. The subsystems were integrated using point-to-point interfaces across military-specific interconnects, such as MIL-STD-1553B.

Often different processing solutions were used for different portions of the system, in order to have optimal support for each subsystem. For very high performance applications, such as radar signal processing, the processors were special purpose implementations. For general purpose applications, such as mission processing, the processors were military-standard computers, such as the Navy standard AYK-14. The software was written in a mixture of military-specific languages, such as CMS-2 for the Navy, and assembler. Operating systems were either custom, or military-specific.

In the 1980's and early 1990's, the military moved towards commercially based processors, such as the MIPS R4000, and commercially based open interconnects, such as VME and Futurebus+, for system upgrades. In the mid to late 1990's, military systems are seriously considering open standards and commercial components (processors, operating systems, interconnects, languages, and even application components) at all levels of integration. **Open standards** and **mainstream commercial availability** are key requirements.

The system must maximize opportunities for inclusion of commercial technology at all levels, including processors, interconnects, operating systems, application support services, and even application components. Where current commercial technology is not suitable for the system, future upgrades should be able to incorporate new commercial technology as it becomes available.

The use of open standards allows customers to select among several implementations, seeking one that is closest to the system requirements. Open standards also provide the ability to swap in a newer, more capable implementation, at a later time. The obvious benefits of commercial technology are the ability to leverage commercial products, reducing time to deployment, and leveraging a large commercial base of users to reduce cost and increase reliability.

In order for open standards and commercial technology to be usable for these systems, the standards and products must support embedding (with the associated constraints on size, weight, and power utilization), hard and soft realtime, a high level of assurance of correct operation, and fault tolerance. Although all of these requirements are found in commercial applications, they are considered "niche" portions of the market by most vendors.

In the 1990's and beyond, another important thrust in avionics systems is towards an Integrated Core Processing system, that executes most of the on-board applications. The F-22 platform is an example of this type of system [3]; the Joint Avionics Tactical Fighter (JAST) will also likely use this approach [2].

In these systems, although most applications are moved into the core processor set, some applications with very stringent requirements, such as safety critical weapons management and vehicle management, are executed in their own processors in order to guarantee that their requirements are met. Thus, some of the original federated characteristics of the system remain.

An integrated system eliminates the problems of having dozens of different on-board processing systems, and can realize significant savings in size, weight, and power. One of the issues raised by Integrated Core Processing are how to support application diversity, providing the various applications / subsystems with near-optimal environments, while maintaining an integrated processing system.

3. Why is Object Technology Attractive and What Special Benefits does CORBA offer?

Support for Application Diversity: One approach to application diversity is a heterogeneous architecture The level of heterogeneity is an open issue. Some new systems use a common operating system and language, but permit a few types of processors for different application needs. The F-22 is such a system. One approach that is being considered for future systems would extend this to permit multiple languages and operating systems, with middleware providing interoperability between different domains.

Treating applications as objects, or collections of objects, is one way to "hide" the internal implementation of an application from other applications. As long as there is a standardized approach to application interactions, the underlying support systems, and even the programming paradigm used in the applications, can be different.

When used for system integration, Object Technology has the potential to operate seamlessly across different languages, operating systems, processors, and even programming paradigms. This allows different applications, or groups of applications, to have execution environments and processing resources that are near-optimal for their needs.

CORBA already supports multiple languages, platforms, and operating systems. CORBA language bindings include C++, C, and Ada, with a JAVA binding in process. Commercial implementations already exist for a wide range of platforms, from PC's and embedded processors to workstations and mainframes. These implementations are hosted on a variety of operating systems. A significant additional advantage of CORBA is that Version 2 of the standard [1] provides interoperability, even between implementations from different vendors.

With this range of interoperable implementations, a system integrator can configure a system that provides near-optimal processors, languages, and operating systems for a variety of applications.

Interface Management: Without arguing the inherent benefits of Object Technology, one clear benefit of using Object Technology for application interfaces is a standardized approach to the design of the interfaces. This facilitates communication among contractors. It also permits development of common tools for interface management and testing that can be used throughout the system.

The additional benefit of CORBA is that language for interface definition (the Interface Design Language, IDL) is defined in the standard, and the interface definitions are compiled. This provides rigor to the interface definition process, as well as more opportunities for automation.

Seamless technology insertion: CORBA is middleware, and was purposely defined to be independent of any particular operating system or processing system. By design, CORBA allows a new operating system, processor, or even processing architecture to be swapped into the system without impacting the application interface.

System Evolution: An object-based approach to application interfaces allows the internal details of each application to be "hidden" behind the interface. This allows one application to be modified, and even extended with minimal impact on other applications. This also permits the addition of new applications, which can utilize the services of the existing components through the object interfaces of those existing components.

Distribution: Because objects are logically "loosely coupled" from one another, they offer an ideal component for distribution. CORBA provides excellent support for transparent object distribution. A client object only needs to know the "name" of the implementation service that it wishes to access, and CORBA does the rest. In a realtime system, the actual location of some objects is be critical due to timing; even in this case, the ability to experiment with the system configuration during development is useful in arriving at a workable configuration.

Reconfiguration: Because the actual locations of objects are "transparent" in CORBA, reconfiguration of the system can be supported transparently. In fact,

CORBA is intended to support the dynamic addition, deletion, and relocation of object. Furthermore, the object paradigm is suitable for replication of objects, for fault tolerance. Degraded modes of operation can occur through deletion of replicates, which would be transparent to other objects. Even where timing-critical operations cannot take full advantage of these capabilities, the capabilities are useful in "tuning" the system configuration to arrive at an optional solution.

Open Standard: CORBA represents a set of fully open standards [1], owned and published by the Object Management Group (OMG). Any company or organization can participate in the standards process, and no single company dominates or controls the contents of the standards.

Mainstream Commercial Availability: The OMG standardization process guarantees that there will be a commercial implementation for any OMG standard within one year of its adoption. Commercial implementations are usually available much sooner than that. Commercial acceptance of CORBA is high [4] [5], and the commercial availability of CORBA implementations is excellent, when compared to other open software standards. Implementations are already available for some popular embedded processors, using commercial realtime operating systems.

In addition to its favorable commercial availability, CORBA is likely to have a long life due to the number of groups currently involved in the OMG (600+ and counting), and the growing commercial popularity of the underlying principle of Object Technology.

For all these reasons, Object Technology is attractive as a general approach to system-level integration, and CORBA a particularly good fit.

4. Special Requirements of Embedded Avionics

Embedded military systems often stress the available technology in the following ways:

- For avionics systems, size, weight, and power are still critical factors. Although advances in commercial memory and processors have narrowed the gap between what is needed to solve the problem and what can fit into the platform, some general pressure to conserve memory and processing power remains. This mainly places stress on the software, particularly the onboard software that supports the execution of applications.
- For some applications, such as signal processing, the inherent data volume still stresses the memory capacity of currently available processors, leading to a constant pressure for applications and the onboard support software to conserve memory.
- Many avionics applications, including signal processing, require very high performance. This leads to a need for good to excellent performance in supporting software, including operating systems, communications systems, and middleware. These requirements are significantly beyond what is needed by most commercial systems.
- Control applications, such as control of flight surfaces and engines, are hard realtime applications, meaning that hard deadlines in time must be met for the safety of the aircraft. The support software for these applications must be able to deliver predictable performance for each service offered, including both overhead and latency. The need for predictability can impact both the design of this software, and the level of documentation that must be provided to the application designers.
- Since many subsystems are "mission critical," they are designed to tolerate and recover from faults. For some of these subsystems, it is acceptable to operate in a degraded mode; some functions or levels of redundancy can be eliminated in order to keep operating when parts of the system fail. Other applications, such as applications that control flight and weapons release cannot be permitted to produce an incorrect result. The application software and supporting software for these subsystems must meet the highest standards of reliability and must support fault-free operation.

These requirements can be summarized as a need to support for embedding (with the associated constraints on size, weight, and power utilization), hard and soft realtime (requiring predictability and good performance), a high level of assurance of correct operation, and fault tolerance.

5. What does this Mean for CORBA?

This section lists characteristics of a CORBA implementation that would support embedded military systems. This is an initial list, that will require additional prototyping and analysis to determine the relative importance of each characteristic, and to determine how to fit these requirements into the OMG specifications with minimal impact to the commercial base of CORBA-related products.

Suitable underlying support: Note that the supporting Operating System and Communication Services must meet the requirements levied on the CORBA implementation. This paper does not address the issues with respect to the underlying operating systems and communication systems.

Configurability: The technology should support vendor implementations that can be configurable into a small, and in some cases a static implementation. The implementation should provide just those services that a particular application needs. This supports embedding the software into a minimal amount of memory, increases predictability by removing sources of dynamic variations, and is helpful in cases where a formal proof of correctness is required.

Predictability: For realtime applications, system designers must be able to verify, and in some cases to actually prove, that the application deadlines can be met. In order to perform this analysis, the time and resources needed for application transactions, and for functions provided by the support software, must be known in advance. In addition, all conditions that could lead to unbounded blocking of a client object must be known. These are primarily issues of documentation.

One common technique to eliminate non-determinism is to use a static configuration of the system. This could impact the design approach used for the CORBA implementation.

Performance characterization: The technology must have high performance, both in terms of actual system overhead and in terms of latency. This implies that the implementation itself, any underlying Operating System, and the communication protocols must offer suitable performance. It is usually acceptable to eliminate capabilities in order to gain performance and predictability, although the limitations must be documented.

Some standardized set of performance measures would allow comparisons between implementations, and would permit customers to select implementations based on which performance characteristics were most important to their application.

Assurance of Correctness: Two common approaches to this requirement are formal verification (proving in advance that the system cannot fail) and redundancy (running two or more different versions of the software and voting among them). The approach used usually depends on the complexity of the software (the more software there is, the less likely that a formal proof can be performed) and the availability of the resources needed to run more than one version of the software.

When a formal proof approach is used, the implementation would be "striped down" to minimize the amount of software to be analyzed.

Fault Tolerance: Briefly, most approaches to fault tolerance use some form of redundancy or replication, sometimes of identical components and sometimes of diverse components. This is coupled with other techniques to detect and correct fault conditions (reasonableness checks, error detecting and correcting codes, fault tolerant communication protocols and the like). This is a huge topic, and only the likely impacts on a technology like CORBA will be addressed.

One of the open architectural issues is how much fault management is performed by the application software, by some form of system-level software that rides above the operating system and middleware, and automatically by the operating system and middleware. In current systems, a combination of all three approaches is used. We can assume that this will continue to be the case, and the standards should permit all three approaches.

Ideally, the CORBA implementations themselves will provide tolerance for some faults, such as lost messages. In many cases, this could be provided by the underlying operating system and communication protocols. Even in these cases, the implementation must be prepared to provide fault statistics to a higher level fault manager so that trend analysis can be performed. These interfaces will impact the CORBA API.

Higher levels of software, either applications or some form of application support services, will need to be able to manage faults. This implies that the CORBA implementation needs to able to pass all fault and error notifications upwards. The architecture and definition of these interfaces will impact the standard.

Realtime Issues: The application, or in some cases a set of application support services, must be able to manage time and system resources. Real time, often some form of mission time that is more accurate than calendar time, will be used to communicate deadlines and time stamps. Time-outs may be in microseconds.

The application will need to be able to find out the status of system resources, such as available dynamic memory, so as avoid having these resources are not depleted. In some cases, an application will "pre-reserve" resources to avoid resource shortages during operation.

It is often unacceptable for an application to be suspended while waiting for a transaction or service to complete. Either some form of asynchronous request or time-outs is required for all services, possibly both.

Prioritization Issues: Substantial work has been done in analyzing the application of Ada to realtime systems, especially using Rate Monotonic Analysis [6]. Much of this work is applicable to CORBA as well. Significant findings are:

- All queuing and dispatching within the system should be based on priority;
- The system should be designed to eliminate priority-inversion;
- Thread-safe implementations should be available.

6. Conclusion

Object Technology, particularly CORBA, offers significant benefits to embedded military applications. In order to serve these applications, the technology needs to address the requirements of realtime, embedded, high assurance and fault tolerant systems. The specific implications for the CORBA standard need further analysis and investigation.

References

[1] *The Common Object Request Broker: Architecture and Specification*, Revision 2, July 1995, OMG TC Document, Object Management Group, 492 Old Connecticut Path, Framingham, MA 01701

[2] *Joint Advanced Strike Technology Program, Avionics Architecture Definition*, Version 1.0, LtCol Chuck Pinney, JAST Avionics Lead, available via FTP from ftp.jast.mil.

[3] Juris Ozols, "Avionics Processing System Trends and Implications", Digital Avionics Systems Conference (DASC) Proceedings, 11/95.

[4] Kurt C. Wallnau and John Rice, "ORBS In the Midst: Studying a New Species of CASE Integration Mechanism," Proceedings of CASE '95.

[5] George Lawton, "Object Efforts Gravitate", Software Magazine, March 1996

[6] Lui Sha and John B. Goodenough, "Read-Time Scheduling Theory and Ada", April 1989, Technical Report CMU/SEI-89-TR-14, ESD-TR-89-22.

Real-Time Programming in the Concurrent Object-Oriented Language TAO

S. E. Mitchell and A. J. Wellings
The University of York, UK

Abstract

This paper introduces the concurrent object-oriented language TAO and then considers how real-time support can be added. The paper describes how concurrent objects can be mapped to sporadic and periodic tasks through inheritance within the concurrent object-oriented language TAO.

1. Introduction

As part of our research we have developed a model for concurrent object-oriented programming called TAO (Threaded Active Objects). The primary motivation for this work arose from the observation that the current COOL's possessed weak synchronisation schemes and also suffered from the inheritance anomaly [5]. For this work, we noted that contemporary languages could not express a wide range of application problems because they failed to comply with existing knowledge on the necessary features of a synchronisation mechanism [4]. We thus wanted to develop an expressive concurrent object-oriented model.

The TAO model has not been designed to consider real-time issues since the the primary motivation was to develop an efficient and expressive concurrent object-oriented language. However, just as language support for concurrency greatly facilitates its use in a concurrent environment so does language support for real-time issues facilitates its use in a real-time environment. Consequently, after a brief introduction to the TAO model, this paper explains how the real-time features which are currently lacking in the TAO model could be added.

2. The TAO Model

Concurrency in TAO arises from the instantiation of active classes each with one or more bodies (termed threads). When an "active" class is created the bodies commence execution after the constructor. Inter-object concurrency is controlled by deontic guards attached to methods. Intra-object concurrency is controlled by the use of "protected" instance variables (which are essentially variables encapsulated so access is only via guarded methods). This paper extends the TAO model further by assigning a priority to each thread (to allow for real-time scheduling policies based on task priority).

Details of an implementation of the TAO model as an extension to C++ are contained in [6].

2.1. Synchronisation in TAO

TAO [6] uses a synchronisation scheme involving a novel combination of boolean guards and deontic logic [1]. Using the synchronisation primitives available it is possible to fully express any combination of Bloom's criteria, eliminate (or alleviate) problems associated with the inheritance anomaly, use inheritance to exploit libraries of concurrent synchronised objects etc.

Deontic Logic allows synchronisation constraints to be specified in terms of the number of requests (deontic operator `req`), activations (`act`), and terminations (`fin`) of methods. TAO extensions allow for synchronisation based on local state and instance variables. In TAO, deontic guards are attached to the method they synchronise. As an example, a bounded buffer holding 10 elements could be implemented as follows:

```
class BB {
  int size;
public:
  Item get() [ guard(size>0) ];
  void put(Item i) [ guard(size<10) ];
};
```

3. Supporting Real-Time Programming in TAO

Clearly it is essential to introduce the notion of time to TAO. A real-time language requires the ability to declare both cyclic and sporadic tasks and then the tasks to be dispatched such that deadlines are met. We assume that the scheduling of tasks is performed by some external object

0-8186-7570-5/96 $5.00 © 1996 IEEE

(such as a real-time kernel) and is thus not considered further. We also assume that schedulability analysis can be performed on the task set [2].

In TAO, the unit of concurrency is a thread in an active object. The system proposed in the subsequent sections shows how generic periodic and sporadic tasks can be defined using a TAO active class and then specialised through inheritance to implement the required functionality. This occurs through the use of deferred methods (in C++ these are called pure virtual methods). The TAO model makes extensive use of such methods to build concurrent class libraries – an example for a periodic class is given in [6]. We now consider how such a class can be made to more closely reflect real-time issues and also how sporadic classes can be created.

3.1. Periodic Tasks

The bodies in each active object are periodic tasks specialised from a generic periodic base class. We use the following model of a periodic task:

```
loop {
  wait_for_next_cycle;
  // task code
}
```

The wait_for_next_cycle call delays the execution of the task loop till the next cycle commences, after which the task code is executed and the process repeats. The start point of the cycle is determined by an external scheduler.

In Real-Time TAO, each periodic task inherits from a `Periodic` class from which the periodic behaviour is derived. The `Periodic` class inherits from the `Scheduler` class which provides the `wait_for_next_cycle` command. In the current C++ implementation of TAO this inheritance is `virtual` so each class hierarchy will only ever have a single copy of class `Scheduler`. The inheritance hierarchy for two periodic tasks if illustrated in Fig. 1 and the following code fragment shows the definition of the base class.

```
class Periodic : public virtual
                          Scheduler {
  thread worker(int period) {
    Scheduler::register(PERIODIC,
                        period);
    do {
      Scheduler::wait_for_next_cycle();
      action();
    } while(1);
  };
public:
```

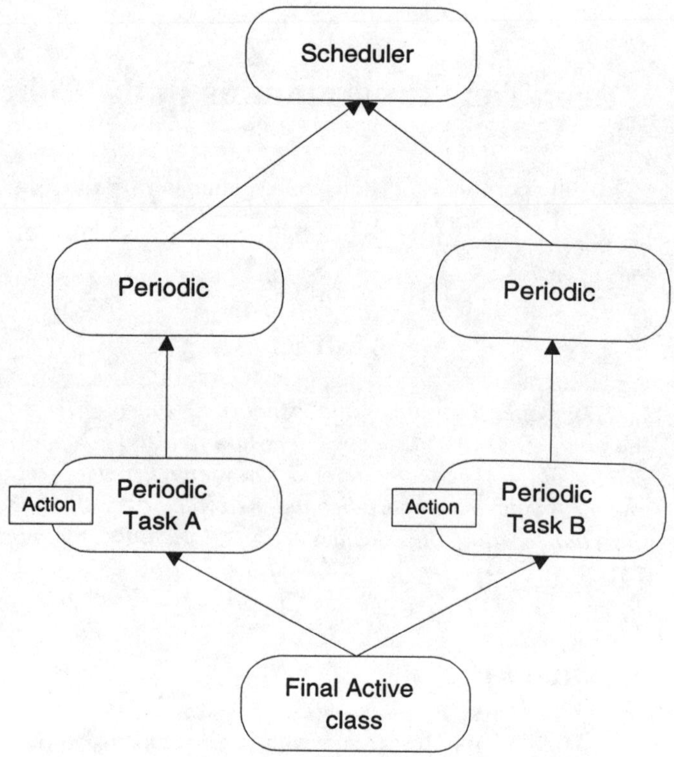

Figure 1. Inheritance hierarchy for two periodic tasks within an active object.

```
  virtual void action() = 0;
  void Periodic(int period, ...) :
                        worker(period);
};
```

The first action of the thread is to register the periodic task with the scheduler. This call sets up the task's period, priority and WCET and is used by the scheduler when deciding which task to release. Note also the use of constructors to establish the period etc (for clarity, only the period is shown in the above fragment).

The `Periodic` class is then specialised as below:

```
class Test : public Periodic {
public:
  void action() {
    // Perform work
  };
  void Test : Periodic(2);
};
```

3.2. Sporadic Tasks

To implement a sporadic task we use a complimentary approach to that taken for periodic tasks. The task thread initially makes an introspective call to `wait` whose deontic guard only permits execution in alternation with the `start` method. Thus the thread is blocked until the `start` method is called at which point the deferred method `action` is invoked. As with periodic tasks, the minimum inter-arrival time of the sporadic is registered with the scheduler so that it can make the relevant release decisions. In the following code fragment, the deontic guards on `wait` and `start` enforce alternating execution.

```
class Sporadic :
     public virtual Scheduler {
private:
  thread worker(int arrival_time) {
    Scheduler::register(SPORADIC,
                        arrival_time);
    do {
      wait(this);
      action();
    } while(1);
  };
  void wait() [act(wait)<act(start)] {
    Scheduler::release(this);
  };
public:
  void start() [act(wait)>=act(start)];
  virtual void action() = 0;

  void Sporadic(int iat, ...) :
                     worker(iat);
};
```

3.3. The Role of the Scheduler

Since there is only a single instance of the `Scheduler` in each class hierarchy is is possible to easily change the scheduling policy used – for example from fixed priority scheduling to earliest deadline first. The above examples use a scheduler which implements four functions:-

- `register` – record details of a periodic/sporadic task.
- `wait_for_next_cycle` – returns at the start of the next period.
- `wait_release` – delay a particular until released by a call to `start` in a sporadic object. The sporadic task is identified by the object's address.
- `release` – release a specified sporadic task.

The scheduler is able to ensure that sporadic tasks do not run more often than has been specified in the call to `register`.

3.4. Accessing Shared Data

In TAO, shared data is protected from concurrent access by encapsulating it in a class with data access only through methods with deontic guards. We propose that such "protected objects" will have an immediate ceiling priority [3] which is as high as any caller. This procedure will bound the blocking on such an object.

4. Conclusion

This paper has shown how the concurrent object-oriented language TAO could be extended to support real-time programming. This work is still very much in its infancy but we believe shows promise as a clear concurrent real-time object-oriented language.

References

[1] C. Atkinson. *Object–Oriented Reuse, Concurrency and Distribution: An Ada based approach.* Addison-Wesley, 1991.

[2] N. Audsley, A. Burns, M. Richardson, K. Tindell, and A. Wellings. Applying New Scheduling Theory to Static Priority Pre-emptive Scheduling. *Software Engineering Journal*, 8(5):284–292, Sept. 1993.

[3] T. Baker. Stack-Based Scheduling of Realtime Processes. *Real-Time Systems*, 3(1), 1991.

[4] T. Bloom. Evaluating Synchronisation Mechanisms. In *Proceedings of the Seventh Symposium on OS Principles*, pages 24–32, Dec. 1979.

[5] S. Matsuoka and A. Yonezawa. Analysis of Inheritance Anomaly in Object-Oriented Concurrent Programming Languages. In G. Agha, A. Yonezawa, and P. Wegner, editors, *Research Directions in Concurrent Object-Oriented Programming*, pages 107–150. MIT Press, 2nd edition, 1993.

[6] S. Mitchell. *TAO – A Model for the Integration of Concurrency and Synchronisation in Object-Oriented Programming.* PhD thesis, University of York, UK, Mar. 1995. FTP://ftp.cs.york.ac.uk/reports/YCST-95-009.ps.Z.

ON REAL-TIME EXTENSIONS TO OBJECT REQUEST BROKERS: A PANEL POSITION PAPER

Bhavani Thuraisingham*
Peter Krupp*
Victor Wolfe**

***The MITRE Corporation, Bedford, MA**
**** The University of Rhode Island, Kingston, RI**

ABSTRACT

This position paper describes some of the issues that need to be investigated in order to develop real-time data processing extensions to CORBA. In particular, issues on extensions to the object model, ORB, and IDL are discussed. Finally a discussion of some real-time services and applications hosted on real-time ORBs will be provided.

1. INTRODUCTION

With the advent of increasingly complex applications that depend on timely execution (such as automated manufacturing, vehicle control, medical patient monitoring, and multi-media), many distributed computer systems must support components capable of real-time data processing. In a real-time system, the data as well as the transactions need to meet temporal constraints. For example, the data may have temporal consistency constraints stating how "old" it can be and still be considered valid. Transaction execution may have to meet start timing and deadline constraints. Recently, the object-oriented approach to developing real-time systems has gained a lot of attention. In this approach, the data along with a set of operations on the data (aka, methods) are regarded as objects. The temporal constraints are associated with the attributes and the methods of the objects.

While the real-time systems community has been conducting research and development activities on the object-oriented approach to designing and developing real-time systems and software [THUR94,PRIC94,WOLF94], the Object Management Group (OMG) has been active in specifying standards for the interoperability of heterogeneous computing systems. Notable among the efforts of OMG is the Common Object Broker Architecture (CORBA). As stated in [CORB92], CORBA is structured to allow integration of a wide variety of object systems. The components of CORBA provide mechanisms by which clients transparently make requests of objects and receive responses so that interoperability between applications on different machines in heterogeneous distributed environments is enabled.

Integration of CORBA technology and real-time systems technology is essentially an unexplored area. The interoperability of real-time systems via CORBA requires real-time extensions to CORBA. This position paper describes some of the issues that need to be investigated in order to develop real-time data processing extensions to CORBA. An overview of CORBA is provided in Section 2. Characteristics of real-time systems are discussed in Section 3. Real-time data processing extensions for CORBA are discussed in Section 4. A discussion of some real-time services to be hosted on real-time ORBs will be given in section 5. The paper is summarized in Section 6. Some of our previous papers on real-time extensions to CORBA include [KRUP94], [THUR95], and [WOLF95].

2. OVERVIEW OF CORBA

Three major components of CORBA are the object model, the Object Request Broker (ORB) and object adapters, and the Interface Definition Language (CORBA-IDL). We discuss the essential points of each component.

The object model describes object semantics and object implementation. Object semantics describe the semantics of an object, type, requests, object creation and destruction, interfaces, operations, and attributes. Object implementation describes the execution model and the construction model.

The ORB essentially enables communication between a client and a server object. As stated in [CORB92], a client is an entity that invokes an operation on the object, and the object implementation is the code and data that actually implements the object. The ORB, together with object adapters, provides all the mechanisms to find the object implementation for the request, to prepare the object implementation to receive the request, to communicate the data making

0-8186-7570-5/96 $5.00 © 1996 IEEE

up the request, to activate and deactivate objects and their implementations, and to generate and interpret object references.

IDL is the language used to describe the interfaces that are called by client objects and provided by object implementations. IDL is a declarative language. Note that clients and object implementations are not written in IDL. The IDL grammar is a subset of ANSI C++ with additional constructs to support the operation invocation mechanism. An IDL binding to the C language has been specified; other language bindings are being processed.

3. CHARACTERISTICS OF REAL-TIME SYSTEMS

As stated in Section 1, in a real-time system the data and the transactions have timing constraints associated with them. The data may have temporal consistency constraints associated with them; e.g. a piece of sensor data in a real-time control application may be valid for a certain number of seconds. The data may also be assigned criticality values to specify their importance. The transactions have to meet timing constraints. For example, a transaction in a real-time control application may have to execute within a specific deadline in order for the result to be considered correct, or, in some other case, to prevent a robot from walking into the wall.

The origin of timing constraints in most real-time systems lies in the environment. To survive in the environment the robot must do the right thing at the right time. Crossing a busy street would be a good example of such a task. Another point that should be considered is that timing constraints are not necessarily satisfied by "faster" processing as in the above example. A third point to consider is that if timing constraints can be met reliably in a system, this provides a means of synchronization of system activities that may require less communication activity (exchange of messages). This may enhance the performance of highly concurrent distributed and parallel systems by reducing the communication workload. A final point in the context of this example is that "speed" is not what is important but "predictability" is. For more information see Stankovic's Tutorial [STAN88].

In addition to timing considerations, criticalities are assigned to transactions. Criticalities are used to ensure that for example, when there is a scheduling conflict, the landing gear on the airplane will be put down before the air conditioning is adjusted. This is a general problem in real-time systems. We cannot economically design for the worst case by providing sufficient resources to meet all timing constraints all the times. Thus, we need to provide a means for "graceful" degradation when demands for resources exceed the resources available. Especially important when interacting with the environment is the requirement that critical activities, such as lowering the landing gear, are given preference over non-critical activities.

Much of the research in real-time systems has been focusing on developing schedules where the transactions meet timing and criticality constraints. The schedules depend on whether the timing constraints are hard or soft. In the case of hard real-time constraints, it is critical that the transactions meet the timing constraints. In this case, the resources utilized by the transactions have to be known ahead of time. In the case of soft real-time constraints, there may still be some value in executing transactions even if they have missed the deadlines.

We also recognize that dividing timing constraints into "soft" and "hard" categories may be overly simplistic. There are other models of timing constraints in which the utility of a result depends on the time at which it is delivered and "hard" and "soft" represent extreme limiting cases. This is seen in operating systems such as ALPHA [CLAR92]. There is also the notion of imprecise computation in which the constraint may be "hard" but the precision of the result depends on the time available to compute and improves montonically as a task executes [CHUN90].

Real-time operating systems, communication systems, and some aspects of real-time database systems are now commercially available. Research on the object-oriented approach to designing real-time systems is also in progress [THUR94,PRIC94].

4. REAL-TIME EXTENSIONS TO CORBA

In this section, we discuss the real-time extensions to the object model, the ORB and object adapters, and to IDL.

4.1 REAL-TIME OBJECT MODEL

We propose that attributes and/or objects have timestamps and related timing information associated with them. In addition to the timestamping of time-sensitive data, information on the time-decay of short-lived data is needed. This timing information is used for determining temporal consistency of a set of attributes and for managing the data. For interoperability among ORB vendors, this requires a new CORBA data type capable of expressing time, preferably "global" time. Another important reason for a global time service is simply that system time will have little meaning without the notion of global time.

We also propose that object operations provide information about their expected execution time and required resources so that a schedule of request activations can be set up, either statically or dynamically, to accomplish the required tasks. This schedule will be produced by a real-time scheduler but will need to interact with the ORB and object adapters.

We also propose the assignment of criticality values to the objects in CORBA, as in [THUR94]. Criticality values specify the importance of the object. During time-constrained query and transaction processing, the objects with higher criticality values could be accessed first. If there is still time remaining, objects with lower criticality values would then be accessed.

4.2 REAL-TIME ORB AND OBJECT ADAPTER

The major issues on developing a real-time ORB and object adapter involve the time constraints and criticalities attached to requests. That is, when clients request services from object implementations, time constraints are attached to the requests, such as start times, start events, deadlines, and periods, to be imposed on the execution of the operations. These ideas have been proposed in the RTSORAC model of [PRIC94]. Techniques to handle these timing constraints have to be developed. This is a heavily-analyzed problem in real-time scheduling which needs to be evaluated for applicability to the ORB environment. For instance, if an object server is to have some responsibility in controlling the order of execution of its requests, the entire request queue needs to be available for reading and writing, not just the top of the queue. Also, real-time communication is also a major issue that needs to be addressed for a real-time ORB.

An example of how CORBA can be used to advantage in a real-time problem: an operation may have multiple implementations which are differentiated by the accuracy of their computed result and their time to execution (worst-case, average, etc). When one implementation is too time-consuming to meet the required deadline, the best implementation which is feasible can be selected by the ORB/object adapter.

4.3 REAL-TIME IDL

Real-time extensions to IDL will depend on the extensions proposed for the object model and the ORB. In general, extensions to IDL will include specifying time constraints and temporal consistency constraints on IDL constructs such as operation, and attribute declarations.

Recently several efforts on real-time extensions to programming languages such as C and C++ have been proposed (see for example, [WOLF93,ISHI92]). These extensions include syntax and semantics for expressing start times, deadlines, periods and exception handling for time constraint violations. The synthesis of these extensions with C++ should facilitate using them as a basis for real-time extensions to IDL.

5. REAL-TIME SERVICES AND APPLICATIONS

While a real-time ORB described in section 4 is intended to provide the underlying mechanisms for time constrained communication between different objects, services such as real-time scheduling for period, aperiodic, and sporadic tasks have to be built on top of real-time ORBs. That is, it is desirable top keep the ORB as simple as possible and to hosted the various services. One approach here is to develop a real-time infrastructure library to encapsulate and extend the use of the operating system and ORB real-time features. For example class libraries can be developed to extend the functionality of threads by allowing them to have their priorities set by a server that calculates the priority based on the period of a task, have timing constraints based on CPU usage or other timing criteria enforced, and to be controlled by events generated by other processes or threads. The library may support features that are important that are directly supported by operating systems or ORBs. Examples may include semaphores, mutexes and condition variables which support priority inheritance. This enables for greater portability of application code that uses the features provided by the library.

Real-time data manager is a example of an application hosted on an ORB. A real-time data manager manages persistent and temporal data, ensures that the queries and transactions meet the timing constraints, enforces temporally correct serializable schedules, and combines techniques from data management and real-time scheduling. Some implementations of real-time data managers relevant to the discussion here is given in [WOLF94] and [BENS95]. In [WOLF94], a discussion of real-time extensions to Texas

Instruments' Open Object-Oriented Database System (Open OODB), which is currently stated to be consistent with CORBA specifications, is given. These extensions follow the RTSORAC model (see Section 4.1) and thus would be consistent with our proposed changes to the Object Model. In [BENS95], the design and implementation of an infrastructure and a data manager, which is based on a real-time object model, for next generation command and control systems is described. Future plans with this latter effort is to use a real-time ORB to integrate the infrastructure services, the data manager, and the application.

6. FUTURE CONSIDERATIONS

In this paper we have provided an overview of CORBA, the characteristics of real-time systems, and discussed possible extensions to CORBA to support real-time data processing. Finally we briefly discussed issues on hosting real-time services and applications on real-time ORBs. One of the major considerations here is to keep the ORB as simple as possible and build real-time services on top of it.

With the advent of computer control in complex real-time applications, the integration of real-time processing components in a CORBA system should be encouraged. The ideas discussed in this paper need to be examined further and a detailed set of requirements to be imposed on CORBA for real-time data processing need to be formulated. The next step is to meet with other interested parties to further explore and prototype real-time processing extensions to CORBA. The goal is to have a set of workable real-time extensions incorporated into the OMG standards and available for production real-time applications. This paper provides the initial step towards this goal.

REFERENCES

[BENS95] Bensley, E. et al, Evolvable Systems Initiative for Real-time C3, Volume II: Design and Implementation of the Infrastructure, Data Management, and Application, MTR 95B 0000116, September 1995

[CLAR92] Clark, R., E. D. Jensen, and F. Reynolds, An Architectural Overview of the Alpha Real-Time Distributed Kernel, Proceedings of the USENIX Workshop on Microkernels and Other Kernel Architectures, Apr.1992

[CORB92] The Common Object Request Broker Architecture, Published by OMG and X/Open, 1992

[ISHI92] Ishikawa, Y, H. Tokuda, and C. Mercer, An Object-Oriented Real-Time Programming Language, IEEE Computer, 25(10),66-73, Oct.1992.

[CHUN90] Chung, J.-Y., J. W.-S. Liu, and K.-J. Lin, Scheduling Periodic Jobs that Allow Imprecise Results, IEEE Computer, 39(9),Sep.1990

[KRUP94] Krupp P., et al. On Real-time Extensions to CORBA, presented at the OOPSLA 94 Conference Workshop on CORBA, Portland, OR, 1994.

[PRIC94] Prichard et al, RTSORAC: A Real-time Object-Oriented Database Model, Proceedings of the International Conference on Database and Expert System Applications, 1994

[STAN88] Stankovic, J., Real-Time Computing Systems: The Next Generation, Hard Real-Time Systems Tutorial, Feb. 1988

[THUR94] Thuraisingham, B., and A. Schafer, RT-OMT: A Real-time Object Modeling Technique for Designing Real-time Database Applications, Proceedings of the IEEE Real-time Systems Applications Workshop,July 1994

[THUR95] Thuraisingham, B., Real-Time Processing and CORBA, Proceedings of Object World West , San Francisco, CA August 1995.

[WOLF93] Wolfe, V. F., S. Davidson, and I. Lee, RTC: Language Support For Real-Time Concurrency, Real-Time Systems Journal, 5 63-87; Spring 1993.

[WOLF94] Wolfe, V. et al, The Design of Real-time Extensions to the Open Object-Oriented Database System, Proceedings of the IEEE Workshop on Object-Oriented Real-time Dependable Systems, Oct. 1994.

[WOLF95] Wolfe, V. et al, Real-time Methods Invocations in Distributed Environments, Proceedings of the IEEE Conference on High Performance Computing, December 1995.

Using Object Frameworks to Enable Real-Time and Dependability in a Modular Operating System

Douglas M. Wells
Open Software Foundation Research Institute
Eleven Cambridge Center
Cambridge, MA 02142, USA
dmw@osf.org
http://www.osf.org/ri

Abstract[1]

The Open Software Foundation (OSF) Research Institute (RI) has developed a multiserver, microkernel-based operating system that provides performance competitive with commercial systems and enables reliable, distributed real-time applications. Use of object frameworks within the operating system allows configuration choices including desktop workstation systems and real-time dependable systems from one source code base. Initial implementation of the networking framework is complete and in daily use; the processor scheduling framework is in progress. This paper reports on the performance and capabilities of those frameworks, and comments on some improvements to be included in a follow-on operating system.

1. Introduction

The Open Software Foundation (OSF) Research Institute (RI) is engaged in a program to improve the capabilities of open systems by addressing the needs of diverse applications within a single operating system source base. Our overall goal is to develop a new generation of systems technology that enables the flexible configuration of system level services, thereby producing a suite of operating systems that are far more extensible and customizable than existing systems. Our immediate goal is to demonstrate the scalability of the OSF MK [13][2] Mach-based[1] operating system in the contexts of numbers of processors and in real-time requirements.

2. Operating System Program

The RI methodology incorporates the development of full-function prototypes. By synthesizing the ideas and code contributed by universities and leading edge companies with our own research into complete systems, the RI is able to reveal the effects of various architectural changes in realistic benchmark environment and in external demonstrations.

In addition to general objectives of portability, scalability, and interoperability, the particular objectives of the current OSF RI Operating System Program are to enhance OSF MK to provide:

- a more maintainable, more extensible systems technology foundation that enables rapid innovation
- local and distributed real-time capabilities that enable both multimedia and mission-critical applications
- a modular framework that enables support for multiple personalities, such as POSIX, OS/2, and specialized network servers
- a scalable foundation that provides support for uniprocessors, SMP, MPP, and clusters

The OSF RI is also developing OSF MK++, a configuration version that incorporates the OSF MK goals listed above in a high trust operating system that provides much higher assurance.

[1] This work was supported in part by the Advanced Research Projects Agency (ARPA) and the Rome Laboratory of the Air Force Materiel Command (AFMC).

0-8186-7570-5/96 $5.00 © 1996 IEEE

1. OSF MK

OSF MK is a modular operating system comprising several components, including the OSF Microkernel, an operating system personality server, external pagers, and various stand-alone servers. The standard OS personality is derived from OSF/1, a commercial quality version of UNIX that complies with SVID, POSIX, XPG3, XPG4, and numerous other standards. An alternate server, OSF AD[14], is designed for massively parallel systems and provides a single system image UNIX on supercomputers, such as the Intel Paragon and the Convex Exemplar.

Although there are myriad reasons why a microkernel-based system offers benefits in the areas of maintenance and innovation, any significant level of performance degradation would not be acceptable to commercial vendors of UNIX-based desktop workstations and servers. Therefore, the RI constantly monitors the performance of various OSF MK configurations and initiates development projects to maintain performance parity with equivalent commercial products. Currently, for example, the desktop workstation configuration of OSF MK provides performance within 2% of that of the fastest commercial UNIX systems on Intel 486 platforms.[2]

The RI is also incorporating advanced real-time capabilities into OSF MK. Recent projects have been oriented toward microkernel mechanisms. There are several reasons for this orientation. First, a real-time server is useless when used with a non-real-time microkernel. Second, the OSF Microkernel is significantly smaller and less complex than the OSF/1 server, allowing us to develop a complete, full-function real-time microkernel utilizing limited engineering resources. Also, many real-time applications are already designed to operate using a very limited set of system primitives, such as the System Programming Interface (SPI) provided by the OSF Microkernel.

2. Use of object frameworks

In order to provide real-time performance while maintaining flexibility and configurability, we are using object frameworks to provide low-overhead, structured interfaces. An object framework as used herein is an architectural support environment within which a developer can create alternate and/or variant services. It is a way of separating policy and mechanism.

[2] These performance measurements have been taken using the AIMIII benchmark with 8 users. AIMIII is widely used as a benchmark for UNIX systems.

Historically, frameworks have been used in operating systems to simplify the design and implementation of device drivers, file systems, and networking stacks. The framework mechanism typically includes a switch construct that directs higher-level invocations to a selected policy object. The framework also includes a set of service objects that hide the details of lower levels of the system. Policy objects that are "plugged into" the framework adhere to a standard set of invocable interfaces as defined by the framework.

In early UNIX systems, for example, device drivers were implemented within a framework. Application calls to read from a character-oriented device, for example, were directed through an indexed jump table (*cdevsw*) to a particular device driver. The device driver used a set of lower level routines to manipulate hardware maps and to manage physical memory. There was no compiler support to assist the programming in managing the interfaces.

In OSF MK, we are extending the use of object frameworks in the areas of networking, dependability, scheduling, interprocess communication, and memory management.

2.1. Networking framework

CORDS (Communication Objects for Real-Time Dependable Systems)[12] is a networking framework and a set of objects that operate within that context to provide a set of services that enable high levels of reliability and availability within a distributed real-time application. A companion paper[11] at this workshop describes an extension to the framework interface to allow explicit management of system resources within a distributed real-time application.

The networking framework is derived from the x-kernel[6], developed at the University of Arizona. Within this framework, primitive building blocks known as *micro-protocols* and *virtual protocols* are configured into structures referred to as *protocol graphs*. (See Figure 1). Each protocol object provides a standard interface to invocations from above and depends upon a standard interface from the protocol objects and framework service objects that it invokes.

The networking framework, including the paths mechanism, is implemented and available in current OSF MK and OSF AD releases.

2.2. Dependability framework

The networking framework is particularly interesting because it, in turn, is used to implement the higher level

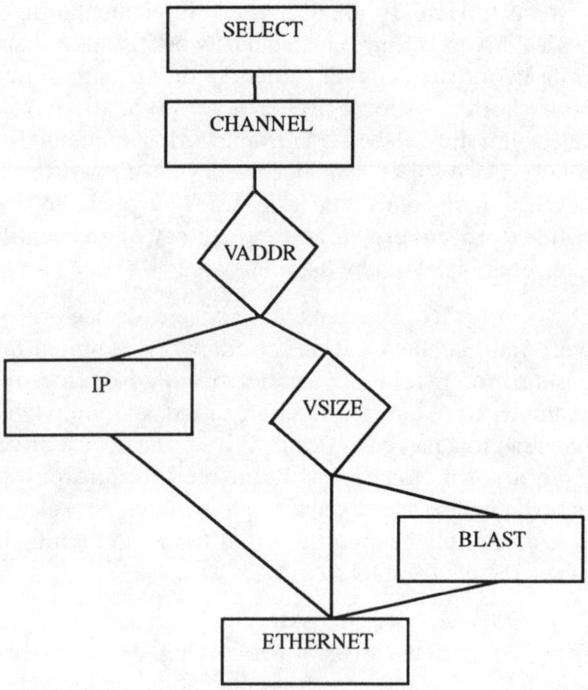

Figure 1. An example x-kernel protocol graph. The rectangles represent micro-protocols, the diamonds represent virtual protocols.

abstraction of a dependability framework. This framework provides standard interfaces for services, such as failure detection, group membership, and distributed clock synchronization., that are required to provide highly available services.

Applications with requirements for high availability can use the dependability framework to reduce or eliminate dependencies on particular algorithms. As an example, synchronized global time can be provided in many ways: there may be a common backplane hardware clock; GPS (Global Positioning System) receivers might be provided for each node; or the synchronization might be implemented in software. The particular mechanism has no effect on the application as long as the mechanism satisfies the requirements of the application for such parameters as accuracy and jitter.

Implementation of the dependability framework is underway. Experimental versions of some CORDS protocol objects are included in recent releases of OSF MK and OSF AD.

2.3. Scheduling framework

CPU scheduling is another area in which OSF RI is developing a framework. The OSF MK Scheduling Framework[5] is derived from one originally developed by the Center for High Performance Computing (CHPC) at Worcester Polytechnic Institute. The OSF MK framework allows two or more scheduling policies to coexist on the same node. Typically, one scheduling policy might be an advanced real-time scheduler, such as rate-monotonic[7] or Best-Effort[8], and the other would be the POSIX-compliant OSF MK time-sharing scheduler or a simpler, fixed priority scheduler.

The scheduling framework is a good example of another benefits of the use of object frameworks. Often, the abstraction maintained by the policy object is significantly simplified within the framework as is shown in Figure 2.

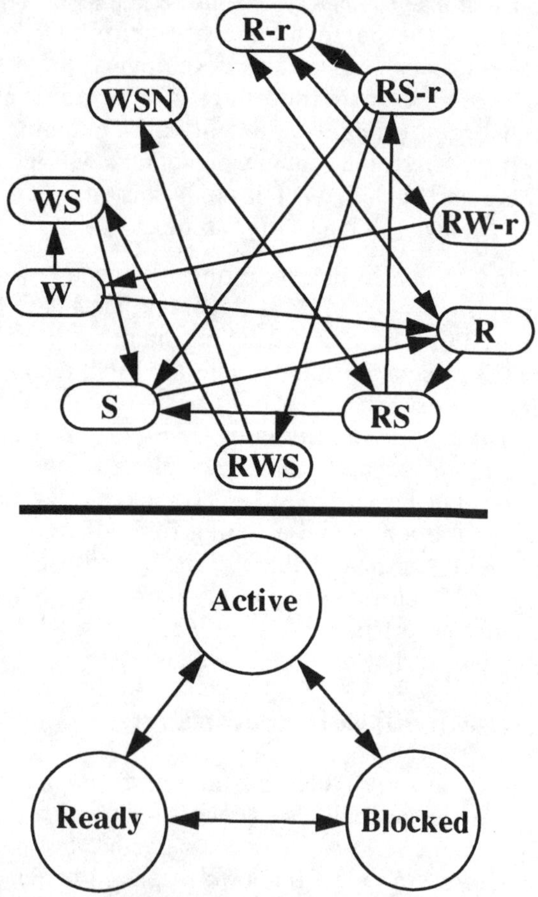

Figure 2. OSF MK internal scheduling state machine (top), and state machine visible to scheduling policy object (bottom).

Implementation of the scheduling framework is underway. An initial version, including implementations of POSIX-compliant scheduling policy objects, will be available in Summer. Advanced scheduling policy objects, such as Earliest Deadline First (EDF) and Best-Effort will be available later. In addition, external users of OSF MK are interested in implementing special purpose schedulers, such as a real-time gang scheduler for the Honeywell Embedded Touchstone, which is a ruggedized version of the Intel Paragon.

2.4. Interprocess communication framework

The OSF RI, in conjunction with the Scalable Systems Division of Intel, has designed the KKT framework[9] for distributed IPC within a group of OSF MK systems. This framework is used within the OSF Microkernel to implement Mach IPC across nodes, including capabilities for message passing between nodes, and distributed port right management.

The KKT framework has been implemented and is available in recent releases of OSF MK and OSF AD. Policy objects have been implemented for widely disparate interconnects, such as the Intel Paragon backplane, where data throughput has been measured at over 120 MB/sec; server clusters, where data is transferred at a utilization rate of 98% over a fast, wide SCSI interconnect. and Ethernet, which utilizes CORDS networking support.

2.5. Virtual memory

Support of virtual memory is another area in which the RI has investigated the use of an object framework. OSF MK includes pagers to implement virtual memory and takes advantage of virtual memory concepts, such as copy-on-write when useful. Not all application environments require or even desire virtual memory. Most current embedded real-time applications, for example, can not tolerate the delay encountered in transferring date from disk to primary memory. In fact, many such environments do not even include secondary storage hardware.

The OSF RI implemented an experimental system[3] in which paging was disabled: virtual memory tricks such as copy-on-reference and copy-on-write were replaced by eager copies, and page faults were reported to the application process as faults. This work will be continued and incorporated into a future version of OSF MK.

2.6. Task abstraction framework

In Mach-based systems, such as OSF MK, resources are associated with tasks. Resources include port rights, address spaces, and threads. Investigations into the sources of performance loss associated with microkernel-based systems have shown that the primary cause is due to the need to context switch from one task to another one when executing UNIX system calls. Marshalling of message date, invalidation of protection caches, and establishment of an execution environment in the operating system server all add overhead that is not present in traditional kernel architectures. In addition, many modern processor architectures include special privileged instructions for accessing data in user space, as in required by the UNIX system functions *copyin* and *copyout*, which are used to implement the system calls *read* and *write*, for instance. These instructions are unusable by the server component of a microkernel-based operating system when executing in a separate task.

The major effort in this area has been the development of server collocation[4]. (See Figure 3.)

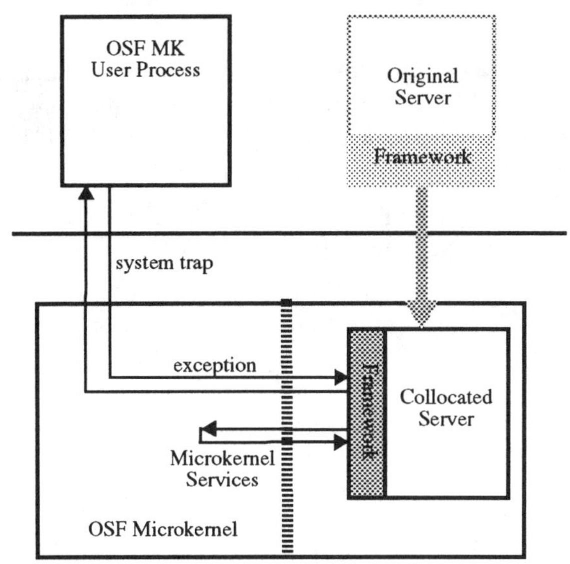

Figure 3. Server collocation into the OSF MK microkernel.

The collocation mechanism allows one server to be dynamically loaded into, or collocated with, another server or the OSF Microkernel while retaining its ability to separately acquire and merge system resources. Ancillary tools allow use of (machine-dependent) hardware optimization operations and the reduction of much of the cross-domain overhead function.

Concurrent enhancements to the server support tools, such as the Mach cthreads multitasking package, allow the identical server image to execute either collocated or as a traditional server executing in a separate task.

The implementation of collocation has required the separation of the task resource management facilities from the address space used by a task. A framework construct now exists such that association and translation of port rights occurs separately from system traps. The OS personality server (or an application) now performs the role of policy object.

3. OSF MK++

The RI is also developing OSF MK++, a high assurance configuration of OSF MK. This microkernel, which targets the B3 and F5/E5 levels of trust, is being redesigned as a strictly layered system. Retaining compatibility at both the source and binary levels with the UNIX API of OSF MK, this new implementation takes advantage of object-oriented development techniques in order to create a highly modular, easily understandable release of OSF MK. Written in C++, the OSF MK++ microkernel uses C++ inheritance to guarantee correct use of parameters within the invocation interface.

We are investigating the use of compiler techniques to detect inconsistencies and to understand abstract requirements, such as support of high availability techniques.

A single-node version of OSF MK++ has been released and is entering formal evaluation.

References

[1] M. Acceta, R. Baron, W. Bolosky, D. Golub, R. Rashid, A. Tevanian, and M. Young, "Mach: A New Foundation for UNIX Development," *Proceedings of the 1986 Summer USENIX Conference*, Atlanta, GA, 1986.

[2] F. Barbou des Places, *et al.*, "Architecture and Benefits of a Multithreaded OSF/1 Server," *OSF Research Institute Symposium*, OSF, Cambridge, MA, February, 1992.

[3] P. Bernadat, D. L. Black, "Real Memory Mach," *OSF Research Institute Collected Papers, Vol.* 1, Open Software Foundation, Cambridge, MA, March, 1993.:

[4] M. Condict, "The Server Co-Location Project," *OSF Research Institute Symposium*, OSF, Cambridge, MA, October, 1993.

[5] R. Haydt, J. CaraDonna, F. Reynolds, "Mach Scheduling Framework," *OSF Research Institute Collected Papers, Vol.* 3, Open Software Foundation, Cambridge, MA, April, 1994:

[6] N.C. Hutchinson, and L.L. Peterson, "The x-kernel: an Architecture for Implementing Network Protocols", *IEEE Trans on Software Eng., vol 17, no. 1*, January, 1991.

[7] J.P. Lehoczky, L. Sha, and Y. Ding, "The rate-monotonic scheduling algorithm: Exact characterization and average case behavior," Carnegie Mellon University, Pittsburgh, PA, 1987.

[8] C.D. Locke, Best-Effort Decision Making for Real-Time Scheduling, Department of Computer Science, Carnegie-Mellon University, Pittsburgh, PA, Ph.D. dissertation, May 1986.

[9] S. Sears, M. Dominijanni, A. Langerman, D. Black, "Kernel to Kernel Transport Interface for the Mach Kernel," *OSF Research Institute Collected Papers, Vol.* 3, Open Software Foundation, Cambridge, MA, April, 1994:

[10] D. Swartzendruber, "OSF Real-Time Mach: Making the Mach Kernel Preemptible," *OSF Research Institute Operating Systems Collected Papers, Vol. 2*, OSF, Cambridge, MA, October, 1993.

[11] F. Travostino, E. Menze, F. Reynolds, "Paths: Programming with System Resources in Support of Real-Time Distributed Applications," to appear in *Proceedings of the Second International Workshop on Object-oriented Real-time Dependable Systems (WORDS 96)*, Laguna Beach, CA, 1996.

[12] F. Travostino, F.D. Reynolds, "An O-O Communication Subsystem for Real-Time Distributed Mach," *Proceedings of the First International Workshop on Object-oriented Real-time Dependable Systems (WORDS 94)*, Dana Point, CA, 1994.

[13] D. Wells, "A Trusted, Scalable, Real-Time Operating System Environment," *1994 Dual-Use Technologies and Applications Conference Proceedings*, pp. II-262/270, Utica, NY, May, 1994.

[14] R. Zajclew, *et al.*, "An OSF/1 UNIX for Massively Parallel Multicomputers," *Proceedings of the 1993 Winter USENIX Conference*, San Diego, CA, January, 1993.

Trademarks

—◊—

Session 5

Real-Time System Design

Session Chair:
Jeff Steinman
JPL, USA

An Object-Based Model for Dependable Real-Time Distributed Systems

Yeong-Jia Chen Daniel Mossé Shi-Kuo Chang
Department of Computer Science
University of Pittsburgh, Pittsburgh, PA 15260
{chang,yjchen,mosse}@cs.pitt.edu

Abstract

A systematic way to introduce fault-tolerant and time-dependent properties into a complex software system is presented. The approach is based on an extended Petri Net model, called the G-Net, enhanced with a deterministic timing scheme. In Timed G-Net (TGN) model, exception handling and time-out mechanisms are used to introduce the fault-tolerant and real-time properties. The basic idea is to avoid the timing error propagation from a faulty object to an object that interacts with it via a time-out mechanism, and also to allow an object to treat a timing error by raising an exception and triggering some corrective actions. Using TGNs as basic building block, we further refine the model into Smart Object (SMO) model. In each SMO, an object can associated with a knowledge structure, which enables the object to dynamic change its behavior to react the changes of environment. We use examples to show that a simplified real-time distributed Air Traffic Control System can be modeled effectively using the SMO model.

1 Introduction

The increased use of computers and our increased reliance on them in every aspect of modern life have led to a need for highly reliable computer systems. There are many areas where computers perform life-critical tasks. Some examples are air traffic control systems, hospital patient monitoring systems, missile guidance and control systems, and aerospace control system. In these systems, failure of components can lead to catastrophe, maybe even loss of human life. There are other application areas that critically depend on computers, where failure can cause great financial loss, or loss of opportunity. The prime examples of this are banking and stock markets. Clearly, from the nature of applications, highly dependable systems are needed now, and that the need for these system will continue to grow.

Dependability is a general concept and, depending on the application, different attributes can be emphasized. The most significant attributes of dependability are *reliability, availability, safety* and *security* [5]. For complex dependable *real-time* (RT) hardware systems there are well-defined methodologies to address these issues. For their software counterparts the same techniques cannot be easily applied because it is difficult to predict all possible changes that can occur in the environment interacting with the software. Another important aspect is that errors can occur due to unpredictable or hostile environments, for example, erroneous data from the environment, undetected hardware failures and/or undetected design errors in the software components.

For such complex RT systems a compositional and/or modular approach that introduces fault-tolerant properties into the design of a component must be considered. This approach is a natural outgrowth of the object-based design methodology, which can be applied in many areas, ranging from the development of RT control systems to the reuse of software components.

In what follows, we propose a modular approach to systematically introduce time-dependent properties into the components, and incorporate the desirable and necessary fault-tolerance features into a complex system. This approach uses the object-oriented model to adjust the object's behavior dynamically (when possible), without violating the timing and resource constraints. Such behavior adjustment takes place according to the environment and execution history. The worst-case timing requirements are important because the systems receive input from an essentially uncontrollable environment, and not meeting the timing requirement even once may have catastrophic consequences. Essentially, we consider hard RT systems.

The paper is divided into six sections. In Section 2, the G-Net model and the notations are introduced. In Section 3, the formal definitions and behavior of the *Time G-Net model* are described. *Timed G-Net model* supports standby redundancy in RT system design as shown in Section 4. In Section 5, the concept of *smart object* is introduced and illustrated with an example. In Section 6, we conclude with some areas of future research.

0-8186-7570-5/96 $5.00 © 1996 IEEE

2 G-Nets and G-Nets Systems

G-Net is a Petri net based multi-level executable specification model which incorporates the concepts of module and system structuring [2]. A *G-Net system* is a composition of several *G-Nets*. Since the *G-Net* framework uses object based concepts, the interaction between *G-Nets* in a *G-Net* system is realized through well-defined interfaces.

A *G-Net*, G, is composed of two parts: a special place called *Generic Switch Place (GSP)* and an *Internal Structure (IS)*. The *GSP* provides the abstraction of the module, and serves as an interface between the *G-Net* and other modules. The internal structure is a modified Petri net, and represents the detailed internal realization of the modeled application. Among other features this notation allows the user to indicate termination and communication among *G-Nets*. The notation for *G-Nets* is shown in Figure 1.

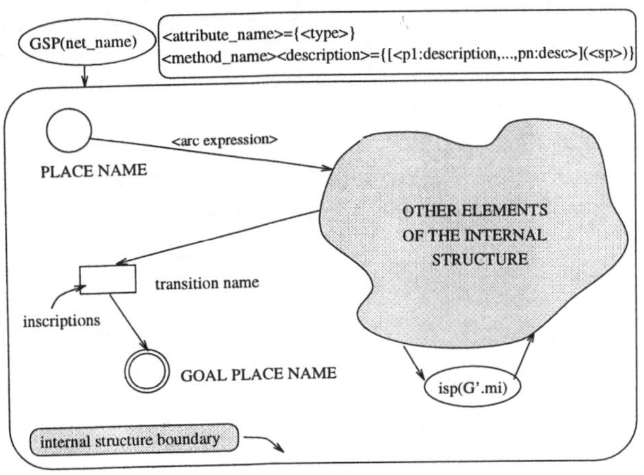

Figure 1. Notations used to represent a *G-Net*

The internal structure of the net (IS) is enclosed by a rounded corner rectangle. The GSP is indicated by the ellipse in the left upper corner of the IS boundary, and its inscription *net_name* defines the G-net id. The rounded corner rectangle in the upper right corner of the IS boundary is used to identify the methods and attributes for the net, where: $\langle attribute\_name \rangle = \{\langle type \rangle\}$ defines the attribute for the net, $\langle attribute\_name \rangle$ is the name of the attributes, and $\langle type \rangle$ is a type for the attribute, restricted to the set of non-negative integers; $\langle method\_name \rangle$ is the name for a method, $\langle description \rangle$ is a description for the method. $\langle p1 : description, \cdots, pn : description \rangle$ is a list of arguments for the method, and $\langle sp \rangle$ is the name of the initial place for the method. A circle represents a normal place. An ellipse in the internal structure represents an *instantiated switching place (isp)*. The *isp* is used to provide *inter-G-Net* communication. The inscription $isp(G'.m_i)$ indicates the invocation of net G' with method m_i. Executing the *isp* primitive implies invoking G' (by sending a token to G') based on the specified method. This token contains the parameters needed to define the tokens for the initial marking of the invoked net. A rectangle represents a transition, that may have an inscription associated with it. This inscriptions may be either an attribute or a firing restriction. A double circle represents the termination place or *goal place*. Places and transitions are connected through arcs that may carry an expression.

In the internal structure, places represent *primitives*, and transitions, together with arcs, represent connections or relations among the primitives. A set of special places called *Goal Places* represents the final state of the execution, and the results (if any) to be returned. Details regarding the dynamic behavior of a *G-Net* system as well as the firing rule can be found in [2].

The G-Net model follows an object-based approach for the design of a software system, in which models of entities are constructed as self-contained components. The system's behavior is defined by the interaction among these self-contained components. The object-based design emphasizes the concepts of information hiding and abstract data types. A detailed definition of G-Nets appears in [3, 4].

G-Nets and G-Nets systems have certain characteristics that make possible the incorporation of fault-tolerant aspects: (a) G-Nets are amenable to formal verification of certain properties; (b) G-Nets are executable; (c) it is possible to systematically identify critical points in the design that may generate a failure when implementing the specification. With G-Nets, it is possible to provide a unified, uniform, and comprehensive framework to model the hardware, software and human behavioral aspects of a complex system by combining these three aspects within one model.

3 Timed G-Net Model

Since the G-Net model does not consider the timing aspect of an application, in order to allow a modular timing analysis for complex real-time distributed systems, a timing scheme needs to be integrated into G-Net model. In the following, we describe how the timing can be embedded into the G-Net model in a deterministic fashion. We assume that clocks in the distributed systems are synchronized to some arbitrary precision. This means that in our real-time distributed model, timing depends only on local clock. Furthermore, we assume that tokens can carry as much information as necessary.

In distributed systems, it is often difficult to precisely characterize the time required for communication and computation due to the varying loading on the networks and servers, respectively. A capacity reservation and enforcement mechanism isolates modules from the timing and execution characteristics of other modules in the same way that a memory protection system isolates programs from memory access by other programs. Therefore, *our modeling strategy is to specify the timing constraint of a specific requesting method (operation) by reserved-time.*

In real-time distributed systems, a scheduling policy is needed to satisfy timing constraints. Moreover, each object attempts to survive, to minimize its loss, and to prevent the propagation of timing faults. These properties are achieved by *time polymorphic invocation* and *timing constrained method.* To implement the former, we associate a timing constraint with a transition. Each object contains several (alternative) methods with different durations $T_i(i = 1..n)$. The method $mtd_k(k = 1..n)$ is executed if timing constraint T_k on the transition is satisfied. In the *timing constrained method*, each method declared in a real-time distributed object has associated with it some timing information (typically maximum execution time), and an exception handling routine for the case that timing constraint cannot be satisfied.

In the *Timed G-Net (TGN)* model, every method is invoked with a timing constraint. The method is expected to provide either the expected and desired service within the timing constraint, or a contingency service (expected, but less desired) if the result cannot be delivered within the timing constraint. The latter may signal an exception after having aborted the current activated method, and the caller of the method can declare a specific handler for coping with a signaled exception. Therefore, in TGNs, we enhance the original G-Net model with two extra parameters in method specification: timing constraint (reserved-time **Rsv**) and timing exception handler **(teh)**. Below, we formally define the TGN, as an extension of G-Net model [4].

Definition 1 (Timed G-Net Definition)
A Timed G-Net is a tuple (GSP,IS), where GSP is the external interface, IS is the internal structure which defines the behavior of the object. GSP is defined as a triple (Gid,AS,MS), where
(1) Gid is a unique logical identifier of the net.
(2) AS = $\{v_i | i=1..n\}$ *is a set of attributes.*
(3) MS = $\{mtd_i | i=1..m\}$ *is a set of methods (operations) defined on the instances of Timed G-Net, where each* mtd_i*, i= 1..m, is defined as mid(In,Out,***Rsv***,imarking,goalp,* **teh***); mid is the name of the method (unique in the net);* In *is a list of input parameters of the method;* Out *is a list of output parameters of the method;* imarking *is the initial marking of the method;* goalp *is the goal place of the method;* **Rsv** *is the reserved-time for the method (operation) invoked; and* **teh** *is the timing exception handler of the method.*

If the timing constraints specified by **Rsv** cannot be met (e.g., due to a fault or transient overload) then the calling G-Net should activate an alternative or recovery operation specified by **teh**. Such activation takes the form of a new initial marking in the invoking G-Net for handling the timing exception. It can be done prior to invoking a method or after receiving a response or a time-out.

In Timed G-Net, each normal place P_i is associated with a primitive, with known worst case execution time (wcet) C_i. P_i can assist in enabling P_i's output transitions on precisely C_i time units after firing its input transition. The waiting time W_i of a place is the elapsed time from the time it enables the output transitions to the time of the firing of an output transition.

In our approach, when a client A requests a service from server B with reserved-time (say, $Rsv = 30$ ms), the charge for invocation and communication overhead is applied to the reserved-time by setting the watchdog timer of A to 30 ms when the request is issued. This provides a mechanism to time-out, preventing the client A from waiting indefinitely and server B from executing forever on behalf (and on the account!) of client A. For accounting and timing purposes, the message arrival time is recorded by the receiver based on the local time, as soon as the message arrives. After the reserved-time is passed to the server B by the token, there might exist some delay (say, 10 ms) from the arrival of a message to the beginning of the execution. The remaining reserved-time (or *remaining time* for short) is denoted by Tr and can be easily derived (in this case 20 ms). Server B uses the remaining time Tr carried by the tokens to decide whether the deadline can be met. Moreover, the timing information carried by the tokens can be used by the scheduler, for example, to activate a specific method.

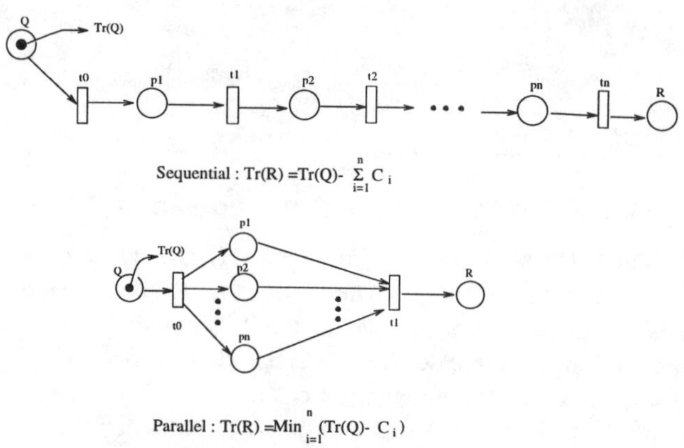

Figure 2. Remaining-Time Calculation

When a transition t fires, the remaining time carried by a token in $O(t)$ depends on the remaining times carried by tokens in $I(t)$. To make guarantees for hard real-time applications, we must consider the worst case execution time in each execution path. This translates into defining the critical path from the GSP to goal place (gp) of a G-Net. In the following, we will discuss how to obtain the remaining time of the token in different cases[1]. Using Figure 2, let the worst case execution time of each primitive denoted by C_i.

Sequential path: When a token is deposited into place R,

[1] In this preliminary work we restrict ourselves to fork/join applications

the remaining time will be $Tr(Q) - \sum_{i=1}^{n} C_i$.

Synchronized parallel path: The remaining time of token deposited in place R is $Min_{i=1}^{n}(Tr(Q) - C_i)$. In this case there may exist some waiting times in each execution path due to the synchronization point.

Sequential/parallel path: In the hybrid case, all the possible execution paths can be segmented, and sequential/parallel case can be applied as described.

In this model each token carries a remaining time which is defined after firing a transition. The dynamic behavior of a TGN depends not only the presence of the suitable tokens in their input places, but also on particular time condition associated with each transition. The merit of using remaining time *Tr* as firing condition is that the timing error will not propagate along the execution path, and can be reported immediately. Moreover, a designer has freedom to choose any alternative action s/he wants by specifying a proper time interval associated with each transition.

Definition 2 (Timed G-Net Firing Rules)

(1) *A transition t is* enabled *if: (a) every $P_i \in$ I(t) holds enough tokens and its contents satisfies the condition defined by I(P_i,t); (b) e(t) = true; (c) the number of tokens in O(t) is smaller than the capacities of the places; (d) the timing condition below must be satisfied : $a \leq Min_{i=1}^{k}(Tr(P_i) - C_i) \leq b$, where the time interval [a,b], $0 \leq a \leq b$, is associated with the transition, and C_i is the worst execution time of $P_i \in$ I(t), where $1 \leq i \leq k$.*

(2) Default condition: *If the time interval of a transition is not specified, then a transition t is enabled if it satisfies 1(a)-1(c), and the transition always can fire as soon as the tokens in its input places are ready. But two cases may occur, depending on the resultant remaining time $Tr = Min_{i=1}^{k}(Tr(P_i) - C_i)$ as follows : (a) if $Tr > 0$ then after firing the transition, the tokens are deposited into the output places of the transition; and (b) if $Tr = 0$, and after firing the transition the goal place is reached, then the execution thread terminates normally.*

(3) *Regardless whether the time interval is specified or not, if $Tr = 0$, and goal place is still not reached after firing the transition or $Tr < 0$, then a timing error is detected. The transition is forced to be fired, all the tokens are sent to the error place, from which an exception signal is sent to the invoking G-Net, and the current activated method is aborted* [2].

[2] The time conditions and the remaining time carried by tokens can influence the choice of the next transition to fire. This reflects the time polymorphic invocation and time-out situation as we will explain later.

(4) *As a result of the firing, a token carrying the new remaining time is assigned to each output place (except for case 3). This new remaining time is computed as a function of: (a) the token's remaining time in current input place, denoted by $Tr(P_i)$. (b) the worst execution time associated with current input place, denoted by C_i.*

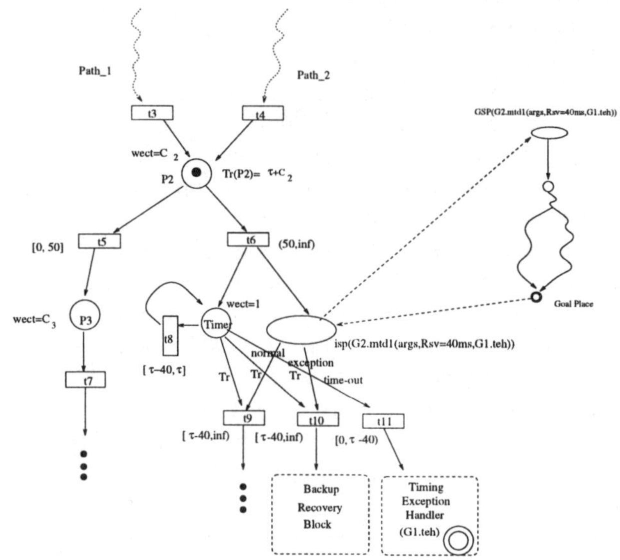

Figure 3. Embedding the timing into the G-Net Model

Example

In the following, we will give an example to help the reader better understand the TGN Model. As illustrated in Figure 3, after the execution along the path $path_1$ or $path_2$, a token with a remaining time $\tau + C_2$ will be deposited into place P_2. We must note that τ is a variable, because the execution times for the two paths might be different. After the primitive associated with place P_2(with wcet=C_2) is executed, the resultant remaining time will be equal to τ. The TGN model provides a mechanism for time polymorphic invocation depending on the remaining time. As shown in this example, if $\tau \in$ [0,50] then transition t_5 is enabled and the smaller-precision local computation P_3 is executed. If $\tau > 50$ then transition t_6 is enabled, and a more precise computation is done by invoking G-Net G_2 via the *isp*.

In a distributed computation environment, there are unpredictable factors such as the communication delay, server response time, etc, that force the client to decide on a time-out value based on a per-case basis. As shown in Figure 3, when the transition t_6 is fired, a request is issued through an *isp* (oval place) and a local timer is also initiated at the same time. The transition t_8 is associated with the time interval $[\tau - 40, \tau]$, since the remaining time is τ when the timer and isp are initiated and the time-out value for the invoked isp is 40 ms (in this example). The main idea is analogous to a hardware watchdog that counts down and signals an exception when the deadline is missed. This mechanism solves the state explosion problem due to infinite loops, since after looping 40 times the transition will not be enabled any more according to firing rule. It is important to note the flexibility in the granularity of the clock: it can be set to arbitrary values, and in this example it is set to 1 ms.

As the timer clicks, three cases might happen : (1) the token returns from the isp with remaining time $Tr > \tau - 40$ (i.e., within the timing constraint), in which case transition t_9 is enabled, and processing continues normally; (2) the token returns within the timing constraint but with an exception (i.e., the request is aborted): transition t_{10} is enabled, and depending on the application design, a backup recovery block may be invoked; (3) the token does not return within the timing constraint: the timer counts down and remaining time becomes $Tr < \tau - 40$, so transition t_{11} fires and the timing exception handler is invoked. Note that transitions t_8, t_9, t_{10} are thereafter disabled since $Tr < \tau - 40$.

The TGN specifications are directly executable in a distributed environment by multiple computation agents, and a TGN is a logical unit for distribution and execution [4]. Therefore, TGNs can be used as a prototyping tool for object-based distributed systems with temporal requirements. In Figure 3, TGNs G_1 and G_2 reside on machines A and B respectively. The isp in G_1 specifies an interaction between an instance of G_1 and an instance of G_2.

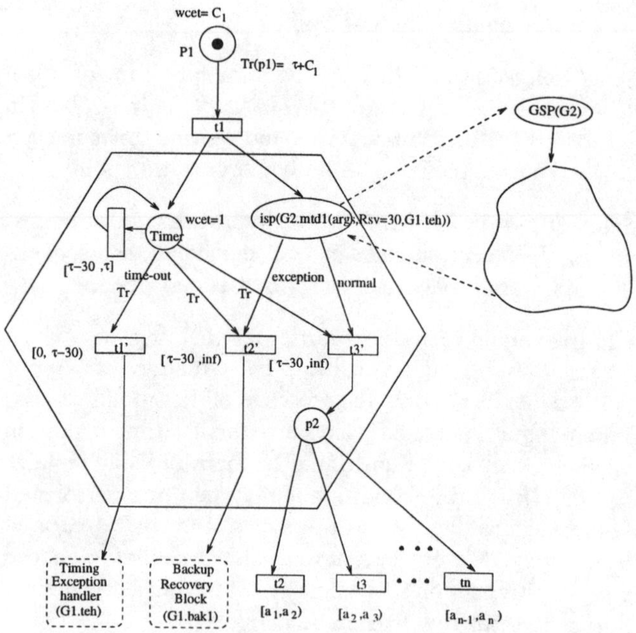

Figure 4. Abstraction for adding the timing constraint to isp(G2)

4 Including Fault Tolerance In Timed G-Nets

In the following, we will use the term *Tdisp module* for the hexagon shown in Figure 4, which depicts the abstraction for an isp with timing constraint and describes how to model software redundancy in real-time systems using TGN. The firing of transition t_1 causes the invocation of G-Net G_2 using method *mtd1* with reserved-time Rsv=30 (the maximum time allowed for execution the requested operation). The TGN specifications capture the three types of results: *normal, exception, time-out.* The system designers only need to use the high level *Tdisp* abstraction, and fill in the application-dependent methods or exception handlers specified in their designs. The system will generate the detailed TGN automatically, reducing the complexity in designing a practical system.

Example: Dynamic Redundancy

In systems employing Primary/Backup redundancy, if a processing module fails, it is replaced by a "backup" module which theoretically performs the same function. The non-redundant software system is thus supplemented by a set of backup modules and an audit module which decide if a fault occurred and performs the switching function (see Figure 5).

Processing Module: In the TGN model, a processing module can be modeled by an instance of G-Net with a timing-constraint. We say the execution of TGN is *successful* if it reaches its goal place within the deadline. Otherwise, the execution has *failed.* This is exemplified by TGN G_1 in Figure 5.

Audit Module: As shown in Figure 5, when *G.mtd* is invoked, the input data is put into the saved buffer (place BF). As the processing module is invoked, the timer of the audit module is also initiated (Figure 4). The audit module monitors the time used by a processing module and a time-out occurs if a certain time limit has reached. As illustrated in Figure 5c, arcs labeled *exception* or *time-out* from the audit module invoke the backup module G_2. When the processing module returns normally within the timing constraint, t_1 is enabled and a functional acceptance test (AT) is conducted at place AT. If the AT returns "correct", then t_3 is fired and *gp* is reached. Otherwise, t_2 is fired, and the backup module G_2 is invoked.

Backup Module: After the system status has been reestablished by using the data saved in the saved buffer (BF), the processing part of the backup module can be initiated. This module may be simpler, shorter and more stable than the corresponding primary module,

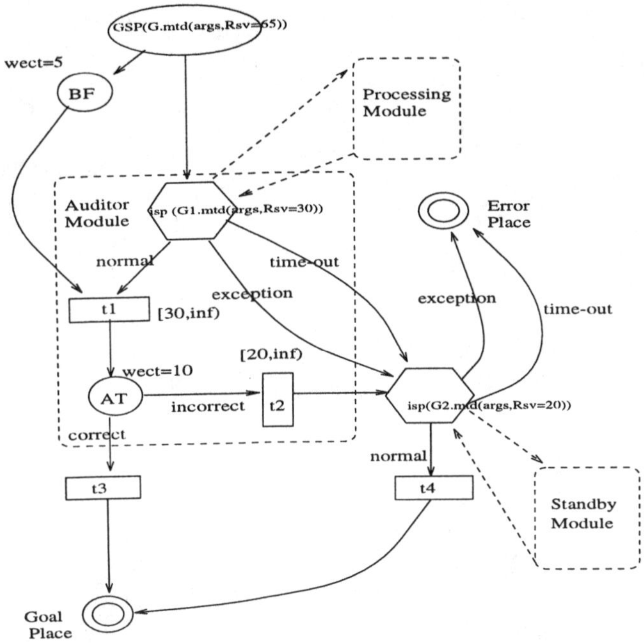

Figure 5. Standby Redundancy.

since we aim to provide the *minimum essential service* to maintain the system in operation.

We presented an example to show that the dynamic redundancy in real-time distributed systems that can be modeled using the *Timed G-Net model.* With the formalization of the *Timed G-Net*, the correctness of the fault-tolerant real-time system design can be analyzed. Using TGNs as building block, we further refine the model into *Smart Object* model. In the following section, we will discuss the methodology to incorporate the concept of *Smart Object* [1] into the dependable real-time system model.

5 The Concept of Smart Objects

We intend to develop a domain-independent model of hierarchical dependable real-time system design and construction, based on building-block technologies exploiting large-scale reusable interchangeable components (objects).

The system accepts a TGN specification, it will become a basic constituent raw object after analysis and redesign. Also, the related parameters such as the worst case execution time, reliability measure, should be embedded along with the TGN specification. Recursively, a higher layered object might consists of several lower layered objects, with related fault specification and parameters.

Since a TGN corresponds to an encapsulated object, by adding redundancy to make it into an resilient TGN, it is possible to make the important objects reliable. Furthermore, we embed a monitoring entity into the original TGN. Attached to each monitor there is an error handler, which can be methods within the original TGN or methods in a separate TGN. These monitors/error handlers (*meh*) carry out error detection/recovery in a user-specified manner.

The monitoring entity and the error handlers can be organized to form a hierarchy that facilitates information exchange, controlling the information flow. Such mechanism facilitates the enhancement of the monitoring for particular critical subsystems, while enabling the encompassing system to control the flow of information. The enhancement can be designed by the user in a customized form, or by drawing from a library of default monitors and error handlers. Among the techniques implemented by the monitoring and error handlers are recovery blocks, forward recovery, software, and timing fault tolerance.

In our framework we develop the concept of *smart objects*, which are objects capable of adjusting their behavior (within permissible temporal and resource boundaries) according to the environment. For example, a TGN can support one or several methods and for each method, it might have several implementations with different properties (e.g., timing and reliability). Thus, a smart object can be used to represent a TGN with time-polymorphism that can realize the best effort to meet the timing constraint. The amount of information required by the object specification depends on how time-critical or fault-sensitive the system is.

A smart object is an object with an associated knowledge structure, which supports the decision-making process on what actions to take based on the internal parameters (e.g., time and resiliency), external interaction with other objects, and environmental changes. The environment may include system hardware and software states, history of previous interactions, exceptional conditions reported by the hardware/software monitors, etc. The information available prior to runtime influences the latency and temporal predictability of the actions when a change of state occurs.

A smart object is a complex object represented by a pair (O, K) where O is the *raw object* and K is its associated *knowledge structure.* The raw object can be either the object itself or a pointer to the object. In general, when a new instance of the raw object is created, the unresolved addresses are resolved dynamically, allowing for adaptive behavior of the smart object, thus facilitating the fulfillment of the fault tolerance objectives. Note that, although the smart object can do dynamic binding and address resolution on the fly, such ability should be curtailed or at least limited for time-critical applications in order to preserve temporal predictability. The knowledge structure consists of A, P, and R, where A is the set of attribute specifications, P is the protocol, and R is the set of links to other smart objects.

The attributes A comprise the characterization of the objects in terms of fault-tolerant requirements and performance requirements, such as number of hardware failures allowed, number of software failures allowed, total allowed time, the resource consumption, the timing properties, and the software alternatives to use in case software redundancy is required. Some of these attributes are independent of the platform such as the timing properties and the reliability degree. Other attributes depend on the system parameters, architecture, and system components, such as the MTTF and the component reliability functions.

The protocol P specifies how the raw object can be extended to be a *resilient object* [6, 7]. It describes how to use the attributes and the monitor specifications to carry out the transformations that will enhance the raw objects in order to meet fault-tolerance and performance criteria.

R defines a set of links to other smart objects, most importantly to the meh's. Since these meh's are themselves smart objects, links can again be added, to relate the meh's to second-level meh's, and so on, thus forming the monitor hierarchy. We associate an meh with TGNs.

5.1 Example: A Simplified Air Traffic Control System

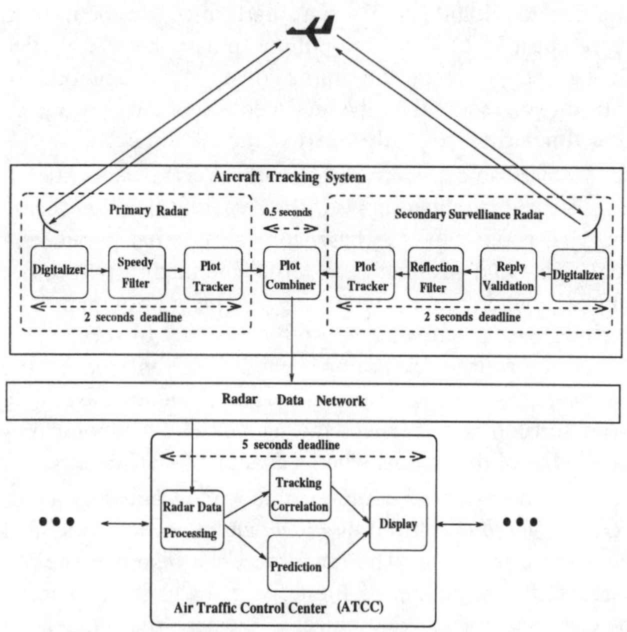

Figure 6. A Simplified ATC System

A simplified air traffic control system is presented as an example of SMO modeling thereby enabling characteristic concerns to be illustrated. The structure of an air traffic control system is shown in Figure 6, the interacting objects include: aircraft, aircraft tracking systems, and Air Traffic Control Centers (ATCCs).

Each object in the ATC system can be modeled by a TGN, which interacts with others as shown in Figure 7a. By adding a monitoring structure as illustrated in Figure 7b, and associate each monitor with knowledge, we support the dynamic behavior adjustment of smart objects. Due to the limitation of space, we will only model the tracking system, composed of a primary and a secondary surveillance radar (SSR). The plot combiner also can be modeled by a TGN $G(combiner)$, which receives plots from two radar modules: primary radar object modeled by $G(P\_R)$ (see Figure 8) and SSR object modeled by $G(S\_R)$. The output monitors of both radar modules are linked to the input monitor of $G(combiner)$.

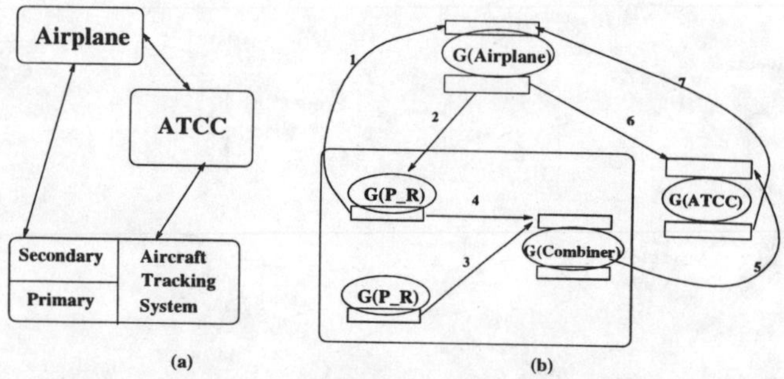

Figure 7. (a) Interaction between objects (b) Monitoring structure of ATC System

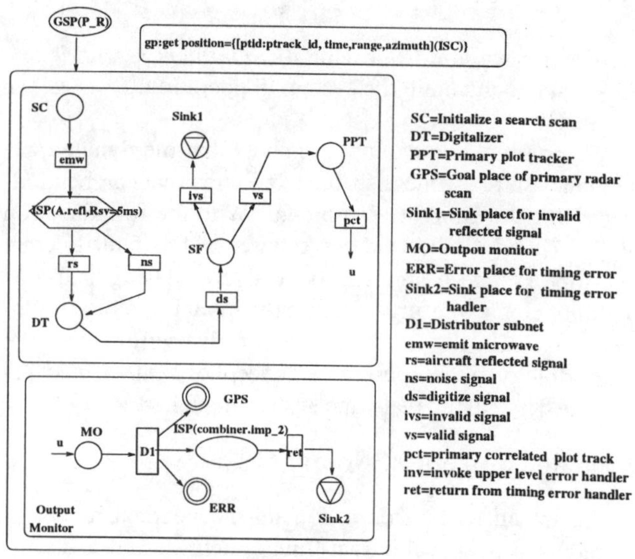

Figure 8. TGN representing the PR module

As transition emw fires, a TGN $G(A\_refl)$ is invoked. That is, as a pulse of microwave energy is emitted, it will be reflected by the aircraft back to the rotating radar antenna (represented by transition rs) thereby enabling the range and azimuth relative to the radar to be determined, or return no/noise signal (represented by transition ns). The analog signals are digitalized and processed. Fourier techniques enable stationary and slow flying flocks of birds to be filtered out, leaving discrete radar returns known as plots. Plots are tracked and correlated over successive rotations of radar antenna so that spurious single returns can be discarded. The places DT, SF, and PPT represent process digitalizer, speedy filter, and primary plot tracker, while corresponding transitions represent the resultant output signals. The sink place is depicted here by a reversed triangle inside a circle. When a token is deposited into a sink place of TGN G, the G is terminated, and token will not return to its invoking TGN. The invoking TGN can only use timeout mechanism to ensure the liveness property. The output monitor MO contains conditions to be evaluated, which are a function of the data, state, and time-stamp. After its execution, the token flows to a distributor, which is actually a subnet (bold transition D_1, see Figure 8) containing a decision net to control the information flow depending upon the value of the result-attribute. Such output monitor provides fail-stop mechanism, which can stop data from being sent when a fault is detected.

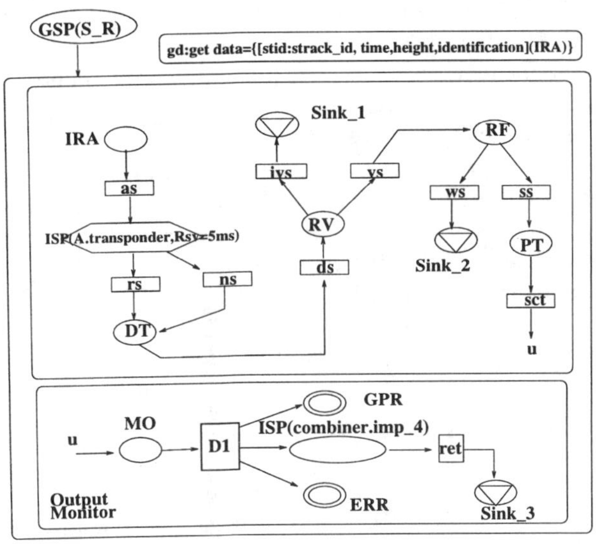

IRA=Initialize interrogation with aircraft
DT=Diditalizer
RV=Reply Validation
RF=Reflection Filter
PT=Plot Tracker
Sink_1=Sink place for invalid signal
Sink_2=Sink place for weak reflected signal
Sink_3=Sink place for timing error handler
MO=Output monitor
ERR=Error place for timing error
D1=Distributor subnet

as=ask signal
rs=returned signal
ns=noise signal
ds=digitized signal
vs=valid signal
ivs=invalid signal
ws=weak reflected signal
ss=strong reflected signal
sct=secondary correlated plot track
inv=invoke upper level error handler
ret=return from upper level error handler

Figure 9. TGN representing the SSR Module

Secondary surveillance radar (SSR) is an active system, which is modeled by TGN $G(SSR)$ illustrated as Figure 9. The SSR radar emits group of microwave pulses which are decoded by an SSR transponder on the aircraft which returns a train of pulses encoding an aircraft identity or a readout from the aircraft altimeter. Again these signals are digitalized for processing. Transponder replies have to be correlated with interrogations by the SSR concerned. Replies triggered by other SSRs are ignored. Plots are correlated over successive scans as for primary.

The input monitor of *G(combiner)* will take the appropri-

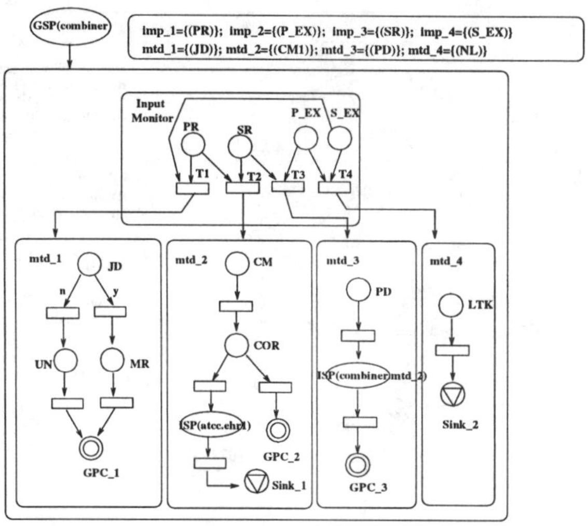

PR=Buffer for primary plot track
SR=Buffer for secondary plot track
P_EX=Primary track exception event
S_EX=Secondary track exception event
JD=Judge if an existing previous track
UN=Unidentified plot track processing
MR=Merge with existing 2ary plot track
CM=Combine primary and 2ary plot track
COR=Correlate previous combine track

PD=Predict the position of aircraft
LTR=Lost plot track processing
GPC_1=Goal place for combining method 1
GPC_2=Goal place for combining method 2
GPC_3=Goal place for combining method 3
MO=Output monitor
ERR=Error place for timing error
Sink_1=Sink place for combining method 2
Sink_2=Sink place for combining method 4

Figure 10. TGN Model for Combine Module with Input Monitor

ate action depending on the arriving time of plots which sent by two radars. As shown in Figure 10, we embed the design knowledge in the input monitor of the combiner module. When a transition satisfies all the input conditions in a marking (i.e., when the input places contain enough tokens to enable the transition), the transition is considered enabled.

As the transition T_i $(i = 1, 2, 3, 4)$ fires, the method mtd_i is activated, representing the occurrence of one of the 4 following mutual exclusive cases: (1) When the deadline is reached, the primary radar plots are ready, but the other is absent. This case is modeled by firing transition T_1, and mtd_1 is invoked. If the previous credible track exists then merge with the old secondary plot track, otherwise mark it as an unidentified track. (2) Before or on the time that deadline is reached, both are present. This case is modeled by firing transition T_2, and mtd_2 is invoked. Two plot tracks are combined, and correlated with the previous tracks. If the error is within a certain bound, send out the track; otherwise, raise an exception to notify the ATCC operator. (3) When the deadline is reached, the secondary radar plots are present, the other is absent. The current position of aircraft is predicted, and then mtd_2 is invoked. (4) When the deadline is reached and both plots are absent. Modeled by firing T_4, the lost tracking method (mtd_4) is invoked.

An aircraft tracking system TGN model is shown in Figure 11. The radar head is rotating and the microwave is

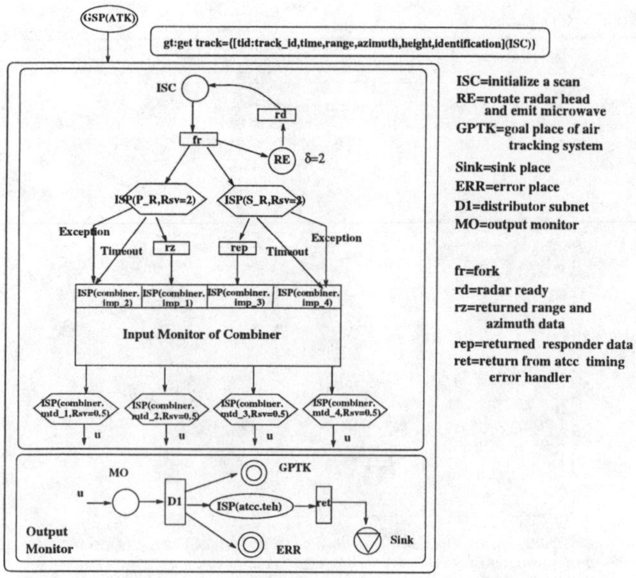

Figure 11. TGN Model for Aircraft Tracking System with Output Monitor

emitted periodically, such behavior is modeled by place RO with time delay $\delta = 2s$. Since the execution of the TGN $G(ATK)$ is a looping task, we must ensure that place ISC is safe (i.e., at most one token for any marking reachable), such that the TGN $G(ATK)$ can always run. The timing constraints for primary/secondary radar plot processing modules can be derived from the physical nature of the radar systems. In this example, the two TGNs are invoked with timing constraint $Rsv = 2s$. After the two processed plots are combined into one plot message, the output monitor of the tracking system will validate, timestamp, and send out the plots through the radar data network to ATCCs. There is a $2.5s$ end-to-end deadline for each scan of radar plots to be sent out. So, there is an intermediate $0.5s$ deadline for each plot combine method, which can modeled by a TGN with time parameter $Rsv = 0.5s$.

In real-time dependable systems, objects interact with each other in various ways, and they also need to reflect the changes of environment. The monitor acting as the knowledge of the SMOs, which enable the objects to dynamic change their behavior in pre-design fashion. In general, a TGN is the basic building block of our model, a TGN can be designed and analyzed for a specific condition. The use of SMO methodology will alleviate the design complexity of real-time dependable applications.

6 Discussion

The Timed G-Net (TGN) methodology offers a systematic approach to incorporate fault-tolerant, fail-safe and vital features into a real-time distributed system. Since each TGN corresponds to an encapsulated object, by adding the smart object concept and the fault-tolerant subnet it is possible to make objects reliable. The meh hierarchy allows the introduction of redundancy at different levels. For example, if a certain part of the system is critical, we can add redundancy and meh's to each level of that part of the system. The remainder of the system can have a simpler fault-tolerance mechanisms added at the higher level of the object hierarchy, thus alleviating the system overhead.

The introduction of the fault tolerant scheme based on smart objects leads to an important characteristic: *by combining the encapsulation of TGNs with the monitoring structure, a designer systematically inserts fault-tolerance and intelligence in a module, without changing its interface.*

Moreover, the meh's play a vital role in the detection and handling of timing faults, which can be rolled forward. The timing constraints of these resilient object structures can be specified. Although it did not receive any attention in this paper, a resource allocation phase is necessary to accommodate all the requests to resource prior to runtime (in order to ensure temporal predictability).

In terms of real-time system analysis, our next step will be to study the properties of TGN by a combination of simulation and analytical approach. We will further investigate how to use the timing analysis results to generate automatic redesign feedback. We also plan to develop a reliability analyzer to use with the redesign feedback mechanism.

References

[1] S. Chang, Y. Chen, and D. Mosse. Smart objects for dependable real-time systems. *Workshop on Object-oriented Real-time Dependable Systems, Dana Point Resort, CA*, Oct. 1994.

[2] Y. Deng, S. Chang, J. de Figueiredo, and A. Perkusich. Integrating software engineering methods and petri nets for the specification and prototyping of complex software systems. In M. Ajmone Marsan, editor, *Application and Theory of Petri Nets 1993*, volume 691 of *Lecture Notes in Computer Science*, pages 206 – 223. Springer-Verlag, Chicago, USA, June 1993.

[3] Y. Deng and S. K. Chang. A framework for the modeling and prototyping distributed information systems. *International Journal of Software Engineering and Knowledge Engineering,2(3)*, Sept. 1991.

[4] Y. Deng and S. K. Chang. Unifying multi-paradigms in software system design. *In 4th Intl. Conf. on Software Engineering and Knowledge Engineering, Capri, Italy*, June 1992.

[5] J. C. Laprie. Dependable computing and fault tolerance : Concepts and terminology. *15th International Symposium on Fault-Tolerant Computing Systems*, June 1985.

[6] D. Mossé. Extending Existing Models to Construct Resilient Computations in Real-Time Systems. Technical Report 93-03, University of Pittsburgh, 1993.

[7] A. Perkusich, J. de Figueiredo, and S. K. Chang. Embedding Fault-Tolerant Properties in the Design of Complex Systems. *Journal of Systems and Software*, November, 1993.

Object-Oriented Implementation of an Infrastructure and Data Manager for Real-time Command and Control Systems

Edward Bensley, Peter Krupp, Ruth Ann Sigel, Michael Squadrito, Bhavani Thuraisingham, Thomas Wheeler

The MITRE Corporation, Burlington Road, Bedford, MA 01730

ABSTRACT

MITRE's Evolvable Real-Time C3 (command, control, and communications) project attempts to develop an approach that would enable current real-time systems to evolve into the systems of the future. The project has chosen AWACS (Airborne Warning and Control System) as an example to test out the concepts and architectures to be developed. In this paper, we describe an object-oriented implementation of an infrastructure and data manager, for next generation real-time command and control systems.

1. INTRODUCTION

Between now and the early part of the next century, significant portions of today's real-time command, control, and communication (C3) systems will become either functionally inadequate or logistically insupportable. Furthermore, due to the continuing budget reductions, new developments of next generation real-time C3 systems may not be possible. Therefore, current real-time C3 systems need to become easier, faster, and less costly to upgrade in capability and easier to support. What is needed is an approach to evolve current real-time C3 systems into the extensible systems required for the future.

MITRE's Evolvable Real-Time C3 project attempts to develop an approach that would enable current real-time systems to evolve into the systems of the future. The candidate evolution approach is to leverage off near-term system upgrade and/or P3I (Pre Planned Product Improvement) activity to put a new architecture framework in place. The emphasis is on transitioning to open architectures which are modular and free from proprietary or unnecessarily complex software designs. The open framework can also accommodate new upgrades more easily. Availability of a suitable software architecture is key for this approach to succeed. The investment plan would continue incremental transition of current systems into more flexible systems. The extensible system architecture would ultimately replace the current hardware and software architecture. The project has chosen AWACS (Airborne Warning and Control System) as an example to test out the concepts and architectures to be developed. Its centralized design is a closed architecture with monolithic custom software. It does not take advantage of state-of-the-art hardware. Processing upgrades to the system is time-consuming and expensive. The project has chosen the multi-sensor integration (MSI) function as a starting point for transitioning AWACS to an open architecture. This is because MSI supports important combat identification capabilities. Also, MSI function's impact on data and display processing provides a thorough test of the concept. The technical challenge is to identify open software technology applicable to AWACS and other real-time C3 systems. The successful execution of this project would substantially reduce the risk of transition to open systems.

In an earlier paper [BENS94], we discussed issues on developing the infrastructure requirements. In particular, we first examined the features of current real-time C3 systems and then determined the systems of the future. Next, we defined various candidate architectures for future systems. Finally, we developed requirements for the infrastructure. The main focus was on operating systems, data management systems, and communication systems.

In this paper we describe the object-oriented implementation of the infrastructure and the data manager. Our infrastructure has focused on the operating system and its associated extensions. Note that while the data manager is part of the infrastructure as far as the application is concerned, it can be regarded as an application by the operating system. Since an evolvable design was a major consideration, we were influenced by object-oriented design and implementation approaches as well as our initial investigation of real-time issues for distributed object management systems. Our infrastructure is essentially a collection of objects interacting with each other. This way existing applications as well as new applications could be encapsulated as objects. The features supported by our real-time infrastructure (RTIS) include real-time scheduling and constraint enforcement, real-time predictability, admission control, overload management, real-time tasks and threads, interprocess/interthread communication, and fault-tolerance and group communication. These features address many of the problems that we have observed in existing C3I systems. The most important problem is the cost of maintenance (evolution and extension). A significant factor driving the cost is the use of the cyclic executive. It results in a global, system-wide dispersal of the knowledge of real-time constraints and behavior of the system. We believe that some of the proposed techniques described in the form of RTIS design requirements will ameliorate this problem significantly. Through our prototyping of a RTIS, we will be able to test our hypothesis.

Our real-time data manager (RTDM) is an application hosted on the RTIS and makes use of the features offered by the RTIS. RTDMs have to meet additional requirements that do not apply to traditional data managers. Data is time-stamped and transactions are time-constrained; they have a deadline and they have a value to the system which varies depending on whether or not they met that deadline. In some cases, missing the deadline may be catastrophic. Other new aspects which must be considered when scheduling transactions are: what is the expected execution time, and is the required data sufficiently up-to-date? While temporal data

0-8186-7570-5/96 $5.00 © 1996 IEEE

managers also contain time-stamped data, these data managers do not require the kind of scheduling that is necessary to meet strict deadlines, as required for functionally complete RTDMs. In addition, an RTDM must also provide the facility to represent the entities, relationships, and constraints of the application, be able to analyze the application for data dependencies, and determine potential inconsistencies.

One of the early tasks was to choose the implementation platform, COTS (Commercial Off the Shelf) products, and the language for implementation. This selection was also constrained by the resources that were available to us. After examining the various options, we decided to carry out the implementation of the RTIS on the Lynx platform [LYNX]. The Lynx operating system provides some of the basic support required by the RTIS. Since and object-oriented approach influenced much of our design, we chose to implement the RTIS in C++.

For the implementation of the data manager, we initially chose the ZIP-RTDBMS (product of DBx Inc. [ZIP]) and encapsulated it as a C++ object. However, this product does not support some of the essential features such as transaction management and concurrency control. Therefore, we implemented a shared memory data manager with concurrency control support ourselves, and ZIP-RTDBMS is being used for storage purposes.

Note that in this paper we describe only the implementation. We have also carried out a detailed design of the infrastructure and data manager. Details are given in [BENS95].

2. INFRASTRUCTURE IMPLEMENTATION

2.1 APPROACH

Some of the critical requirements that have influenced the design include: real-time scheduling/constraint enforcement, fault tolerance/redundancy management, object management, and distributed systems support. We started initially with a real-time distributed object management reference model illustrated in figure 2-1. In this model, applications and resources are encapsulated as objects. The real-world constraints that influenced this approach were the following. Most software/hardware technology does not directly support real-time processing, although there is some support for POSIX 1003.4 (Real-Time Standard) [IEEE] in emerging operating systems such as LynxOS. There is inadequate transparent distribution support in current operating systems. Furthermore, current operating systems do not support heterogeneity (machines, languages, etc.). No operating system supports timing-constraint enforcement directly (timing violation detection and recovery). Any near-term, open, real-time infra-structure must address all these constraints and issues.

Our design is based on real-time objects which combine C++ objects and Posix 1003.4 threads. Objects are treated as common real-time system components/entities. Polymorphism is used to map real-time operations supported by multiple implementations onto their correct run-time implementation. Abstract classes represent common characteristics of real-time activities (deadline, schedulability, workload, etc.). Concrete classes represent realizations (threads with hard deadlines invoked periodically, message streams with periodic arrivals and hard deadlines, sporadic tasks, etc.). Container classes are used to encapsulate non-real-time components.

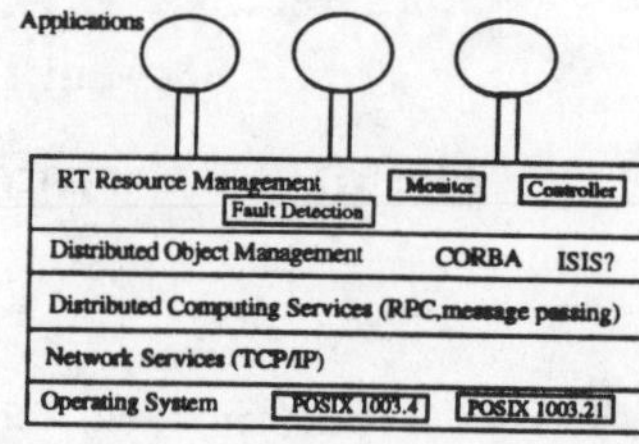

Figure 2-1 Real-time Distributed Object Management Reference Model

The RTIS Application Programming Interface (API) consists of a C++ class library. This is needed to encapsulate the creation of schedulable units of work (processes and threads), provide abstractions for application development, and to utilize polymorphism and inheritance to reduce real-time system complexity. API functions (within a single infrastructure object) include create and destroy constrained threads, create and destroy synchronization objects, provide interprocess/thread communication, time constraint enforcement and exception processing, register and unregister threads, and synchronization of objects with priority service. Priorities are assigned as directed by a priority service.

2.2 REAL-TIME INFRASTRUCTURE LIBRARY

We have chosen LynxOS as our operating system platform for the development of the Real-Time Infrastructure Library. Lynx gives us the following features:

- Real-Time pre-emptable kernel.
- POSIX compliance for most of the POSIX real-time extensions.
- POSIX threads.
- Support for Priority Inheritance and Ceiling.
- Multi Hardware Platform availability.
- Availability of per-thread timers.

We created the Real-time Infrastructure Library to encapsulate and extend the use of the POSIX and Lynx Real-time features. Using POSIX threads and the LynxOS Real-time features can be confusing and requires a large quantity of setup code before using the features provided. We have eliminated this set up code with the class library. We have also extended the functionality of threads by allowing them to have their priorities set by a server that calculates the priority based on the period of a task, have timing constraints based on CPU usage or other timing criteria enforced, and to be controlled by events generated by other processes or threads.

The library encapsulates the POSIX and real-time portions of the operating system. The library also emulates some features we believe are important that are not yet in the standard or are not directly supported by Lynx. Examples of these are POSIX threads which are not yet in the standard. Lynx POSIX threads (pthreads) are based on Lynx system threads and as such do not quite meet the proposed standard. Semaphores, mutexes and condition variables which support priority inheritance are a part of the

POSIX standard but the implementation of them by Lynx is not POSIX compliant. Our library classes for these synchronization objects makes up for this lack of support.

This encapsulation enables the user to concentrate on what he/she needs to do rather than on how the operating system works. Additionally this allows for greater portability of application code that uses the features provided by the library. Much of POSIX has not yet been implemented in the various vendor versions of UNIX. Therefore it is important that users be able write their code without concern to standards, operating systems, or hardware. As operating systems and the standard further develop only the classes of the library will need to be modified, not the applications themselves.

2.3 ASSOCIATED TOOLS

1. System Monitor (watch)

This standalone program has been developed to provide information about processes and threads running on the system. It may be used as a more complete version of ps command. The UNIX ps command reports information regarding processes, the watch command reports information for both processes and the individual threads they contain. The program accesses the kernel data structures for all existing processes and their "child" threads. Information found in the data structures includes, the state of a thread (i.e., running, on ready queue, etc.), name and id of the parent process, owner of the thread, priority, pointers to and owner of any semaphore or mutex that the thread is waiting on, etc.

Since this program currently runs in user space it does have the limitation of only seeing itself as the active process. We need to further investigate whether running this in the kernel, as a pseudo device driver, will remove this limitation.

2. Priority Server

The priority server is a stand-alone RPC program. Although the client and server stubs and the server itself are not part of the Real-Time C++ infrastructure library, there is a class within the infrastructure library to interface with the server.

Any thread that is created using the Infrastructure Library and that requires a priority other than the system default will have its priority calculated and assigned by the Priority Server. This calculation is based on the periodic rate at which the thread runs as well as various system wide parameters. Technical details and interface information on using this server can be found in the Infrastructure Implementation Appendix.

Processes and threads that do not use the Infrastructure Library may use the server directly.

NOTE: We assume that all users of the system will set priorities, other than system default, through the use of this server. It is essential, to ensure the schedulability of threads and processes, that all processes use the server. There is no way to enforce the use of the priority server by a process and/or threads unless the program is using the Infrastructure Library for all of its thread code. The user must never make operating system calls to change priorities either from within the code or from their terminal.

3. CPU Time Device Driver (timdrv)

The device driver, timdrv, monitors CPU usage on a per-thread basis for all threads on the system. At every clock tick (currently every 10 msecs), the driver increments a time count for the thread id that is currently running. All threads, whether created by the library or directly by the user will be timed by this driver. Threads that use the Infrastructure Library will get CPU usage counts (in ticks) by querying the driver. We need to further investigate whether the granularity of this timing method is fine enough for our purposes or that this method of timing CPU usage is accurate. This will be examined during fiscal year '96.

2.4 REAL-TIME INFRASTRUCTURE LIBRARY CLASSES

NOTE: At present, there are three groups of classes: miscellaneous, threads, and synchronization. The Interface, interaction, and usage of the classes are described in the Infrastructure Implementation Appendix.

The terms base class and abstract base class which are used below are two variants of classes which may be inherited by other classes. A base class contains functions and data which are completely defined and although the class may not be complete, may be used directly. An abstract base class is one in which one or more member functions are declared but not defined. This forces any classes which derives from them to implement a function with the exact name and calling sequence.

2.4.1 Miscellaneous Classes

EVENT_SIG

Class EVENT_SIG handles all POSIX user defined events to be sent to processes or individual threads. This does the following:

- Setting up POSIX events
- Waiting for an event.
- Sending an event.

MUTEX_ATTR

Class MUTEX_ATTR is the base class from which the standard Lynx mutex is derived. It handles the initialization and deletion of the Lynx data structures for a Lynx mutex. This class is not visible to the user and is to be used only by derived classes.

PTHREAD_ATTR

Class PTHREAD_ATTR is the class which handles setting up all of the attributes of a thread. It is a requirement of Lynx that all threads have a set of attributes created and initialized before the thread is created. This class will never be used directly, it is accessed only by Class THREAD. The class implements the following attribute features for a thread.

- Set stacksize
- Set inheritance policy (does the thread inherit the attributes of its parent or does it have its own).
- Set scheduling policy.
- Set priority.
- The class allows queries which will return the various values of attributes to the user.
- Deletion of the attributes from the operating systems tables when a thread exits.

PRIORITY_SVC

The class PRIORITY_SVC enables the user of the Real-time Infrastructure library to interface to the priority server without knowing about the RPC server or client. The user does not have to understand how RPC works, only what arguments and return values are expected by the PRIORITY_SVC class. The class supports the following priority server functions:

- Assigning priority
- Releasing priority
- Calculating priority

THREAD_ERR

Class THREAD_ERR is nothing more than a place holder that will be replaced when a more robust fault tolerant error handler has been designed. At this point in its development it accepts a defined constant describing the type of error and whether the error is fatal or not. The class prints out an error message and exits or returns. During the next phase of development the implementation will be changed to handle errors in a manner appropriate for Real-Time threads and fault tolerant systems.

2.4.2 Thread Classes

Figures 2-2 and 2-3 illustrate the thread classes. A description of these classes is given below.

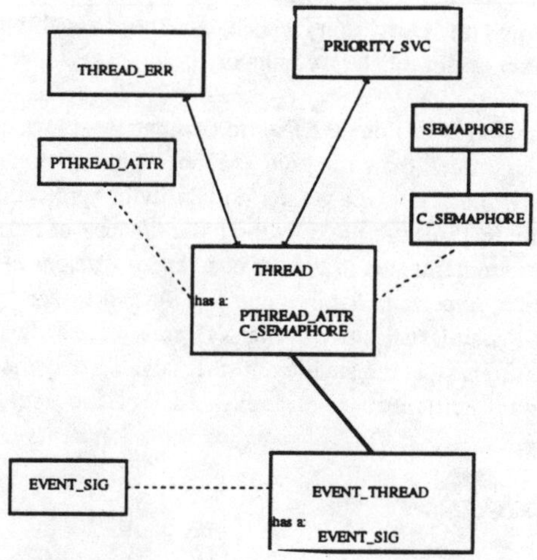

Figure 2-2 Threads-I

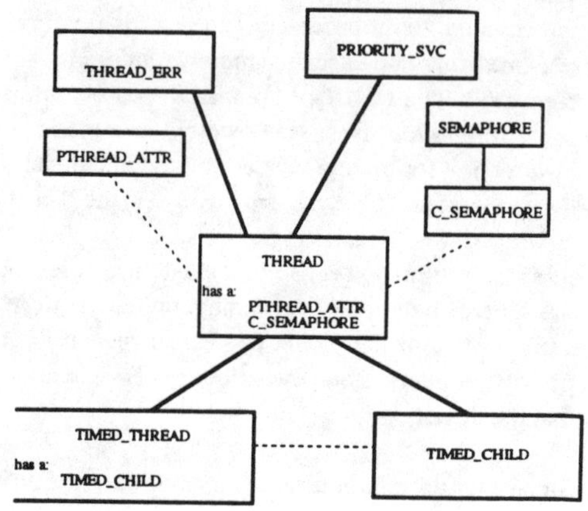

Figure 2-3 Threads-II

THREAD

Class THREAD is an abstract base class which can be used only when defining a new class which is derived from it. This class is the highest level class from which all thread class types are derived. The class supports the following features:

- POSIX thread creation
- Thread Exit
- Cancel Properties: Ability to cancel thread synchronously or asynchronously. Synchronous cancellation allows the thread to prohibit cancellation while it is executing critical sections of code.
- Query thread's CPU time
- Push and pop cleanup routines on the threads stack which can be executed before the thread exits.

EVENT_THREAD

Class EVENT_THREAD is the abstract base class for all threads which require and are controlled by an event handler. This allows a thread's execution to be controlled by events sent to it by other processes or threads. In addition to the attributes inherited from threads, this class supports:

- Creating an event driven thread
- Sending an event to a thread.

TIMED_THREAD

Class TIMED_THREAD is the basic class of threads used in the real-time infrastructure library. It may be used as implemented, or it may be used as a base class. In addition to the functionality of a basic thread it encapsulates the properties of real-time constraints and deadlines. The thread manages a series of thread instances which perform the required actions on a periodic basis. The priority server assigns a priority to this thread based on its period/rate. Optionally, it can implement a maximum time that each thread can run before it is forced to exit. Before the thread is forced to exit it is required to cleanup any locked resources that it may be holding. The class supports the following:

- Creation of a periodic timed thread.
- Creation of a periodic timed thread with an execution deadline.
- Specification of an abort routine if a thread has missed or will miss its proscribed deadline.

TIMED_CHILD

Class TIMED_CHILD is a thread class that can only be created by a TIMED_THREAD instance. This creates the actual periodic thread which is constrained by the periods and deadlines given to the TIMED_THREAD at creation. The abort routine specified to the TIMED_THREAD is used when this thread misses its deadline to ensure that all system resources are freed before the thread exits. Without an appropriate abort routine the thread might leave synchronization objects locked, causing the system to behave unpredictably.

2.4.3 Synchronization Classes

Figure 2-4 illustrates the synchronization class hierarchy. A description of the classes are given below.

SEMAPHORE
Class SEMAPHORE is an abstract base class from which standard Lynx counting semaphores and Lynx fast (priority inheritance) semaphores are derived. This class contains the virtual functions of wait and post that all semaphores must have. This class can never be directly used by the programmer.

FA_OBJ
Class FA_OBJ is another support class found in the library. This class does the set up in the Lynx data structures for Lynx fast semaphores, mutexes, and condition variables used to support priority inheritance.

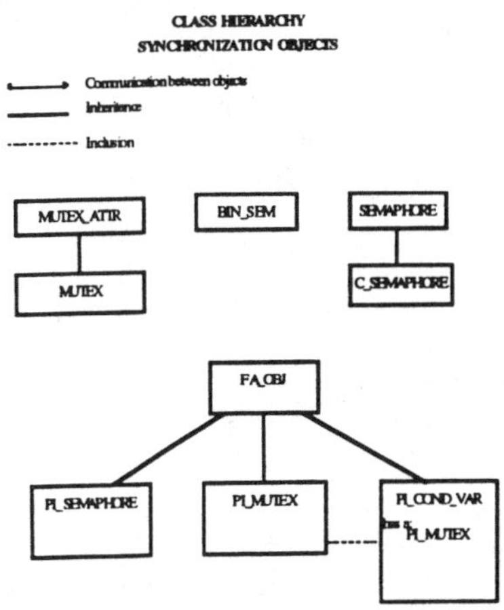

Figure 2-4 Class Hierarchy Synchronization

Note: This class also supports priority ceiling. Currently there is a limitation with the use of this class and those derived from it. We have not as yet been able to make priority ceiling work when the synchronization object is a class object. Priority ceiling, however, does work when the implementation is done outside of a class implementation.

C_SEMAPHORE
Class C_SEMAPHORE supports standard Lynx counting semaphores. These are ordinary UNIX semaphores which allow an initial count to be set when the semaphore is created. In addition to wait and post, which it inherits from class SEMAPHORE, it adds the functionality of try_wait. Try_wait allows the thread to do something else if it can not get the semaphore.

PI_SEMAPHORE
Class PI_SEMAPHORE supports priority inheritance semaphores for use within a single process. They appear to the user as ordinary semaphores but have the property of priority inheritance to prevent cases of priority inversion on the semaphore.

BIN_SEM
Class BIN_SEM supports named priority inheritance semaphores which are capable of use across processes running on the same CPU.

MUTEX
Class MUTEX supports standard Lynx mutexes. The mutex will work within a process. Unlike the semaphore, a mutex has the concept of an owner. Only the thread which locked a mutex may unlock it. Care must be exercised when a thread is killed or does a normal exit that the synchronization objects that it holds have been released.

PI_MUTEX
Class PI_MUTEX supports priority inheritance through inheritance of FA_OBJ and works in a similar fashion to the PI_SEMAPHORE. It cannot be released except by the thread that locked it. This may be used in conjunction with the PI_COND_VAR.

PI_COND_VAR
Class PI_COND_VAR supports condition variables, which must be used in conjunction with a PI_MUTEX. It supports condition variable functions such as:

- Wait,
- Signal,
- Signal and Unlock the associated mutex,
- Broadcast,
- Broadcast and Unlock the associated mutex,
- Lock associated mutex.

3. DATA MANAGER IMPLEMENTATION

3.1 APPROACH

A real-time data manager manages persistent and temporal data, ensures that the queries and transactions meet the timing constraints, enforces temporally correct serializable schedules, and combines techniques from data management and real-time scheduling.

Our goal was to implement a two-level data manager; shared memory data manager to manage the shared memory database needed for real-time applications; persistent storage data manager to manage persistent database. The persistent database is larger than the shared memory database and is needed for long-term storage. Interaction between the two data managers is as follows: The shared memory data manager manages the data for the operation of the application subsystem. If data is not available, the shared memory data manager communicates with the persistent storage data manager to retrieve data. Caching data may be performed in advance. At every interval, the shared memory data manager sends all updates to the persistent storage data manager. Figure 3-1 illustrates the two data managers.

Our initial plan was to use ZIP-RTDBMS for both data managers and make the necessary changes. However we found that implementing transaction management and concurrency control with ZIP-RTDBMS could not be accomplished within the resources allocated to the project. Therefore, we used ZIP-RTDBMS as a persistent storage manager and encapsulated it as a C++ class. This implementation is discussed in section 3.4.2. We implemented the shared memory data manager. This implementation is discussed in sections 3.4.3 - 3.4.6. In particular, the implementation of the data manager and the associated C++ classes are discussed. Note that we have not implemented the interaction between the shared memory data manager and the persistent storage data manager.

Figure 3-2 is a representation of the current Data Manager implementation design. The unshaded boxes are those portions which have been implemented or obtained as a COTS product.

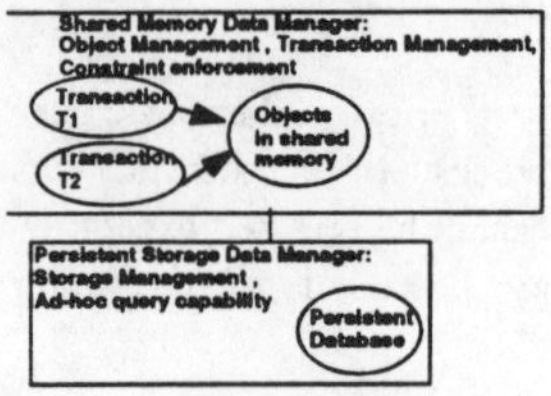

Figure 3-1 Implementation Approach

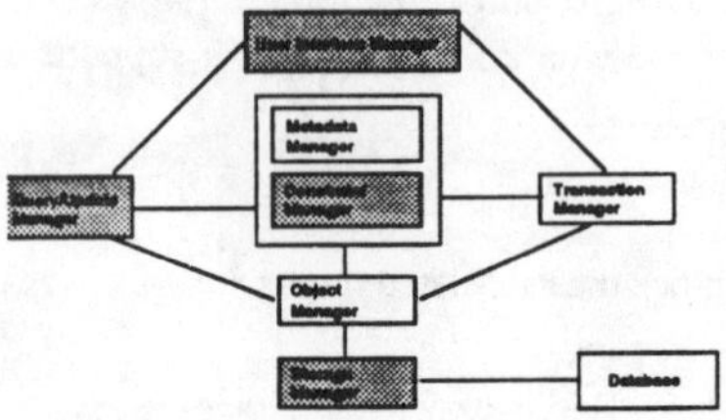

Figure 3-2 Functional Modules of the Data Manager

3..2 ENCAPSULATED Zip_RTDBMS

Zip_RTDBMS is a COTS Real-Time database management system. Typically a user acting as a server will create a database for use by themselves and other client users who may connect to the database. A database is instantiated by inputting a schema file into Zip which specifies all relations in the database. Queries are written using Zip 'C' functions and compiled. All data manipulation language (DML) update and select queries must be written in post-fix form. The encapsulation was done to make Zip more user friendly and evolvable. The encapsulation consisted of C++ wrapper classes and an SQL to Zip parser.

The Zip Wrapper Classes

Three C++ classes were implemented to hide various Zip features.

1) Class **Zip_rtdbms** contains all of the DML functions which Zip offers, such as insert and select. Each wrapper function corresponds to a Zip function, i.e., wrapper function **Zip_rtdbms::insert_into** calls the Zip function **insertDB**. Each wrapper function also handles errors returned by the Zip functions, which simplifies the users programs. This class also establishes the connection to the Zip server database when a **Zip_rtdbms** object is instantiated.

2) Class **relation** contains and initializes all of the attributes of a given Zip relation. The user creates a **Zip_rtdbms** object which has as members, **relation** objects that are contained within the database. Since the database is instantiated with the schema file, a client or server may elect to only specify those relations specific to their current queries.

3) Class **attr_t<class Type>** hides the Zip information required for each attribute (column) in a given relation. A user of a Zip database is required to specify four fields for each attribute, the connection id of the database, the character name of the column in the relation, a pointer to the actual variable being used as the buffer, and a pointer to a structure containing the size and type information of the variable. Class **attr_t<class Type>** contains all of the information that Zip needs for a users buffer. All that is required is to instantiate an attribute of the required type, i.e., attr_t<int> column1_buffer.

To use these classes a server or client would derive their database from the Zip_rtdbms class and specify any relations that are going to be accessed by their DML, as shown below.

```
#include "relation.h"
#include "zip.h"

class stu_adv_server : public Zip_rtdbms {
 public:
  stu_adv_server() : Zip_rtdbms("/tmp/student_advisor.schema"),
                     student(masterDBid) { }

  student_t  student;
};
```

When the stu_adv_server class is instantiated, the variable masterDBid is assigned a connection address which is used by the relation class to initialize the attributes. The class student_t is derived from the class relation and contains all attributes required by the users DML queries, as shown below.

```
class student_t : public relation {
 public:
  student_t(DBId_t * masterDBid) :
     relation(masterDBid, "student", ROLLING, 1000),
     id(this, "id"),
     name(this, "name"),
     advisor(this, "advisor"),
     grad_date(this, "grad_date")    { }

  attr_t<byte4_t>   id;
  attr_t<byte_t>    name;
  attr_t<ubyte_t>   advisor;
  attr_t<date_t>    grad_date;
};
```

The SQL to Zip Parser

The C++ wrapper encapsulation hides many Zip specific details but the user is still required to express the DML in terms of Zip functions and in post-fix notation. The SQL to Zip parser enables the user to express the DML in embedded SQL and use the parser to convert the query into Zip wrapper function calls which can then be used by a C++ compiler. The parser also generates the appropriate derived **relation** classes from the data definition language (DDL) schema file. An example SQL query and its parsed Zip wrapper equivalent are shown below.

Example SQL SELECT query :

```
EXEC SQL DECLARE tld1 CURSOR FOR
        SELECT *
        FROM student
        WHERE ( advisor      = :advar
        OR      advisor      = 'Vic' )
        AND     grad_date    <= 'Dec' ;
```

Output of previous SQL query in post-fix form :

```
cursor_t tld1;

strcpy(student.advisor.value, advar);
declare_predicate(student.advisor.name, EQ, &student.advisor.value,    &student.advisor.type);

strcpy(student.advisor.value, "Vic");
declare_predicate(student.advisor.name, EQ, &student.advisor.value,    &student.advisor.type);

log_op = OR;
declare_logop(log_op);

strcpy(student.grad_date.value, "Dec");
declare_predicate(student.grad_date.name, LEQ, &student.grad_date.value,
&student.grad_date.type);

log_op = AND;
declare_logop(log_op);
```

3.3 CONCURRENCY CONTROL

Before discussing the other major components of the Data Manager, it is necessary to discuss the various techniques that are and will be used to manage multiple executing transactions. The following classes make use of the basic infrastructure concurrency control classes, which include the class **BIN_SEM** and the class **PI_COND_VAR**.

Preemptable Mutex / Condition Variable Class

Derived from the infrastructure **PI_COND_VAR** class, the preemptable class **PREEMPT_PI_MUT_CV** allows a transaction to be interrupted and aborted without leaving any data structures in an inconsistent state. It provides the ability to lock and unlock a mutex, which may be used in conjunction with a condition variable. A condition variable is used to signal other users of the mutex that a particular condition has been satisfied and that the mutex is available for use. After a mutex is unlocked or a condition variable is signaled, a test is done to find out if an abort signal was issued to the transaction while the mutex was locked. The test function is a built-in LYNX feature. Only one user may lock a mutex at any time.

Readers / Writer (Latch) Class

This class allows many readers or one writer to lock an object. It can use either the binary semaphore class or the priority inheritance mutex / condition variable class, which are both available in the Real-time infrastructure class library. When a user performs a read lock, a count is incremented to indicate a reader is accessing the guarded data structure. More readers are allowed to access the data structure, but writers must wait until the reader count is zero. Only one writer is allowed access at any time. This class also provides the ability to upgrade a lock from a read to a write.

Semantic Locking Mechanism (SLM)

This implementation was acquired from the University of Rhode Island and ported from the Solaris operating system to Lynx OS version 2.2.2. The SLM allows object locking on the method level, affording more concurrency than basic object read / write locking. The SLM is actually a family of C++ classes which operate together. Each object must be specified in a predetermined manner, deriving from a **Base** class. Objects must use the **Attr** class to specify each attribute that will be manipulated and accessed by users of that object. Each method of the object will access the attributes through the Attr member functions **Read** and **Write**.

The object is parsed to determine the read / write affected sets for each method. The read affected set is the set of attributes read by a method and the write affected set is the set of attributes written by a method. From the read / write affected sets, a compatibility matrix is defined which identifies which methods can run concurrently with other methods of the same object. Each method pair has a compatibility function which is used to determine if the methods are compatible. The user is allowed to manipulate the compatibility functions, therefore creating specific semantics under which the methods may run.

Read/Write Affected Set Priority Ceiling Protocol

This algorithm is an extension of the Basic Priority Ceiling Protocol and the Read/Write Priority Ceiling Protocol. Both of these protocols operate on the object locking level and provide prevention from deadlock and bound priority inversion to one lower priority transaction. Priority inversion occurs when lower priority transactions hold locks which prevent a higher priority transaction from running.

The Read/Write Affected Set Priority Ceiling Protocol provides the same benefits of the previous protocols, except that it works at the method locking level. This allows for greater concurrency than the previous protocols, while at the same time offering the same benefits of no deadlock and a bounded priority inversion of one.

Although this algorithm has not been fully implemented, the algorithm for determining the priority ceilings of the methods has been implemented in the form of a parser as long as the objects are in conformance with SLM objects (as described above).

3.4. MEMORY MANAGERS

We first describe the dynamic shared memory manager which we also refer to as the object manager. To implement the Meta Data and Transaction Managers, it was also necessary to implement some other memory managers. We also describe these other memory managers.

Dynamic Shared Memory Manager

This implementation was acquired from the University of Rhode Island and ported from the Solaris operating system to Lynx OS version 2.2.2. It allows dynamic allocation and deallocation of objects in shared memory. This **SharedHeap<int m>** class is currently being used as the Object Manager in a limited capacity. The Transaction Manager has access to the objects through the **SharedHeap**, however, since there is no Storage Manager, there is no connection to the Zip Database.

To use this class a process first creates a new shared memory heap or attaches an existing one to it's virtual address space. A process allocates objects using an overloaded C++ new operator. The process then registers an object id with the Shared Memory Manager (done through the Transaction Manager). Any thread running within that process can retrieve an object from shared memory if it knows the object id.

Static Memory Manager

This class was created to take advantage of the *a priori* information in a Real-Time system. The class may be used with or without shared memory. The user pre-allocates the desired type and quantity. Once allocated, transactions can use an overloaded C++ new operator to retrieve a block of the desired type. This class cannot retrieve an object by 'id'.

Hash Table Class

This class was needed to keep track of transaction id's in the Meta Data Manager. The class uses a template so it will work with any user defined type. The only requirement is that the user type contain a member named 'id' which is used by the hash function.

The class may be used with or without shared memory and uses a **PREEMPT_PI_MUT_CV** to allow concurrent and preemptable thread access to the Hash Table structures. The class can currently only hash *int* 'id' values, but through inheritance, any 'id' type may be hashed.

3.5 METADATA MANAGER

This implementation was based on [BILI95]. In its current form the Metadata Manager **TXN_META_DATA** class is simply a storage device. Any manipulation of the metadata contained within it is performed by external classes. Currently the only friends which use the Metadata Manager are the Transaction Manager and Transaction Clean-Up class. The Metadata Manager contains three major structures for storing information :

1) The Transaction Descriptor (TD) structure contains information for a transaction. A transaction is a single thread within a process. The status of the transaction is recorded, and can be either initiated, running, committed, or aborted. The TD stores the transaction id, which is used to locate a TD in the hash table. The TD id is the same as the thread id that is executing the transaction. A pointer to the infrastructure **THREAD** object which was instantiated for the transaction is also stored, so when the transaction status is changed from initiated to running, the thread's function may be executed. Finally, the TD keeps a list of all locked objects for the transaction.

2) The Object Descriptor (OD) structure contains information for an object. It contains the object's id, supplied by the user, and the object's size. This structure also contains a list of granted and pending locks for transactions on the object. Although the term object is being used, it actually could refer to any lockable entity, such as a method of an object. This means that this structure could be used with any of the concurrency techniques that were previously discussed.

3) The Lock Request Descriptor (LRD) structure contains information for a lock on a particular object. This structure contains a pointer to the TD structure of the transaction who owns the lock, a pointer to the OD structure of the locked object, whether the lock is a read or a write, and whether the lock is granted, pending, or released. This structure also contains a shadow copy of the locked object, which is used by the transaction until that transaction is committed.

The Meta Data Manger may be used with or without shared memory. This class uses the Static Memory Manager class to preallocate all of the required structures in shared memory and the Hash Table class to locate the TD structure for a particular transaction . It also contains a **PREEMPT_PI_MUT_CV** class to allow concurrent and preemptable thread access to the Meta Data Manager structures.

3.6 TRANSACTION MANAGER

This implementation was based on [BILI95]. Through the **TXN_MAN** class transactions manipulate the Metadata Manager to log their lock requests. This class supports both periodic and non-periodic Real-Time infrastructure threads.

Periodic Threads

The **TIMED_TXN_THREAD** class is derived from the infrastructure class **TIMED_THREAD**. It has as a **TXN_CLEANUP** class as a member.

Non-Periodic Threads

The **TXN_THREAD** class is derived from the infrastructure class **THREAD**. It has a **TXN_CLEANUP** class as a member.

Transaction Clean Up

Since both types of threads need to deallocate Meta Data Manager structures when they are destroyed, the **TXN_CLEANUP** class was made to centralize the destructor function without having to use multiple inheritance, which seemed unnecessary for this simple function. When a periodic or non-periodic thread is deleted, the destructor for the **TXN_CLEANUP** class is called to deallocate any Meta Data Manager structures still being used by the transaction.

Use of Transaction Manager

Transactions are 'initiated' to allocate resources. At this time the appropriate thread class is instantiated, and the TD structure is placed in the hash table. If initiated, the transaction is executed by a call to the function 'begin'. The transaction manager changes the transaction's status to running and starts the threads function. The transaction manager presently issues read and write locks on objects stored in shared memory. Deadlock is avoided by explicitly ordering the locks. This could be changed in the future by implementing the read/write affected set priority ceiling protocol to allow locking on the method level with no deadlock.

The Transaction Manager uses two-phase locking to maintain serializability. The user may make as many locks as they wish, but cannot make any more locks after a lock has been released. The locks may be released one at a time throughout the transaction and/or released all at once when a commit or abort is performed. Releasing the locks individually allows more concurrency with incompatible transactions.

When a transaction finishes executing it can commit the changes to the database and release any remaining locks. If a periodic transaction does not meet it's deadline, a signal is issued and the transaction is aborted. An abort releases any locks the transaction may have had.

Another function related to the commit/abort functions is the 'wait' function. The function wait returns a true value if the transaction in question commits. It may be used to express dependencies between transactions. A transaction will wait to commit until a certain transaction or group of transactions commits. Depending on the logical expression used, any number of dependencies may be expressed. For example:

Transaction self will commit if (T1 commits) OR
(T2 AND T3 commit).

```
if ( wait(T1) OR (wait(T2) AND wait(T3) )
    commit(self);
else
    abort(self);
```

Transaction Issues

A transaction either commits and updates the shared memory or aborts and leaves the shared memory untouched, therefore satisfying the atomic requirement of the ACID (Atomicity, Consistency, Isolation, Durability) property for databases. This Transaction Manager also fulfills the consistency requirement, since any changes are not reflected in the database unless a transaction commits. The isolation requirement is also satisfied because each transaction has it's own version of an object. Therefore, a transaction works on an object in isolation from other transactions. The durability requirement currently cannot be met completely at this time. There are several aspects to durability, which concern the state of the database in the event of a failure of some form.

Since the data resides in shared memory, if the system were to lose power, the data would be lost. This situation will be remedied when a persistent store is added to the Data Manager. Another situation of failure is when the process running the transactions is terminated abruptly. When this occurs, a single process can be started to traverse all active TD structures and finish aborting and committing based on the current state of the TD status flags. Once the objects have been placed in a consistent state, the TXN_META_DATA may reinitialized to reset the structures and replace any mutexes that were left in a locked state.

4. CURRENT STATUS

In this paper we have described our object-oriented implementation of the infrastructure and data manager for real-time command and control systems. In addition to the work described in this paper, we have also carried out a detailed design of the infrastructure and data manager which includes several additional features not described in the implementation. One of the major contributions of this work is a concurrency control protocol which incorporates the semantic locking mechanism described in [WOLF94] with the [priority ceiling protocol,. described in [RAJK89]. We have also completed an object-oriented implementation of the MSI application. This implementation was carried out in the SGI [SGI] environment. Details of the design and implementation are given in [BENS95].

Our current work is proceeding in the following directions. We are porting the application implemented in the SGI environment to the Lynx platform. The application will be hosted on the infrastructure and will make used the data manager for real-time data management. We are also investigating real-time issues for distributed object management systems such as OMG's (Object Management Group's_ CORBA (Common Object Request Broker Architecture. We have described some preliminary ideas in [KRUP94]. Finally, we are investigating the use of an Object Request Broker (ORB) to be included in the implementation. The ORB will be a glue for the infrastrucutre, data manager, and the application.

REFERENCES

[BENS94] Bensley, E. et al, Evolvable Systems Initiative for Real-time C3, Volume I: Infrastructure Requirements,
MTR 94B 0000116, September 1994, Unclassified, NOFORN; A version also published in the Proceedings of the IEEE Complex Systems Conference.

[BENS95] Bensley, E. et al, Evolvable Systems Initiative for Real-time C3, Volume II: Design and Implementation of the Infrastructure, Data Management, and Application, MTR 95B 0000127, September 1995

[BILI95] A. Biliris, S. Dar, N. Gehani, H. Jagadish, K. Ramamritham, "ASSET: A System for Supporting Extended Transactions," SIGMOD, 1995.

[IEEE] IEEE POSIX 1003.4 Working Group, *POSIX Std 1003.1b-1993 (P1003.4) Real-Time Extension*. 1994, IEEE.

[KRUP94] Krupp P., et al. Real-Time Processing and CORBA, presented at the OOPSLA 94 Conference Workshop on CORBA, Portland, OR, 1994.

[LYNX] Lynx Operating System Reference Manual

[RAJK89] Rajkumar, R. , *Synchronization in Real-Time Systems: A Priority Inheritance Approach*. Real-Time Systems, ed. J.A. Stankovic. 1991, Boston: Kluwer Academic Press.

[SGI] SGI Reference Manual

[WOLF94] V. Wolfe, J. Prichard, L. DiPippo, J. Black, J. Peckham, and P. Fortier, "The RTSORAC Real-Time Object-Oriented Database Model and Prototype," Technical Report, University of Rhode Island; a version also published in the 1993 IEEE Real-time Systems Symposium.

[ZIP] ZIP-RTDBMS Reference Manual

ACKNOWLEDGMENTS

This work was carried out at the MITRE Corporation during FY95 under contract F19628-94C-0001. We gratefully acknowledge the Air Force Mission Oriented Investigation and Experimentation Program for supporting the work described in this report. We thank Thomas Lawrence of Rome Laboratory for monitoring the project, the AWACS program office and MITRE members of the AWACS project for their support, various groups from Industry and Academia for the interactions and information provided to us, and MITRE management for their support. We also thank Larry Fisher, Gary Gengo, Jim Houchens, Arkady Kanevsky, Soohee Kim, and Alice Schafer for their inputs into this project. Finally, we thank Prof Victor Wolfe of the University of Rhode Island and his real-time systems group (John Black in particular) for many useful suggestions, comments, discussions, and also for supplying the code for part of the data manager implementation.

Object-Oriented Method for Real-Time Systems Based on Timed Automaton

Satoshi Yamane
Department of Computer Science, Shimane University, Japan

ABSTRACT

It is useful to introduce object-orientation into analysis and design methodology for real-time systems[1]. We have proposed the object-oriented methodology for real-time systems, which is integrated with timed automaton[2]. But using proposed method, the verification cost is very large. In this paper, we focus on the verification of dynamic model and propose two effective verification methods as follows.

(1)explicit enumeration method

We represent state transitions as list structures. Using depth's first search algorithm, we search SCC(Strongly Connected Components)[3] satisfying acceptance conditions and test whether the SCC is reachable from initial states. At the same time, we check timing constraints.

(2)implicit enumeration method

We represent state transitions as BDD(Binary Decision Diagram)[4]. Using inverse image computation, we search SCC satisfying acceptance conditions and test whether the SCC is reachable from initial states. At the same time, we check timing constraints.

We have developed the verification system and shown our method effective.

1. Introduction

Real-time systems consist of many concurrent processes and many states, and timing constraints are strict. It is necessary to model and specify ,verify real-time systems including timing constraints. There have been many development methods and specification methods such as real-time structured analysis and timed Petri net, timed automaton, timed process algebra, real-time temporal logic [1]. But there are many problems in these methods, such as lack of modeling technique and intuitonic understandability, documentability.

On the other hand, object-orientation is promising in database and information processing bacause of modeling technique and information hiding, polymorphism, inheritance[5]. Moreover, in real-time systems, object-orientation is effective in following points.

(1)encapsulation

As real-time systems have many processes and states, system specification has huge numbers of states. But as system states are divided into local states of each object, these can be encapsulated. From this point, the large system can be specified.

(2)concurrency

Real-time systems consist of many concurrent processes. As object is computed by message communication and logically autonomous, object is suitable for concurrency.

As object-oriented analysis/design method for real-time systems, there are many methods such as OMT(Object Modeling Technique)[6] and Schlaer/Mellor method[7], Booch method[8], Buhr method[9], ROOM(Real-tome Object-Oriented Modeling)[10], Objectchart[11]. There are various problems in these methods and specification languages as follows.

(1)There is no methodology in order to formally and explicitly specify and verify timing conditions.

(2)There is no methodology in order to clarify relations between object model and dynamic model, functional model.

In order to solve problems, we have proposed our object-oriented methodology for real-time systems as follows[2].

(1)We represent dynamic models using timed automaton and verify it using language inclusion algorithm.

(2)In order to clarify relations between object model and dynamic model, functional model, we define the relations between associations of object model and other models.

But in (1), the verification cost is very large. In this paper, we focus on the verification of dynamic model and propose two effective verification methods as follows.

(1)explicit enumeration method

Using depth's first search algorithm, we search SCC (Strongly Connected Components)[3] satisfying acceptance conditions and test whether the SCC is reachable from initial states. At the same time, we check timing constraints.

(2)implicit enumeration method

We represent state transitions as BDD(Binary Decision Diagram)[4]. Using inverse image computation, we search SCC satisfying acceptance conditions and test whether the SCC is reachable from initial states. At the same time, we check timing constraints.

The structure of the rest of the paper is as follows. The real-time object-oriented methodology is proposed in Section 2. The timing verification method of dynamic model is proposed in Section 3. The method is shown effective by automobile example and implemented tools in Section 4. The conclusion is in Section 5.

0-8186-7570-5/96 $5.00 © 1996 IEEE

2. Real-time object-oriented methodology

This section explains real-time object-oriented methodology. This section assumes that class is equal to object and object has no creation /destruction.

2.1 Real-time object-oriented specification

Real-time object-oriented sepcification consists of object model and dynamic model, functional model based on OMT as follows.

(1)object model

Object model represents object and its association. Object's definition consists of name and attributes, operations, and associations ,which consists of inheritance and aggregation, refer/use.

(2)dynamic model

Dynamic model represents object behaviors by timed statechart[12,13], and consists of states and events, timing constraints, functions. Each dynamic model keep balances of events all over the system.

(3)functional model

Functional model represents functions by dataflow diagram, and consists of dataflows and functions, stores. Each functional model keep balances of data all over the system.

(4)relations between three models

The relations are defined according to associations as follows.

(a)Case of aggregation

Aggregation maps dynamic model into concurrent statechart, and maps functional model into concurrent program.

(b)Case of inheritance

Inheritance maps dynamic model into hierarchical statechart, and maps functional model into differential program.

(c)Case of refer/use

Refer/use maps dynamic model into sequential statechart, and maps functional model into sequential program.

(5)other relations

Attributes and operations of object model depends on following factors of dynamic model and functional model.

(a)Attributes correspond to states of dynamic model and stores of functional model.

(b)Operations correspond to functions of dynamic model and functional model.

2.2 Syntax of specification language

Syntax of real-time object-oriented specification language consists of twelve definition parts such as class name and attribute, operation, etc. Syntax is defined in Fig.1. In Fig.1, State transition definition part (9) has notable feature as follows.

(9)State transition definition part

Dynamic model is based on timed statechart combining timed Muller automaton and statechart.

Here, we formally define timed statechart, safety conditions.

[Definition 1](timed statechart)

Specification language consists of nine tuples T=(Σ ,S, s0, C, E, Inv, F, ρ , ψ).

where

Σ : a finite set of events

S : a finite set of states

$s0 \subseteq S$: a finite set of start states

C : a finite set of clocks

$E \subseteq 2^S \times 2^S \times \Sigma \times 2^C \times \Phi(C)$: a set of transitions

$Inv \in S \rightarrow \Phi(C)$: an invariant assignment mapping states to clock constraints

$F \subseteq 2^S$: an acceptance family

ρ : $S \rightarrow 2^S$: hierarchical function. ρ determines substates of each state.

ψ : S→{AND, OR} : type function. ψ determines type of each state.

For a set C of clocks, the set $\Phi(C)$ of clock constraints δ is defined inductively by

$\delta ::= x \leq d \mid d \leq x \mid \neg \delta \mid \delta_1 \wedge \delta_2$

where x is a clock variable in C and d is a constant in integer. An edge(2^S , $2^{S'}$, a, λ, δ) represents a transition from 2^S to state $2^{S'}$ on input symbol $a \in \Sigma$. The set $\lambda \subseteq C$ gives the clocks to be reset with this transition. A run r of T is an accepting run if $inf(r) \in F$. If ψ(S)=AND, ρ (S) is AND-decomposition(concurrency). If ψ(S)=OR, ρ (S) is OR-decomposition(hierarchy). □

2.3 Semantics of specification language

This semantics is mapping relations in Fig.2. Fig.2 is described as follows.

[Definition 2](Aggregation)

Syntax definition is defined as follows.

Class O_1 ;

Association aggregation : O_2, O_3 ;

Here as $\psi(S_1)$=AND, $\rho(S_1)=\{S_2, S_3\}$.

$S_1=S_2 \parallel S_3$

In dynamic model, S_1 is concurrent statechart consisting of S_2 and S_3.

In functional model,

Function $F_1=F_2 \parallel F_3$

In functional model, F_1 is concurrent program consisting of F_2 and F_3 .

```
specification language:=  |object definition part|

object definition part:=Class name definition part ;          (1)
                        Attribute definition part ;           (2)
                        Operation definition part ;           (3)
                        Association definition part ;         (4)
          [             State definition part ;               (5)
                        Event definition part ;               (6)
                        Initial state definition part ;       (7)
                        Acceptance state definition part ;    (8)
                        State transition definition part ;    (9)
                        Data definition part ;                (10)
                        Store definition part ;               (11)
                        Function definition part ;    ]       (12)
                        Classend ;

(1)Class name definition part     :=Class  class name ;
(2)Attribute definition part      :=Attribute  attribute name : type definition
                                               |, attribute name : type definition|  ;
   where type definition :=integer | real | bool | char | set(a:b) | range(a:b)
(3)Operation definition part      :=Operation  operation name  |,  operation name |  ;
(4)Association definition part    :=Association  association : association name [attribute]  (class name)
                                  |,association : association name [attribute] (class name)|  ;
   where association :=aggregation | inheritance | refer/use
(5)State definition part          :=State state name   |,state name|  ;
(6)Event definition part          :=Event  event name : type definition  [(class name)]
                                        |,event name : type definition  [(class name)] |  ;
   where class name which inputs/outputs events
(7)Initial state definition part     :=state name    |,state name |  ;
(8)Acceptance state definition part:=state name    |,state name |  ;
(9)State transition definition part:=Transition  state name→event, timing constraint/function→next state name
                          |,state name→event, timing constraint/function→next state name |  ;
   where timing constraint δ :=x<d | x>d | ¬ δ | δ1∧δ2 | x:=0
          x : clock variable      d : time constant
(10)Data definition part          :=Dataflow  data name:type definiton  [(class name)]
                                        |,data name:type definiton  [(class name)] |  ;
(11)Store definition part         :=Store  store name : type definition   [(class name)]
                                        |,store name : type definition   [(class name)] |  ;
(12)Function definition part :=Function  data name | store name   |, data name | store name|
                            →function→data name | store name   |, data name | store name|
                          | ,data name | store name   |, data name | store name|
                            →function→data name | store name   |, data name | store name|  |  ;

   where  [] and || , | are BNF notation
```

Fig.1 Definition of syntax of specification language

Here

‖ : concurrent operator

S_i : statechart of class O_i

F_i : function of class O_i (i=1,2,3) □

[Definition 3](Inheritance)

Syntax definition is defined as follows.

Class O_1 ;

Association inheritanc : O_2, O_3 ;

Here as $\psi(S_1)$=OR, $\rho(S_1)=\{S_2, S_3\}$.

$S_1=S_2$ OR S_3

In dynamic model, S_1 is hierachical statechart consisting of S_2 and S_3.

In functional model,

Function $O_2=F_2 \triangle F_1$;

Function $O_3=F_3 \triangle F_1$;

In functional model, Function O_2 consists of F_2 and F_1, and F_2 is differencial program .

Here

△ : differencial operator □

[Definition 4](refer/use)

Syntax definition is defined as follows.

Class O_2 ;

Association refer/use : O_1, O_3 ;

Here following specification can be described.

Class O_1 ;

Association refer/use : O_2 ;

Event I_1 :bool(O_2) ;

or

Dataflow D_1 : interger(O_2) ;

(store)

and

Class O_3 ;

Association refer/use : O_2 ;

Event I_3 : bool(O_2) ;

or

Dataflow D_3 : interger(O_2) ;

(store) □

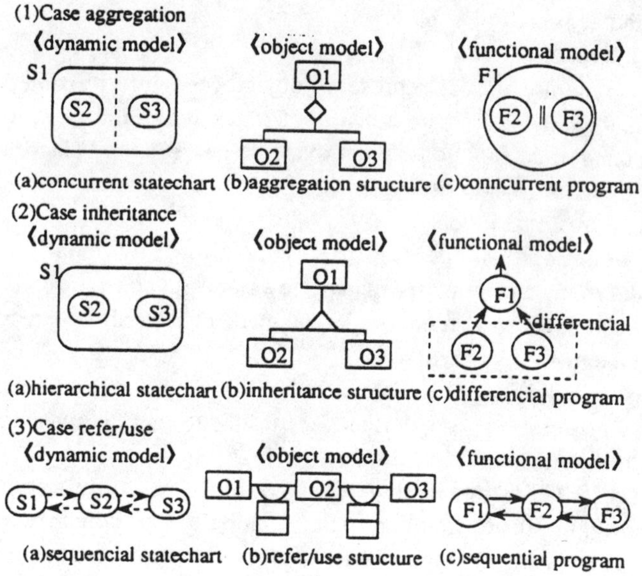

Fig.2 relations of models

3. Timing verification method

This section explains timing verification method of dynamic model. In verification method, verification property specification language and verification algorithm are important. In this paper, formal verification method is based on language inclusion problem .

3.1 Verification specification language

In order to unify both specification language and verification property specification language, I decide that verification property specification is described in timed automaton. In formal language theory, there is language inclusion algorithm as verification algorithm. And verification property specification language is described in deterministic timed Muller automaton. Because deterministic timed Muller automaton is more expressive than deterministic timed Buchi automaton and closed under complementation[14]. We extend deterministic timed Muller automaton with safety timing constraints in a set of states.

[Defunition 5](safety deterministic timed Muller automaton)

Deterministic timed Muller automaton consists of seven tuples (Σ, S, s0, C, E, Inv, F).

where

Σ : a finite set of events

S : a finite set of states

$s0 \subseteq S$: a start state

C : a finite set of clocks

$E \subseteq S \times S \times \Sigma \times 2^C \times \Phi(C)$: a set of transitions

$Inv \in S \rightarrow \Phi(C)$: is an invariant assignment mapping states to clock constraints

$F \subseteq 2^S$: an acceptance family

It has only one start state, and on any input symbol there is exactly one transition possible from any given state. □

3.2 Verification algorithm

If system specification and verification specification are described in automaton, verification problem reduces to language inclusion problem[14].

[Defunition 6](language inclusion problem)

The definition of language inclusion problem is as follows.

$L(M1) \subseteq L(M2)$

$L(M1) \subseteq L(M2)$ is equal to $L(M1) \cap \overline{L(M2)} = \phi$.

where

M1 : timed automaton of system specification

L(M1) : language accepted by M1

M2 : timed automaton of verification specification

L(M2) : language accepted by M2 □

Firstly, it is necessary to convert timed statechart into timed Muller automaton[12].

[Defunition 7](to convert timed statechart into timed Muller automaton)

Timed statechart is AND/OR tree, which has unique root. In order to convert timed statechart into timed Muller automaton, following operation are repeated from root to bottom.

(1)If ψ(S)=AND at i level, intersect automata ρ(S) and product automaton become at i+1 level.

(2)If ψ(S)=OR at i level, S is converted to ρ(S) at i+1 level. □

[Defunition 8](verification algorithm)

The verification algorithm of language inclusion problem $L(M1) \cap \overline{L(M2)} = \phi$ reduces to checking emptiness of $L(M1) \cap L(M2)$. The verification algorithm consists of following procedures.

(1)Constructing the product automaton $L(M1) \cap \overline{L(M2)}$.

(2)Searching for a cycle meeting all the desired acceptance conditions by Tarjan' depth-first search algorithm[15].

(3)Checking timing conditions of cycles by geometric region method[16,17,18].

(4)If there is no cycle satisfy timing conditions, verification property issatisfied.

(5)If cycles satisfy timing conditions, verification property is not satisfied because of $L(M_1) \cap \overline{L(M2)} \neq \phi$. If cycles do not satisfy them, verification property is satisfied. □

Here, we explain geometric region.

[Defunition 9](geometric region method)

Geometric region method consists of following procedures.

(1)Constructing DBM(differences bound matrix) per each state

(2)Constructing canonical DBM by Floyd-Warshall algorithm

(3)Analizing forward reachability of state transitions by the intersection of DBMs

(4)If there is a negative-cost cycle, there is no state transition because of unsatisfiable timing conditions. If there is no negative-cost cycle, there is state transitions. □

But the algorithm in [Defunition 8](2) takes very large cost. In order to reduce the verification cost, we propose explicit enumeration method and implicit enumeration method.

4. Effective timing verification method

We propose effective timing verification methods using explicit enumeration method and implicit enumeration method. In this paper, verification method is based on geometric region verification in order to avoid the state-space explosion problem. But in usual verification method, the verification algorithm checks for a cycle with several desired properties from an initial state in system specification and verification property specification. For this reason, the verification cost is very large. we propose two following methods.

(1)Explicit enumeration verification

Computing SCC(Strongly Connected Components) satisfying acceptance conditions with checking timing constraints.

(2)Implicit enumeration verification

Computing SCC by image computation based on BDD (Binary Decision Diagram) with checking timing constraints.

4.1 explicit enumeration method

We define explicit enumeration verification methods as follows.

[Definition 10](Explicit enumeration verification)

Explicit enumeration verification consists of following procedures.

(1)First, intersect system specification with the complement of deterministic timed Muller automaton.

(2)Next, search SCC satisfying an acceptance family using the algorithm in [3] and check whether the SCC is reachable from an initial state.

(3)If there is no SCC reachable from an initial state, verification property is satisfiable. If there is a SCC reachable from an initial state, test whether state sequences including a SCC reachable from an initial state satisfy timing constraints or not. If the SCC satisfies timing constraints, verification property is unsatisfiable. If it does not satisfy it, verification property is satisfiable. □

[Example 11](Example of explicit enumeration verification)

Firstly, we show the verification example by explicit enumeration verification. Specification and verification property are shown in Fig3. The set of accepting states of specification is {{s0,s1,s2,s3,s4}}. The acceptance family of verification property is {{v0,v1}}. As the complementation of the acceptance family of verification property is {{v0},{v1}}, the acceptance family of product automaton(generated by intersecting system specification with the complement of deterministic timed Muller automaton) is {{s0v0,s1v0,s2v0,s3v0,s4v0}, {s0v1,s1v1,s2v1,s3v1,s4v1}} shown in Fig.4.

If we compute SCC,we can find two SCCs as follows : s0v0→s1v0→s3v0→s4v0→s2v0 and s0v1→s1v1→s3v1→4v1→s2v1. We show the reachability analysis of s0v1→s1v1→s3v1→s4v1→s2v1 as follows. Firstly, we compute DBM. For example, we show DBM about s3v1→s4v1. In s3v1, t-t0≦7 and u-to≦8 are computed from t≦7∧u≦8 and DBM D3 is generated. DBM D4 is computed by the same method. Next, we test whether s4v1 is reachable from s3×v1 by computing the intersection D3∩D4 DBM. In this case, there no negative-cost cycle in the intersection D3 ∩D4 DBM, s4v1 is reachable from s3v1. We can test other transitons. Next, we test whether SCCs are reachable froman initial state.

In this case, SCCs are reachable from an initial state satisfying timing constraints. We can show this verification property is unsatisfiable in specification. □

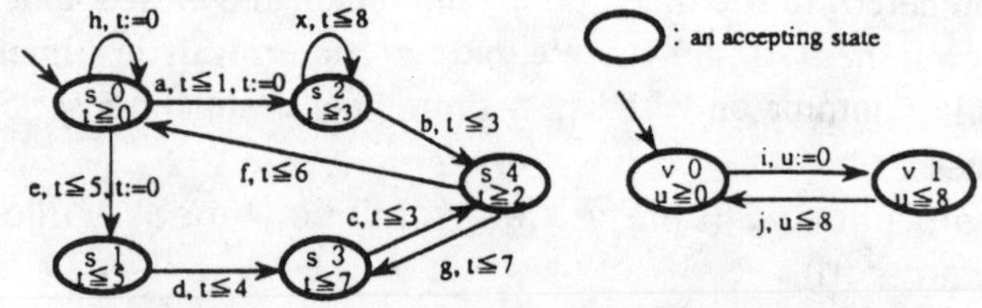

(1)specification (2)verification property

Fig.3 specification and verification property

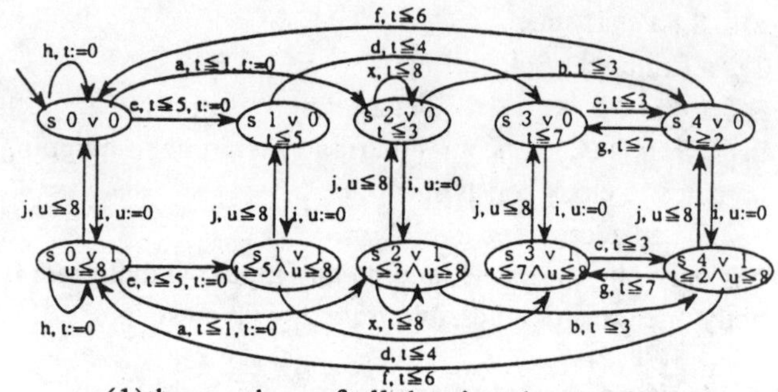

(1)the product of all the timed automata

At state s3v1, from t≦7∧u≦8

```
          * * *                     * * *
DBM D3=7 * *           canonical DBM=7 * *
          8 * *                     8 * *
```

At state s4v1, from t≧2∧u≦8

```
          * -2 *                    * -2 *
DBM D4=* * *           canonical DBM=*  *  *
          8 * *                     8  6  *
```

At state s3v1→state s4v1, from t≦7∧u≦8 to t≧2∧u≦8.

```
                        * -2 *                   6 -2 -2
intersection D3∩D4 DBM=* * 0      canonical DBM=8  6  0
                        8 6 *                    8  6  6
```

(2)Testing for timing constraints

Fig.4 Example of testing for timing constraints

4.2 implicit enumeration method

We define implicit enumeration verification methods as follows.

[Definition 12](Implicit enumeration verification)

Implicit enumeration verification consists of following procedures.

(1)First, intersect system specification with the complement of safety deterministic timed Muller automaton.

(2)Next, compute BDD of state transitons R(x,y)[19] and represent each state as (Si ,Zi)[20]. Si is represented as BDD and Zi is timing constraints and represented as DBM.

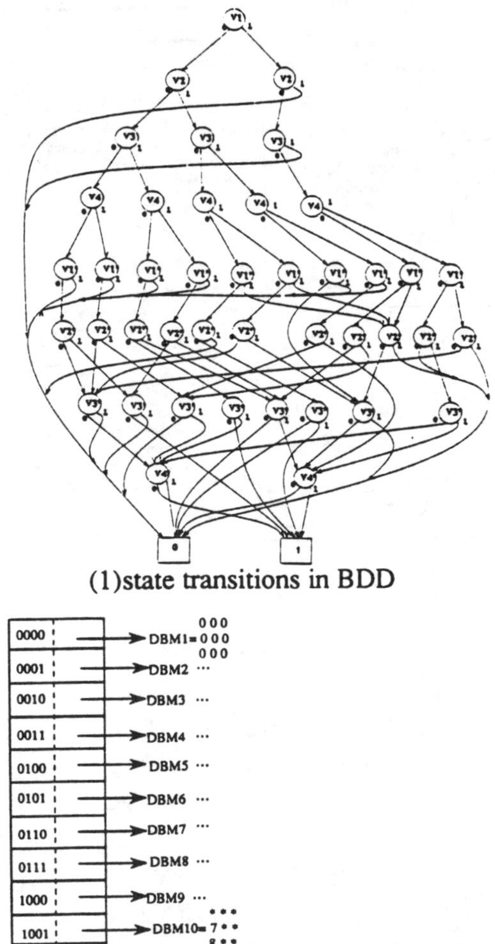

(1)state transitions in BDD

(2)structure (S i, Z i)

Fig.5 state transitions in BDD and (S i ,Z i)

(3)Next, compute BDD of state transitons R(x,y) reachable from initial states S(x) using

$R(x,y)=R(x,y)\wedge S(x)$

(4)Next, compute the transitive closure T(x,y) satisfying an acceptance family using

$T(x,y)=\exists z.\ [R(x,z)\wedge R(z,y)\wedge A(x)\wedge A(y)]\vee R(x,y)$,

where A(x) and A(y) are acceptance family represented by BDD.

(5)Next, compute BDD of states being included in the SCC satisfying an acceptance family using

$SCC(x)=\exists(T(x,y)\wedge T(y,x))$[21].

(6)If SCC(x) is equal to A(X),and SCC(x) is reachable from an initial state satisfying timing constraints, verification property is unsatisfiable. If it is not so, verification property is satisfiable. □

[Example 13](Example of implicit enumeration verification)

We show implicit enumeration verification using Fig.3. Firstly, we compute BDD of state transitons R(x,y) and represent each state as (Si ,Zi) as shown in Fig.5. Next, compute the set of reachable state S(x) from an initial state based on image computation by BDD. Next, compute the transitive closure T(x,y). Next, compute the SCC satisfying an acceptance family and test whether SCC(x) is equal to A (x) or not. In this case, verification property is unsatisfiable because SCC(x) is nonempty. □

5. Example of specification and verification

5.1 Example of specification

Automobile consists of accel and engine, transmission, brake. Automobile communicates with driver and external environment. Automobile is an aggregation of accel and engine etc. Brake is inherited to normal brake and ABS (Antilock Brake System). Normal brake is operated by driver ,and ABS is computer-controlled in order not to lock wheels. Both normal brake andABS control oil pressure and operate brake.

5.1.1 object model

Object model has following feature as shown in Fig.6.

(1)System is an aggregation of accel and engine, transmission, brake.

(2)Normal brake and ABS are an inheritance of brake.

(3)Accel and engine are combined by refer/use association of 「control valve」.Transmmision and brake are combined by refer/use association of 「admit gear R」 Data and events are communicated by associations.

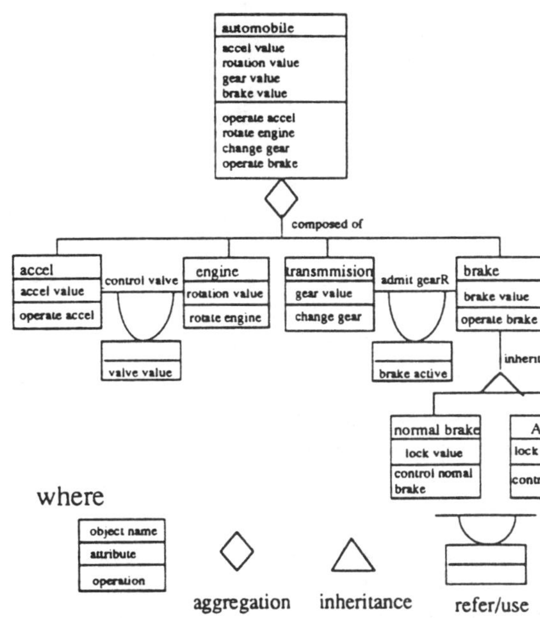

Fig.6 object model

5.1.2 dynamic model

Dynamic model has following feature.

(1)Concurrent statechart consists of accel and engine, transmission, brake.

(2)Hierachical statechart consists of normal brake and ABS.

(3)Transmission and brake communicate by 「brake active」 event.

This statechart is timed statechart consisting of concurrent and hierachical states based on timed ω-language. As soon as brake locks, ABS operates. But if ABS operates for more 7 time, ABS has trouble. Dynamic model is shown in Fig.7.

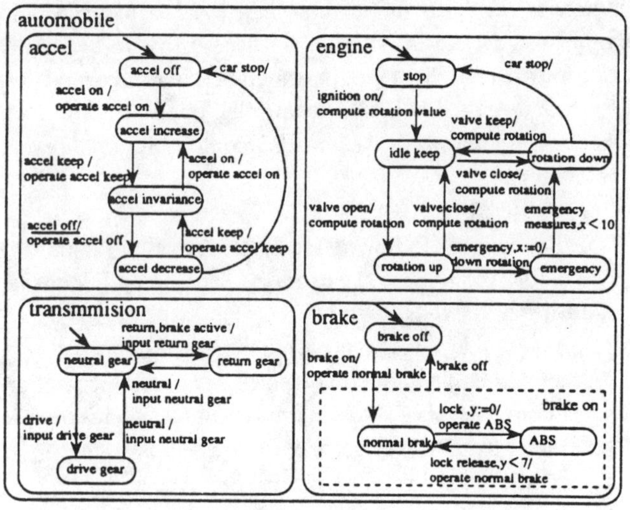

Fig.7 Dynamic model

5.1.3 functional model

Functional model is shown in Fig.8 and has following feature.

(1)Accel and engine, transmission, brake concurrently behave.

(2)Normal brake and ABS are described as differencial program for inheritance.

Like real-time structured analysis, following rule makes computational semantics clear.

(1)Active process continues to execute during state transition.

(2)As process continues to be activated, data and event continue to flow.

(3)Process can not transact event, but process outputs event by data condition.

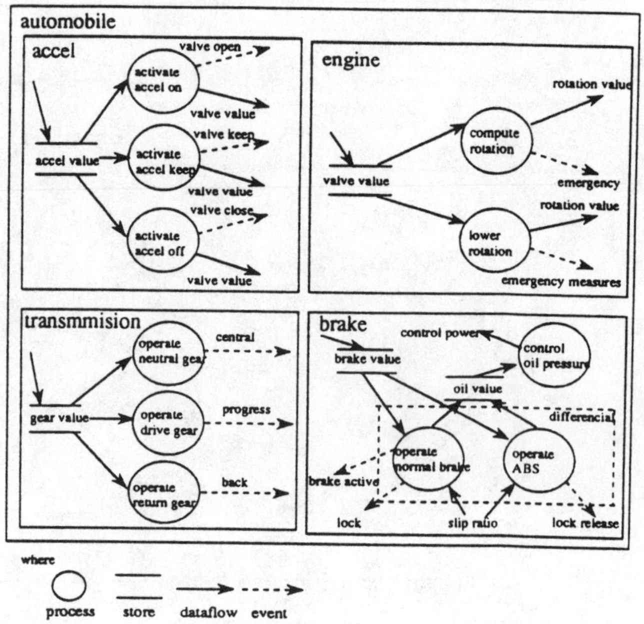

Fig.8 Functional model

5.1.4 verification specification

Verification specification is shown in Fig.9. (1) is satisfiable, but (2) is unsatisfiable. It is clear.

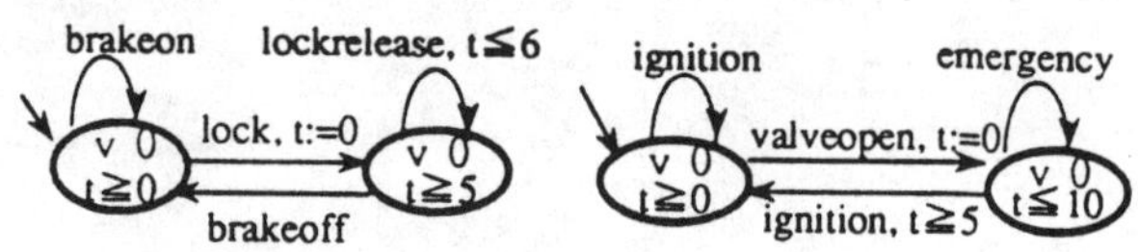

(1)satisfiable verification property (2)unsatisfiable verification property

Fig.9 specification and verification property

5.2 Verification system

We have developed verification systems as shown in Fig.10. It consists of about 6000step in C language on SUN4/IP(12MB). We have used SBDD library made of KYOTO University[22]. We have verified using explicit enumeration method and implicit enumeration method, previous methods such as region graph method and depth-first search method.

Using previous methods, we could not verify more than 200 states because of being lack of memory or not enough memory. We show the verification evaluation of our proposed methods in Table1 and Fig.11. Using explicit enumeration method, we could verify real-time systems by smaller time of complexity. Using implicit enumeration method, we could verify real-time systems by smaller space of complexity.

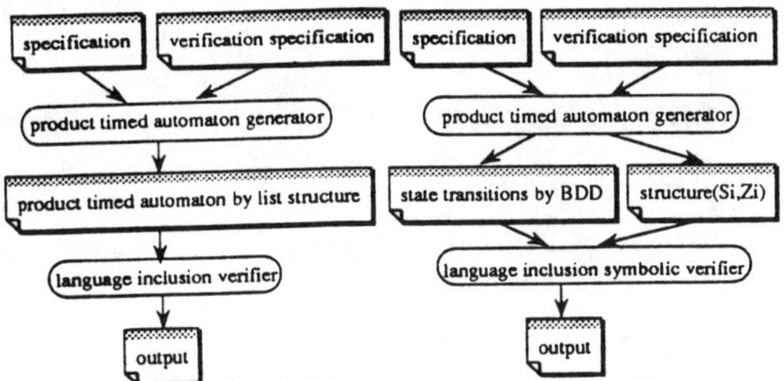

(1)explicit enumeration verifier (2)implicit verifier

Fig.. 10 Configuration of verification system

Table1 Evaluation of space and time complexity

specification states	specification transitions	true/ false	explicit enumeration method CPU time	explicit enumeration method memory	implicit enumeration method CPU time	implicit enumeration method memory
16	54	true	0.2sec	0.4Mb	0.9sec	2.8Mb
16	51	false	0.2sec	0.4Mb	1.1sec	2.8Mb
80	329	true	0.5sec	0.7Mb	6.7sec	2.8Mb
80	315	false	0.5sec	0.7Mb	9.2sec	2.9Mb
350	1926	true	3.4sec	2.4Mb	73.1sec	2.8Mb
350	1845	false	3.7sec	2.5Mb	111.2sec	2.9Mb
560	3141	true	7.9sec	3.9Mb	32.9sec	2.8Mb
560	3037	false	8.7sec	3.9Mb	400.8sec	2.9Mb
640	3710	true	9.9sec	4.3Mb	307.5sec	2.8Mb
640	3592	false	11.0sec	4.4Mb	447.5sec	2.9Mb
1024	6840	true	25.0sec	4.5Mb	532.6sec	2.8Mb
1024	6704	false	26.3sec	4.5Mb	587.2sec	2.9Mb
1400	9622	true	47.9sec	4.5Mb	772.8sec	2.8Mb
1400	9250	false	50.1sec	4.5Mb	1084.3sec	2.9Mb
1750	12248	true	73.7sec	4.6Mb	1171.6sec	2.8Mb
1750	11795	false	77.8sec	4.5Mb	1800.4sec	2.9Mb
2000	14305	true	97.4sec	4.7Mb	1177.6sec	2.8Mb
2000	13795	false	101.6sec	4.6Mb	1819.3sec	2.9Mb

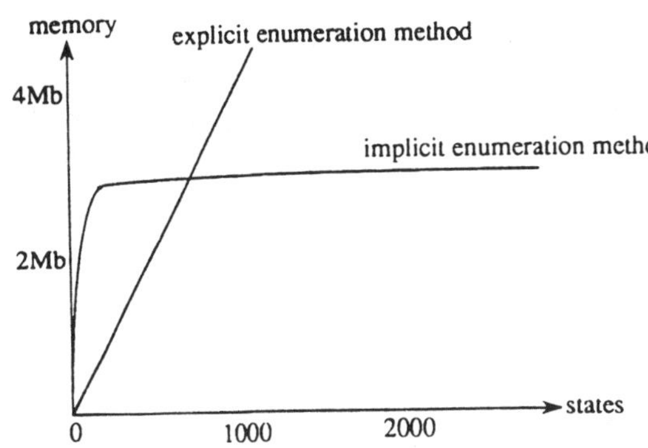

Fig.11 the relation between memory and states

5. Conclusion

In this paper, object-oriented method for real-time systems is proposed and it is shown effective by example of automobile. Real problem is that complex processes are multi-tasking and each process has timing conditions. If this method is applied to real problem, process van be specified as object and this method can be applied.

Reference

[1]Kavi K. M.: Real-timeSystems,Abstraction,Languages, and Design Methodologies, P.660,IEEE Computer Society (1992)

[2]Yamane S. :"Object-oriented specification and verification", Intensional programming 1,World scientific (1995)

[3]Aho A. Hopcrofft J.E. Ullman J.D.:The design and analysis of computer algorithms,Addison-Wesley(1974)

[4]Bryant R.E.:"Graph-Based Algorithms for Boolean Function Manipulation",IEEE Transactions on Computers, Vol.C-35,No.8,pp.677-691(1986)

[5]Kim W. Lochovsky F.H.: Object-Oriented Concepts, Databases, and Applications,P.602,ACM PRESS(1989)

[6]Rumbaugh J. et-al : Object-Oriented Modeling and Design,P.500,Prentice Hall (1991)

[7]Shlaer S. Mellor S.J.:OBJECT LIFECYCLES Modeling the World in States,Englewood Cliffs(1992)

[8]Booch G.:Object-Oriented Design with Applications, Benjamin/Cummings(1991)

[9]Buhr R.:"Architectures with Pictures",OOPSLA 92, pp.466-483ACM PRESS(1992)

[10]Selic B. Gullekson G. Ward P.T.:Real-Time Object-Oriented Modeling,P.525,John Wiley&Sons(1994)

[11]Coleman D.Hayes F.Bear S.:"Introduction Objectcharts or How to Use Statecharts in Object-Oriented Design", IEEE Trans. on SE,Vol.18,No.1,pp.9-18(1992)

[12]Yamane S.:"Method of verification of real-time system based on timed statechart",Trans. of IPSJ,Vol.35,No.12, pp.2640-2650(1994)

[13]Harel D.:"Statecharts:A Visual Formalism for Complex Systems",Science of Computer Programming 8,pp.231-274(1987)

[14]Alur R. Dill D.:"The theory of Timed automata",LNCS 600,pp.45-73(1992)

[15]Tarjan R.,"Depth-first search and linear graph algorithms",SIAM J. of Computing1(2),pp.146-160(1972)

[16]Dill D.:"Timing assumptions and verification of finite-state concurrent systems",LNCS 407,pp.197-212(1989)

[17]Alur R. et-al:"An Implementation of Three Algorithms for Timing Verification Based on Automata Emptiness", Proc. RTSS,pp.157-166(1992)

[18]Rokicki T.G. Myers C.J.:"Automatic Verification of Timed Circuits",LNCS 818,pp.468-480(1994)

[19]McMillan K.L.:Symbolic Model Checking,Kluwer, P.194(1993)

[20]Dill D. et-al:"Verification of real-time systems by successive over and under approximation",LNCS939, pp.409-422,1995

[21]Touatic H. et-al.:"Testing language contaiment for ω-automata using BDDs",Info. and comp.118,pp.101-109 (1995)

[22]Minato S. et-al.:"Shared Binary Decision Diagram with Attributed Edges for Efficient Boolean Function Manipulation",Proc. 27th DAC,pp.52-57(1990)

Design and Optimization of Complex Real-Time Dependable Systems*

Leyuan Shi
Dept. of Industrial Engineering
University of Wisconsin-Madison
Madison, WI 53706

Phillip Q. Hwang
Acquisition & Technology
Defence Mapping Agency
Bethesda, MD 20816-5003

Abstract

In this paper we propose a Rapid Iterative Nested Design Optimization algorithm (RINDO) for optimizing multi-criteria complex systems. We first use utility function to formulate multi-criteria stochastic optimization problems. Ordinal Optimization algorithm is then applied to utility function for obtaining a set of good solutions. Finally, the Gradient Surface Method is used to search for the optimal solution iteratively. An especially appealing characteristic of RINDO is: it can identify a set of good designs very rapidly. The key thrust of this approach is its integrated, nested (hierarchical), system level design capabilities. The framework proposed here is generic and application independent.

1. Introduction

This paper presents a methodology for optimizing the design of large, complex, real-time systems in distributed and parallel processing environments. When a poor design decision is made early in the design phase, the cost for correcting it may far exceed the planned budget. Therefore, one key issue in designing a real-time, large-scale, complex system is how to optimize and assess the design, based on multiple competing requirements, **early** and throughout the design process. Thus, a large number of design problems is essentially selection problems: from a discrete set of alternative design options, select a subset of satisfactory designs (i.e., meeting certain performance requirement).

The main difficulties in designing large, complex, computer-based systems are (a) the complex, multi-attribute nature of these systems tending to be overwhelming to analysts and designers, for example, the designs that tend to optimize one attribute (or objective) (such as real-time performance) generally will degrade the other required behavior (such as system dependability), (b) the combinatorial explosion of alternatives normally leading to NP-complete optimization problems, and (c) the fact that no analytical expressions relating performance criteria to design parameters are available. In this paper we propose a Rapid Iterative Nested Design Optimization algorithm (RINDO) for optimizing multi-criteria complex systems. The optimization emphasizes system design structuring and function to resources allocation.

This paper is divided into seven sections. The multi-objective stochastic optimization problem is formulated via utility functions in Section 2. In Section 3 Ordinal Optimization techniques are applied to the utility functions to obtain a set of good solutions . Section 4 and Section 5 provide brief discussions on Gradient Surface Method and Perturbation Analysis techniques, respectively. Section 6 is on Rapid Iterative Nested Design Optimization Algorithm. Section 7 presents some concluding remarks and suggestions for future research.

2. Problem Formulation

The multi-objective stochastic optimization problem can be formulated as follows:

$$\textit{Find } \theta \textit{ in } \Theta \textit{ to minimize } ((J_1(\theta), \ldots, J_M(\theta)) \quad (2.1)$$

where θ is the design parameters , $J_i(\theta) = E[L_i(\theta)]$ for $i = 1, \ldots, M$.

Unlike traditional optimization problem with a single objective function, an optimal solution in the sense of one that minimizes all the objective functions simultaneously does not necessarily exist in multiobjective optimization problems, and hence, we are troubled with conflicts among objectives in decision making problems with multiple objectives. Therefore, a new problem of value judgment called value trade-off

*This work is supported by Naval Surface Warefare Center under contract N60921-95-M-0706.

arises. One approach to evaluation modeling is to find a scale-valued function $U(\mathbf{J}) = U(J_1, \ldots, J_M)$ representing the decision maker's preference. A preference function in decision making under uncertain is called a utility function. The theory regarding existence, uniqueness, and practical representation of such a utility is called the utility theory. Utility theory is one of the well-accepted mathematical frameworks for studying multi-objective decision making [Canada & Sullivan 1989]. The utility $U(\mathbf{J}) = U(J_1, \ldots, J_M)$ of any combination of outcomes $(J_1, \ldots, J_M)$ for M attributes $(A_1, \ldots, A_M)$ can be expressed as either (1) an additive or (2) a multiplicative function of individual attribute utility functions $U_1(J_1), \ldots, U_M(J_M)$.

- *Additive Utility Model*: the utility $U(\mathbf{J})$ of any combination of outcomes for M attributes $(A_1, \ldots, A_M)$ can be expressed as:

  $$U(\mathbf{J}) = \sum_{i=1}^{M} k_i \cdot U_i(J_i), \quad \sum_{i=1}^{M} k_i = 1$$

  where k_i is weight (scaling factor) and $U_i(J_i)$ is utility of the outcome J_i for ith attribute.

- *Multiplicative Utility Model*: the utility $U(\mathbf{J})$ of any combination of outcomes for M attributes $(A_1, \ldots, A_M)$ can be obtained from the solution to the following equation:

  $$U(\mathbf{J}) = \frac{\prod_{i=1}^{M}[K \cdot k_i \cdot U_i(J_i) + 1] - 1}{K}$$

  K is a scaling constant that is a solution to

  $$1 + K = \prod_{i=1}^{M}(1 + K \cdot k_i)$$

  and must be found iteratively.

Further condition and explanations for utility model are:

1. U is normalized by $U(J_1^0, \ldots, J_M^0) = 0$ and $U(J_1^*, \ldots, J_M^*) = 1$. (Note: J_i^0 means worst outcome of J_i and J_i^* means best outcome of J_i.)
2. U_i is a conditional utility function of J_i normalized by $U_i(J_i^0) = 0$ and $U_i(J_i^*) = 1$ for $i = 1, 2, \ldots, M$ attributes.

To formulate multi-objective stochastic optimization problem, we can define each utility function U_i as a function of J_i. For example, we may define:

$$U_i = \frac{J_i^* - J_i}{J_i^* - J_i^0} \quad \text{for all} \quad i, \tag{2.2}$$

or

$$U_i = 1 - (1 - \frac{J_i^* - J_i}{J_i^* - J_i^0})^2 \quad \text{for all} \quad i, \tag{2.3}$$

or some other forms.

Note that in both cases we have $U_i(J_i^0) = 0$ and $U_i(J_i^*) = 1$. In the first case, U_i is a linear function of J_i and in the second case U_i is a concave function of J_i. There are many ways to define each utility function. In this paper, we adopt *Additive Utility Model* and use only utility function (2.2) to formulate multi-objective stochastic optimization problems.

3. Ordinal Optimization

Traditional cardinal optimization techniques are based on the absolute values of the objective function. Ordinal optimization, on the other hand, is derived by the relative order of the estimates of the objective function-not their absolute values.

Ordinal optimization technique can quickly obtain an ordinal (or a good) optimal design based on the order statistics. To see how ordinal optimization techniques can be applied to multi-objective stochastic optimization, we use the utility function given in the previous section to demonstrate our approach.

Assume that no analytical formula is available for each J_i's and the additive utility model is used and each utility function is defined as

$$U_i = \frac{J_i^* - J_i}{J_i^* - J_i^0} \quad \text{for} \quad i = 1, \ldots, M \tag{3.1}$$

Note that when we define the above utility functions in (3.1), we need to obtain J_i^* and J_i^0 first. However, finding J_i^* and J_i^0 themselves is a difficult and time consuming process if detailed cardinal optimization method is used (such as hill climbing type of algorithms). Using Ordinal Optimization technique, we can quickly obtain the largest and smallest order statistics for estimating J_i^* and J_i^0. Let us elaborate on the technical approach by focusing the discussion on the utility functions given above.

Consider a set of N experiments at $\theta = \theta_1, \ldots, \theta_N$ which are randomly chosen. Assuming we are minimizing, we order the sample results of these N experiments for each objective function as $\hat{J}_i^{[1]} < \hat{J}_i^{[2]} < \ldots < \hat{J}_i^{[N]}$ for $i = 1, \ldots, M$. Replace J_i^* and J_i^0 by $\hat{J}_i^{[1]}$ and $\hat{J}_i^{[N]}$, respectively and define a new utility function for each attribute as :

$$\hat{U}_i(\theta) = \frac{\hat{J}_i^{[1]} - J_i(\theta)}{\hat{J}_i^{[1]} - \hat{J}_i^{[N]}} \quad \text{for} \quad i = 1, \ldots, M \tag{3.2}$$

Then with additive utility model the utility function will be

$$J(\theta) = k_1\hat{U}_1(\theta) + k_2\hat{U}_2(\theta) + \ldots, +k_M\hat{U}_M(\theta) \quad (3.3)$$

Assume that each k_i can be specified via a given procedure. We can obtain a set of N values for the utility function $J(\theta)$ at $\theta = \theta_1, \ldots, \theta_N$. We order the N values as $J^{[1]} < J^{[2]} <, \ldots < J^{[N]}$. θ^k which corresponds to $J^{[N]}$ is chosen as an ordinal optimal value for the current iteration, because in this case, the utility function is maximized.

This technique is to explicitly tackle the combinatorial explosion of system design alternatives that normally leads to computationally intractable problems. Using Ordinal Optimization approach, we can evaluate many designs in parallel and to develop optimization techniques that can work in concert with parallel simulation and learning approaches. Not only does this approach provides us with a global view of the response surface and permits quick localization of promising search regions, but it also allows us to take maximal advantage of parallel computing architectures.

4. Gradient Surface Method

Now the problem under consideration is:

$$\min_{\theta \in R^m} J(\theta) = \min_{\theta \in R^m} E_\omega L(\theta, \omega) \quad (4.1)$$

where θ is the design parameter,L is some performance functional of the sample path of the system under the design, and ω is all random effects of the problem.

We assume that $J(\theta)$ is not available in closed form, and thus are faced with an iterative stochastic optimization problem. Under certain smoothness and convexity conditions on $J(\theta)$, we can find the minimum by solving instead the problem

$$\nabla J(\theta) \equiv dJ(\theta)/d\theta = 0 \quad (4.2)$$

i.e., to optimize $J(\theta)$ is equivalent to finding the zero of $\nabla J(\theta)$. The problem now is to determine the zero point of $\nabla J(\theta)$ without a closed form expression for $\nabla J(\Theta)$, under stochastic conditions. In this paper, we use a linear function to estimate $\nabla J(\theta)$ directly (i.e., fitting $J(\theta)$ by a quadratic surface).

Denote the gradient surface fitting function $F(\theta) = A\theta + b$, where A is an $m \times m$-dimensional matrix and b is an m-dimensional vector. It is very natural for us to use the following criteria:

$$\min_{F(\theta)} \|\nabla J(\theta) - F(\theta)\|_2 = \min_{[A,b]} \|\nabla J(\theta) - (A\theta + b)\|^2 \quad (4.3)$$

Assuming that (A^*, b^*) is the optimal solution of (4.3), i.e.,

$$\|\nabla J(\theta) - (A\theta^* + b^*)\|^2 = \min_{[A,b]} \|\nabla J(\theta) - (A\theta + b)\|^2 \quad (4.4)$$

it is then obvious that any quadratic response surface will yield gradient solutions at most as good as $F^*(\theta) = A^*\theta + b^*$, but no better. Furthermore, fitting a quadratic surface will require one more parameter. Note also that the use of linear function for the estimated gradient surface is for convenience. In principle, more complex surfaces can be used.

Now we are in position to discuss the Gradient Surface Method [Ho & Shi, etc. 1992]. Assume that we have $n(n \geq m)$ estimates of the gradient $z(\theta_i) = \nabla J(\theta_i) + E_i$, where E_i is a random vector. The least-squares method says that, based on these n observations, the best estimate of (A^*, b^*) is given by minimizing

$$\sum_{i=1}^{n} \|A\theta_i + b - z(\theta_i\|^2 \quad (4.5)$$

Denote the solution of (4.5) to be (A_{n+1}, b_{n+1}). The approximate gradient surface is $F_{n+1}(\theta) = A_{n+1}\theta + b_{n+1}$. We take the zero point of $F_{n+1}(\theta)$ as a candidate of the next iteration point, θ_{n+1}, i.e.,

$$\theta_{n+1} = -A - n + 1^{-1} b_{n+1} \quad (4.6)$$

It is important to emphasize that minimization (4.5) (or surface fitting) is accomplished based on the information of all n points, i.e., in terms of the global behavior of the response surface. As a result, a small change in the estimation error at a particular point will have less effect on the result of surface fitting and its zero point (4.6) than ordinary gradient methods which can be also regarded as a special case of surface fitting in which only one point is used to define the gradient surface.

5. Perturbation Analysis

The fundamental problem we are facing here is that the function $J(\theta)$ is simply unknown due to the complexity of the system we are designing and evaluation. Solving the problem (4.1) typically involves repetitive simulation in order to estimate $J(\theta)$ over the parameter range θ. We can: **(a)** try to obtain $J(\theta)$ and then determine the "best" value of θ, i. e. the one yielding the smallest value over θ, or **(b)** use gradient-based approaches, by starting with an initial value θ, and then use gradient information to iteratively improve performance [Robbins & Monro 1951]. Since $J(\theta)$ is unknown we need to either estimate $J(\theta)$ over the parameter range θ (or at lease some subset) or the gradient $dJ/d\theta$ at every new point selected by the optimization algorithm used. At first sight, this can only be done through trial-and-error experiments on a real system (obviously inefficient) or by simulation.

The need to efficiently obtain gradient estimates of complex and dynamic systems has motivated the development of techniques which can provide such estimates from a single observed sample path of the system. In other words, one sample path under θ can provide not only $L(\theta)$ but $L(\theta+\Delta\theta)$ as well, for some small $\Delta\theta$, and in some case $dL/d\theta$. Moreover, this can often be done with little added computational cost. This approach, known as Perturbation Analysis (PA) was pioneered by Ho and co-workers in the early 1980's [Ho & Cao 1992].

Perturbation analysis (PA) was initiated by Ho et al. (1979) in the context of production lines in manufacturing systems. Its infinitesimal version, infinitesimal perturbation analysis (IPA), was first introduced by Ho and Cao (1983) for the sensitivity analysis of throughput in closed queueing networks. Since then, it has been successfully applied to many different systems.

It needs to point out that PA has come to embody a whole class of different techniques. Infinitesimal Perturbation Analysis (IPA) is a well-defined technique which is easy to derive and implement with very little computational overhead. Intuitively, it is the estimator resulting from assuming that small changes cause no event order changes. Here we provide a simple example to illustrate IPA technique.

Consider a single first come, first served G/G/1 queue. Let C_n denote job n in the arrival stream, with service time and system time denoted by S_n and T_n, respectively. Denote the interarrival time between C_n and C_{n+1} by A_{n+1}. We assume that the system starts with empty and C_0 arrives at the system at time 0. Let S be a generic service time with p.d.f. $F(\theta; S)$, where θ is the parameter of interest, say, $\theta \in \Theta = (a, b)$. The quantity we are concerned with here is the mean steady-state system time of a job. We want to estimate its derivative.

The system time of a customer for such a system satisfies the well-known recursive Lindley equation:

$$T_{n+1} = \begin{cases} T_n + S_{n+1} - A_{n+1} & \text{if } T_n - A_n \geq 0 \\ S_{n+1} & \text{otherwise} \end{cases}$$

Then the IPA estimator of the derivative of T_n, is defined recursively by

$$\frac{\partial T_{n+1}}{\partial \theta} = \begin{cases} \frac{\partial T_n}{\partial \theta} + \frac{\partial S_{n+1}}{\partial \theta} & \text{if } T_n - A_n \geq 0 \\ \frac{\partial S_{n+1}}{\partial \theta} & \text{otherwise} \end{cases}$$

where we have under mild conditions [Glasserman 1991]

$$\frac{dS}{d\theta} = -\frac{dF(\theta; S)/d\theta}{dF(\theta; S)/dS}$$

For example, for scale parameters, such as if θ is the mean of an exponential distribution, we have $dS/d\theta = S/\theta$. Using above recursion, the IPA estimator for the derivative of mean system time is given by

$$\left(\frac{dT}{d\theta}\right)_{IPA} = \frac{1}{N}\sum_{i=1}^{N}\frac{dT_i}{d\theta}$$

For instance, for M busy periods in an M/M/1 queue with N customers served, we would use

$$\left(\frac{dT}{d\theta}\right)_{IPA} = \frac{1}{N}\sum_{m=1}^{M}\sum_{i=1}^{n_m}\sum_{j=1}^{i}\frac{dS_{j,m}}{d\theta}$$

where n_m is the number of customers served in the mth busy period, $N = \sum_{m=1}^{M} n_m$is the total number of customers served, and the (j, m) subscript denotes the jth customer in the mth busy period. Thus, implementation of the gradient estimator involves the addition of accumulators; *no alternation of the underlying simulation is required.* A number of subsequent developments in the PA field, related techniques, and applications can be found in [Ho & Cao 1992, Cassandras 1991]. In [Shi 1994 (a)], we have provided a general framework to derive the gradients for structure or threshold type of problems such as the admission control problem.

6. Rapid Iterative Nested Design Optimization algorithm (RINDO)

Now we are in a position to present the Rapid Iterative Nested Design Optimization (RINDO) algorithm. The proposed RINDO methodology includes four basic components: (1) a formulation procedure; (2) an adaptive search strategy; (3) an ordinal optimization technique and (4) a fine-tuning (or local search) technique. These component are linked together into an iterative method that constructs a feasible design one at a time at the early design stage and then feeds the design to the fine-tuning procedure at the final stage of the design process. At any time a solution for the optimization problem can be obtained. An especially appealing characteristic of RINDO is: it can identify a set of good designs very rapidly. The key thrust of this approach is integrated, nested (hierarchical), system level design capabilities. These capabilities include capture, decomposition, recomposition, and evaluation of the system at different level representations. At each level the same design engine (RINDO) is used. Thus, we can quickly decompose and recompose the system in a consistent and easy to communicate form.

The flow chart on the following page demonstrates the RINDO algorithm.

Rapid Iterative Nested Design Optimization Algorithm

Assume that all attributes should be minimized.

Initialize

0. Given N (initial number of points), M (number of attributes), MSN (maximum switching number).
   Set $J_i^* = \infty, J_i^0 = -\infty, S = 0$.

Ordinal Optimization Phase (Using Utility Functions)

1. Randomly choose N initial points $\theta_1, \theta_2, \ldots, \theta_N$.
2. Order the N sample results for each of the M attributes.
$$J_i^{[1]} < J_i^{[2]} < \cdots < J_i^{[N]}, i = 1, \ldots, M$$
3. Update the best and worst outcome for each attribute, if necessary.
$$\text{If } J_i^* > J_i^{[1]} \text{ , let } J_i^* = J_i^{[1]}, i = 1, \ldots, M$$
$$\text{If } J_i^0 < J_i^{[N]}, \text{ let } J_i^0 = J_i^{[N]}, i = 1, \ldots, M$$
4. Define an utility function for each attribute
$$U_i = \frac{J_i^* - J_i}{J_i^* - J_i^0} \quad \text{for } i = 1, \ldots, M$$
5. Determine scaling constants $k_i, i = 1, \ldots, M$; using designer feedback.

6a. If using additive utility function model define
$$U(\mathbf{J}) = \sum_{i=1}^{M} k_i \times U_i(J_i)$$

6b. If using the multiplicative utility function model define
$$U(\mathbf{J}) = \frac{\prod_{i=1}^{M}[K \cdot k_i \cdot U_i(J_i) + 1] - 1}{K}$$
where K is the solution to
$$1 + K = \prod_{i=1}^{M}(1 + K \cdot k_i)$$

7. Order the utility function values
$$U^{[1]} < U^{[2]} < \cdots < U^{[N]}$$
8. Check if simulation budget is used up. If so, pick the design corresponding to $U^{[N]}$ as the best design. STOP.

Gradient Surface Method (GSM) Phase I

9. Pick the $(k-1)$ best designs as starting points for GSM algorithm ($k < N$ is the window size).
10. Use IPA algorithm to find the corresponding derivatives $z(\theta_i), i = 1, \ldots, k-1$
11. Use least squares method to find the GSM function F_k:
$$F_k(\theta) = A_k\theta + b_k.$$
where (A_k, b_k) is the solution to
$$\min_{A,b} \sum_{i=1}^{n} \|A\theta_i + b - z(\theta_i)\|^2.$$
12. Find the zero point of F_k as $\theta_k = A_k^{-1} b_k$.
13. If θ_k is not acceptable (e. g. does not satisfy some constraints), discard θ_k. Randomly choose $N + 1 - k$ points $\theta_k, \theta_{k+1}, \ldots, \theta_N$ and go to step 2.

GSM Phase II

14. Use IPA algorithm to find the derivative $z(\theta_k)$.
15. Remove the worst point $\theta_\omega = \underset{\theta_i}{\text{argmax}}\{U(\theta_i)\}$ If $\omega = k$, let $S = S + 1$, go to step 20.
16. Use least squares to find the GSM function F_{k+1}:
$$F_{k+1}(\theta) = A_{k+1}\theta + b_{k+1}$$
where (A_{k+1}, b_{k+1}) is the solution to
$$\min_{A,b} \sum_{i=1}^{n} \|A\theta_i + b - z(\theta_i)\|^2.$$
17. Find the zero point of F_{k+1} as $\theta_{k+1} = A_{k+1}^{-1} b_{k+1}$.
18. Set $\theta_{k+1} = (1 - \rho_k)\theta_k + \rho_k \underline{\theta}_{k+1}$.
19. If stopping rule satisfied, STOP. Otherwise $k = k + 1$, go to step 14.

GSM Phase III

20. Use one point to determine $\underline{\theta}_k = \theta_{k-1} - z(\theta_{k-1})$. Set $\theta_k = (1 - \rho_k)\theta_{k-1} + \rho_k \underline{\theta}_k$.
21. Is $S > MSN$? If no, go to step 14.
22. Is stopping rule satisfied? If yes, STOP. Otherwise $k = k + 1$, go to step 20.

7. Conclusions and Future Work

To summarize, we submit that we can very quickly narrow the scope of the search for the optimal design. The framework proposed here is generic and application independent. The nesting of ordinal and cardinal optimization will be extended to further explore global optimization.

References

[1] J. Canada and W. Sullivan. *Economic and Multi-attribute Evaluation of Advanced Manufacturing systems.* Prentice Hall, 1989.

[2] C. G. Cassandras. *Discrete Event Systems: Modeling and Performance Analysis.* Irwin, 1993.

[3] A. A. Gaivoronski, L. Shi, and R. S. Sreenivas. Augmented infinitesimal perturbation analysis: An alternate explanation. *Journal of Discrete Event Systems*, 2(2):121–138, 1992.

[4] P. W. Glynn. Likelihood ratio gradient estimation: an overview. *Proc. Winter Simulation Conf.*, pages 366–375, 1987.

[5] Y. C. Ho. Ordinal optimization of deds. *Journal of DEDS*, 2(2):61–88, 1992.

[6] Y. C. Ho and X. R. Cao. *Discrete Event Dynamic Systems and Perturbation Analysis.* Kluwer Academic, 1991.

[7] Y. C. Ho, L. Shi, L. Dai, and W.-B. Gong. Optimization of discrete event dynamic systems via gradient surface method. *Journal of DEDS*, 2(2):99–120, 1992.

[8] P. Korhonen. Multiple criteria decision support: The state of research and future directions. *Computers Ops. Res.*, 19:549–551, 1992.

[9] H. Kushner and Clark. Stochastic approximation algorithms for the local optimization of functions with nonunique stationary points. *IEEE Trans. Auto. control*, AC-17(5):646–654, 1972.

[10] P. L'Ecuyer. A unified view of the ipa, sf, and lr gradient estimation techniques. *Management Science*, 36(11):364–1383, 1990.

[11] J. L. Ringuest. *Computers Ops. Res.*, 19, 1992.

[12] H. Robbins and S. Monro. A stochastic approximation method. *Annals of Mathematical Statistics*, 22:400–407, 1951.

[13] L. Shi. Perturbation analysis with discontinuous sample performance functions. *IEEE Trans. on AC*, 1994(a).

[14] L. Shi. On the consistency of high order derivative estimators for the m/g/1 queue. *Proceedings of Thirty-second Annual Allerton Conference on Communication, Control, and Computing*, 1994(b).

[15] L. Shi. Sample path derivative and its variance properties. *Proceedings of Hong Kong International Workshop on New Directions of Control and Manufacturing*, 1994(c).

Rapid Iterative Nested Design Optimization(RINDO) Algorithm

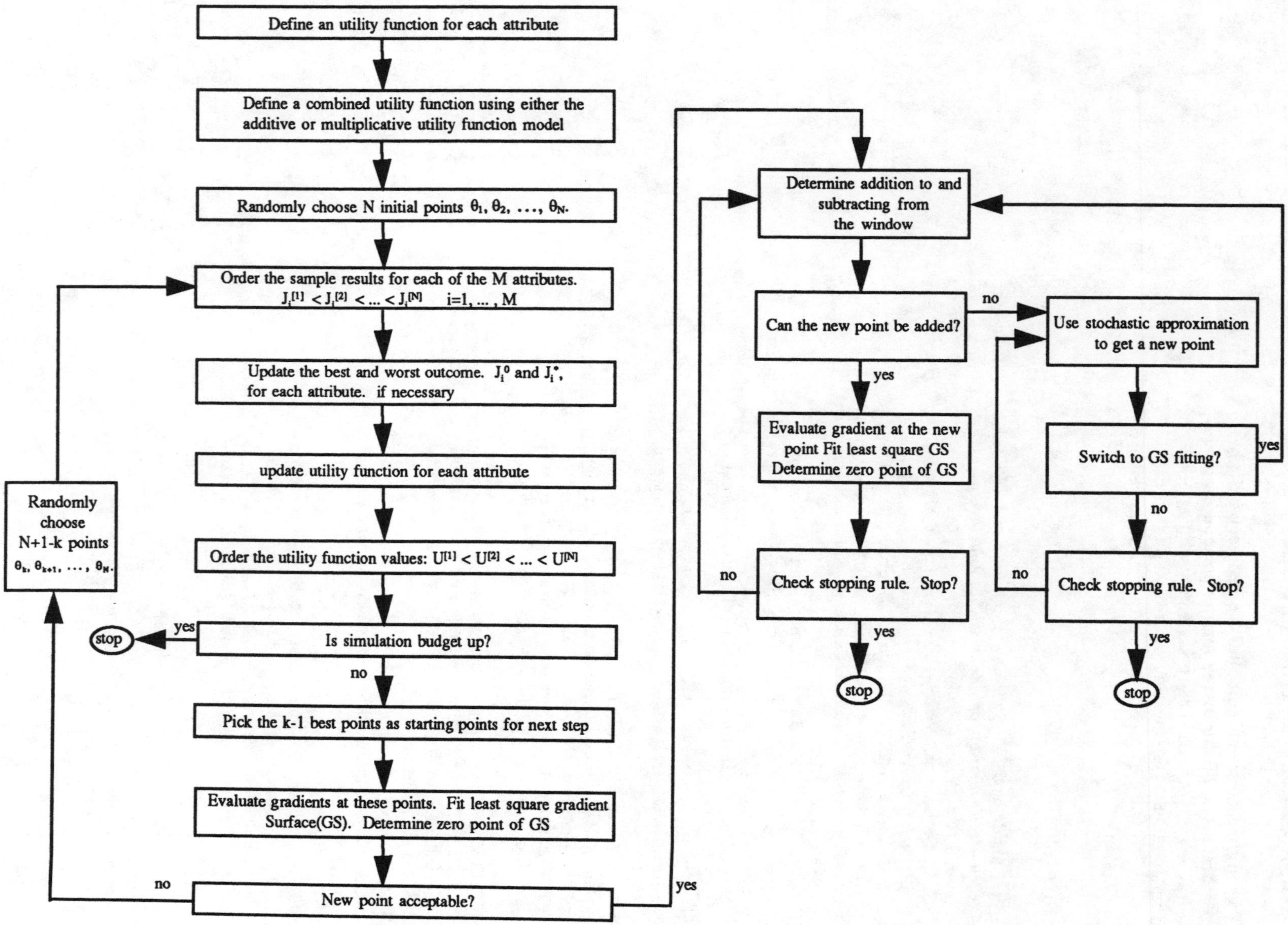

Session 6

Simulation Technologies and Applications

Session Chair:
Charlie Jung
IBM Austin, USA

Fault-Tolerant Real-Time Synchronous Collaboration Environment using WWW*

Soomyung Park[+], Insook Chung[+], Geonyou Jo[+], Sunyoung Han [+], Kwan-Ho Song[++]

[+] Department of Computer Science
Konkuk University
93-1 Mojin-Dong Kwangjin-Gu
Seoul 133-701, Korea
E-mail:{smpark,syhan}@galaxy.konkuk.ac.kr
{ischung,gyjo}@pluto.konkuk.ac.kr
Tel: +82-2-450-3537, 3534
Fax: +82-2-444-1458

[++] IT Standards Division
National Computerization Agency
168, Jukjon-Ri, Suji-Myun, Yongin-Gun
Gyonggi-Do 449-840, Korea
E-mail: khsong@nca.or.kr
Tel: +82-331-289-1501
Fax: +82-331-284-2752

Abstract

Drastic advancement in computer industry has brought today's complicated and various computer technologies. Especially two can be noticed with respect to the area of computing environment. One is the technology to have enabled video conferencing, which handles multimedia data on Mbone. And World-Wide Web(WWW) provides environment which easily locates various kinds of information user wants.

It has been suggested that growing interest in video-conferencing has drawn researchers' attention to Real-Time Synchronous Collaboration Environment.

The paper aims to present the development of synchronous collaboration environment which can deal with multimedia data between multi-users in real-time using WWW systems. The developed environment can be characterized with the two features, multiplication of transfer channels which support realtime multimedia data transfer and fault tolerancy. Multiplication of transfer channels amplifies the ability to transfer multimedia data and fault tolerency gives its users reliability.

1. Introduction

Synchronous collaboration environment means that users in the environment can share and transfer multimedia information in real-time to each other.

The synchronous collaboration environment generally utilizes virtual network, Mbone [6][11][14] as its sublayer in order to achieve multicasting ability, and uses RTP[2](developed by IETF) for real-time transfer.

The research is to develop real-time synchronous collaboration for multimedia conferencing[1][15] using WWW's ability to share information between multi-users. The users in the environment can join a conference simply using WWW[8] browser such as Netscape, Mosaic, and etc. The synchronous collaboration environment consists of three functions which are session management, real-time group conferencing and user tools for the group conferencing.

The functions for session management[9] are mainly to register a session or manage the process necessary when a new comer joins already registered session. To do so, it uses the ability to share information between multi-users, which WWW provides. Functions for real-time group conferencing need to support multicasting[6] of multimedia data. Thus it employs multicasting IP and RTP developed by IETF AVT-WG[2][3], and hires the multi-channels mechanism to improve the ability of the multimedia data transfer. User tools for the group conferencing are applications for the convenience of users joining a conference. They support functions of audio, video and whiteboard. Lastly there are more than one Domain Session Manager per each domain to manage session. To give fault-tolerance(when a Domain Session Manager downs), Arjuna developed by Newcastle Univ. was hired to a Domain Session Manager.

2. Related researches

The following explains related subjects:

* This research is supported by the University Basic Research Grant funded by the Ministry of Information and Communications, Korea.

0-8186-7570-5/96 $5.00 © 1996 IEEE

Mbone[11][14] for multicasting[6], RTP[2] for real-time data transfer, tools for audio, video, and whiteboard functions[4][9][10][12][13] on Mbone, and tools supporting fault tolerance[16][17].

2.1 Mbone (Multicast Backbone)

Mbone is a virtual network that has been in existence since 1992. It was originated from the effort to multicast audio and video from meetings of the Internet Engineering Task Force. Today it is actively used in media-on-demand, live music concert as well as audio, video conferencing. Useful tools for Mbone-based audio/video conference often provided in a public domain.

2.2 RTP: A Transport Protocol for Real-Time Applications

AVT WG(Audio/Video Transport Working Group) is a working group of IETF which experimentally develop a transfer protocol for real-time applications. This working group has developed the RTP and now improved it up to RTPv2. RTP itself consists of RTP(data transfer protocol) and RTCP(RTP control protocol). RTP data transfer protocol normally runs on UDP but can run on ST-II or IP. RTCP allows monitoring of data transfer so as to provide quality of service, and transfering information on the people joining a session.

2.3 Audio/Video Tool

For audio-side tool, vat is used. LBL(Lawrence Berkeley Laboratory) has devised it for audio conferencing. Since vat uses RTP and it is a tool for multiparty conferencing, it should be connected to Mbone.

IVS, audio/video conferencing tool, was developed by INRIA in France and now is being employed in seminars and meetings of MICE(Multimedia Integrated Conference for Europe).

NV is another Internet video conferencing tool. It has been told that Xerox PARC(Palo Alto Research Center) was motivated by workstation's high-rising computing power and that led them to devise a video tool which is purely softwares not requring and hardware.

2.4 Tool for Session Management

SD(Session Directory) is provided as a tool for session management. SD lists session information(session description packets) under multicasting environment which provide by Mbone. With the help of SD, users can create a conference and join it. Without SD, they may waste their time to give all the information such as address, and port of conferencing tool. SD automates such tiresome arrangement so as to give users more convenience in registering and joining a session.

2.5 Tool for WhiteBoard

WhiteBoard such as Xpaint or Pbrush allows several users to share or exchange graphic images or texts simultaneously. And WhiteBoard is more effective on a viewpoint of a bandwidth since it uses the concept of objects not just bitmap. Wb is a tool for WhiteBoard and was developed by LBL(Lawrence Berkeley Laboratory). It runs on various platforms such as Alpha, Ultrix, Sun, SGI, etc.

2.6 Fault-tolerant Distributed Processing Tool

Arjuna[16][17] was developed by Newcastle University in British as a fault-tolerant distributed object-oriented system. The system provides a tool set for fault-tolerant distributed application. Arjuna allows a nested atomic action in order to allow application to be designed and configured on it. Each Arjuna object becomes a main storage for the information of system state. Those objects generated may be either atomic or persistent atomic object . Operations on the objects can be divided into two types: read and write under control of an atomic action. The principle that Arjuna only allows persistent object and atomic action on it guarantees integrity in object even though there may be some fault including loss of message and node damage on the network.

2.7 Collaborative Tool

Shared-Mosaic is developed by NCSA and is an extended version of the WWW browser,
Mosaic. Since it utilizes Mbone for its multicasting, Shared-Mosaic clients can transfer audio data to each other through a WWW server.

Needless to say, Shared-Mosaic can unicast as well as multicast. Unicasting simply means point-to-point communication between two clients.

3. The Environment, design, and implementation of the Synchronous Collaboration

The Figure 1. shows the environment of real-time synchronous collaboration which supports fault-tolerance. The relations between DSMS and WSBS, and between DSMSs are depicted in detail in Figure 2.

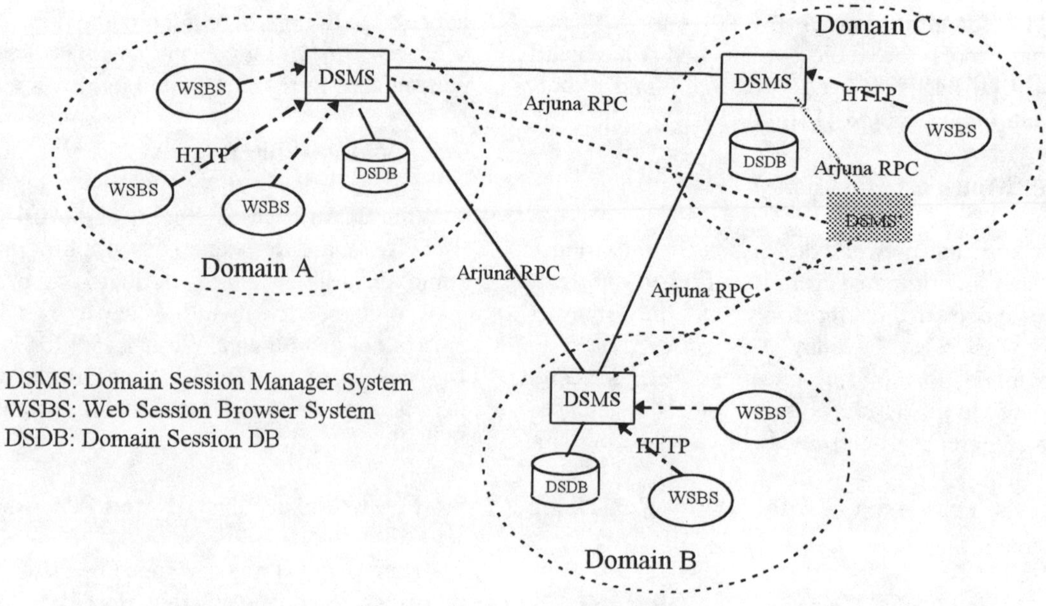

<Figure 1. Synchronous Collaboration Environment>

The synchronous collaboration environment consists of three parts, which are session management, real-time group conferencing and user tools for the group conferencing.

3.1 Session Management

There are more than a Domain Session Manager System(DSMS) which takes the function of Web server, and Web Session Browser(WSBS) for the users willing to join or register a session per domain.

WSBS and DSMS are communicated via HTTP. Each DSMS updates its session information, that is, multicasting routing information using Arjuna RPC(⑥). Using WSBS, one registers a new session to the DSMS in the domain where he belongs to(①). Then the updated session information belongs to the DSMS is notified to other DSMS connected to each other such as a transaction via Arjuna RPC(②).

This process guarantees that every DSMS has identical session information.

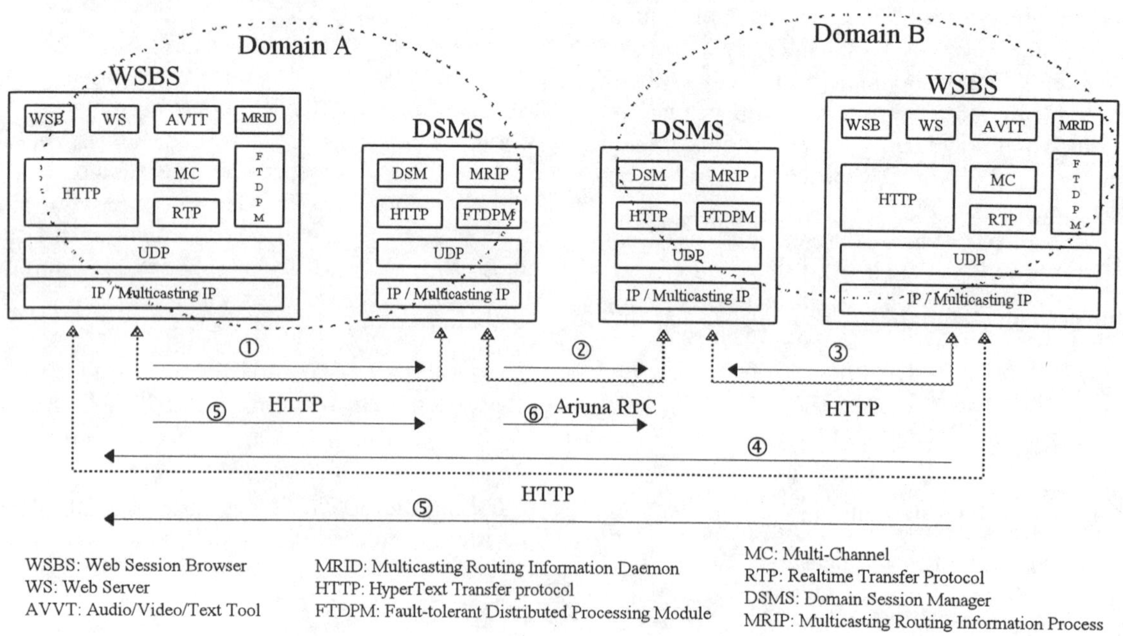

<Figure 2. Relation between DSMS and WSBS>

Each user willing to join a session firstly contacts the DSMS in his domain using WSBS in order to find the session information he/she wants. Then he/she takes session information from the DSMS(③). And, user contacts to the web server system, to join the registered session he/she wants(④). The web server which has received join request from the WSBS updates the acceptance or rejection of the session about him on its session joining home page.

User can find the acceptance or rejection of the session about him/her through the session joining home page which the web server provides(⑤). In the acceptance case, the web server transfers the information related the comer to the DSMS in the domain where the web server belongs to(⑤). In the other case, there is nothing.

In the Figure 1, there are two DSMSs in domain C. WSBS can connect either of them. When the DSMS which a WSBS has connected is down, the WSBS connects and communicates with the other. So, it can run fault-tolerantly.

joining user's information from his WSBS in the domain user updates its multicasting routing information, and informs it to DSMSs in another domains as well as the other DSMS in the same domain at the same time(①). The updated multicasting routing information is notified to other DSMS connected to each other such as a transaction via Arjuna RPC. This information is actually delivered by MRIP via Arjuna RPC to daemon process, so called MRID, which updates multicasting routing information AVTT refers(②). Each MRID then performs AVTT on the system with that information(③).

Via RTP, AVTT of the user links to AVTT of others. Then group conferencing is enabled(④). Each AVTTs are dynamically controlled by MRID which receives multicasting routing information from MRIP.

To improve transfer ability, Multi-channel has been inserted between AVTT and RTP, as Figure 4 shows. And practically AVTT consists of three conferencing tools: Audio, Video and Text tools.

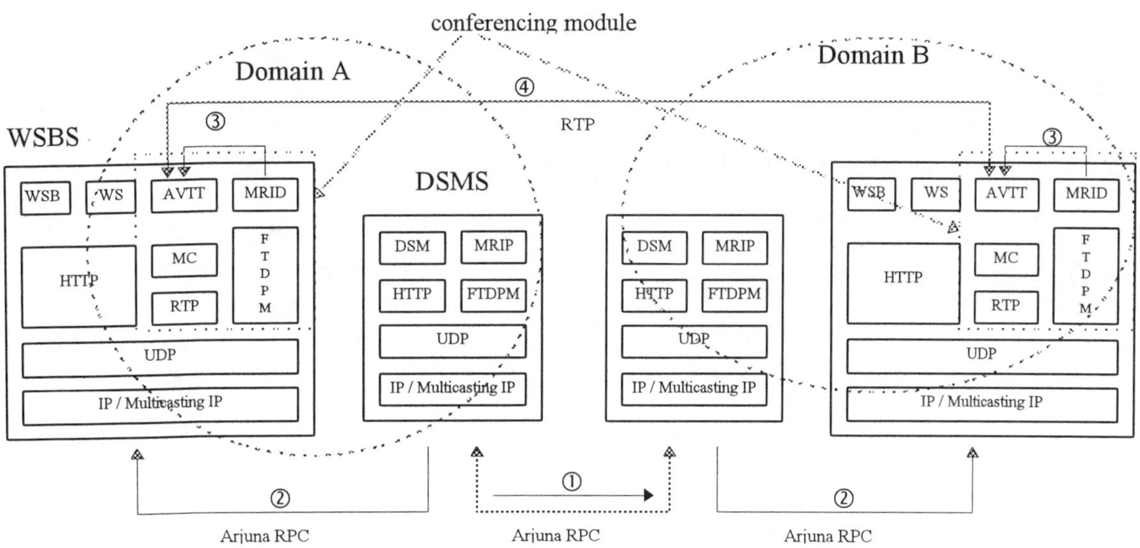

<Figure 3. relations of WSBS, DSMS, the conferencing module>

This kind of fault tolerance is achieved by Arjuna's distributed processing environment. Also, DSMS has the function of storing session's contents. If the type of session is such as a lecture, DSMS can store the session's contents(voice and image) to DSDB. So, users can retrieve session's contents which DSMS have had.

3.2 Real-Time Group Conferencing

Figure 3 shows the relations of WSBS, DSMS, and the conferencing module necessary for practical conferencing in the synchronous collaboration environment.

The DSMS in the same domain which received a session

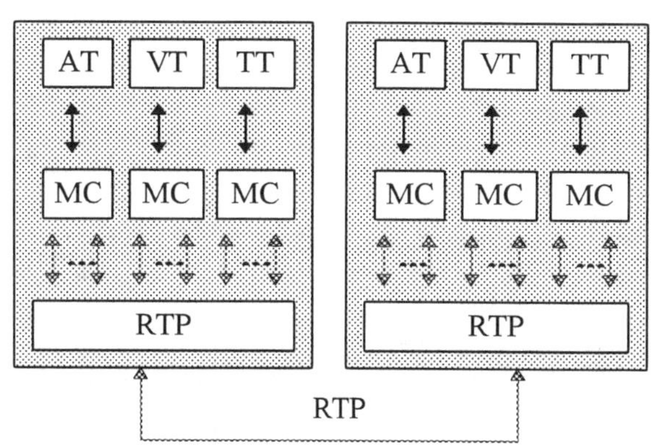

<Figure 4. Multi-Channel Mechanism>

3.3 Tools for group conferencing

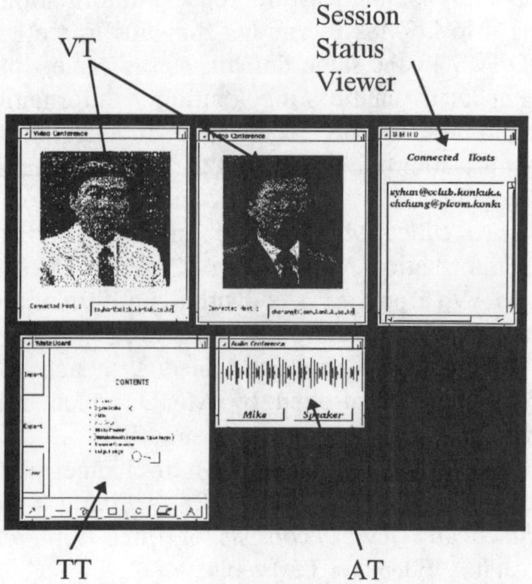

<Figure 5. AVTT running on the X-Windows>

WSBS is a tool for registering and joining a session and AVTT(AT for audio, VT for video, TT for text data) performed by MRID with multicasting routing information is user tool for group conferencing. Figure 5 shows those tools which pop up on X-Window.

WSBS(Web Session Browser System) takes care of registration of a session to its DSMS and it also gets service on the registered session information. Figure 6 shows WSBS. To register a session, user should fill out proper information such as session name, subject, system information hosting the session. This information is then registered to the session list home page of the DSMS. The registered session list is also useful for the people who want to join the session.

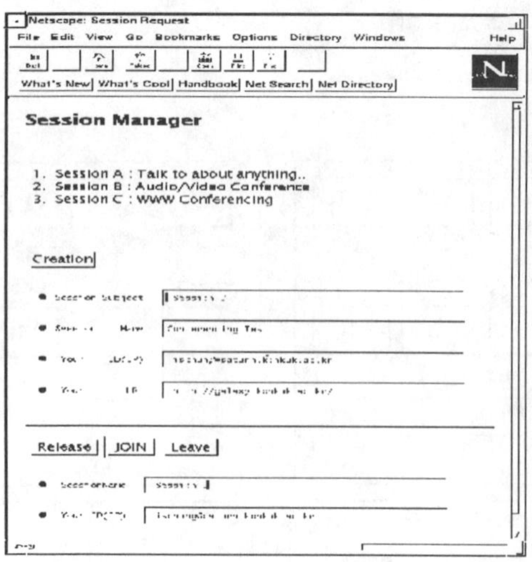

<Figure 6. WSBS>

VT(Video tool) performed via RTP. Video data browser shows each of participants who join the session. AT (Audio tool)performed via RTP. Audio data browser gives audio data from each participant. VT performed via RTP. This gives a brief look of how personal document data to be treated with the help of WhiteBoard Implementation Environment

Aforementioned, Fault-tolerant Real-Time Synchronous Collaboration Environment is implemented as extension of WWW on MBone. The following is the environment of the implementation.

- System: Sun Sparc Station 20 supporting Multimedia Kit
- OS: Solaris 2.x
- GUI: X-Window & Motif
- Compiler: GNU C compiler 2.6.3
- Fault-Tolerance tool kit: Arjuna

4. Conclusion

The research aims to develop Real-Time Synchronous Collaborative Environment which can deal with multimedia data using WWW. The environment developed here introduces idea of planting multi-channel mechanism between AVTT and RTP in order to improve transfer ability. Since the environment enables DSMS, which take care of session information and session contents, to be fault-tolerant, it also can be said that the environment gives reliability necessary in synchronous collaboration.

References

[1] T. K. Woo, M. J. Rees, "A Synchronous Collaboration Tool for World-Wide Web", 1994. 5.

[2] H. Schulzrinne, S. Casner, R. Frederick, C.D. Schulz and V. Jacobson, "RTP: A transport protocol for real-time applications", INTERNET-DRAFT, 1994.6.

[3] R. Braude, S. Zabel, "RFC 1458: Requirements for Multicast Protocols", 1993. 5.

[4] R. Freerick "NV manual pages", Xerox PARC, 1992.

[5] Robert W. Scheifler, James Gettys, "The X Window System, third edition", Digital Press, 1992.

[6] S. Deering, "RFC 1112: Host Extention for IP multitasking", 1989. 8.

[7] The FAQ can be viewed on the World Wide Web at address: http://www.research.att.com/mbone-faq.html, 1994. 8.

[8] Thierry Turletti, Jean-Chrysostome Bolot, "Issues with multicast video distribution in heterogeneous packet networks", INRIA, 1993.

[9] V. Jacobson, "SD manual pages", Lawrence Berkeley Laboratory (LBL), 1993.3.

[10] Van Jacopson, Steve McCane, "vat-LBNL Audio Conferencing Tool", 1995. 10.

[11] Jean Bunn, "MBONE (Multicast Backbone)", Geneva

University, 1994. 11.
[12] Thierry Turletti, "The INRIA Videoconferencing System", 1994. 10.
[13] Van Jacobson, Steve McCanne, "Using the LBL Network Whiteboard", 1993.2.
[14] Steve Caster, "Frequently Asked Question(FAQ) on the Multicast Backbone(MBONE)", 1994. 8.
[15] Martina A. Sasse, Claus-D Schulz, "Remote Seminars through Multimedia Conferenceing: Experiences from the MICE prpject", 1994.
[16] Dixon, G. N, G. D. Parrington, S. K. Shrivastara, S. M. Wheater, "The Treatment of Persistent Objects in Arjuna", 1989. 7.
[17] S. K. Shrivastava, G. N. Dixon, G. D. Parrington, "The Design and Implementation of Arjuna", 1989. 3.

A Simulation Study of Real-time MPEG Traffic Using Forward Error Control Scheme

Lee Chuan Hu and Kwei-Jay Lin
Department of Electrical and Computer Engineering
University of California, Irvine
Irvine, California 92717

Abstract

MPEG coding dramatically reduces the data amount for video. On the other hand, its bursty and cyclic nature creates a challenge for transmission in ATM networks. Various methods have been proposed to transmit MPEG streams, such as priority coding, dynamic bandwidth allocation and smoothing. This paper suggests another scheme to the problem. The original MPEG streams are split into n substreams and the substreams are skewed a fixed amount of time from each other. The substreams are transmitted with the forward error correction code. It provides a deterministic bound on the delay and has the implementation simplicity needed for the high-speed video applications. We have tested the scheme on five real MPEG-coded video/movie stream traces. It significantly reduces the peak cell losses at the price of longer delay.

1 Introduction

MPEG coding was originally designed for digital video storage. It dramatically reduces the total amount of data in a video stream. However, when transmitting an MPEG traffic over an ATM network for real-time applications like on-line video and tele-conferencing, it is difficult to guarantee the required quality of service (QOS) due to the bursty nature of MPEG on the network utilization. A real-time MPEG stream has the real-time requirements of delay and jitter constraints. The high-speed nature of data streams (w.r.t. today's processors) puts tight limits on the complexity of algorithms used in switches and network interfaces. In the mean time, since MPEG is a variable bit rate (VBR) traffic with cell loss restrictions, we need to allocate an adequate network bandwidth. A connection based on the peak data rate gives a low network utilization, while a connection with too low a bandwidth may result in high cell losses (caused by switch buffer overflow) and introduces extra network delays and jitters.

There are three approaches to the problem: coding-based scheme, network based scheme and real-time scheme. [1-3] The coding-based scheme prioritizes the cells based on the importance of the content. In case of a congestion, low priority cells are dropped to maintain the minimum QOS. The network based scheme tries to dynamically adjust the allocated network bandwidth to accommodate the predicted upcoming bursts. Both schemes require a significant amount of real-time processing and/or network support. On the other hand, real-time approach uses the real-time characteristics of the streams (not the content of the data) to maximize the network utilization (or minimize the required bandwidth) while maintaining the QOS.

In this paper, we propose a scheme using the forward error correction technique that falls into the real-time scheme category. We observe that a real-time MPEG stream has several properties. First, the bandwidth may not be well used. A given channel capacity may not be enough at bursty peaks while the channel is under-utilized for the rest of the time. The time constraint adds some complexity in using the unused bandwidth since data can not be arbitrarily delayed. Secondly, when we keep the cell loss rate within a specified range (within QOS), the bounded delay and jitter are more important than a good average behavior. Third, multimedia data streams like those in video and audio are different from the traditional data in computers. The traditional computer data are "precise" data which we cannot lose even a single bit. On the other hand, multimedia data are similar to those in analog systems. The human perception determines the satisfactory degree of a system. We propose a scheme tailored to these properties. The scheme bounds the

This work was supported in part by contracts from the Office of Naval Research N00014-94-1-0034 and N00014-95-1-0262, Hughes Aircraft/UC MICRO 94-082 and 95-097, and the US Navy NSWCDD N60921-94-M-1261 and N60921-94-M-2714.

0-8186-7570-5/96 $5.00 © 1996 IEEE

delay and jitter instead of cell losses. It utilizes the unused inter-frame channel capacity and redundant codes to minimize the cell loss. In our scheme, the fault-tolerant computing technique is not used to provide a reliable computation, but to reduce the overall cell losses. The interesting result is that even we add extra forward error correction (FEC) code to the data stream, the actual required bandwidth is reduced for the same error rate. In addition, it has the implementation simplicity needed for high-speed applications.

In this paper, we consider both video playback and interactive multimedia streams as those in the video conference. The jitter is more critical in the former case while both jitter and delay are important for the latter. This is an application of the scheme introduced in [3]. The rest of the paper is organized as follows: Section 2 describes the background and relevant works. Section 3 presents our scheme. Section 4 shows the simulation results. The paper is concluded in Section 5.

2 Background

For encoding frames, MPEG has three types of frames (solid line in Figure 1): I-frames for intra coded frames (peaks above 150,000 bits/frame), P-frames for predictive coded frames (small peaks less than 100,000 bits/frame), and B-frames for bi-directionally predictive coded (interpolative) frames (valleys). One I frame is interleaved with a certain number of P and B frames. Since P and B frames have the information of the difference among the frames they are not independent frames themselves. P and B frames usually have much smaller size (fewer bits) than that of I frames. From the standpoint of recovery cell loss, the fewer I frames are inserted in a P and B frame stream, the longer it takes to recover from frame losses. The interarrival time of I frames is the worst case time needed for an image to recover. Another important characteristic of MPEG is that after the coding, the scene changes are no longer the highest bursty peaks. Instead, the I frames become the bursty peaks which are deterministically periodic. Figure 1 shows the typical MPEG coded trace which is excerpted from a movie. This one has a frame sequence of "IBBPBBPBBPBBPBB", which can be seen from the solid line plotout.

Various methods have been proposed for the VBR applications, e.g. priority coding, dynamic bandwidth allocation and Δ-smoothing. By identifying some critical components, e.g. header information, in the video streams for playing back, priority coding labels the data with different priorities. During the transmission, in case of queue overflow or traffic congestion, the network sacrifices the low priority data to guarantee the delivery of high priority data. [1,10] This scheme requires sophisticated coding and is video content- and application-dependent (for the purpose of network transmission). Since this method does not change the bursty characteristics of an application, the traffic can experience a high delay and jitter.[5]

The dynamic bandwidth allocation deals with the problem from the network standpoint. Based on their study, [2] indicates that the low-frequency signal captures the slow time-variation of consecutive scene changes (usually more than a second) while the high-frequency signal exhibits the feature of strong frame autocorrelation. The video transmission bandwidth in a finite-buffer system is essentially characterized by the low-frequency signal. Therefore, it is feasible to implement dynamic allocation of video transmission bandwidth using on-line observation and prediction of scene changes. It presents two prediction schemes (recursive least square method and time delay neural network method) to predict future bandwidth requirements of a video stream and dynamically allocates the needed bandwidth. Like in the case of priority coding, this scheme is also video content-related and the efficient bandwidth allocation depends on the network design. This method targets at the slow fluctuation of the bandwidth requirement caused by the video scene change. If the setup time for extra bandwidth is constant, a network may add some jitter for different action speed in the video (content-related). On the other hand, the setup time is likely affected by the network load since the routing and admission control are all related to the network load. Also the fractal (or self-similar) phenomenon may further hinder the effectiveness of the scheme. For a multiple-switched wide area network like ATM, the effectiveness of the scheme is yet to be actually tested.

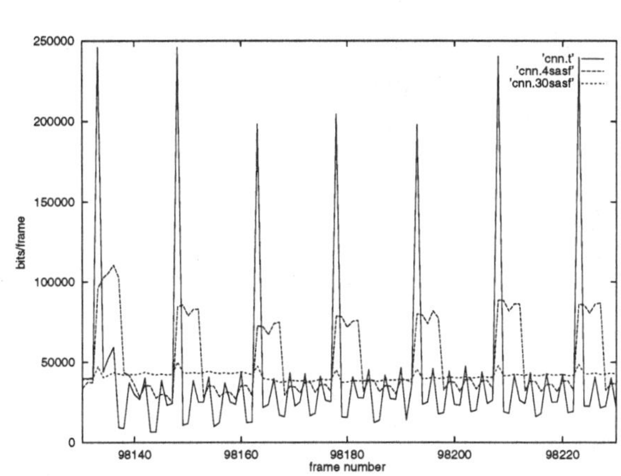

Figure 1 MPEG traces

[9] shows a real-time scheme called Δ-smoothing. Although it is theoretically optimal and has a delay bound, it is based on the knowledge of future bandwidth requirement like that in the case of dynamic bandwidth allocation (It would not be optimal if the forecast is not perfect.) The assumption for the scheme to work is that recent behavior of the traffic stream can be used to approximately forecast near-term future behavior. The forecast process is computationally expensive. The dilemma is that computation is only needed at the points where bandwidth requirement changes but the purpose of the computation is to predict such a change. Video conferencing trace is used in their simulation (which is not as bursty as those in normal movies because of static background.) They calculate the value once per frame. Obviously, this method works only for low frequency scene changes like the case in dynamic bandwidth allocation. Further, because of non-perfect prediction of the future traffic behavior, the bound given is actually non-deterministic.

3 Our Scheme

Our scheme is to exploit the under-utilized bandwidth in neighboring frames. The scheme is deterministic on delay and jitter bounds. Since the data arrival of a video stream is not independent in MPEG streams, the (σ, ρ) model described in [4, 5] is used in our discussion. For a given $\sigma \geq 0$ and $\rho \geq 0$, we say $R \sim (\sigma, \rho)$ if and only if for all x, y with $y \geq x$ the following relation holds

$$\int_x^y R \leq \sigma + \rho(y - x) \qquad (1)$$

where R is a nonnegative function. $\int_x^y R$ is the amount of data from the stream that is transmitted on the link in the interval [x, y]. In general, R(t) is the rate function of a stream and represents the instantaneous rate of traffic at time t. The constant ρ determines the upper bound to the long term average rate of the traffic while the value σ allows the burstiness. The larger σ is, the more bursty a traffic can be.

We can define the nonnegative cell loss function $L_\rho(R, t)$ for any function R and $\rho > 0$ for all x and t as

$$L_\rho(R, t) = \max\left[\int_x^t R - \rho(t - x)\right] \; (\geq 0) \qquad (2)$$

Obviously, for any finite value $\sigma \geq 0$, if $R \sim (\sigma, \rho)$, $L_\rho(R)(t) \leq \sigma$. Zero cell loss is possible if we allow infinite delay (and buffer the finite data) $t - x \to \infty$ with large enough ρ.

From (2) we can see in the case of video playback, by allowing enough delay (and buffering), we can achieve the required cell loss QOS if enough bandwidth is allocated and traffic is bounded by (1). On the other hand, with the low human tolerance for delays, interactive multimedia applications have to find a means to balance the cell loss, delay and jitter. If we want to limit the loss to a specified value L_c, we have the channel capacity C as

$$C = \max\left[\left(\int_{t0}^t R - L_c\right) / (t - t_0)\right] \qquad (3)$$

We can see the tradeoff between delay and cell losses here. If C and L_c are given, it is a cell loss priority delay scheme since the delay could be very significant for VBR applications like that in MPEG streams. The delay is not bounded. In real-time and multimedia applications, later arrival data may have the same effect of cell loss. Moreover, non-constant delay among the frames introduces extra complexity in the processing. So in our scheme, we bound delay instead of cell losses.

3.1 n-split and Skew

By looking at Figure 1, we find that the channel capacity is under-utilized for B frames and at the same time potential cell losses may occur for I frames. We want to use the unused capacity in those small frames by delaying part of the large frames. Moreover, we want to bound the delay associated with the smoothing process. The scheme we propose is split-and-skew (SAS). Figure 2 shows the block diagram of the scheme. Since a frame is packed and repackaged in cells locally, the ordering of cells (and frames) can still be guaranteed.

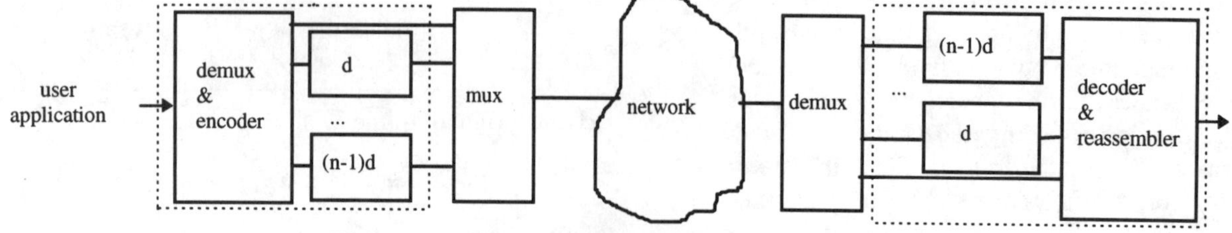

Figure 2 Stream splitting forward error control scheme

The input stream is split into n equal-sized sub-streams S_i, $0 \le i \le n-1$. Each sub-stream is skewed one frame from the previous one. Since we have n sub-streams, the total delay (sum of the skews) is n-1 frames. In our discussion, the frame arrival and delay are all discrete events and the time unit is measured in frame interval (or 33 msec for 30 frames/sec case). If the original traffic function is f(k), where k is the frame number and the time, and the unit delay (one frame time) function is D, for a particular frame, we have

$$f'(k) = f_0(k) + D(f_1(k)) + \ldots + D^{n-1}(f_{n-1}(k))$$

$$= (\sum_{0 \le i \le n-1} D^i(f_i(k))) \quad 0 \le k \le \infty \qquad (4)$$

where $f_i(k)$ is the *ith* equal-divided sub-frame with $|f_i(k)| = \frac{1}{n}|f(k)|$.

Then these f(k)'s are reassembled for transmission. The new data stream becomes

$$S(k) = f_0(k) + f_1(k-1) + \ldots + f_{n-1}(k-(n-1))$$

$$= (\sum_{0 \le i \le n-1} f_i(k-i)) \quad \text{for } k \ge n-1 \qquad (5)$$

Note that this is an average over time (over last n frames). We call this *local averaging*. Local averaging reduces the amplitude of the fluctuation of input streams. The larger the n is, the less bursty the resulted stream will be. If the original stream has a burst peak at $k = k_0$, the amount of reduction can be calculated as

$$\Delta_k = |f(k_0)| - (\sum_{0 \le i \le n-1} |f_i(k_0 - i)|)$$

$$= \frac{n-1}{n}|f(k_0)| - (\sum_{1 \le i \le n-1} |f_i(k_0 - i)|) \qquad (6)$$

For MPEG, since two frames in front of an I frame (the burst frame) are always B frames (further, even a P frame is less than half of the size of an I frame for most cases), the reduction is significant. As the later part of the paper shows, if we allow n to be large enough (as in the case of video playback application), we can get very smooth stream for transmission.

We define the I frame and P frame cycle times as T_I and T_P, respectively and let $R \sim (\sigma, \rho)$ for $t - x \le T_I$. For an MPEG pattern of "IBBPBBPBBPBBPBB" (the one used in the trace of CNN news), T_I and T_P are 15 and 3 (in number of frames). For our discussion, σ can be represented in two parts σ_I and σ_P, which represent the bursty parts of I and P frames respectively. We have

$$L_\rho(R,t) = \max[\int_x^t R - \rho(t - x)] \qquad (7)$$

$$\le \max[\sum_{j=k_0}^{k} (\sum_{i=0}^{n-1} |f_i(j-i)|) - \rho(k-k_0)] \qquad (8)$$

$$= \max[\frac{1}{n} \sum_{i=0}^{n-1} \sum_{j=k_0}^{k} |f_i(j-i)| - \rho(k-k_0)] \qquad (9)$$

$$\le \frac{1}{n}[\lceil (k-k_0)/T_I \rceil \sigma_I + (\lceil (k-k_0)/T_P \rceil - 1)\sigma_P + n\rho(k-k_0)] - \rho(k-k_0) \qquad (10)$$

$$= \frac{1}{n}[\lceil (k-k_0)/T_I \rceil \sigma_I + (\lceil (k-k_0)/T_P \rceil - 1)\sigma_P] \qquad (11)$$

where $k = \lceil t \rceil$ and $k_0 = \lfloor x \rfloor$. Without loss of generality, we limit our discussion to within one I frame cycle or $n \le T_I$. Then the worst case burstiness becomes I and P frame burst averaging over the delay period n. On the other hand, we can see this from (5):

$$|S(k)| = (|f_0(k)| + |f_1(k-1)| + \ldots + |f_{n-1}(k-(n-1))|$$

$$= \frac{1}{n} \sum_{i=0}^{n-1} |f(k-i)|$$

$$\le \frac{1}{n}(\sigma_I + (\lceil (k-k_0)/T_P \rceil - 1)\sigma_P + \rho(k-k_0)) \qquad (12)$$

This means the burstiness is reduced to $\frac{1}{n}$ of the σ_I and the number of σ_P in the region. Due to the repetitive nature of the MPEG, for a split and skew more than T_I the benefit gained is reduced. The improvement of this can only be for the case σ_I fluctuation over the different sections which is not the main objective of the scheme.

The buffer requirement is another issue. At any time, the buffer needed for the current frame is $\frac{(n-1)}{n}|f(j)|$ and for the last frame is $\frac{(n-2)}{n}|f(j-1)|$, and so on

$$B(j) = \frac{(n-1)}{n}|f(j)| + \frac{(n-2)}{n}|f(j-1)| + \ldots + \frac{1}{n}|f(j-(n-2))|$$

$$= \sum_{1 \le i \le n-1} \frac{n-i}{n}|f(j-i+1)| \qquad (13)$$

For example, if n = 4 and f(j) = 200,000 , f(j-1) = 20,000 , and f(j-2) = 20,000 , all in bits, the $B(j) \approx 20$ Kbytes. So the delay associated buffer requirement is not significant (it is usually less than a single I frame.)

3.2 Forward Error Correction

In our proposed scheme, forward error correction (FEC) is used to further reduce cell losses.[3][6-8] We use parity + checksum to detect erroneous cells and correct them if possible. The method is similar to the diversity coding. But instead of using spatial diversity channels, to keep the temporal relation among the cells and to reduce the reassembling complexity, we diversify the connection in terms of time (since the splitting and reassembling a new stream is done in the sender node under a single local clock, to have tight timing control is not difficult). In this way, the stream can resist to longer bursty interference or overflow cell losses.[3] This is suitable for switching networks like ATM. Temporary buffer overflow and switch blocking can cause this kind of bursty errors. It is a case of temporary erasure channel problem. For example, we can look at the checksum to decide if a section of data is corrupted.

Let the original frame j: $f_j = f_0^j + f_1^j + \ldots + f_{n-1}^j$, where the subscripts indicate the part which will be assigned to corresponding substreams. We can form the (n+1)th sub-frame (so does the sub-stream) by

$$c_i(f_0^j) = b_i(f_0^j) \oplus b_i(f_1^j) \oplus \ldots \oplus b_i(f_{n-1}^j) = \bigoplus_{l=0}^{n-1} b_i(f_l^j) \quad (14)$$

where $b_i(\bullet)$ means taking the bit i of that sub-frame and $\bigoplus_{l=0}^{n-1}$ indicates doing logical exclusive-OR of b_0, b_1, ..., b_{n-1}. c_i is the corresponding bit in the forming parity frame. Then, in case we lose one sub-frame k, we can recover the sub-frame (so does the whole frame) by

$$\tilde{b}_i = c_i(f_n^j) \oplus \bigoplus_{\substack{l=0 \\ l \neq k}}^{n-1} b_i(f_l^j) \quad (15)$$

This can be seen by expanding $c_i(\bullet)$ in (15) by (14). Since we skew the sub-frames one frame from its previous sub-frame, it produces the effect of the temporal diversity coding, which means by including the (n + 1)th sub-stream we can recover burst loss up to one sub-frame in length for the worst case. Another interesting thing is to selectively drop the cells from different substreams when a burst occurs. This can result in correcting a burst length up to one newly formed "frame" or $\sum_{i=0}^{n} f_k^{j+i}$, where $0 \leq k \leq n$. As the simulation shows, although the parity adds extra data to the traffic, the overall loss is actually decreased. But a large n does have a better performance since it both reduces the overhead and the burstiness.

The associated overhead for FEC has a different impact than other overheads for fault tolerance. For VBR traffic, we compare cell losses under the same given bandwidth to the case without. It has no other negative effect associated with it. The logic needed to implement the FEC is very simple and has no significant extra delay.

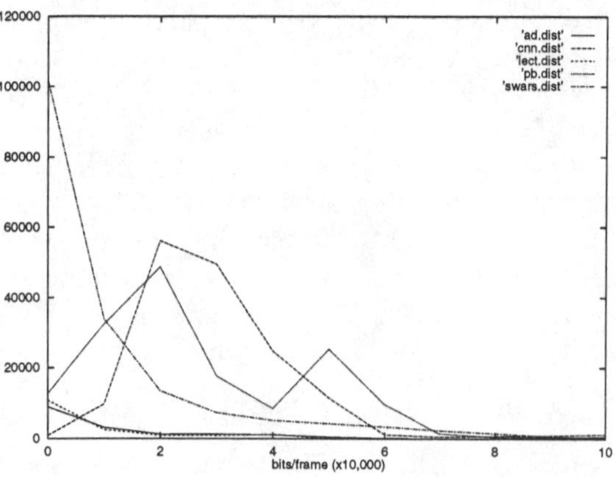

Figure 3 Frame size distributions of 5 videos

4 Simulation Result

We have studied five real MPEG video traces which are coded by three different MPEG coding implementations. These MPEG streams represent a class of video VBR traffic supported in ATM. Star Wars from Bellcore is coded in MPEG I with "IBBPBBPBBPBB". The original video was 500x400 frame size and 24 frames/sec and is a complete movie running about two hours. The rest of the traces are from UC-Berkeley. Advertisements and Lecture are software coded with a resolution of 160x120 and 30 frames/sec. and each runs about 10 minutes. They have a pattern of "IBBPBB." CNN news and Princess Bride are both coded by a commercial hardware MPEG coder with a resolution of 320x240 and 30 frames/sec. Each of them is about 90 minutes and have a pattern of "IBBPBBPBBPBBPBB." Table 1 gives a brief statistics about the original traces.

From the table we can see that there is a great disparity between the maximum frame and the minimum sizes. The maximum frame sizes are 5 - 11 times of the mean frame sizes. The standard deviations are also very large. These can be expected from Figure 1. On the other hand, a particular implementation can have strong influence over the resulted streams. This characteristic can overweight the impact of a specific video content.

In the simulation, we take the original stream given in (4) and do splitting and skew process by using (5). For the case with error correction, we also form a parity sub-stream according to (14). Afterwards, the new stream is transmitted under a given bandwidth. During the transmission, if an overflow happens, it is checked to see if it is recoverable. If it is a non-recoverable error (loss), the case is recorded. The relation between the loss rate and the channel capacity is plotted out. We have simulated the following cases:

1. n = 4 split with an extra sub-stream (the 5th) for FEC (4-SASF). It has a worst case delay of 5 frames.
2. n = 4 split without the forward error correction (4-SAS). It has a delay of 4 frames.
3. n = 6 split with forward error correction and 7 frame delay in worst case (6-SASF).

In all the cases above, the total delays are less than 250 msec and usually considered as tolerable to human perception for interactive applications. On the other hand, as indicated in [5], the better regulated stream may actually decrease the network delay so the net delay increase can be smaller. To see the longer averaging effect, which can be used in video playback, we also run a case of n = 30 frames (about 1 second delay for 30 frames/sec or 1.25 sec. for 24 frames/sec case) with error correction (30-SASF). Tables 2 - 5 list the corresponding mean, maximum, minimum and standard deviation of frames. The impact of FEC is included in all simulations.

Figure 1 is an excerpt from CNN news. The long and short dashed lines show two cases of SAS processed streams. The corresponding original stream is in solid line. The long dashed line gives the stream with 4-split and skew and FEC coding (4-SASF). Its new peaks are less than 50% of original ones. This means that even including 25% overhead for FEC the bandwidth requirement is decreased over 50%. The short dashed line (30-SASF) is plotted for reference. With 30-split and skew and FEC, the stream almost becomes a constant bit rate stream and the bandwidth requirement decreases down to one fifth of original peak. At the same time, the buffer requirement is less than 200 kbytes, a practical size. Although a one second delay is not feasible for interactive application, it can still be used for on-line video playback.

Figure 3 is the distribution of frame size from the five MPEG streams. From the curves we can see the impact of a particular implementation. The traces of Advertisement and Lecture are both coded by a software coder and are almost identical although the contents are very different. They will show similar characteristics in the simulation. On the other hand, CNN news and Princess Bride are closer to each other than to, say, Star Wars. The statistics in Table 1 also reveals the same thing. This coder-implementation-dependent feature is useful: in bandwidth allocation, we may not have to test a particular video for the allocation. Instead, we get the characteristics of the coder through *any* video and based on that information for future bandwidth allocation. But one question remains: Is the difference caused by sub-frame coding or by the I, P and B frame interleaving?

In Tables 1-5, we can see that SAS has the most significant impact over CNN news and Princess Bride. The maximum frame size in 4-SAS with FEC (4-SASF) is less than 45% of the original. It is further down to 35% in 6-SASF case and 20% in 30-SASF. In 30-SASF case, the maximum is very close to the mean frame size, which indicates a very smooth stream (the standard deviation is only 3% of the mean, down from 28% in the original stream.) All the traces show a similar trend when SAS is applied. But they are not as significant as that shown in CNN news and Princess Bride. We believe this is because of T_I cycle. Longer T_I gives more room to exploit the unused bandwidth among the frames.

Table 1 Basic Statistics of the Original Traces

Movie	mean frame size (bits)	maximum frame size (bits)	minimum frame size (bits)	standard deviation (bits)	# of frames	frame cycle time T_I (# of frames)
Advertisement	15235	82560	1152	10808	16316	6
Lecture	11242	50304	1920	9443	16316	6
CNN news	40039	246624	3584	11356	164748	15
Princess Bride	40038	243552	3584	14829	167766	15
Star Wars	15598	185267	476	10391	174136	12

Table 2 Statistics of 4-split with FEC

Movie	mean (bits)	max. size (bits)	min. size (bits)	std. dev. (bits)
Advertisement	19041	85632	2112	8555
Lecture	14051	27168	4320	4085
CNN news	50049	110592	22016	10745
Princess Bride	50048	107776	19200	12244
Star Wars	19495	119858	2972	8933

Table 3 Statistics of 6-split with FEC

Movie	mean (bits)	max. size (bits)	min. size (bits)	std. dev. (bits)
Advertisement	17771	79808	2688	7548
Lecture	13113	27264	4736	3238
CNN news	46710	87037	22355	8684
Princess Bride	46709	85160	16613	10386
Star Wars	18194	114859	2101	7611

Table 4 Statistics of 4-split without FEC

Movie	mean (bits)	max. size (bits)	min. size (bits)	std. dev. (bits)
Advertisement	15234	71136	1728	7239
Lecture	11241	24864	3744	4046
CNN news	40039	100568	17280	7239
Princess Bride	40038	97752	14976	11251
Star Wars	15594	110291	2467	7986

Table 5 Statistics of 30-split with FEC

Movie	mean (bits)	max. size (bits)	min. size (bits)	std. dev. (bits)
Advertisement	15714	63104	1638	6386
Lecture	11596	19560	947	2235
CNN news	41356	50564	5012	1239
Princess Bride	41355	50296	3322	1518
Star Wars	16102	58811	420	5598

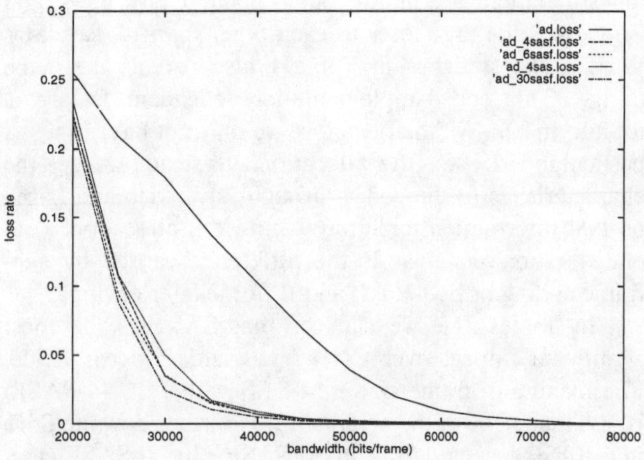

Figure 4: Loss rate vs. Bandwidth for Advertisements

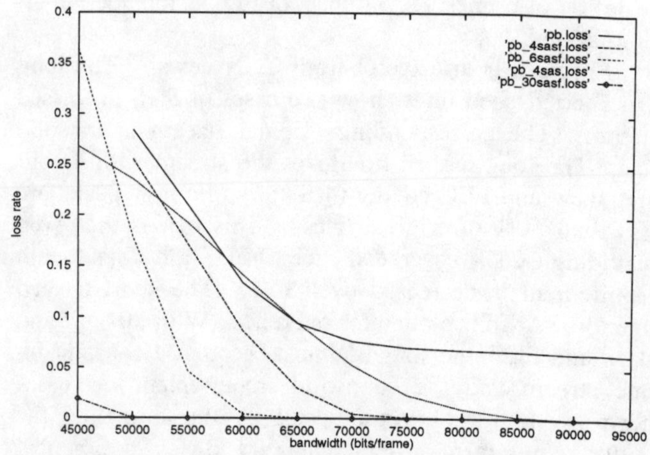

Figure 7: Loss rate vs. Bandwidth for Princess Bride

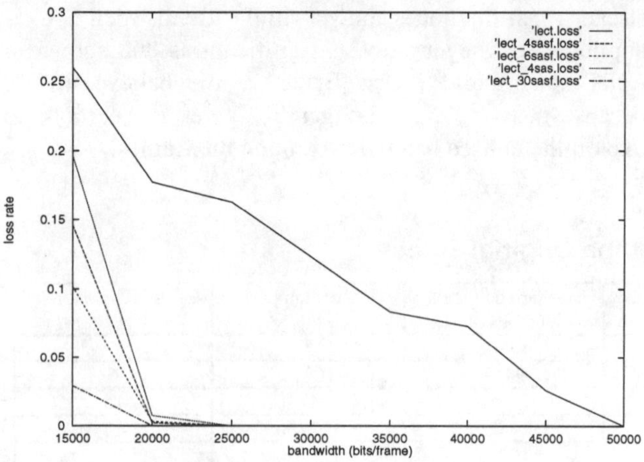

Figure 5: Loss rate vs. Bandwidth for Lecture

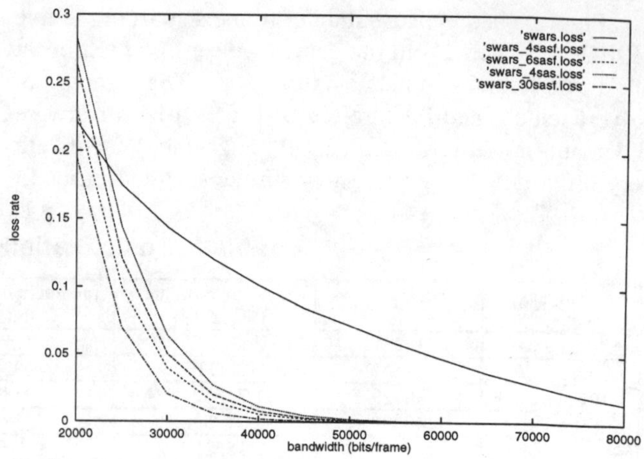

Figure 8: Loss rate vs. Bandwidth for Star Wars

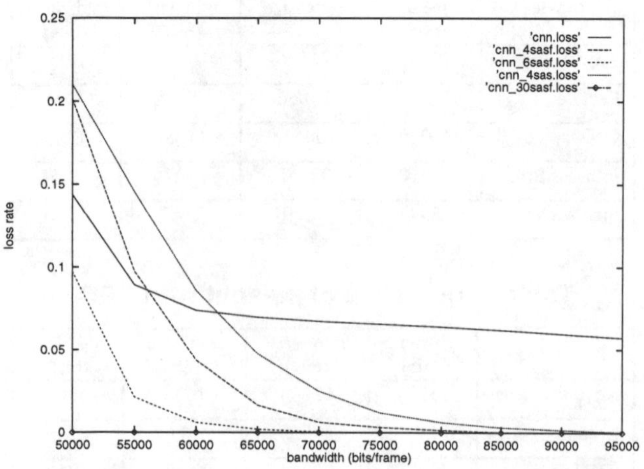

Figure 6: Loss rate vs. Bandwidth for CNN News

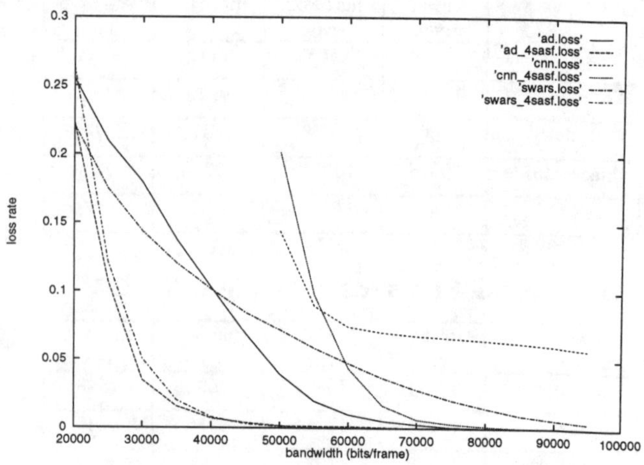

Figure 9: 4-SASF traces from 3 diff. coders

Figures 4 through 8 give the simulation result of the loss rate vs. the bandwidth. The similar characteristics discussed above can be seen from the plotouts. SAS with or without FEC reduces the bandwidth requirement under a given loss rate consistently across all the traces. But at the same time, there are some differences. To show the characteristics of the implementations from different coders, we put three traces from three different coders in the same plotout (Figure 9). From the simulations, we can see:

1. Stream splitting and skew significantly reduces the cell loss or the required bandwidth consistently across all MPEG traces. The implementation and video contents do not change this result.

2. Although FEC adds extra data to the original streams, i.e. increasing the average traffic, it still reduces the overall loss rate comparing to split and skew only situation. So the overhead associated is less of an issue than, say, that in storage situation.

3. The splitting and skew have the effect of re-conditioning the distribution of frame size in a favorable way. In Figures 6-8, the loss rate of the original streams does not go down quickly as the given bandwidth increases. But after SAS, the new streams are more responsive to the bandwidth increase. This is understandable since after the local averaging (so smoother traffic) the differences of the frame sizes are reduced and the frame size is more linearly distributed comparing to the original streams with clear distinction of I, P, and B frames. As a result, the loss rate is more responsive to the bandwidth increase.

4. When the split and skew exceed T_I, the benefit becomes less significant. Since the scheme mainly targets at shifting I frame data to unused bandwidth of B (and, less significantly, also P) frames, once the second I frame arrives, the data of the first I frame can not be absorbed further. At the same time, the associated long delay is not desirable for applications and buffering concern. So the coding implementation actually determines the number of splitting and skew allowed. For small T_I like in the trace of advertisement video, although there is some improvement, the performance of the different splitting is very close. On the other hand, long T_I streams like CNN news can be benefited from the scheme much more by deep splitting.

5. The different coding implementations have a certain impact to the effectiveness of the scheme. Among the traces, the software coder produces the traces which have similar characteristics of that from Star Wars coder (with the exception of the Lecture trace, which has very static background, presumably.) They require about three times of the mean frame size to have a less-than-one-percent loss rate. On the other hand, the traces generated by the hardware coder can have the same loss rate by a bandwidth of one and one half of the mean size with a 6-split FEC scheme. Since we do not have enough excerpts from different movie coded by the same coder to have a statistics, we segment the traces to find the mean variations. The results show that the means vary very little among the sections. It may suggest that the mean is more related to a coder instead of the content (for normal content. The lecture case should be very special with almost non-background changes.) If this is the case, a bandwidth allocation scheme can be devised based on a coder, which is known ahead of time.

6. The method works less well for long continuous burst frames with abnormal short cycle time (less than assumed T_I). But even these occasional burst patterns (IBBIBB...) last less than a second in all these traces. We think they are the highly action part of the movie and may not be perceived by human well originally. Some cell losses may have less significance in these sections. Fortunately, after the burst, normal video can be resumed by itself within a few frames because of the interleaved I frames.

5 Conclusion

This paper proposes a scheme to real-time MPEG transmissions on ATM. Some unique properties of the scheme are that 1) it provides a deterministic bound on the delay and jitter. 2) it has the implementation simplicity to accommodate the high-speed video applications. 3) it has built-in mechanism for dependable transmission. 4) it exploits the unused bandwidth among the frames which leads to higher channel capacity utilization and, because of resulted less bursty traffic, reduces network delay. The scheme can be seen as a case of temporal diversity coding. The redundant code used in the scheme is to reduce the overall cell losses without any extra bandwidth requirement. This study shows that it significantly reduces the peak cell losses, which is the major cause of QOS degradation in ATM, at the price of acceptable extra delay. This method can also be used in other real-time VBR applications.

References

[1] P. Pancha and M. El Zarki, "Prioritized Transmission of Variable Bit Rate MPEG Video," *Proc. IEEE Globecom 1992*

[2] S. Chong, S.Q. Li and J. Ghosh, "Predictive Dynamic Bandwidth Allocation for Efficient Transport of Real-Time VBR Video over ATM," *IEEE J. Select. Areas Commun., Jan. 1995*

[3] L. C. Hu and K.-J. Lin, "A Forward Error Control Scheme for Real-Time VBR Traffic in ATM," *Proc. of*

Pacific Rim International Symposium for Fault-Tolerant Systems, Newport Beach, Calif., Dec. 1995

[4] R.L. Cruz, "A Calculus for Network Delay, Part I: Network Elements in Isolation," *IEEE Trans. on Infor. Theory*, Vol. 37, No. 1, Jan. 1991, pp. 114-131

[5] R.L. Cruz, "A Calculus for Network Delay, Part II: Network Analysis," *IEEE Trans. on Infor. Theory*, Vol. 37, No. 1, Jan. 1991, pp. 132-141

[6] E. Ayanoglu, C-L I, R.D. Gitlin, and J.E. Mazo, "Diversity Coding for Transparent Self-Healing and Fault-Tolerant Communication Networks," *IEEE Trans. on Commun.*, vol.41, no.11, Nov. 1993

[7] E. Ayanoglu, C-L I, R.D. Gitlin, and J.E. Mazo, "Diversity Coding: Using Error Control for Self-Healing in Communication Networks," *IEEE Infocom'90*, pp. 95-104, 1990

[8] N. Shacham and P. McKenney, "Packet recovery in High-Speed Networks Using Coding and Buffer Management," *IEEE Infocom'90*, pp. 124-131, 1990

[9] T. Ott, A. Tabatabai, and T.V. Lakshman, "A Scheme for Smoothing Delay Sensitive Traffic Offered to ATM Networks," *Proc. IEEE Infocom'92*, May 1992

[10] B. DeCleene, P. Pancha, M. El Zarki, and H. Sorensen, "Comparison of Priority Partition Methods for VBR MPEG," *Proc. of IEEE Infocom'94*, vol.2, pp.689-696, June 1994

Author Index

Notes

Notes

Notes

Notes